VOLUME ONE HUNDRED AND FORTY

ADVANCES IN CANCER RESEARCH

Sphingolipids in Cancer

VOLUME ONE HUNDRED AND FORTY

ADVANCES IN CANCER RESEARCH

Sphingolipids in Cancer

Edited by

CHARLES E. CHALFANT

Professor and Chairman,
Department of Cell Biology, Microbiology, and Molecular Biology, University of South Florida, Florida, USA

PAUL B. FISHER

Professor and Chairman,
Department of Human & Molecular Genetics, Director, VCU Institute of Molecular Medicine, Thelma Newmeyer Corman Chair in Cancer Research, VCU Massey Cancer Center, Virginia Commonwealth University, School of Medicine, Virginia, USA;
Emeritus Professor, Columbia University, College of Physicians and Surgeons, New York, USA

ELSEVIER

ACADEMIC PRESS
An imprint of Elsevier

Academic Press is an imprint of Elsevier
125 London Wall, London EC2Y 5AS, United Kingdom
525 B Street, Suite 1650, San Diego, CA 92101, United States
50 Hampshire Street, 5th Floor, Cambridge, MA 02139, United States
The Boulevard, Langford Lane, Kidlington, Oxford OX5 1GB, United Kingdom

First edition 2018

ISBN: 978-0-12-814223-3
ISSN: 0065-230X

Publisher: Zoe Kruze
Acquisition Editor: Zoe Kruze
Editorial Project Manager: Shellie Bryant
Production Project Manager: James Selvam
Cover Designer: Alan Studholme

Typeset by TNQ Technologies

CONTENTS

CONTRIBUTORS

Erhard Bieberich
Department of Physiology, University of Kentucky, Lexington, KY, United States

Myles C. Cabot
Department of Biochemistry and Molecular Biology, East Carolina University, Brody School of Medicine, and the East Carolina Diabetes and Obesity Institute, Greenville, SC, United States

Christopher J. Clarke
Department of Medicine and Cancer Center, Stony Brook University, Stony Brook, NY, United States

Pedro Costa-Pinheiro
Department of Pathology, University of Virginia, Charlottesville, VA, United States

Giovanni D'Angelo
Institute of Protein Biochemistry, National Research Council of Italy, Naples, Italy

Kelly Drews
Department of Pathology, University of Virginia, Charlottesville, VA, United States

Ahmed Elsherbini
Department of Physiology, University of Kentucky, Lexington, KY, United States

Antonio Gomez-Muñoz
Department of Biochemistry and Molecular Biology, Faculty of Science and Technology, University of the Basque Country (UPV/EHU), Bilbao, Spain

Kristen A. Jeffries
Nutrition Research Institute, UNC Chapel Hill, Kannapolis, NC, United States

Mark Kester
Department of Pharmacology, University of Virginia, Charlottesville, VA, United States; University of Virginia Cancer Center, University of Virginia, Charlottesville, VA, United States

Natalia I. Krupenko
Nutrition Research Institute, UNC Chapel Hill, Kannapolis, NC, United States; Department of Nutrition, UNC Chapel Hill, Chapel Hill, NC, United States

Clayton S. Lewis
Department of Internal Medicine, University of Cincinnati, Cincinnati, OH, United States

Chiara Luberto
Department of Physiology and Biophysics, Stony Brook University, Stony Brook, NY, United States

Sitapriya Moorthi
Department of Physiology and Biophysics, Stony Brook University, Stony Brook, NY, United States

Samy A.F. Morad
Department of Pharmacology, Faculty of Veterinary Medicine, South Valley University, Qena, Egypt

Rose Nganga
Department of Biochemistry and Molecular Biology, and Hollings Cancer Center, Medical University of South Carolina, Charleston, SC, United States

Mariana Nikolova-Karakashian
University of Kentucky College of Medicine, Department of Physiology, Lexington, KY, United States

James S. Norris
Department of Microbiology and Immunology, Medical University of South Carolina, Charleston, SC, United States

Besim Ogretmen
Department of Biochemistry and Molecular Biology, and Hollings Cancer Center, Medical University of South Carolina, Charleston, SC, United States

Natalia Oleinik
Department of Biochemistry and Molecular Biology, and Hollings Cancer Center, Medical University of South Carolina, Charleston, SC, United States

Falak Patel
Department of Physiology, University of Kentucky, Lexington, KY, United States

Logan Patterson
Department of Pathology, University of Virginia, Charlottesville, VA, United States

Jeremy Shaw
Department of Pathology, University of Virginia, Charlottesville, VA, United States

Charles D. Smith
Apogee Biotechnology Corporation, Hummelstown, PA, United States

Stefka D. Spassieva
Department of Physiology, University of Kentucky, Lexington, KY, United States

Sarah Spiegel
Department of Biochemistry and Molecular Biology, Virginia Commonwealth University School of Medicine, Richmond, VA, United States

Christina Voelkel-Johnson
Department of Microbiology and Immunology, Medical University of South Carolina, Charleston, SC, United States

Hong-Gang Wang
Department of Pediatrics, Pennsylvania State University College of Medicine, Hershey, PA, United States

Shai White-Gilbertson
Department of Microbiology and Immunology, Medical University of South Carolina, Charleston, SC, United States

Megan M. Young
Department of Pediatrics, Pennsylvania State University College of Medicine, Hershey, PA, United States

FOREWORD

The pathways of sphingolipid metabolism regulate the formation of many biologically active molecules (bioactive sphingolipids) that exert important biochemical and cellular functions. Indeed, bioactive sphingolipids have been recognized for over 30 years now, and their connections to cancer-related cellular responses were noted from the earliest studies on functions of sphingosine in inhibiting protein kinase C as well as in the initial studies on the sphingomyelin cycle and regulation of leukemia cell differentiation. This was followed by the early appreciation of induction of ceramide levels by chemotherapeutic agents and by stress stimuli, thus casting ceramide as a tumor suppressor lipid. On the other hand, sphingosine 1-phosphate (S1P) has emerged as a tumor promoting lipid due to its varied cellular and organismal functions promoting cancer cell growth, angiogenesis, inflammation, and tumor cell migration. Ceramide 1-phosphate (C1P), a relatively newer addition to the group of bioactive sphingolipids, has also emerged as an enhancer of tumor cell growth and inflammatory responses.

This field of research has attracted many prominent investigators and continues to attract highly capable and creative scientists who are devoted to advancing the overall field of bioactive sphingolipids and more specifically their many roles in cancer biology and therapeutics. Indeed, this volume brings together an elite group of sphingolipidologists. In fact, most of the authors, with the exception of a couple of original investigators of bioactive sphingolipid, may be considered the "second generation" of sphingolipidologists, led by Dr. Chalfant, who did his training with me as an exceptional postdoctoral fellow. Sphingolipid investigators 2.0 are on the right track!

At this juncture, it appears that all/most major cancer attributes (i.e., proliferation and escape from senescence, sensitivity/resistance to apoptosis, angiogenesis, reprogrammed metabolism, inflammation, invasion and metastasis, immune evasion, and genomic instability) are regulated in some key fashion by one or more bioactive sphingolipid. Given these roles, it should be obvious that no satisfactory understanding of any of these major cancer attributes would be possible without invoking sphingolipids. That is, one would expect that bioactive sphingolipids would be on the radar screen of all cancer biologists. Yet, this is not the case. So, one question is what are the issues that have resulted in this serious deficiency?

First and foremost, the study of lipids at a biochemical and biological level is rather impenetrable to the "general" cell biologist. These molecules and their enzymes are mostly hydrophobic molecules that are difficult to study and manipulate. This field attracts the intrepid! Second, sphingolipids are difficult to measure and analyze. Only with the increasing availability of "lipidomic" mass spectrometry have investigators been more exposed to important changes in bioactive lipids. Third, the actual pathways of sphingolipids are quite complex. They are compartmentally restricted, and they are interconnected, diverting one metabolite into one or more "derivatives," that often exert opposing effects. For example, formation of ceramide in one case may be coupled to the generation of its metabolite sphingosine which in turn can be phosphorylated to S1P. Thus, what initially starts as a tumor suppressor pathway is "converted" to a tumor promoting one. Fourth, defining the biochemical underpinnings of sphingolipid metabolism has lagged significantly when compared with more "traditional" biochemical pathways such carbohydrate or nucleic acid metabolism.

Luckily, many of these obstacles have been falling under the persistent studies on bioactive sphingolipids, many of which are illustrated in the various chapters of this volume. These walk us through the functions and mechanisms of the key bioactive sphingolipids: ceramide, S1P, and C1P, including roles in cell death, autophagy, exosome formation, and RNA splicing. Other chapters focus on enzymes that are emerging as key regulators of important cancer biologies such as sphingomyelin synthases, neutral sphingomyelinase, glucosyl ceramide synthase, acid ceramidase, and sphingosine kinase in tumor pathogenesis. Four chapters are dedicated to the study of bioactive sphingolipids in the action of chemotherapeutics, drug resistance, and side effects.

Given all the impressive research presented and discussed here, one has to wonder, what are the remaining challenges in thrusting bioactive sphingolipids more into the heart of cancer biology research? There are some deficiencies in the field. First, none of the enzymes of sphingolipid metabolism have been studied for either inactivating mutations as candidate tumor suppressors or for activating mutations as oncogenes and driver genes. Although not all drivers or suppressors need be mutated, this criterion is helpful to distinguish one class of cancer regulating genes from the myriad genes that are either overexpressed or suppressed in various cancers. Notwithstanding, evidence appears to be forthcoming. Stay tuned! Second, with the exception of S1P for which high affinity receptors have been conclusively identified, no such targets have been identified for ceramides,

C1P, or other candidate bioactive sphingolipids. "At the end of the day," most non lipidologists are more comfortable with protein and gene components to study rather than lipids. Witness the extensive study of PTEN, PI kinases, and Akt in the PI3P pathway with modest study of PI3P itself. Last but not least, the development of cancer therapeutics based on enzymes and pathways of sphingolipid metabolism lags compared with other pathways, including other lipid-mediated pathways. Many such evolving targets are discussed in this volume, and advancing this goal would provide a central stage for bioactive sphingolipids in cancer therapeutics.

This is a timely volume that consolidates the premise that bioactive sphingolipids are key mediators and that their enzymes can emerge as important targets in cancer therapy. The chapters are well presented and make for accessible and highly informative reading.

Yusuf A. Hannun, MD

PREFACE

Johann Ludwig Wilhelm Thudichum (August 27, 1829–September 7, 1901) is considered "The Father of Sphingolipids" as he is credited with the identification and characterization of the first sphingolipids, sphingomyelin and sphingosine, from brain extracts. In 1884, he described his research in a publication titled "A Treatise on the Chemical Constitution of the Brain" and initially named the compound *sphingosine*, due to its enigmatic nature. Indeed, the root word, *Sphingo,* is linked to the mythological story of the Greek Sphinx, and for specific English translation means "to strangle" or "to kill" in ancient Greek. In Greek mythology, the Greek Sphinx was the harbinger of death and the arbiter of bad luck. In the legend, a traveler would come across the Sphinx, who would ask them to solve a riddle: "What is the creature that walks on four legs in the morning, two legs at noon and three legs in the evening?". If not answered correctly, this harbinger of Death would strangle the life from the traveler. It was not until Oedipus that the riddle was solved: "man, crawls on all fours as a baby then walks on two feet as an adult, and then walks with a cane in old age." The Sphinx then committed *suicide* by throwing herself from her high rock and died.

The above story of the naming of sphingolipids and the nature of the legend in Greek Mythology are interestingly linked to a later historical coincidence, when the first biological function for a sphingolipid was reported, essentially solving the "riddle" of mechanisms regulated by this puzzling class of lipids discovered by Dr. Thudichum. Unambiguously and after being overlooked for approximately 100 years, Dr. Yusuf A. Hannun, considered by many as the "Modern Father of Sphingolipids" and author of the Forward for this book, demonstrated that two particular sphingolipids, ceramide and sphingosine, were bioactive molecules. Indeed, his initial findings were historically congruent with the Sphinx committing suicide as soon as her riddle was solved. More specifically, treatment of cells with these sphingolipids "shutdown" oncogenic kinase pathways and induced apoptosis, also referred to as cellular *suicide*. This book begins with Chapter 1, authored by Rose Nganga, Natalia Oleinik, and Besim Ogretmen, which since these early findings by Hannun and coworkers discusses in detail the mechanisms described by which ceramide "drives" cell death.

For another historical coincidence, sphingein, means to bind tightly in ancient Greek, is linked to more recent work on the strong interaction of specific sphingolipids with protein targets. This thematic *Advances in Cancer Research* article explores the role of these interactions as well as the additional opposing roles of bioactive sphingolipids in the induction of cell survival, inflammation, and oncogenesis. In Chapter 8, for example, Dr. Antonio Gomez-Muñoz discusses the role of the bioactive sphingolipid, ceramide-1-phosphate, binding to cytosolic phospholipase $A_2\alpha$, acid sphingomyelinase, and cellular receptors in the context of inflammation and cell survival.

From these initial and more recent findings that are linked historically to the naming of sphingolipids, the field of sphingolipids has exploded with roles for numerous sphingolipids in cellular stress pathways, oncogenesis, inflammation, and immune responses. For example, sphingosine-1-phosphate (S1P) can be produced intracellularly to bind specific protein targets and illicit epigenetic modifications or it can be secreted from the cell to act in an autocrine/paracrine manner via the interaction with specific S1P receptors to drive G protein–mediated cellular signaling. In regard to the latter effect, a sphingolipid analog of sphingosine, FTY720, was found to antagonize this system acting to suppress immune responses and later to suppress both colon and breast tumors in preclinical animal studies. This compound, also called Fingolimod, was later marketed under the name Gilenya by Novartis and was FDA-approved to treat multiple sclerosis reducing the rate of relapses by approximately 50%. This is just one example of how basic research into bioactive sphingolipids has led to translation into the clinic. Drs. Cabot, Smith, Kester, and Voelkel-Johnson in Chapters 9 through 12 discuss, in depth, in this thematic volume, along with their coauthors, as to how our expanded knowledge of sphingolipid biosynthetic pathways and cell signaling cascades targeted by these bioactive lipids is translating to the clinic as new therapeutics targeting these pathways. Dr. Smith, for example, in Chapter 11 discusses the preclinical and clinical efficacy of a new generation of inhibitors that block the function of the S1P-generating enzymes, sphingosine kinase 1 and 2. Moreover, one of the major hurdles in targeting the sphingolipid pathways for therapeutics has been efficient delivery of rather insoluble molecules. In Chapter 12, Drs. Mark Kester and Sarah Spiegel along with their colleagues discuss new delivery systems for sphingolipids and sphingolipid analogs.

Overall, there is a new era of sphingolipid pharmacology that is emerging and is expected to have a major impact on numerous disease states. One aspect important to discuss as it is not overtly discussed in this thematic volume of *Advances in Cancer Research* is the catalyst for the explosion in our knowledge of sphingolipid biosynthetic pathways and cellular signaling pathways. In this regard, sphingolipidomics has emerged in the last decade as a major "tool" to examine in appropriate detail the generation of specific forms of sphingolipids, thereby explaining conundrums in the field caused by the use of biochemical and biophysical techniques that could not delineate these specific forms of sphingolipids, such as different chain lengths of ceramide. For example, some initial studies did not observe increases in the sphingolipid, ceramide, prior to or concomitant with the induction of apoptosis. Many scientists during this time concluded that ceramide synthesis was not involved in the biological process of programmed cell death. The conclusion was further confounded in the field as inhibitors of ceramide synthesis blocked apoptosis induced by a number of apoptosis-inducing agonists. With the emergence of sphingolipidomics, which utilizes mass spectrometric analysis of various sphingolipids, individual ceramide species could be delineated and measured in a complex biological sample. On revisiting early studies, specific forms of ceramide differing by the length of the acyl chain were found to be increased by apoptotic agonists at the expense of other forms, thereby demonstrating that whereas total ceramide was not increased, specific forms were being increased by these agonists. This simple finding along with a decade of enhanced mass spectrometric technologies and techniques has now led to many studies showing that different forms of ceramide are generated in specific cellular locations by specific enzymes called ceramide synthases. Now targeted therapies can be developed as our understanding of the biochemistry of these enzymes and signaling cassettes expands. Drs. Wang, Luberto, Bieberich, Clarke, Nikolova-Karakashian, and Krupenko in Chapters 2 through 7 cover, along with their coauthors, various sphingolipid biosynthetic pathways that are emerging as major regulators of biological systems, such as autophagy and exosome biosynthesis as well as the relation of sphingolipid biosynthesis to various diseases' states, such as cancer and metabolic disorders.

This focused thematic volume of *Advances in Cancer Research* concludes with Chapter 13 in which Dr. Spassieva discusses an overlooked and understudied area of sphingolipid biology; the role of sphingolipids in the side effects observed for existing therapies for disease states such as cancer. Indeed, there is emerging evidence that novel and overlooked sphingolipids, deoxysphingolipids, are highly stable, exhibit long half-lives, and accumulate with genotoxic stress agents. This accumulation has now been linked to side effects such as peripheral neuropathy induced by conventional chemotherapies. Thus, this chapter discusses the "double-edged sword" of targeting the sphingolipid biosynthetic pathways for disease therapeutics, which awareness is recommended and required for the development of future therapeutics. More importantly, the work discussed in this chapter demonstrates that greater understanding of sphingolipid biosynthesis may allow for simple oral supplementations to ameliorate the production of deoxysphingolipids and, thus, unwanted side effects linked to some chemotherapies.

Because of the extraordinary developments in recent technologies and the rapid increase in our knowledge of cell signaling pathways, biochemical mechanisms, and disease states mediated by sphingolipids, this thematic volume is timely in covering these new developments, which are of a very large breadth. This volume can serve as a compendium of our current knowledge serving as a foundation for the field of sphingolipids. By employing ever-expanding newer technologies to identify sphingolipid binding proteins and new abilities to manipulate and "trap" specific sphingolipids new discoveries will be forthcoming. As discoveries of specific cell functions linked to specific forms of ceramides explained scientific conundrums in the field, the hope is that in the future by building on the foundations explored in this thematic volume, it will be possible to explain current conundrums and fill knowledge gaps discussed in all chapters throughout this volume. Indeed, we "throw down the gauntlet" and, hereby, challenge the next generation of sphingolipid scientists to solve the next riddle of this novel and important class of lipids in both cancer and other cellular functions and disease states.

Charles E. Chalfant, PhD
Paul B. Fisher, MPH, PhD, FNAI

CHAPTER ONE

Mechanisms of Ceramide-Dependent Cancer Cell Death

Rose Nganga, Natalia Oleinik and Besim Ogretmen[1]
Department of Biochemistry and Molecular Biology, and Hollings Cancer Center, Medical University of South Carolina, Charleston, SC, United States
[1]Corresponding author: E-mail: ogretmen@musc.edu

Contents

Advances in Cancer Research, Volume 140
ISSN 0065-230X
https://doi.org/10.1016/bs.acr.2018.04.007

Abstract

Mechanistic details for the roles of sphingolipids and their downstream targets in the regulation of tumor growth, response to chemo/radiotherapy, and metastasis have been investigated in recent studies using innovative molecular, genetic and pharmacologic tools in various cancer models. Induction of ceramide generation in response to cellular stress by chemotherapy, radiation, or exogenous ceramide analog drugs mediates cell death via apoptosis, necroptosis, or mitophagy. In this chapter, distinct functions and mechanisms of action of endogenous ceramides with different fatty acyl chain lengths in the regulation of cancer cell death versus survival will be discussed. In addition, importance of ceramide subcellular localization, trafficking, and lipid-protein binding between ceramide and various target proteins in cancer cells will be reviewed. Moreover, clinical trials from structure-function—based studies to restore antiproliferative ceramide signaling by activating ceramide synthesis will also be analyzed. Future studies are important to understand the mechanistic involvement of ceramide-mediated cell death in anticancer therapy, including immunotherapy.

1. INTRODUCTION

Sphingolipids are structural molecules of biological membranes, maintaining barrier function and fluidity (Hannun & Obeid, 2008). Ceramide, ceramide 1-phosphate (C1P), glucosylceramide, sphingosine, and sphingosine 1-phosphate (S1P) also exert signaling functions to regulate various biological processes within the cancer cell signal transduction network (Hannun & Obeid, 2008). Identification of key metabolic enzymes, transporters, downstream targets (by sphingolipid-protein binding), and specific receptors of sphingolipids has revealed various distinct signaling roles of these molecules for the regulation of numerous aspects of cancer biology (Dressler, Mathias, & Kolesnick, 1992; Ogretmen & Hannun, 2004). For example, ceramide and sphingosine generation induced by chemotherapy, radiation, and/or oxidative stress mediates cell death, senescence, and/or cell cycle arrest. However, conversion of ceramide to S1P promotes cancer cell growth, inhibits apoptosis, and enhances cancer cell—host communication, inducing tumor metastasis (Cuvillier et al., 1996; Ogretmen, 2018). Importantly, these multiple signaling nodes of sphingolipid metabolic enzymes and/or receptors provide novel targets to develop mechanism-based novel anticancer intervention strategies. Moreover, reconstitution of ceramide stress in cancer cells using small-molecule ceramide analog drugs or enzyme inhibitors modulates antitumor T cell response, regulates myeloid-derived suppressor cell function, and inhibits tumor growth and/or metastasis (Cuvillier et al., 1996; Ogretmen, 2018; Ogretmen & Hannun, 2004).

In this chapter, recent mechanistic studies with significant impact on various aspects of cancer biology and therapeutics, including ceramide-mediated cell death with regard to antitumor chemotherapy and immunotherapy, will be discussed.

2. SPHINGOLIPID METABOLISM AND CANCER

Cellular stress induces the generation of sphingosine and/or ceramide (Fig. 1) by the activation of de novo synthesis, sphingomyelin (SM) hydrolysis, or salvage pathway to mediate cancer cell death via apoptosis (Obeid, Linardic, Karolak, & Hannun, 1993), necroptosis (Saddoughi et al., 2013), mitophagy (Dany et al., 2016; Sentelle et al., 2012), and/or endoplasmic reticulum (ER) stress (Swanton et al., 2007). In contrast, many tumors exhibit increased ceramide catabolism mainly by acid ceramidase (AC) and/or sphingosine kinase (SphK) activities, increasing the generation of prosurvival S1P (Cheng et al., 2013; Liang et al., 2013).

2.1 Structure and Metabolism of Sphingolipids

Ceramide consists of a sphingosine long-chain base (LCB) and amide-linked fatty acyl chain varying in length from C_{14} to C_{26}. Endogenous ceramides are synthesized via the de novo pathway, which begins with serine and palmitoyl CoA condensation to form 3-ketosphinganine by serine palmitoyltransferase (SPT). Reduction of 3-ketosphinganine to generate sphinganine is catalyzed by 3-ketosphinganine reductase. This is followed by the synthesis of dihydroceramide by dihydroceramide synthases (dhCerS1−6) (Laviad, Kelly, Merrill, & Futerman, 2012; Tidhar et al., 2012) (Fig. 1). Dihydroceramide is desaturated to yield ceramide by dihydroceramide desaturase (DES), via insertion of a double bond between carbons 4 and 5 in the LCB (Rahmaniyan, Curley, Obeid, Hannun, & Kraveka, 2011). CerS1−6 are specialized for the generation of selective ceramides with different fatty acyl chain lengths: CerS1 and CerS4 mainly generate $C_{18}-C_{20}$-ceramides, CerS5 and CerS6, $C_{14}-C_{16}$-ceramides, and CerS2 is selective for $C_{22}-C_{24}$-ceramides (Pewzner-Jung, Ben-Dor, & Futerman, 2006; Raichur et al., 2014). CerS3 is involved in the synthesis of ultra-long ceramides, distinct for testes and skin tissues (Jennemann et al., 2012). Ceramide is also generated by sphingomyelinases (SMases), which hydrolyze SM, or cerebrosidase, and glucosyl/galactosyl-ceramidase (Fig. 1), which break down GlcCer and/or galactosylceramide (GalCer) (Dressler et al., 1992; Hannun & Bell, 1987; Hannun & Obeid, 2008; Ogretmen & Hannun, 2004). The salvage pathway

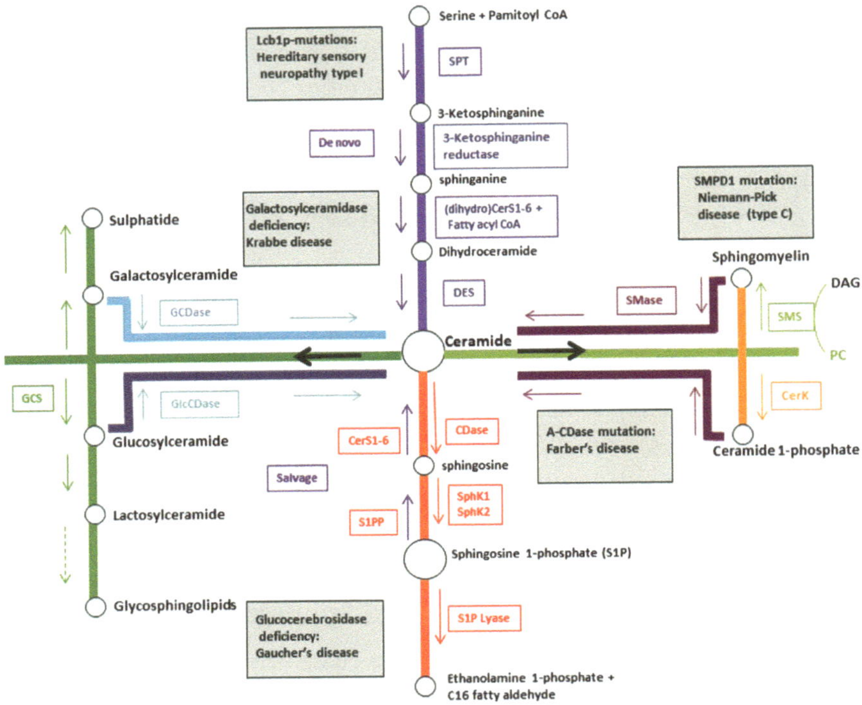

Figure 1 ***Pathways of sphingolipid metabolism and key enzymes.*** Ceramide is the central molecule in sphingolipid metabolism and can be synthesized de novo (*blue arrows* and *boxes*) by the functions of serine palmitoyltransferase (SPT), 3-ketosphinganine reductase, (dihydro)ceramide synthases (CERS1–6) and dihydroceramide desaturase (DES). Ceramide can also be generated by the hydrolysis of complex sphingolipids such as sphingomyelin (SM) by the action of sphingomyelinases (SMases) (*purple boxes*). Conversion of sphingosine to ceramide is catalyzed by the salvage pathway via CERS1-6 (*red boxes*); the same enzymes involved in de novo ceramide synthesis. Ceramide is metabolized to generate sphingosine 1-phosphate (S1P) by the functions of ceramidases (CDases), and sphingosine kinase 1 (SphK1) and SphK2. S1P is further hydrolyzed by S1P lyase to exit the sphingolipid metabolic cycle. The metabolism of ceramide to generate complex sphingolipids such as glycosphingolipids or gangliosides requires the synthesis of glucosylceramide (GlcCer) by glucosylceramide synthase (GCS) (*green boxes*), which is a precursor for synthesis of complex sphingolipids (*dotted arrows*). Sphingomyelin synthase (SMS) enzymes convert ceramide to SM via insertion of a choline moiety into ceramide as a head group using phosphatidylcholine as a donor, yielding free diacylglycerol, and ceramide kinase (CERK) converts ceramide to ceramide 1-phosphate (C1P) (*orange boxes*). Ceramide is hydrolyzed by ceramidases (CDases), including acid (A)-CDase (A-CDase), to release free fatty acids and sphingosine. Ceramide can be synthesized by the hydrolysis of SM by SMases, including SMPD1. Mutations in enzymes involved in sphingolipid metabolism have been associated with various lysosomal storage disorders (*gray boxes*), including Farber's, Gaucher's, Niemann-Pick, and Krabbe diseases. *S1PP*, S1P phosphatase.

entails the recycling of sphingosine and the generation of endogenous long-chain $C_{14}-C_{26}$-ceramides from exogenous short-chain C_2-C_6-ceramides by CerS1−6. These ceramides exert various signaling functions in cancer cells including inhibition of c-Myc and telomerase (Ogretmen et al., 2002).

Ceramide kinase (CERK) (Wijesinghe, Lamour, Gomez-Munoz, & Chalfant, 2007) or SM synthase (SMS) (Huitema, van den Dikkenberg, Brouwers, & Holthuis, 2004) generates ceramide 1-phosphate (C1P) and SM, respectively (Fig. 1). Ceramide is also utilized by glucosylceramide synthase (GCS) as a precursor for the generation of GlcCer (Watters et al., 2013). GlcCer is then transported from early Golgi to distal Golgi compartments by four-phosphate adaptor protein (FAPP2) for the synthesis of glycosphingolipids, determining the lipid composition of the plasma membrane (D'Angelo et al., 2013). Nonvesicular C1P trafficking between membranes is regulated by a transporter protein GLTPD1 (renamed as C1P transporter protein, CPTP), which then controls prostaglandin signaling and inflammation (Simanshu et al., 2013).

Ceramide is hydrolyzed by ceramidases (CDases) to yield sphingosine, which is phosphorylated by sphingosine kinase (SphK) 1 or 2 (SphK1 or SphK2), to generate sphingosine 1-phosphate (S1P). S1P engages with S1P receptors (SIP_{1-5}/S1PR1−5) in an autocrine/paracrine manner for oncogenic signaling (Maceyka & Spiegel, 2014). S1P is usually rapidly (within 10−20 min in cells/tissues) metabolized by S1P phosphatases (Huang et al., 2016) or S1P lyase (Zamora-Pineda, Kumar, Suh, Zhang, & Saba, 2016), yielding ethanolamine 1-phosphate and hexadecanol (C_{16}-fatty aldehyde) (Fig. 1).

2.2 Enzymes of Ceramide Metabolism and Cancer

The abundance of sphingolipid molecules is highly regulated by metabolic enzymes (Fig. 1), whose altered expression or activity plays key roles in the regulation of cancer signaling and/or therapeutic treatment.

2.2.1 Serine Palmitoyltransferase

Mammalian trimeric SPT consists of two of three large subunits (SPTLC1, SPTLC2, or SPTLC3) and a small subunit (SPTssa or SPTssb) localized in the ER (Han et al., 2009). Specific missense mutations in the human SPTLC1 gene, encoding Lcb1p subunit, result in the synthesis/accumulation of 1-deoxysphingoid bases, modified sphingolipids, which are associated with hereditary sensory neuropathy type I, an autosomal dominant disease (Bode et al., 2016). SPT-generated 1-deoxysphingoid bases have been linked to chemotherapy-related peripheral neuropathy (Kramer et al., 2015), a major dose-limiting toxicity for paclitaxel or cisplatin treatment in some cancer patients.

Identification of biological regulators of metabolic enzymes has important implications. For example, the ORM1-like proteins (ORMDLs) have been identified as key suppressors of SPT function (Breslow et al., 2010). Phosphorylation of SPTLC1 by Bcr-Abl kinase at Tyr164 inhibited its activity, promoting CML (chronic myeloid leukemia) cell survival (Taouji et al., 2013). However, whether ORMDLs play any roles in the regulation of P-SPTLC1 and CML cell survival remains unknown.

2.2.2 Ceramide Synthases 1–6

Discovery and cloning of CerS1–6 have been key to understand the roles of ceramides with different fatty acyl chain lengths in cancer cell death versus survival. For example, in head and neck (HN) cancer cells, histone deacetylase 1 and microRNA-574-5p repressed CerS1 expression, inhibiting C_{18}-ceramide generation (Meyers-Needham et al., 2012). Activation of CerS1/C_{18}-ceramide induced cell death and tumor suppression in HNSCC (head and neck squamous cell carcinoma) (Koybasi et al., 2004; Senkal et al., 2007). Genetic loss of CerS2 resulted in pheochromocytoma in some mice due to possible apoptotic defects (Park et al., 2015). Moreover, in lung cancers expressing wild-type p53, CerS6 was shown to be a transcriptional target of p53, activated by nongenotoxic folate stress, inducing cell death (Fekry, Jeffries, et al., 2016). In colon cancer cells, CerS6 increased TRAIL sensitivity via inducing nuclear translocation of caspase-3 (White-Gilbertson et al., 2009). In response to irradiation, CerS6-dependent ceramide synthesis induced Bax-mediated apoptosis (Lee et al., 2011). Furthermore, mitochondria-associated antiapoptotic protein Bcl2L13 directly inhibited ceramide synthesis by blocking formation of homodimeric and heterodimeric CerS2/CerS6 complex (Jensen et al., 2014). These studies support the concept that CerS-generated ceramides with different fatty acyl chains play distinct roles in inducing cancer cell death.

2.2.3 Dihydroceramide Desaturase (DES or DEGS1)

Exogenous dihydroceramide was considered to be an inactive sphingolipid molecule possibly due to poor solubility. However, molecular knockdown or pharmacological inhibition of DES using fenretinide or C_8-cyclopropenyl-ceramide resulted in increased accumulation of endogenous dihydroceramides, delaying G1/S transition of neuroblastoma cell cycle (Rahmaniyan et al., 2011). These data suggest that endogenous dihydroceramides suppress cancer cell growth without inducing apoptosis. It remains unknown whether targeting DES inhibits apoptosis due to decrease in ceramide generation.

2.2.4 Sphingomyelinases

In addition to de novo synthesis, ceramide can be generated by the hydrolysis of SM by SMases. Based on their optimal activity at different pH conditions, there are three types of SMases: acid, neutral, and alkaline. Crystal structure of acid SMase demonstrated that the catalytic domain contains a calcineurin-like fold associated with two zinc ions, and this domain also accommodates a hydrophobic track leading to the active site (Gorelik, Illes, Heinz, Superti-Furga, & Nagar, 2016). Induction of acid SMase increases ceramide generation, resulting in apoptotic cell death (Hannun & Obeid, 2008). Carpinteiro et al. showed that acid SMase (SMPD1−/−)-deficient mice were protected from lung and spleen metastasis of melanoma cells. Remarkably, transplantation of wild-type platelets to SPMD1−/− mice reactivated tumor metastasis (Carpinteiro et al., 2015). The melanoma cells enhanced platelet secretory acid SMase, inducing C_{16}-ceramide formation, activating clustering of alpha5-beta1 integrin on melanoma cells and promoting cancer cell adhesion and metastasis. These data suggest that while induction of acid SMase in cancer cells results in apoptosis, systemic acid SMase activation in platelets leads to enhanced melanoma metastasis, supporting the importance of the localization of ceramide signaling in inducing cell death versus increasing survival/migration.

Ceramide generation by nSMase2 mediated doxorubicin-induced cell cycle arrest via p53 (Shamseddine et al., 2015). Trajkovic et al. (2008) demonstrated that cells' exosome release is dependent on nSMase2/ceramide-induced intra-endosomal membrane transport and exosome formation. Purified exosomes were enriched in ceramide, and their release was decreased in response to nSMase2 inhibition. However, whether nSMase2-dependent ceramide-enriched exosome secretion plays any roles in the induction of cancer cell death and/or growth inhibition remains unknown.

In mice, alkaline SMase (NPP7) deficiency resulted in increased tumor size and number in AOM/DSS (azoxymethane/dextran sodium sulphate)-induced colon cancers, possibly due to reduced ceramide generation (Chen et al., 2015). These data support that hydrolysis of SM by SMases to generate ceramide mediates cancer cell death and tumor suppression.

2.2.5 Sphingomyelin Synthases and Ceramide Transporter

Metabolic conversion of ceramide for SM synthesis, which is known to reduce ceramide-mediated cell death, is catalyzed by SMS1 and SMS2 (Huitema et al., 2004). Synthesis of SM is regulated by ceramide transport from ER to the Golgi via a nonvesicular mechanism by ceramide transporter (CERT) (Hanada et al., 2003; Wang et al., 2009). Heering et al. (2012) reported that loss of CERT augmented EGFR (epidermal growth factor receptor) signaling in triple-negative breast cancer cells through SM alterations at the plasma membrane. In addition, Lee et al. (2012) reported that CERT (COL4A3BP) knockdown sensitized ovarian, colorectal, and Her2+ breast cancer cells to paclitaxel due to induction of ceramide-mediated ER stress or LAMP2-dependent autophagic flux. Targeting CERT by 3-chloro-8-beta-hydroxycarapin-3,8-hemiacetal (CHC) was effective to induce ceramide-dependent cancer cell death (Hullin-Matsuda et al., 2012).

2.2.6 Ceramide Kinase

CERK phosphorylates ceramide to generate C1P, which induced eicosanoid synthesis and cell migration via the activation of cPLA (2)-alpha (Wijesinghe et al., 2014). Increased CERK/C1P signaling directly reduced apoptotic ceramide accumulation and promoted breast cancer cell survival along with mammary tumor recurrence after Her2/neu inhibition (Payne, Pant, Pan, & Chodosh, 2014). CERK inhibitor NVP-231 suppressed breast and lung cancer cell proliferation by inducing ceramide-mediated cell cycle arrest and apoptosis (Pastukhov et al., 2014).

2.2.7 Glucosylceramide Synthase

Ceramide is a substrate for GCS to produce glucosylceramide. In fact, GCS and P-glycoprotein expression was associated with poor prognosis in patients with oral cavity cancers (Kim et al., 2016). Molecular or pharmacological inhibition of GCS attenuated resistance to chemotherapeutic drugs in HN cancer or hepatocellular carcinoma (Roh, Kim, Park, & Kim, 2015; Stefanovic et al., 2016). Recently, Liu et al. (2011) reported that in ovarian cancer cells expressing mutant p53, inhibition of GCS restored ceramide-regulated p53-dependent apoptosis. Conversion of ceramide to GlcCer by GCS was key for the maintenance of breast cancer stemness (CD44+/ESA+/CD24−), which was predicted to be associated with glycosphingolipid (globotriosylceramide, Gb3) accumulation (Gupta et al., 2012).

2.2.8 Ceramidases

There are three classes of CDases (acid, neutral and alkaline), which are identified based on their optimal pH for enzymatic activity to cleave ceramide. AC was found to be upregulated in various cancers, including acute myeloid leukemia (AML), melanoma, and prostate cancers (Eliyahu, Park, Shtraizent, He, & Schuchman, 2007; Realini et al., 2016; Tan et al., 2016). Cheng et al. (2013) demonstrated that radiation-induced AC mediates resistance to irradiation, resulting in tumor relapse. It was also shown that in prostate cancer cells AC promoted Crm1-dependent nuclear export of PTEN (phosphatase and tensin homologue) through S1P-mediated AKT (also known as protein kinase B, PKB) signaling. Pharmacological inhibitor of lysosomal AC (LCL521) induced ceramide-dependent cancer cell death and prostate tumor suppression (Beckham, Cheng, et al., 2013). A recent study revealed that there was a positive feedback mechanism between CerS6 and AC activation, as induction of CerS6 enhanced AC expression through JNK activation in colon cancer cells (Tirodkar et al., 2015). Thus, these data suggest that AC provides a metabolic communication between ceramide and S1P synthesis to activate prosurvival S1P signaling in cancer cells.

NCDase (neutral ceramidase), whose crystal structure was solved recently (Airola et al., 2015), is highly abundant in the intestine. Inhibition of NCDase resulted in ceramide-mediated apoptosis and autophagy in colon cancer cells and xenograft-derived tumors in mice (Garcia-Barros et al., 2016). NCDase knockout mice were protected from azoxymethane-derived colon cancers (Garcia-Barros et al., 2016).

There are three alkaline CDases (ACER1, ACER2, and ACER3). Although the roles of ACER1 (Liakath-Ali et al., 2016) in cancer are still unknown, ACER2 activation, due to increased sphingosine accumulation and reactive oxygen species (ROS) signaling, enhanced cancer cell death in response to doxorubicin-induced DNA damage despite reduction in ceramide generation (Mao et al., 2010). ACER3 (selective for catabolizing $C_{18:1}$- and $C_{20:1}$-ceramides) deficiency due to homozygous p.E33G mutation was associated with progressive leukodystrophy in early childhood of Ashkenazi-Jewish patients (Edvardson et al., 2016). ACER3$^{-/-}$ mice exhibited enhanced colitis and colitis-associated colorectal cancers with increased systemic inflammation and hyperactive innate immune system due to accumulation of lactosyl/monohexosyl-ceramides (Wang et al., 2016). In addition to conventional ceramides, accumulation of complex sphingolipids including lactosyl/hexosyl-sphingolipids might also play role in relation between death and survival of cancer cell.

3. CERAMIDE IN CANCER CELL DEATH

3.1 Ceramide and Apoptosis

Obeid et al. (1993) demonstrated that ceramide induces apoptosis in leukemic cells. Roles of ceramide biogenesis in radiation-induced apoptosis via the involvement of ceramide-rich microdomains on mitochondrial membrane for Bax integration (Fig. 2) were shown (Lee et al., 2011; Siskind et al., 2008). The involvement of ceramide channels in the induction of mitochondrial apoptosis by mitochondrial outer membrane permeabilization (MOMP) was also reported (Chang et al., 2015). nSMase-dependent sphingolipid metabolism sensitized MOMP to coordinate BAX/BAK activation, suggesting that BH3-only proteins and specific sphingolipid milieu at the mitochondrial membrane regulate apoptosis (Chipuk et al., 2012).

Lysosome-associated protein transmembrane-4b (LAPTM4B) removes ceramide from late endosomal (LE) organelles. LAPTM4B silencing induced LE-sphingolipid accumulation and destabilized lysosomal membrane, resulting in resistance to caspase-3 activation and apoptosis in response to chemotherapeutic drugs (Blom et al., 2015). These data suggest that compartmentalization of ceramide by LAPTM4B regulates sphingolipid-mediated caspase-3 activation and apoptosis (Fig. 2).

There are studies, however, that have demonstrated a cell death-protective role of ceramide in some cancer cells. For example, CERS6-generated C_{16}-ceramide was important to protect HNSCC cells from ER stress-mediated apoptosis or metastasis (Senkal, Ponnusamy, Bielawski, Hannun, & Ogretmen, 2010; Suzuki et al., 2016). This was also consistent with studies showing protective effect of C_{20}-ceramide and C_{22}-ceramide generated by HYL2-ceramide synthase from anoxia-induced death in *Caenorhabditis elegans* (Menuz et al., 2009). Moreover, studies in HeLa cells revealed that overexpression of CerS2 increased production of C24-ceramide and protected cells from ionizing radiation (IR)-induced apoptosis. However, C_{16}-ceramide, generated by CerS5 and CerS6, was responsible for IR-mediated apoptosis in these cells (Deng et al., 2008; Mesicek et al., 2010). Overall, these studies imply that although ceramide mediates apoptosis in majority of cancer cells via mitochondrial membrane perturbation and/or influencing cell death signaling, it might also play protective roles against apoptosis. These differential roles of ceramides might depend on cell type, the presence and/or absence of downstream ceramide targets, and/or subcellular localization of ceramide.

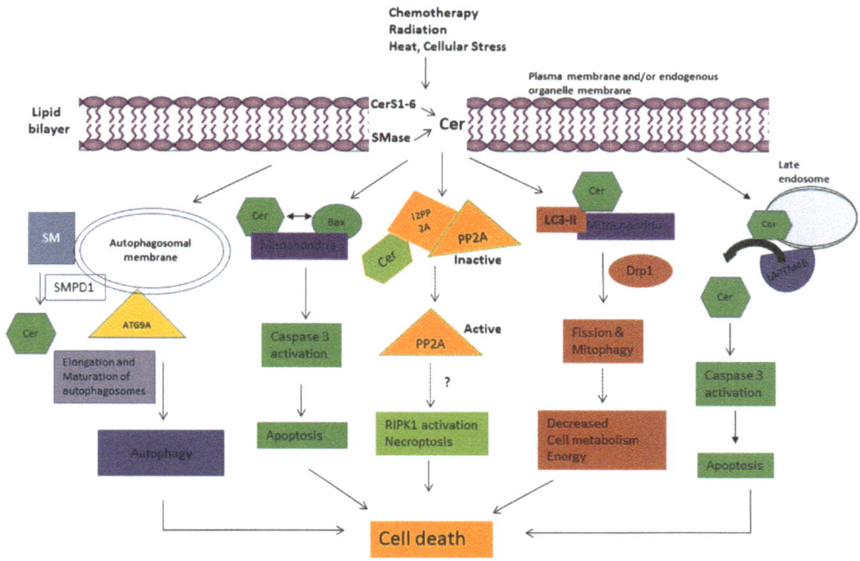

Figure 2 ***Intracellular ceramide signaling in cancer cell death and tumor suppression.*** In response to cellular stress, the increased generation of ceramide via ceramide synthases (CERS1−6) and sphingomyelinases (SMases) results in cancer cell death and tumor suppression, which is regulated by various mechanisms involving both direct and indirect targets of ceramide. Autophagy in cancer cells has been shown to be induced by accumulation of dihydroceramide and/or ceramide in the endoplasmic reticulum (ER) of glioma cells in response to tetrahydrocannabinol, a psychotrophic cannabinoid. Accumulation of dihydroceramide/ceramide in the ER leads to the permeabilization of the autophagosomal and lysosomal membranes, resulting in cathepsin release and apoptotic cell death. Moreover, hydrolysis of SM to generate ceramide is important to promote the elongation and maturation of autophagosomal membranes via recruitment of autophagy-related protein 9A (ATG9A), suggesting a novel role for ceramide in the maturation of early autophagosomal membranes. Ceramide accumulation in the mitochondrial membrane induces BAX recruitment to the mitochondria, resulting in mitochondrial outer membrane permeabilization (MOMP), caspase activation and apoptosis in response to radiation. Activation of protein phosphatase 2A (PP2A) is mediated by direct binding between phosphatase 2A inhibitor (I2PP2A; also known as SET) and ceramide or FTY720, a sphingosine analogue drug, leading to receptor-interacting serine/threonine-protein kinase 1 (RIPK1)−dependent necroptosis in lung cancer cells. Cellular stress in response to chemotherapy results in C_{18}-ceramide trafficking to the outer mitochondrial membrane, which then directly binds the lipidated form of microtubule-associated protein 1 light-chain 3 beta (LC3B-II) to recruit autophagosomes to mediate lethal mitophagy by the activation of dynamin-related protein 1 (DRP1) and mitochondrial fission. In addition, regulation of transport of ceramide from late endosomal organelles by lysosomal-associated transmembrane protein 4B (LAPTM4B) stabilizes lysosomal membranes, leading to ceramide-mediated caspase-3 activation and apoptosis induced by chemotherapeutic drugs.

3.2 Ceramide and Necroptosis

Induction of ceramide signaling was involved in programmed necrosis (necroptosis)-mediated cancer cell death in response to FTY720 (Mukhopadhyay et al., 2009; Saddoughi et al., 2013). Interaction of ceramide or FTY720, but not P-FTY720, with inhibitor 2 of PP2A (I2PP2A/SET oncoprotein) activated tumor suppressor PP2A, inducing necroptosis and inhibition of lung xenograft-derived tumor growth. This pathway was shown to be activated by the kinase domain of RIPK1 independent of RIPK3 (Mukhopadhyay et al., 2009; Saddoughi et al., 2013) (Fig. 2). However, whether FTY720-mediated necroptosis and cell death involves the generation of endogenous ceramide in lung cancer cells is still unknown.

3.3 Ceramide and Lethal Autophagy or Mitophagy

Accumulation of dihydroceramide or ceramide induced autophagy resulting in context-dependent cancer cell death or survival (Hernandez-Tiedra et al., 2016; Sentelle et al., 2012). Tetrahydrocannabinol (THC, the active component of marijuana) triggered dihydroceramide accumulation in the ER of glioma cells, promoting autophagosomal and lysosomal membrane permeabilization, leading to cathepsin release and apoptosis (Hernandez-Tiedra et al., 2016). Accumulation of SM (by silencing of SMPD1) in cancer cells resulted in aggregation of elongated and unclosed autophagic membranes with reduced association of ATG9A (Corcelle-Termeau et al., 2016). Recent data demonstrated that induction of ceramide stress using chemotherapeutic agents or exogenous ceramides increased mitochondrial accumulation of CerS1-generated C_{18}-ceramide at the outer membrane, leading to the recruitment of autophagosomes to mitochondria via ceramide-LC3 binding and mitophagy in HN cancer (Sentelle et al., 2012; Thomas et al., 2017) and AML (Dany et al., 2016) (Fig. 2). It was also shown that ceramide-dependent lethal mitophagy is regulated by the activation of Drp1 and mitochondrial fission, involving PKA inhibition and decreased phosphorylation of Drp1 at S637 in FLT3 + AML cells (Dany et al., 2016). Thus, these data suggest that CerS1-generated ceramide acts as a receptor on mitochondrial membrane for LC3-containing autophagosomes for inducing lethal mitophagy.

3.4 Ceramide and Endoplasmic Reticulum Stress

Induction of ceramide-associated ER stress was reported as an upstream regulator of autophagy (Salazar et al., 2009). Melanoma differentiation associated gene-7 (mda-7) regulated IL-24—induced ER stress, leading to ceramide-mediated lethal autophagy in glioblastoma cells (Yacoub et al., 2010). Methotrexate-induced folate stress in lung cancer cells increased the formation of ER stress aggregates with CerS6 accumulation, which was associated with the activation of p53 (Fekry, Esmaeilniakooshkghazi, Krupenko, & Krupenko, 2016).

3.5 Ceramide and Cell Cycle Arrest

Ceramide, generated via the salvage pathway in response to C_6-ceramide exposure, interacted with phosphatidylinositol 3-kinase C2β (PI3KC2β), suppressing its activation. This resulted in the inhibition of cell motility and tumor metastasis in response to EGF stimulation (Kitatani et al., 2016). This was also consistent with inhibition of tumor metastasis by nanoliposomal ceramide (NLC), which induced lysosomal CD44 degradation, and anoikis activation (Haakenson et al., 2015).

Overall, these studies suggest that various downstream targets of sphingolipids, and their distinct subcellular compartmentalization/trafficking, dictate their roles for induction of cancer cell death and/or growth inhibition.

4. SPHINGOLIPID METABOLISM AND ANTICANCER THERAPY

4.1 Ceramide Generation in the Induction of Chemotherapy/Radiation-Mediated Tumor Suppression

CerS-generated ceramide was linked to daunorubicin-induced apoptosis in various cancer cells (Bose et al., 1995). Vorinostat and sorafenib combination was efficacious in the inhibition of gastrointestinal tumor and pancreatic tumor-derived xenograft growth in mice. The mechanism involved activation of CD95 through the generation of Ca^{+2} and CerS6-dependent ceramide, leading to PP2A and ROS signaling (Park et al., 2010). In a phase II clinical trial in patients with recurrent HN cancers, elevation of serum C_{18}-ceramide was significantly associated with improved response to gemcitabine plus doxorubicin combination therapy, which induced CerS1/C_{18}-ceramide (Saddoughi et al., 2011).

Ataxia telangiectasia kinase regulated target switching to escalating doses of radiation in the intestines via activation of ceramide synthesis in intestinal stem cells, without affecting microvascular response (Ch'ang et al., 2005). Interestingly, neutralizing ceramide signaling by anticeramide antibody prevented gastrointestinal damage in response to radiation, relieving gastrointestinal syndrome mortality (Rotolo et al., 2012). Measurement of plasma ceramide levels during the treatment of 35 patients with hypofractionated stereotactic body radiation therapy showed that elevated total ceramide was associated with positive response (Dubois et al., 2016). Moreover, ABC294640, a pharmacological inhibitor of SphK2, which was shown to also inhibit DES, resulting in the accumulation of dihydroceramides (Venant et al., 2015), has been successfully tested in a phase I clinical trial for the treatment of patients with advanced solid tumors (Britten et al., 2017). Thus, these data suggest that increased (dihydro)ceramide generation in cancer cells in response to chemotherapy or radiation plays important roles in tumor suppression.

4.2 Ceramide Metabolism in Tumor Immunology and Immunotherapy

In addition to its roles in chemo/radiotherapy, ceramide signaling is implicated in the regulation of tumor immunology and immunotherapy. For example, ceramide inhibited growth of myeloid-derived suppressor cells by activation of lysosomal cathepsin B/D. Attenuation of autophagy and ER stress enhanced cytotoxic T cell lymphocyte function for tumor suppression (Liu et al., 2016). In Gaucher's disease patients, who are susceptible to myeloma, clonal immunoglobulin was reactive against lyso-glucosylceramide, which is highly elevated in these patients due to deficiency in glucocerebrosidase (Nair et al., 2016). These data suggested that lyso-glucosylceramide might be involved in the origin of myeloma and/or Gaucher's disease—associated gammopathy (Nair et al., 2016). Interestingly, in Gaucher's disease, complement signaling via C5a/C5aR1 induced GlcCer accumulation and tissue inflammation (Pandey et al., 2017). Recently, Sofi et al. (2017) demonstrated that CerS6/C_{16}-ceramide enhances T cell response to alloantigens during allogeneic hematopoietic cell transplantation (allo-HCT), an effective immunotherapy for various hematologic malignancies. The data revealed that genetic loss of CerS6 prevented graft versus host disease (GVHD), a major toxicity limiting the antileukemic efficacy of allo-HCT (Sofi et al., 2017). Thus, these data suggest that ceramide metabolism/signaling plays key roles in the regulation of anticancer immunotherapy.

4.2.1 Targeting Sphingolipid Metabolism and Signaling for Cancer Therapy

The US Food and Drug Administration (FDA) approved FTY720 (Fingolimod/Gilenya, Novartis, Inc.), which is phosphorylated by SphK2 to generate P-FTY720, a functional antagonist of S1PR1, for the treatment of relapsing MS (multiple sclerosis) (Cohen et al., 2010; Kappos et al., 2010). FTY720 (prodrug) also exhibited tumor suppression via S1PR-dependent (Liang et al., 2013) and receptor-independent (Saddoughi et al., 2013) mechanisms. Receptor-independent mechanism involves inducing necroptosis via inhibition of oncoprotein SET and activation of tumor suppressor PP2A signaling in various cancer types, including lung and colon cancers. Another tumor suppressor mechanism results in inhibition of CML stem cell proliferation/expansion, reversing CML drug resistance (Neviani et al., 2013). These data suggest that sphingolipid metabolism and signaling present a valuable target for cancer therapy in patients.

In addition, FTY720 was used as a lead compound in developing inhibitors for specific ceramide synthases: ST1060 inhibited CerS2, ST1072 inhibited CerS4 and CerS6, whereas ST1058/1074 inhibited CerS2 and CerS4 (Schiffmann et al., 2012). These chemicals provide valuable preclinical tools to distinctly regulate various CerS enzymes and their specific ceramide products in different cancer models. In fact, inhibition of CerS6 by ST1072 reduced T cell activation in response to allogeneic stimulation in mouse and human T cells, preventing GVHD toxicity during allo-HCT for antileukemia therapy (Sofi et al., 2017). Inhibition of GCS by PDMP (D-threo-1-phenyl-2-decanoylamino-3-morpholino-1-propanol) sensitized CML cells expressing mutant Bcr-Abl (T315I) to tyrosine kinase inhibitors by inducing glycogen synthase kinase-3–dependent apoptosis (Huang et al., 2011).

Highly soluble pyridinium-ceramides, selectively accumulating in cancer cells' mitochondria, induced lethal mitophagy (Dany et al., 2016; Sentelle et al., 2012) and tumor suppression in HN, AML, and prostate cancers (Beckham, Lu, et al., 2013; Senkal et al., 2006). Inhibition of survivin by NLC resulted in complete remission of the natural killer type of aggressive large granular lymphocytic leukemia (Liu et al., 2010). NLC is now tested in a phase I clinical trial for the treatment of patients with advanced solid tumors (NCT02834611).

5. DISCUSSION

During the past decade, many novel discoveries helped to delineate the roles and mechanism of action of sphingolipids in cancer signaling. These discoveries required the development of analytical, molecular, structural, genetic, and pharmacologic tools. For example, highly quantitative analytical methodology using mass spectrometry (Bielawski et al., 2009; Shaner et al., 2009) utilized to measure different sphingolipid species, including 1-O-acylceramide (Senkal et al., 2017), in cancer cells, tissues, and/or blood, was key to understand the alterations of these signaling molecules and their metabolism in cancer versus noncancerous cells/tissues. Moreover, identification and cloning of almost all of the key enzymes involved in sphingolipid metabolism helped generate molecular and genetic models to study their roles in cancer biology and/or treatment. These models are now coupled with structure-function–based anticancer drug design and development. For example, mitochondrial accumulation of ceramide, regardless of its fatty acyl composition, was shown to induce cancer cell death by mitophagy (Dany et al., 2016; Sentelle et al., 2012).

In addition, understanding the roles of sphingolipid signaling in the regulation of communication between tumors and host cells, including stromal, endothelial, and immune cells or platelets, will help develop novel anticancer immunotherapy strategies, such as strengthening of antitumor functions of T cells, attenuation of suppressor functions of myeloid-derived suppressor cells, and/or cancer-associated macrophages, leading to cancer cell death (Scheffel et al., 2017).

There are, however, some biochemical and biophysical challenges for defining the roles of sphingolipids in the regulation of cancer cell death versus survival. These challenges involve their rapid metabolic interconversions and trafficking/signaling roles at biological membranes, including primary cilia (Gencer et al., 2017). To this end, genetic loss or shRNA-mediated knockdown of CerS4, reducing C_{18}-C_{20}-ceramides, induced trafficking of TGF-β receptor I/II (TβRI/II) to primary cilia, activating sonic hedgehog-related Smo to enhance cancer cell migration and metastasis. Reconstitution of CerS4 inhibited this process, attenuating tumor metastasis (Gencer et al., 2017). These recent data suggest that alterations of selective enzymes involved in ceramide/sphingolipid synthesis can have major impact to control cancer cell migration and tumor metastasis. However, future studies, which will help determine how enzymes of sphingolipid and general metabolism are interregulated, possibly by protein-protein and/or sphingolipid-protein interactions, are needed. These studies should focus on control of sphingolipid trafficking, metabolic enzyme transport from one organelle to another, such as from ER to mitochondria, receptor aggregation/activation, and/or exosome formation/secretion, which have important implications in cancer cell signaling, resulting in cancer cell death, leading to improved anticancer chemotherapy and/or immunotherapy.

ACKNOWLEDGMENTS

The authors thank the members of the Ogretmen laboratory for their helpful discussions. The authors apologize to those investigators whose publications were not mentioned in this review due to space limitations. This work is supported by NIH research grants (R01-DE16572, R01-CA88932, R01-CA173687 and P01-CA203628) and the SmartState Endowment for Lipidomics and Drug Discovery.

REFERENCES

Airola, M. V., Allen, W. J., Pulkoski-Gross, M. J., Obeid, L. M., Rizzo, R. C., & Hannun, Y. A. (2015). Structural basis for ceramide recognition and hydrolysis by human neutral ceramidase. *Structure, 23*(8), 1482–1491. https://doi.org/10.1016/j.str.2015.06.013.

Beckham, T. H., Cheng, J. C., Lu, P., Shao, Y., Troyer, D., Lance, R., et al. (2013). Acid ceramidase induces sphingosine kinase 1/S1P receptor 2-mediated activation of oncogenic Akt signaling. *Oncogenesis, 2*, e49. https://doi.org/10.1038/oncsis.2013.14.

Beckham, T. H., Lu, P., Jones, E. E., Marrison, T., Lewis, C. S., Cheng, J. C., et al. (2013). LCL124, a cationic analog of ceramide, selectively induces pancreatic cancer cell death by accumulating in mitochondria. *The Journal of Pharmacology and Experimental Therapeutics, 344*(1), 167–178. https://doi.org/10.1124/jpet.112.199216.

Bielawski, J., Pierce, J. S., Snider, J., Rembiesa, B., Szulc, Z. M., & Bielawska, A. (2009). Comprehensive quantitative analysis of bioactive sphingolipids by high-performance liquid chromatography-tandem mass spectrometry. *Methods in Molecular Biology, 579*, 443–467. https://doi.org/10.1007/978-1-60761-322-0_22.

Blom, T., Li, S., Dichlberger, A., Back, N., Kim, Y. A., Loizides-Mangold, U., et al. (2015). LAPTM4B facilitates late endosomal ceramide export to control cell death pathways. *Nature Chemical Biology, 11*(10), 799–806. https://doi.org/10.1038/nchembio.1889.

Bode, H., Bourquin, F., Suriyanarayanan, S., Wei, Y., Alecu, I., Othman, A., et al. (2016). HSAN1 mutations in serine palmitoyltransferase reveal a close structure-function-phenotype relationship. *Human Molecular Genetics, 25*(5), 853–865. https://doi.org/10.1093/hmg/ddv611.

Bose, R., Verheij, M., Haimovitz-Friedman, A., Scotto, K., Fuks, Z., & Kolesnick, R. (1995). Ceramide synthase mediates daunorubicin-induced apoptosis: An alternative mechanism for generating death signals. *Cell, 82*(3), 405–414.

Breslow, D. K., Collins, S. R., Bodenmiller, B., Aebersold, R., Simons, K., Shevchenko, A., et al. (2010). Orm family proteins mediate sphingolipid homeostasis. *Nature, 463*(7284), 1048–1053. https://doi.org/10.1038/nature08787.

Britten, C. D., Garrett-Mayer, E., Chin, S. H., Shirai, K., Ogretmen, B., Bentz, T. A., et al. (2017). A phase I study of ABC294640, a first-in-class sphingosine kinase-2 Inhibitor, in patients with advanced solid tumors. *Clinical Cancer Research, 23*(16), 4642–4650. https://doi.org/10.1158/1078-0432.CCR-16-2363.

Carpinteiro, A., Becker, K. A., Japtok, L., Hessler, G., Keitsch, S., Pozgajova, M., et al. (2015). Regulation of hematogenous tumor metastasis by acid sphingomyelinase. *EMBO Molecular Medicine*, 7(6), 714–734. https://doi.org/10.15252/emmm.201404571.

Ch'ang, H. J., Maj, J. G., Paris, F., Xing, H. R., Zhang, J., Truman, J. P., et al. (2005). ATM regulates target switching to escalating doses of radiation in the intestines. *Nature Medicine, 11*(5), 484–490. https://doi.org/10.1038/nm1237.

Chang, K. T., Anishkin, A., Patwardhan, G. A., Beverly, L. J., Siskind, L. J., & Colombini, M. (2015). Ceramide channels: Destabilization by Bcl-xL and role in apoptosis. *Biochimica et Biophysica Acta, 1848*(10 Pt A), 2374–2384. https://doi.org/10.1016/j.bbamem.2015.07.013.

Cheng, J. C., Bai, A., Beckham, T. H., Marrison, S. T., Yount, C. L., Young, K., et al. (2013). Radiation-induced acid ceramidase confers prostate cancer resistance and tumor relapse. *Journal of Clinical Investigation, 123*(10), 4344–4358. https://doi.org/10.1172/JCI64791.
Chen, Y., Zhang, P., Xu, S. C., Yang, L., Voss, U., Ekblad, E., et al. (2015). Enhanced colonic tumorigenesis in alkaline sphingomyelinase (NPP7) knockout mice. *Molecular Cancer Therapeutics, 14*(1), 259–267. https://doi.org/10.1158/1535-7163.MCT-14-0468-T.
Chipuk, J. E., McStay, G. P., Bharti, A., Kuwana, T., Clarke, C. J., Siskind, L. J., et al. (2012). Sphingolipid metabolism cooperates with BAK and BAX to promote the mitochondrial pathway of apoptosis. *Cell, 148*(5), 988–1000. https://doi.org/10.1016/j.cell.2012.01.038.
Cohen, J. A., Barkhof, F., Comi, G., Hartung, H. P., Khatri, B. O., Montalban, X., et al. (2010). Oral fingolimod or intramuscular interferon for relapsing multiple sclerosis. *The New England Journal of Medicine, 362*(5), 402–415. https://doi.org/10.1056/NEJMoa0907839.
Corcelle-Termeau, E., Vindelov, S. D., Hamalisto, S., Mograbi, B., Keldsbo, A., Brasen, J. H., et al. (2016). Excess sphingomyelin disturbs ATG9A trafficking and autophagosome closure. *Autophagy, 12*(5), 833–849. https://doi.org/10.1080/15548627.2016.1159378.
Cuvillier, O., Pirianov, G., Kleuser, B., Vanek, P. G., Coso, O. A., Gutkind, S., et al. (1996). Suppression of ceramide-mediated programmed cell death by sphingosine-1-phosphate. *Nature, 381*(6585), 800–803. https://doi.org/10.1038/381800a0.
D'Angelo, G., Uemura, T., Chuang, C. C., Polishchuk, E., Santoro, M., Ohvo-Rekila, H., et al. (2013). Vesicular and non-vesicular transport feed distinct glycosylation pathways in the Golgi. *Nature, 501*(7465), 116–120. https://doi.org/10.1038/nature12423.
Dany, M., Gencer, S., Nganga, R., Thomas, R. J., Oleinik, N., Baron, K. D., et al. (2016). Targeting FLT3-ITD signaling mediates ceramide-dependent mitophagy and attenuates drug resistance in AML. *Blood, 128*(15), 1944–1958. https://doi.org/10.1182/blood-2016-04-708750.
Deng, X., Yin, X., Allan, R., Lu, D. D., Maurer, C. W., Haimovitz-Friedman, A., et al. (2008). Ceramide biogenesis is required for radiation-induced apoptosis in the germ line of *C. elegans*. *Science, 322*(5898), 110–115. https://doi.org/10.1126/science.1158111.
Dressler, K. A., Mathias, S., & Kolesnick, R. N. (1992). Tumor necrosis factor-alpha activates the sphingomyelin signal transduction pathway in a cell-free system. *Science, 255*(5052), 1715–1718.
Dubois, N., Rio, E., Ripoche, N., Ferchaud-Roucher, V., Gaugler, M. H., Campion, L., et al. (2016). Plasma ceramide, a real-time predictive marker of pulmonary and hepatic metastases response to stereotactic body radiation therapy combined with irinotecan. *Radiotherapy and Oncology: Journal of the European Society for Therapeutic Radiology and Oncology, 119*(2), 229–235. https://doi.org/10.1016/j.radonc.2016.03.014.
Edvardson, S., Yi, J. K., Jalas, C., Xu, R., Webb, B. D., Snider, J., et al. (2016). Deficiency of the alkaline ceramidase ACER3 manifests in early childhood by progressive leukodystrophy. *Journal of Medical Genetics, 53*(6), 389–396. https://doi.org/10.1136/jmedgenet-2015-103457.
Eliyahu, E., Park, J. H., Shtraizent, N., He, X., & Schuchman, E. H. (2007). Acid ceramidase is a novel factor required for early embryo survival. *The FASEB Journal, 21*(7), 1403–1409. https://doi.org/10.1096/fj.06-7016com.
Fekry, B., Esmaeilniakooshkghazi, A., Krupenko, S. A., & Krupenko, N. I. (2016). Ceramide synthase 6 is a novel target of methotrexate mediating its antiproliferative effect in a p53-dependent manner. *PLoS One, 11*(1), e0146618. https://doi.org/10.1371/journal.pone.0146618.
Fekry, B., Jeffries, K. A., Esmaeilniakooshkghazi, A., Ogretmen, B., Krupenko, S. A., & Krupenko, N. I. (2016). CerS6 is a novel transcriptional target of p53 protein activated by non-genotoxic stress. *Journal of Biological Chemistry, 291*(32), 16586–16596. https://doi.org/10.1074/jbc.M116.716902.

Garcia-Barros, M., Coant, N., Kawamori, T., Wada, M., Snider, A. J., Truman, J. P., et al. (2016). Role of neutral ceramidase in colon cancer. *The FASEB Journal, 30*(12), 4159–4171. https://doi.org/10.1096/fj.201600611R.

Gencer, S., Oleinik, N., Kim, J., Panneer Selvam, S., De Palma, R., Dany, M., et al. (2017). TGF-beta receptor I/II trafficking and signaling at primary cilia are inhibited by ceramide to attenuate cell migration and tumor metastasis. *Science Signaling, 10*(502). https://doi.org/10.1126/scisignal.aam7464.

Gorelik, A., Illes, K., Heinz, L. X., Superti-Furga, G., & Nagar, B. (2016). Crystal structure of mammalian acid sphingomyelinase. *Nature Communications, 7*, 12196. https://doi.org/10.1038/ncomms12196.

Gupta, V., Bhinge, K. N., Hosain, S. B., Xiong, K., Gu, X., Shi, R., et al. (2012). Ceramide glycosylation by glucosylceramide synthase selectively maintains the properties of breast cancer stem cells. *Journal of Biological Chemistry, 287*(44), 37195–37205. https://doi.org/10.1074/jbc.M112.396390.

Haakenson, J. K., Khokhlatchev, A. V., Choi, Y. J., Linton, S. S., Zhang, P., Zaki, P. M., et al. (2015). Lysosomal degradation of CD44 mediates ceramide nanoliposome-induced anoikis and diminished extravasation in metastatic carcinoma cells. *Journal of Biological Chemistry, 290*(13), 8632–8643. https://doi.org/10.1074/jbc.M114.609677.

Han, G., Gupta, S. D., Gable, K., Niranjanakumari, S., Moitra, P., Eichler, F., et al. (2009). Identification of small subunits of mammalian serine palmitoyltransferase that confer distinct acyl-CoA substrate specificities. *Proceedings of the National Academy of Sciences of the United States of America, 106*(20), 8186–8191. https://doi.org/10.1073/pnas.0811269106.

Hanada, K., Kumagai, K., Yasuda, S., Miura, Y., Kawano, M., Fukasawa, M., et al. (2003). Molecular machinery for non-vesicular trafficking of ceramide. *Nature, 426*(6968), 803–809. https://doi.org/10.1038/nature02188.

Hannun, Y. A., & Bell, R. M. (1987). Lysosphingolipids inhibit protein kinase C: implications for the sphingolipidoses. *Science, 235*(4789), 670–674.

Hannun, Y. A., & Obeid, L. M. (2008). Principles of bioactive lipid signalling: Lessons from sphingolipids. *Nature Reviews. Molecular Cell Biology, 9*(2), 139–150. https://doi.org/10.1038/nrm2329.

Heering, J., Weis, N., Holeiter, M., Neugart, F., Staebler, A., Fehm, T. N., et al. (2012). Loss of the ceramide transfer protein augments EGF receptor signaling in breast cancer. *Cancer Research, 72*(11), 2855–2866. https://doi.org/10.1158/0008-5472.CAN-11-3069.

Hernandez-Tiedra, S., Fabrias, G., Davila, D., Salanueva, I. J., Casas, J., Montes, L. R., et al. (2016). Dihydroceramide accumulation mediates cytotoxic autophagy of cancer cells via autolysosome destabilization. *Autophagy, 12*(11), 2213–2229. https://doi.org/10.1080/15548627.2016.1213927.

Huang, W. C., Liang, J., Nagahashi, M., Avni, D., Yamada, A., Maceyka, M., et al. (2016). Sphingosine-1-phosphate phosphatase 2 promotes disruption of mucosal integrity, and contributes to ulcerative colitis in mice and humans. *The FASEB Journal, 30*(8), 2945–2958. https://doi.org/10.1096/fj.201600394R.

Huang, W. C., Tsai, C. C., Chen, C. L., Chen, T. Y., Chen, Y. P., Lin, Y. S., et al. (2011). Glucosylceramide synthase inhibitor PDMP sensitizes chronic myeloid leukemia T315I mutant to Bcr-Abl inhibitor and cooperatively induces glycogen synthase kinase-3-regulated apoptosis. *The FASEB Journal, 25*(10), 3661–3673. https://doi.org/10.1096/fj.10-180190.

Huitema, K., van den Dikkenberg, J., Brouwers, J. F., & Holthuis, J. C. (2004). Identification of a family of animal sphingomyelin synthases. *The EMBO Journal, 23*(1), 33–44. https://doi.org/10.1038/sj.emboj.7600034.

Hullin-Matsuda, F., Tomishige, N., Sakai, S., Ishitsuka, R., Ishii, K., Makino, A., et al. (2012). Limonoid compounds inhibit sphingomyelin biosynthesis by preventing CERT protein-dependent extraction of ceramides from the endoplasmic reticulum. *Journal of Biological Chemistry, 287*(29), 24397–24411. https://doi.org/10.1074/jbc.M112.356733. 10.1074/jbc.M112.344432.

Jennemann, R., Rabionet, M., Gorgas, K., Epstein, S., Dalpke, A., Rothermel, U., et al. (2012). Loss of ceramide synthase 3 causes lethal skin barrier disruption. *Human Molecular Genetics, 21*(3), 586–608. https://doi.org/10.1093/hmg/ddr494.

Jensen, S. A., Calvert, A. E., Volpert, G., Kouri, F. M., Hurley, L. A., Luciano, J. P., et al. (2014). Bcl2L13 is a ceramide synthase inhibitor in glioblastoma. *Proceedings of the National Academy of Sciences of the United States of America, 111*(15), 5682–5687. https://doi.org/10.1073/pnas.1316700111.

Kappos, L., Radue, E. W., O'Connor, P., Polman, C., Hohlfeld, R., Calabresi, P., et al. (2010). A placebo-controlled trial of oral fingolimod in relapsing multiple sclerosis. *The New England Journal of Medicine, 362*(5), 387–401. https://doi.org/10.1056/NEJMoa0909494.

Kim, J. W., Park, Y., Roh, J. L., Cho, K. J., Choi, S. H., Nam, S. Y., et al. (2016). Prognostic value of glucosylceramide synthase and P-glycoprotein expression in oral cavity cancer. *International Journal of Clinical Oncology, 21*(5), 883–889. https://doi.org/10.1007/s10147-016-0973-1.

Kitatani, K., Usui, T., Sriraman, S. K., Toyoshima, M., Ishibashi, M., Shigeta, S., et al. (2016). Ceramide limits phosphatidylinositol-3-kinase C2beta-controlled cell motility in ovarian cancer: Potential of ceramide as a metastasis-suppressor lipid. *Oncogene, 35*(21), 2801–2812. https://doi.org/10.1038/onc.2015.330.

Koybasi, S., Senkal, C. E., Sundararaj, K., Spassieva, S., Bielawski, J., Osta, W., et al. (2004). Defects in cell growth regulation by C18:0-ceramide and longevity assurance gene 1 in human head and neck squamous cell carcinomas. *Journal of Biological Chemistry, 279*(43), 44311–44319. https://doi.org/10.1074/jbc.M406920200.

Kramer, R., Bielawski, J., Kistner-Griffin, E., Othman, A., Alecu, I., Ernst, D., et al. (2015). Neurotoxic 1-deoxysphingolipids and paclitaxel-induced peripheral neuropathy. *The FASEB Journal, 29*(11), 4461–4472. https://doi.org/10.1096/fj.15-272567.

Laviad, E. L., Kelly, S., Merrill, A. H., Jr., & Futerman, A. H. (2012). Modulation of ceramide synthase activity via dimerization. *Journal of Biological Chemistry, 287*(25), 21025–21033. https://doi.org/10.1074/jbc.M112.363580.

Lee, A. J., Roylance, R., Sander, J., Gorman, P., Endesfelder, D., Kschischo, M., et al. (2012). CERT depletion predicts chemotherapy benefit and mediates cytotoxic and polyploid-specific cancer cell death through autophagy induction. *The Journal of Pathology, 226*(3), 482–494. https://doi.org/10.1002/path.2998.

Lee, H., Rotolo, J. A., Mesicek, J., Penate-Medina, T., Rimner, A., Liao, W. C., et al. (2011). Mitochondrial ceramide-rich macrodomains functionalize Bax upon irradiation. *PLoS One, 6*(6), e19783. https://doi.org/10.1371/journal.pone.0019783.

Liakath-Ali, K., Vancollie, V. E., Lelliott, C. J., Speak, A. O., Lafont, D., Protheroe, H. J., et al. (2016). Alkaline ceramidase 1 is essential for mammalian skin homeostasis and regulating whole-body energy expenditure. *The Journal of Pathology, 239*(3), 374–383. https://doi.org/10.1002/path.4737.

Liang, J., Nagahashi, M., Kim, E. Y., Harikumar, K. B., Yamada, A., Huang, W. C., et al. (2013). Sphingosine-1-phosphate links persistent STAT3 activation, chronic intestinal inflammation, and development of colitis-associated cancer. *Cancer Cell, 23*(1), 107–120. https://doi.org/10.1016/j.ccr.2012.11.013.

Liu, F., Li, X., Lu, C., Bai, A., Bielawski, J., Bielawska, A., et al. (2016). Ceramide activates lysosomal cathepsin B and cathepsin D to attenuate autophagy and induces ER stress to suppress myeloid-derived suppressor cells. *Oncotarget, 7*(51), 83907–83925. https://doi.org/10.18632/oncotarget.13438.

Liu, X., Ryland, L., Yang, J., Liao, A., Aliaga, C., Watts, R., et al. (2010). Targeting of survivin by nanoliposomal ceramide induces complete remission in a rat model of NK-LGL leukemia. *Blood, 116*(20), 4192–4201. https://doi.org/10.1182/blood-2010-02-271080.

Liu, Y. Y., Patwardhan, G. A., Bhinge, K., Gupta, V., Gu, X., & Jazwinski, S. M. (2011). Suppression of glucosylceramide synthase restores p53-dependent apoptosis in mutant p53 cancer cells. *Cancer Research, 71*(6), 2276–2285. https://doi.org/10.1158/0008-5472.CAN-10-3107.

Maceyka, M., & Spiegel, S. (2014). Sphingolipid metabolites in inflammatory disease. *Nature, 510*(7503), 58–67. https://doi.org/10.1038/nature13475.

Mao, Z., Sun, W., Xu, R., Novgorodov, S., Szulc, Z. M., Bielawski, J., et al. (2010). Alkaline ceramidase 2 (ACER2) and its product dihydrosphingosine mediate the cytotoxicity of N-(4-hydroxyphenyl)retinamide in tumor cells. *Journal of Biological Chemistry, 285*(38), 29078–29090. https://doi.org/10.1074/jbc.M110.105296.

Menuz, V., Howell, K. S., Gentina, S., Epstein, S., Riezman, I., Fornallaz-Mulhauser, M., et al. (2009). Protection of *C. elegans* from anoxia by HYL-2 ceramide synthase. *Science, 324*(5925), 381–384. https://doi.org/10.1126/science.1168532.

Mesicek, J., Lee, H., Feldman, T., Jiang, X., Skobeleva, A., Berdyshev, E. V., et al. (2010). Ceramide synthases 2, 5, and 6 confer distinct roles in radiation-induced apoptosis in HeLa cells. *Cellular Signalling, 22*(9), 1300–1307. https://doi.org/10.1016/j.cellsig.2010.04.006.

Meyers-Needham, M., Ponnusamy, S., Gencer, S., Jiang, W., Thomas, R. J., Senkal, C. E., et al. (2012). Concerted functions of HDAC1 and microRNA-574-5p repress alternatively spliced ceramide synthase 1 expression in human cancer cells. *EMBO Molecular Medicine, 4*(2), 78–92. https://doi.org/10.1002/emmm.201100189.

Mukhopadhyay, A., Saddoughi, S. A., Song, P., Sultan, I., Ponnusamy, S., Senkal, C. E., et al. (2009). Direct interaction between the inhibitor 2 and ceramide via sphingolipid-protein binding is involved in the regulation of protein phosphatase 2A activity and signaling. *The FASEB Journal, 23*(3), 751–763. https://doi.org/10.1096/fj.08-120550.

Nair, S., Branagan, A. R., Liu, J., Boddupalli, C. S., Mistry, P. K., & Dhodapkar, M. V. (2016). Clonal immunoglobulin against Lysolipids in the origin of myeloma. *The New England Journal of Medicine, 374*(6), 555–561. https://doi.org/10.1056/NEJMoa1508808.

Neviani, P., Harb, J. G., Oaks, J. J., Santhanam, R., Walker, C. J., Ellis, J. J., et al. (2013). PP2A-activating drugs selectively eradicate TKI-resistant chronic myeloid leukemic stem cells. *Journal of Clinical Investigation, 123*(10), 4144–4157. https://doi.org/10.1172/JCI68951.

Obeid, L. M., Linardic, C. M., Karolak, L. A., & Hannun, Y. A. (1993). Programmed cell death induced by ceramide. *Science, 259*(5102), 1769–1771.

Ogretmen, B. (2018). Sphingolipid metabolism in cancer signalling and therapy. *Nature Reviews Cancer, 18*(1), 33–50. https://doi.org/10.1038/nrc.2017.96.

Ogretmen, B., & Hannun, Y. A. (2004). Biologically active sphingolipids in cancer pathogenesis and treatment. *Nature Reviews Cancer, 4*(8), 604–616. https://doi.org/10.1038/nrc1411.

Ogretmen, B., Pettus, B. J., Rossi, M. J., Wood, R., Usta, J., Szulc, Z., et al. (2002). Biochemical mechanisms of the generation of endogenous long chain ceramide in response to exogenous short chain ceramide in the A549 human lung adenocarcinoma cell line. Role for endogenous ceramide in mediating the action of exogenous ceramide. *Journal of Biological Chemistry, 277*(15), 12960–12969. https://doi.org/10.1074/jbc.M110699200.

Pandey, M. K., Burrow, T. A., Rani, R., Martin, L. J., Witte, D., Setchell, K. D., et al. (2017). Complement drives glucosylceramide accumulation and tissue inflammation in Gaucher disease. *Nature, 543*(7643), 108–112. https://doi.org/10.1038/nature21368.

Park, M. A., Mitchell, C., Zhang, G., Yacoub, A., Allegood, J., Haussinger, D., et al. (2010). Vorinostat and sorafenib increase CD95 activation in gastrointestinal tumor cells through a Ca(2+)-de novo ceramide-PP2A-reactive oxygen species-dependent signaling pathway. *Cancer Research, 70*(15), 6313–6324. https://doi.org/10.1158/0008-5472.CAN-10-0999.

Park, W. J., Brenner, O., Kogot-Levin, A., Saada, A., Merrill, A. H., Jr., Pewzner-Jung, Y., et al. (2015). Development of pheochromocytoma in ceramide synthase 2 null mice. *Endocrine-Related Cancer, 22*(4), 623–632. https://doi.org/10.1530/ERC-15-0058.

Pastukhov, O., Schwalm, S., Zangemeister-Wittke, U., Fabbro, D., Bornancin, F., Japtok, L., et al. (2014). The ceramide kinase inhibitor NVP-231 inhibits breast and lung cancer cell proliferation by inducing M phase arrest and subsequent cell death. *British Journal of Pharmacology, 171*(24), 5829–5844. https://doi.org/10.1111/bph.12886.

Payne, A. W., Pant, D. K., Pan, T. C., & Chodosh, L. A. (2014). Ceramide kinase promotes tumor cell survival and mammary tumor recurrence. *Cancer Research, 74*(21), 6352–6363. https://doi.org/10.1158/0008-5472.CAN-14-1292.

Pewzner-Jung, Y., Ben-Dor, S., & Futerman, A. H. (2006). When do Lasses (longevity assurance genes) become CerS (ceramide synthases)?: Insights into the regulation of ceramide synthesis. *Journal of Biological Chemistry, 281*(35), 25001–25005. https://doi.org/10.1074/jbc.R600010200.

Rahmaniyan, M., Curley, R. W., Jr., Obeid, L. M., Hannun, Y. A., & Kraveka, J. M. (2011). Identification of dihydroceramide desaturase as a direct in vitro target for fenretinide. *Journal of Biological Chemistry, 286*(28), 24754–24764. https://doi.org/10.1074/jbc.M111.250779.

Raichur, S., Wang, S. T., Chan, P. W., Li, Y., Ching, J., Chaurasia, B., et al. (2014). CerS2 haploinsufficiency inhibits beta-oxidation and confers susceptibility to diet-induced steatohepatitis and insulin resistance. *Cell Metabolism, 20*(4), 687–695. https://doi.org/10.1016/j.cmet.2014.09.015.

Realini, N., Palese, F., Pizzirani, D., Pontis, S., Basit, A., Bach, A., et al. (2016). Acid Ceramidase in Melanoma: Expression, localization, and effects of pharmacological inhibition. *Journal of Biological Chemistry, 291*(5), 2422–2434. https://doi.org/10.1074/jbc.M115.666909.

Roh, J. L., Kim, E. H., Park, J. Y., & Kim, J. W. (2015). Inhibition of glucosylceramide synthase sensitizes head and neck cancer to cisplatin. *Molecular Cancer Therapeutics, 14*(8), 1907–1915. https://doi.org/10.1158/1535-7163.MCT-15-0171.

Rotolo, J., Stancevic, B., Zhang, J., Hua, G., Fuller, J., Yin, X., et al. (2012). Anti-ceramide antibody prevents the radiation gastrointestinal syndrome in mice. *Journal of Clinical Investigation, 122*(5), 1786–1790. https://doi.org/10.1172/JCI59920.

Saddoughi, S. A., Garrett-Mayer, E., Chaudhary, U., O'Brien, P. E., Afrin, L. B., Day, T. A., et al. (2011). Results of a phase II trial of gemcitabine plus doxorubicin in patients with recurrent head and neck cancers: Serum C(1)(8)-ceramide as a novel biomarker for monitoring response. *Clinical Cancer Research, 17*(18), 6097–6105. https://doi.org/10.1158/1078-0432.CCR-11-0930.

Saddoughi, S. A., Gencer, S., Peterson, Y. K., Ward, K. E., Mukhopadhyay, A., Oaks, J., et al. (2013). Sphingosine analogue drug FTY720 targets I2PP2A/SET and mediates lung tumour suppression via activation of PP2A-RIPK1-dependent necroptosis. *EMBO Molecular Medicine, 5*(1), 105–121. https://doi.org/10.1002/emmm.201201283.

Salazar, M., Carracedo, A., Salanueva, I. J., Hernandez-Tiedra, S., Lorente, M., Egia, A., et al. (2009). Cannabinoid action induces autophagy-mediated cell death through stimulation of ER stress in human glioma cells. *Journal of Clinical Investigation, 119*(5), 1359–1372.

Scheffel, M. J., Helke, K., Lu, P., Bowers, J. S., Ogretmen, B., Garrett-Mayer, E., et al. (2017). Adoptive transfer of ceramide synthase 6 deficient splenocytes reduces the development of colitis. *Scientific Reports, 7*(1), 15552. https://doi.org/10.1038/s41598-017-15791-x.

Schiffmann, S., Hartmann, D., Fuchs, S., Birod, K., Ferreiros, N., Schreiber, Y., et al. (2012). Inhibitors of specific ceramide synthases. *Biochimie, 94*(2), 558–565. https://doi.org/10.1016/j.biochi.2011.09.007.

Senkal, C. E., Ponnusamy, S., Bielawski, J., Hannun, Y. A., & Ogretmen, B. (2010). Antiapoptotic roles of ceramide-synthase-6-generated C16-ceramide via selective regulation of the ATF6/CHOP arm of ER-stress-response pathways. *The FASEB Journal, 24*(1), 296–308. https://doi.org/10.1096/fj.09-135087.

Senkal, C. E., Ponnusamy, S., Rossi, M. J., Bialewski, J., Sinha, D., Jiang, J. C., et al. (2007). Role of human longevity assurance gene 1 and C18-ceramide in chemotherapy-induced cell death in human head and neck squamous cell carcinomas. *Molecular Cancer Therapeutics, 6*(2), 712–722. https://doi.org/10.1158/1535-7163.MCT-06-0558.

Senkal, C. E., Ponnusamy, S., Rossi, M. J., Sundararaj, K., Szulc, Z., Bielawski, J., et al. (2006). Potent antitumor activity of a novel cationic pyridinium-ceramide alone or in combination with gemcitabine against human head and neck squamous cell carcinomas in vitro and in vivo. *The Journal of Pharmacology and Experimental Therapeutics, 317*(3), 1188–1199. https://doi.org/10.1124/jpet.106.101949.

Senkal, C. E., Salama, M. F., Snider, A. J., Allopenna, J. J., Rana, N. A., Koller, A., et al. (2017). Ceramide is metabolized to acylceramide and stored in lipid droplets. *Cell Metabolism, 25*(3), 686–697. https://doi.org/10.1016/j.cmet.2017.02.010.

Sentelle, R. D., Senkal, C. E., Jiang, W., Ponnusamy, S., Gencer, S., Selvam, S. P., et al. (2012). Ceramide targets autophagosomes to mitochondria and induces lethal mitophagy. *Nature Chemical Biology, 8*(10), 831–838. https://doi.org/10.1038/nchembio.1059.

Shamseddine, A. A., Clarke, C. J., Carroll, B., Airola, M. V., Mohammed, S., Rella, A., et al. (2015). P53-dependent upregulation of neutral sphingomyelinase-2: Role in doxorubicin-induced growth arrest. *Cell Death and Disease, 6*, e1947. https://doi.org/10.1038/cddis.2015.268.

Shaner, R. L., Allegood, J. C., Park, H., Wang, E., Kelly, S., Haynes, C. A., et al. (2009). Quantitative analysis of sphingolipids for lipidomics using triple quadrupole and quadrupole linear ion trap mass spectrometers. *The Journal of Lipid Research, 50*(8), 1692–1707. https://doi.org/10.1194/jlr.D800051-JLR200.

Simanshu, D. K., Kamlekar, R. K., Wijesinghe, D. S., Zou, X., Zhai, X., Mishra, S. K., et al. (2013). Non-vesicular trafficking by a ceramide-1-phosphate transfer protein regulates eicosanoids. *Nature, 500*(7463), 463–467. https://doi.org/10.1038/nature12332.

Siskind, L. J., Feinstein, L., Yu, T., Davis, J. S., Jones, D., Choi, J., et al. (2008). Antiapoptotic Bcl-2 family proteins disassemble ceramide channels. *Journal of Biological Chemistry, 283*(11), 6622–6630. https://doi.org/10.1074/jbc.M706115200.

Sofi, M. H., Heinrichs, J., Dany, M., Nguyen, H., Dai, M., Bastian, D., et al. (2017). Ceramide synthesis regulates T cell activity and GVHD development. *Journal of Clinical Investigation Insight, 2*(10). https://doi.org/10.1172/jci.insight.91701.

Stefanovic, M., Tutusaus, A., Martinez-Nieto, G. A., Barcena, C., de Gregorio, E., Moutinho, C., et al. (2016). Targeting glucosylceramide synthase upregulation reverts sorafenib resistance in experimental hepatocellular carcinoma. *Oncotarget, 7*(7), 8253–8267. https://doi.org/10.18632/oncotarget.6982.

Suzuki, M., Cao, K., Kato, S., Komizu, Y., Mizutani, N., Tanaka, K., et al. (2016). Targeting ceramide synthase 6-dependent metastasis-prone phenotype in lung cancer cells. *Journal of Clinical Investigation, 126*(1), 254–265. https://doi.org/10.1172/JCI79775.

Swanton, C., Marani, M., Pardo, O., Warne, P. H., Kelly, G., Sahai, E., et al. (2007). Regulators of mitotic arrest and ceramide metabolism are determinants of sensitivity to paclitaxel and other chemotherapeutic drugs. *Cancer Cell, 11*(6), 498–512. https://doi.org/10.1016/j.ccr.2007.04.011.

Tan, S. F., Liu, X., Fox, T. E., Barth, B. M., Sharma, A., Turner, S. D., et al. (2016). Acid ceramidase is upregulated in AML and represents a novel therapeutic target. *Oncotarget, 7*(50), 83208–83222. https://doi.org/10.18632/oncotarget.13079.

Taouji, S., Higa, A., Delom, F., Palcy, S., Mahon, F. X., Pasquet, J. M., et al. (2013). Phosphorylation of serine palmitoyltransferase long chain-1 (SPTLC1) on tyrosine 164 inhibits its activity and promotes cell survival. *Journal of Biological Chemistry, 288*(24), 17190–17201. https://doi.org/10.1074/jbc.M112.409185.

Thomas, R. J., Oleinik, N., Panneer Selvam, S., Vaena, S. G., Dany, M., Nganga, R. N., et al. (2017). HPV/E7 induces chemotherapy-mediated tumor suppression by ceramide-dependent mitophagy. *EMBO Molecular Medicine, 9*(8), 1030–1051. https://doi.org/10.15252/emmm.201607088.

Tidhar, R., Ben-Dor, S., Wang, E., Kelly, S., Merrill, A. H., Jr., & Futerman, A. H. (2012). Acyl chain specificity of ceramide synthases is determined within a region of 150 residues in the Tram-Lag-CLN8 (TLC) domain. *Journal of Biological Chemistry, 287*(5), 3197–3206. https://doi.org/10.1074/jbc.M111.280271.

Tirodkar, T. S., Lu, P., Bai, A., Scheffel, M. J., Gencer, S., Garrett-Mayer, E., et al. (2015). Expression of ceramide synthase 6 transcriptionally activates acid ceramidase in a c-Jun N-terminal kinase (JNK)-dependent manner. *Journal of Biological Chemistry, 290*(21), 13157–13167. https://doi.org/10.1074/jbc.M114.631325.

Trajkovic, K., Hsu, C., Chiantia, S., Rajendran, L., Wenzel, D., Wieland, F., et al. (2008). Ceramide triggers budding of exosome vesicles into multivesicular endosomes. *Science, 319*(5867), 1244–1247. https://doi.org/10.1126/science.1153124.

Venant, H., Rahmaniyan, M., Jones, E. E., Lu, P., Lilly, M. B., Garrett-Mayer, E., et al. (2015). The sphingosine kinase 2 inhibitor ABC294640 reduces the growth of prostate cancer cells and results in accumulation of dihydroceramides in vitro and in vivo. *Molecular Cancer Therapeutics, 14*(12), 2744–2752. https://doi.org/10.1158/1535-7163.MCT-15-0279.

Wang, X., Rao, R. P., Kosakowska-Cholody, T., Masood, M. A., Southon, E., Zhang, H., et al. (2009). Mitochondrial degeneration and not apoptosis is the primary cause of embryonic lethality in ceramide transfer protein mutant mice. *The Journal of Cell Biology, 184*(1), 143–158. https://doi.org/10.1083/jcb.200807176.

Wang, K., Xu, R., Snider, A. J., Schrandt, J., Li, Y., Bialkowska, A. B., et al. (2016). Alkaline ceramidase 3 deficiency aggravates colitis and colitis-associated tumorigenesis in mice by hyperactivating the innate immune system. *Cell Death and Disease, 7*, e2124. https://doi.org/10.1038/cddis.2016.36.

Watters, R. J., Fox, T. E., Tan, S. F., Shanmugavelandy, S., Choby, J. E., Broeg, K., et al. (2013). Targeting glucosylceramide synthase synergizes with C6-ceramide nanoliposomes to induce apoptosis in natural killer cell leukemia. *Leukemia and Lymphoma, 54*(6), 1288–1296. https://doi.org/10.3109/10428194.2012.752485.

White-Gilbertson, S., Mullen, T., Senkal, C., Lu, P., Ogretmen, B., Obeid, L., et al. (2009). Ceramide synthase 6 modulates TRAIL sensitivity and nuclear translocation of active caspase-3 in colon cancer cells. *Oncogene, 28*(8), 1132–1141. https://doi.org/10.1038/onc.2008.468.

Wijesinghe, D. S., Brentnall, M., Mietla, J. A., Hoeferlin, L. A., Diegelmann, R. F., Boise, L. H., et al. (2014). Ceramide kinase is required for a normal eicosanoid response and the subsequent orderly migration of fibroblasts. *The Journal of Lipid Research, 55*(7), 1298–1309. https://doi.org/10.1194/jlr.M048207.

Wijesinghe, D. S., Lamour, N. F., Gomez-Munoz, A., & Chalfant, C. E. (2007). Ceramide kinase and ceramide-1-phosphate. *Methods in Enzymology, 434*, 265–292. https://doi.org/10.1016/S0076-6879(07)34015-9.

Yacoub, A., Hamed, H. A., Allegood, J., Mitchell, C., Spiegel, S., Lesniak, M. S., et al. (2010). PERK-dependent regulation of ceramide synthase 6 and thioredoxin play a key role in mda-7/IL-24-induced killing of primary human glioblastoma multiforme cells. *Cancer Research, 70*(3), 1120–1129. https://doi.org/10.1158/0008-5472.CAN-09-4043.

Zamora-Pineda, J., Kumar, A., Suh, J. H., Zhang, M., & Saba, J. D. (2016). Dendritic cell sphingosine-1-phosphate lyase regulates thymic egress. *The Journal of Experimental Medicine, 213*(12), 2773–2791. https://doi.org/10.1084/jem.20160287.

FURTHER READING

Senkal, C. E., Ponnusamy, S., Manevich, Y., Meyers-Needham, M., Saddoughi, S. A., Mukhopadyay, A., et al. (2011). Alteration of ceramide synthase 6/C16-ceramide induces activating transcription factor 6-mediated endoplasmic reticulum (ER) stress and apoptosis via perturbation of cellular Ca2+ and ER/Golgi membrane network. *Journal of Biological Chemistry, 286*(49), 42446–42458. https://doi.org/10.1074/jbc.M111.287383.

CHAPTER TWO

Sphingolipids as Regulators of Autophagy and Endocytic Trafficking

Megan M. Young and Hong-Gang Wang[1]
Department of Pediatrics, Pennsylvania State University College of Medicine, Hershey, PA, United States
[1]Corresponding author: E-mail: huw11@psu.edu

Contents

Advances in Cancer Research, Volume 140
ISSN 0065-230X
https://doi.org/10.1016/bs.acr.2018.04.008

Abstract

Macroautophagy (herein referred to as autophagy) is a highly conserved stress response that engulfs damaged proteins, lipids, and/or organelles within double-membrane vesicles called autophagosomes for lysosomal degradation. Dysregulated autophagy is a hallmark of cancer; and thus, there is great interest in modulating autophagy for cancer therapy. Sphingolipids regulate each step of autophagosome biogenesis with roles for sphingolipid metabolites and enzymes spanning from the initial step of de novo ceramide synthesis to the sphingosine-1-phosphate lyase 1–mediated exit from the sphingolipid pathway. Notably, sphingolipid metabolism occurs at several of the organelles that contribute to autophagosome biogenesis to suggest that local changes in sphingolipids may regulate autophagy. As sphingolipid metabolism is frequently dysregulated in cancer, a molecular understanding of sphingolipids in stress-induced autophagy may provide insight into the mechanisms driving tumor development and progression. On the contrary, modulation of sphingolipid metabolites and/or enzymes can induce autophagy-dependent cell death for cancer therapy. This chapter will overview the major steps in mammalian autophagy, discuss the regulation of each step by sphingolipid metabolites, and describe the functions of sphingolipid-mediated autophagy in cancer. While our understanding of the signaling and biophysical properties of sphingolipids in autophagy remains in its infancy, the unique cross talk between the two pathways is an exciting area for further development, particularly in the context of cancer therapy.

1. INTRODUCTION

Autophagy is a catabolic process in which cytoplasmic components, such as damaged organelles, proteins, lipids, or pathogens, are sequestered within double-membrane vesicles called autophagosomes (0.5–1.5 μm diameter) and delivered to lysosomes for degradation and recycling (Fig. 1). While originally thought to be a bulk and nonselective process, it is now understood that autophagy regulates the selective degradation of pathogens (xenophagy) and organelles, including mitochondria (mitophagy), peroxisomes (pexophagy), endoplasmic reticulum (ER) (reticulophagy), and lipid droplets (lipophagy) (Galluzzi et al., 2017). The identification of over 30 conserved autophagy-related (ATG) genes has defined core sets of machinery for the five major steps of autophagy: initiation, nucleation, elongation, closure, and fusion (Fig. 1) (Feng, He, Yao, & Klionsky, 2014; Mercer, Gubas, & Tooze, 2018).

Dysregulated autophagy is a hallmark of cancer (Hanahan & Weinberg, 2011), and thus, a frequent question is whether autophagy should be activated or inhibited for cancer therapy. Unfortunately, the answer to this

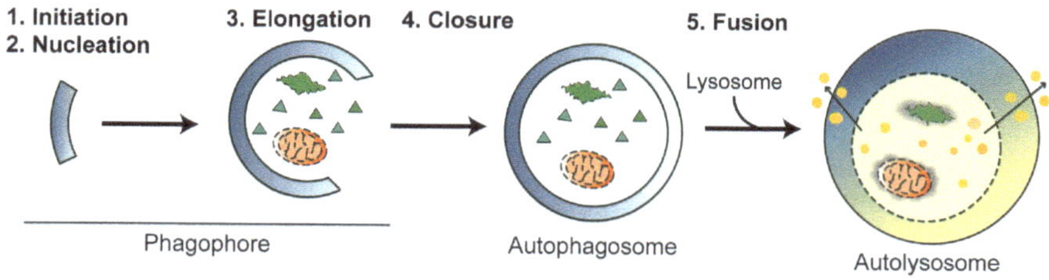

Figure 1 ***Overview of autophagy.*** Autophagosome biogenesis is comprised of five major steps: initiation, nucleation of the phagophore membrane, elongation of the phagophore to engulf autophagic cargo, phagophore closure to generate the completed double-membrane autophagosome, and lysosomal fusion to generate the autolysosome in which cargo is degraded and recycled to the cytosol.

question is complex and context-dependent as autophagy has paradoxical functions as both a tumor suppressor and tumor promoter and can serve to promote or suppress cell death (Levy, Towers, & Thorburn, 2017). During cancer development, autophagy constrains tumor initiation through the removal of dysfunctional organelles, such as mitochondria, and reactive oxygen species that drive genomic instability and malignant transformation (Yang et al., 2011). In contrast, autophagy supports cell survival in established tumors by providing an intracellular source of nutrients to meet the high proliferative and metabolic demands of the hostile, and often nutrient-deprived, tumor microenvironment (Galluzzi et al., 2015; Rebecca & Amaravadi, 2015; Zhi & Zhong, 2015). As high autophagic flux is associated with poor patient prognosis, enhanced metastasis, and resistance to chemotherapy (Lazova et al., 2012; Mikhaylova et al., 2012), inhibiting autophagy may be an effective anticancer strategy. Alternatively, autophagy can serve as a cell death mechanism to kill cancer cells that are resistant to canonical cell death signaling. Here, we will provide an overview of the major steps in mammalian autophagy, discuss the regulation of each step by sphingolipid metabolites, and describe the functions of sphingolipid-mediated autophagy in cancer.

2. THE MOLECULAR MACHINERY OF AUTOPHAGY AND REGULATION BY SPHINGOLIPIDS

2.1 Autophagy Initiation: the Unc-51–Like Kinase Complex

The Unc-51–like kinase 1/2 (ULK1/2) complex contains ATG13, FIP200, and ATG101 and is responsible for initiating autophagosome

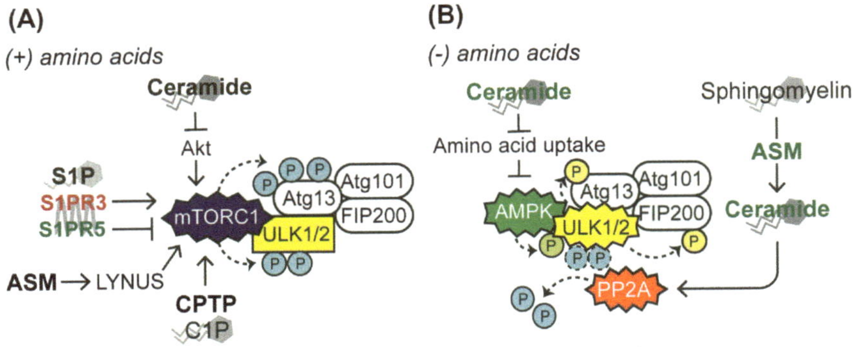

Figure 2 ***Autophagy initiation.*** (A) In the presence of amino acids, mTORC1 inhibits ULK1/2 complex activation to impair autophagy. mTORC1 is regulated by ceramide and sphingosine-1-phosphate (S1P) signaling as well as acid sphingomyelinase (ASM) and ceramide-1-phosphate transfer protein (CPTP). LYNUS, lysosomal nutrient sensing complex. (B) Amino acid deprivation inhibits mTORC1 and activates ULK1/2 via AMPK and PP2A-mediated signaling. Ceramide positively regulates AMPK and PP2A to enhance autophagy initiation. Autophagic regulation by sphingolipids in *red text* has been implicated with toxic autophagy, whereas *green text* represents survival autophagy. *See text for further details.*

biogenesis (Fig. 2) (Hara et al., 2008; Hosokawa, Hara et al., 2009; Hosokawa, Sasaki et al., 2009; Mercer, Kaliappan, & Dennis, 2009). ULK1/2 is regulated by the nutrient and energy-sensing kinases mechanistic target of rapamycin complex 1 (mTORC1) and AMP-activated protein kinase (AMPK). Under nutrient-rich conditions, mTORC1 phosphorylates ULK1 and ATG13 to inhibit autophagy (Kim, Kundu, Viollet, & Guan, 2011; Mizushima, 2010). Amino acid starvation activates AMPK, which directly phosphorylates and inhibits mTORC1 while activating ULK1 (Egan et al., 2011; J. Kim et al., 2011; Lee, Park, Takahashi, & Wang, 2010). Starvation also induces the protein phosphatase 2A (PP2A)—mediated dephosphorylation of ULK1 to enhance its activation (Wong, Feng, Wang, Shi, & Jiang, 2015). ULK1 phosphorylates ATG13 and FIP200 and translocates to subdomains of the ER coinciding with ER—mitochondrial contact sites or autophagy-specific ER exit sites to promote phagophore nucleation (Hamasaki et al., 2013; Karanasios et al., 2016).

2.1.1 Sphingolipids in Autophagy Initiation

Sphingolipid metabolites regulate the ULK1/2 complex via mTORC1, AMPK, and PP2A (Fig. 2). Amino acid starvation, a widely used stimulator of autophagy, activates acid sphingomyelinase (ASM), increases ceramide, and triggers the PP1/PP2A-dependent inactivation of mTORC1 (Taniguchi et al., 2012). Sphingosine kinase activity is also enhanced on nutrient deprivation, and loss of ASM or sphingosine kinase 1 (SphK1) impairs starvation-induced autophagy (Lavieu et al., 2006; Taniguchi et al., 2012). Interestingly, inhibition of ASM under nutrient-replete conditions inhibits mTORC1 and induces autophagy by disrupting the lysosomal nutrient-sensing complex (Justice et al., 2018). Here, ASM inhibition accumulates newly formed long-chain (LC) ceramides (C14:0, C16:0, C18:0) and significantly reduces sphingosine and sphingosine-1-phosphate (S1P) at the level of ceramide hydrolysis, leading the authors to propose that decreases in lysosomal sphingosine, rather than S1P or ceramide, drive autophagy initiation in response to ASM dysfunction. The overexpression of SphK1, which converts sphingosine to S1P, also inhibits mTORC1 for autophagy initiation in a manner that is independent of ceramide synthase (CerS) activity (Lavieu et al., 2006). While cytosolic S1P has been shown to contribute to autophagosome biogenesis (Moruno Manchon et al., 2015), extracellular S1P can modulate autophagy by activating or inhibiting mTORC1 signaling downstream of S1P receptor 5 (S1PR5) (Chang et al., 2009) or S1P receptor 3 (S1PR3) (Taniguchi et al., 2012), respectively. Therefore, the effect of S1P signaling on autophagy appears to occur by intracellular or extracellular pools of S1P as well as S1PR isoform expression.

Exogenous cell-permeable ceramides mimic amino acid starvation by triggering the rapid internalization of nutrient transporter proteins and activation of AMPK for the initiation of cytoprotective autophagy (Guenther et al., 2008). Furthermore, the accumulation of endogenous ceramide by short-chain ceramide analogues, tamoxifen-induced de novo synthesis, or inactivation of glucosylceramide synthase initiates autophagy by inhibiting Akt upstream of mTORC1 (Scarlatti et al., 2004; W. Shen et al., 2014). In contrast, increases in ceramide-1-phosphate (C1P) at the trans-Golgi network on the loss of ceramide-1-phosphate transfer protein (CPTP) inhibit mTORC1 and induce ULK1-dependent autophagy in the absence of significant changes in ceramide (Mishra et al., 2017). In the same study, overexpression of CPTP was shown to impair starvation-induced autophagy by mitigating mTORC1 inactivation on nutrient depletion to support the notion that C1P trafficking regulates ULK1/2 activity.

2.2 Phagophore Nucleation: the Phosphatidylinositol 3-Kinase Complex

ULK1 phosphorylates members of class III phosphatidylinositol 3-kinase (PI3K) complex I to stimulate phosphatidylinositol 3-monophosphate (PI3P) production for phagophore nucleation (Park et al., 2016, 2018; Russell et al., 2013). The class III PI3K complex I consists of Beclin 1, phosphatidylinositol 3-kinase catalytic subunit type 3 (PIK3C3 or VPS34), phosphoinositide-3-kinase regulatory subunit 4 (VPS15), ATG14L (or BARKOR), and nuclear receptor binding factor 2 (Fig. 3) (Itakura, Kishi, Inoue, & Mizushima, 2008; Lu et al., 2014). Autophagy and Beclin 1 regulator 1 (AMBRA1) and ER-resident vacuole membrane protein 1 (VMP1) are positive regulators of VPS34 activity, whereas the

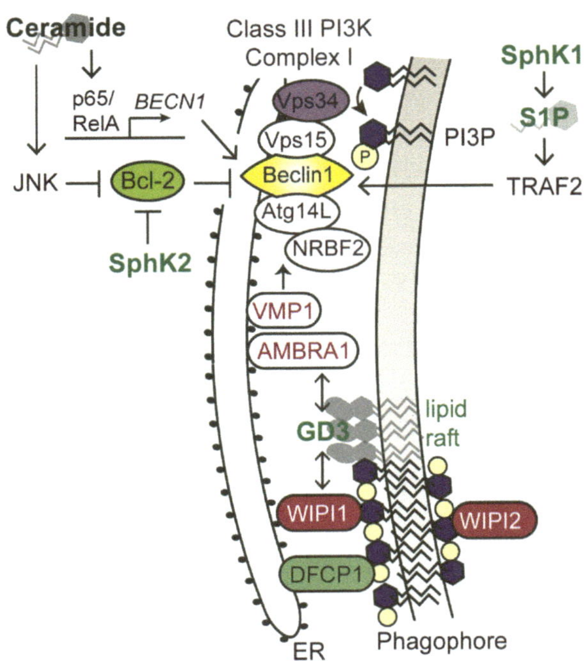

Figure 3 ***Phagophore nucleation.*** The Beclin 1–containing class III PI3K complex I regulates the nucleation of the phagophore membrane via the Vps34-mediated generation of phosphatidylinositol-3-monophosphate (PI3P) and recruitment of DFCP1 and WIPI1/2. Ceramide, SphK2, and the SphK1-dependent production of S1P positively regulate the Beclin 1 complex, whereas the ganglioside GD3 associates with PI3P, AMBRA1, and WIPI1 at mitochondria–ER contact sites and is required for phagophore nucleation. Autophagic regulation by sphingolipids in *green text* has been implicated with survival autophagy. *See text for further details.*

complex is negatively regulated by the Beclin 1 inhibitor Bcl-2 (Fimia et al., 2007; Molejon, Ropolo, Re, Boggio, & Vaccaro, 2013; Pattingre et al., 2005). However, direct regulation of Beclin 1 by Bcl-2 remains controversial (Lindqvist, Heinlein, Huang, & Vaux, 2014). The generation of PI3P recruits PI3P-binding proteins, including downstream ATG proteins, the phagophore marker double FYVE domain—containing protein 1 (DFCP1), and WD-repeat domain phosphoinositide-interacting proteins 1/2 (WIPI1/2) that are required for the nucleation of the cup-shaped isolation membrane or phagophore (Proikas-Cezanne, Takacs, Dönnes, & Kohlbacher, 2015).

2.2.1 Sphingolipids in Phagophore Nucleation

Several sphingolipid metabolites, including GD3, ceramide, and S1P, regulate the Beclin 1—containing PI3K complex (Fig. 3). Most notably, the ganglioside GD3, a core component of lipid rafts, interacts with PI3P and the autophagy proteins AMBRA1 and WIPI1 in immature LC3-II—positive autophagic structures during amino acid deprivation, and knockdown of GD3 synthase (ST8SIA1) is sufficient to impair starvation-induced autophagy (Garofalo et al., 2016; Matarrese et al., 2014). As exogenous GD3 reactivates starvation-induced autophagy in the presence of the CerS-inhibitor fumonisin B1, raft lipids appear to be important regulators of autophagosome biogenesis (Matarrese et al., 2014). Furthermore, lipid rafts are present within ER—mitochondria contact sites (a site of phagophore nucleation) with knockdown of ST8SIA1 impairing ER—mitochondria contact site formation and autophagosome nucleation (Garofalo et al., 2016). These results suggest the assembly of lipid rafts may be critical event that drives changes in membrane curvature and/or fluidity for phagophore nucleation.

Ceramide, S1P, and SphK2 promote phagophore nucleation by regulating Beclin 1. Endogenous ceramide enhances autophagy by the p65/RelA-dependent transcription of Beclin 1 (Copetti, Bertoli, Dalla, Demarchi, & Schneider, 2009) or the liberation of Beclin 1 on JNK1-mediated phosphorylation of the N-terminal loop of Bcl-2 (Pattingre et al., 2005). Interestingly, SphK2 directly interacts with Bcl-2 via a putative BH3 domain (Bcl-2 homology domain 3) to displace Beclin 1 and enhance autophagy independent of its catalytic activity (Song et al., 2017). In contrast, the SphK1-dependent production of S1P enhances TNF receptor-associated factor 2 (TRAF2) and Beclin 1 binding to promote the Lys63-linked ubiquitination of Beclin 1 for autophagy

(Liu, Ma, He, Zhao, & Shao, 2017). Indeed, cytosolic S1P appears to coincide with autophagic response as the overexpression of SphK1 upregulates Beclin 1, whereas expression of dominant-negative SphK1 or the ER-localized S1P-metabolizing enzymes sphingosine-1-phosphate phosphohydrolase-1 (SPP1) or sphingosine-1-phosphate lyase 1 (SGPL1) impair autophagic response (Moruno Manchon et al., 2015).

2.3 Phagophore Elongation: LC3 Lipidation

The elongation of the phagophore requires two ubiquitin-like conjugation systems that conjugate the ubiquitin-like proteins ATG12 and LC3 (consisting of six family members: LC3A, LC3B, LC3C, GABARAP, GABARAPL1, and GABARAPL2) to the substrates ATG5 and phosphatidylethanolamine (PE), respectively (Fig. 4A) (Geng & Klionsky, 2008). The unlipidated, cytosolic form of LC3 is referred to as LC3-I, whereas LC3 that has become conjugated to the amine head group of PE is known

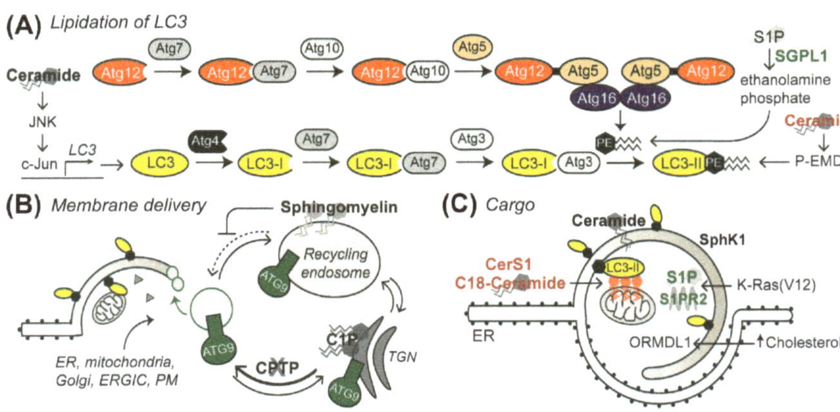

Figure 4 ***Elongation of phagophores.*** (A) The lipidation of LC3 is regulated by two ubiquitin-like conjugation systems that covalently conjugate LC3-I to PE on the phagophore. Ceramide and sphingosine-1-phosphate lyase 1 (SGPL1) promote LC3-II production. (B) ATG9-containing vesicles and several organelles are proposed to contribute membrane to the elongating phagophore. Ceramide-1-phosphate (C1P) promotes ATG9 vesicle formation whereas the accumulation of sphingomyelin within recycling endosomes on the loss of SMPD1 impairs ATG9 trafficking. (C) Ceramide synthase 1 (CerS1) and C18-ceramide directly interact with LC3-II to target mitochondria to the phagophore for degradation. Ceramide and SphK1 have been shown to colocalize with LC3-containing autophagic structures, whereas S1P/S1PR2 and ORMDL1 have been identified as autophagic substrates. Autophagic regulation by sphingolipids in *red text* has been implicated with toxic autophagy, whereas *green text* represents survival autophagy. *See text for further details.*

as LC3-II. The protease ATG4B primes pro-LC3 for lipidation by exposing a C-terminal glycine residue that is critical for conjugation (i.e., LC3-I). Both conjugation reactions utilize a common E1-like activating enzyme, ATG7, but distinct E2-like enzymes (ATG10 for ATG12–ATG5; ATG3 for LC3-II). The E3-like enzyme for ATG12–ATG5 conjugation remains unknown; however, the ATG12–ATG5 conjugate associates with ATG16L to serve as an E3-like enzyme for the transfer of LC3-I to PE (Fujita et al., 2008). Interestingly, the loss of ATG3 impairs LC3-II formation while also markedly suppressing the ATG12–ATG5 conjugate to indicate that the two conjugation systems work in concert with one another (Sou et al., 2008). While the ATG conjugation systems are not strictly required for autophagosome biogenesis, evidence suggests that the lipidation of LC3 contributes to the elongation of autophagosomal membranes, autophagosome closure, lysosomal fusion, and degradation of the inner autophagosomal membrane (Nguyen et al., 2016; Tsuboyama et al., 2016). In addition, LC3-II directly interacts with autophagy receptors (e.g., p62/sequestosome-1) via LC3-interacting region (LIR) motifs to target cargo to the autophagosomal membrane for degradation (Bjørkøy et al., 2005; Pankiv et al., 2007).

LC3-II decorates the inner and outer autophagosomal membranes and is widely used as a marker of autophagy by monitoring LC3-II levels by immunoblotting or LC3 cytoplasmic foci using fluorescence microscopy (Klionsky et al., 2016). While LC3-II on the outer membrane of completed autophagosomes is removed by the ATG4 family, LC3-II on the inner autophagosomal membrane is degraded on lysosomal fusion (Lamb, Yoshimori, & Tooze, 2013). Therefore, an increase in LC3-II can reflect enhanced autophagosome biogenesis or suppressed degradation. To discriminate between the two possibilities, autophagic flux must be examined by comparing LC3-II levels in the presence or absence of inhibitors of autophagosome–lysosome fusion (e.g., the v-ATPase inhibitor Bafilomycin A1) or lysosomal protease activity (e.g., E64d and pepstatin A) (Klionsky et al., 2016; Tooze et al., 2015).

2.3.1 Sphingolipid-Mediated Regulation of LC3 Lipidation

Ceramide and S1P directly regulate the lipidation of LC3 (Fig. 4A). SGPL1 controls the exit from the sphingolipid pathway by cleaving S1P into hexadecenal and ethanolamine phosphate, which can be directed to the synthesis of PE. Neuronal-specific depletion of SGPL1 reduces PE levels and impairs LC3 lipidation to accumulate unclosed phagophore-like structures and autophagic substrates, including p62 and aggregate-prone proteins that promote neurodegenerative disease (Mitroi et al., 2017). The addition of PE to SGPL1-deficient cells restores LC3-II levels and autophagic flux to suggest that S1P regulates PE availability for LC3 lipidation and the elongation of phagophores. Additionally, ceramide enhances the availability of LC3 by activating the JNK/c-Jun-mediated transcription of LC3 (Sun et al., 2011). Exogenous C16-ceramide also triggers autophagy-dependent cell death that is accompanied by the interaction of phosphorylated emerin (EMD) and LC3B (Deroyer, Rénert, Merville, & Fillet, 2014). While the role of the inner nuclear membrane protein EMD in autophagy remains unclear, Deroyer et al. hypothesized that phosphorylated EMD may exist in a complex with ceramide/LC3-II to promote LC3 lipidation for the induction of lethal autophagy.

2.4 Phagophore Elongation: Membrane Delivery and Autophagic Cargo

One of the most intriguing and unresolved events in autophagosome biogenesis is the membrane source for the elongating phagophore. ATG9 is the only multispanning transmembrane ATG protein, which cycles between the Golgi network, recycling endosomes, and LC3-containing phagophores during autophagy in a manner that is dependent on ULK1 and class III PI3K complex I (Fig. 4B) (Takahashi et al., 2011, 2016; Young et al., 2006). Notably, ATG9-containing vesicles only transiently associate with the phagophore to suggest that the vesicles transfer material to the elongating membrane during autophagy (Mari et al., 2010; Orsi et al., 2012). In addition to ATG9-containing vesicles, multiple organelles contribute membrane during autophagosome biogenesis, including the ER, mitochondria, the Golgi, ER–Golgi intermediate compartment, plasma membrane, and recycling endosomes (Shibutani & Yoshimori, 2014). However, the organelle selectivity, lipid composition, and mechanism(s) of bilayer delivery to the expanding phagophore remain unclear.

2.4.1 Sphingolipid Regulation of Membrane Delivery and Autophagic Cargo

C1P and sphingomyelin are important regulators of ATG9 trafficking for phagophore elongation (Fig. 4B). Loss of CPTP is accompanied by an increase in C1P at the trans-Golgi network to trigger Golgi fragmentation similar to that observed during starvation-induced autophagy (Simanshu et al., 2013). Indeed, loss of CPTP triggers ATG9A vesicle trafficking to promote autophagy for autophagy-dependent inflammasome activation (Mishra et al., 2017). As low levels of C1P increase Ca^{2+}-mediated membrane fusion and ceramide kinase promotes phagolysosome formation (Hinkovska-Galcheva et al., 1998), Mishra et al. propose that C1P may aid in ATG9 vesicle formation and/or membrane delivery to phagophore. In addition to C1P, increased sphingomyelin in Niemann–Pick type A patient fibroblasts or SMPD1-depleted cancer cells leads to the accumulation of elongated and unclosed autophagic membranes and impaired stress response (Corcelle-Termeau et al., 2016). Notably, the immature autophagic membranes in SMPD1-deficient cells contain WIPI2, ATG16L1, and LC3B but reduced ATG9 due to the mislocalization of ATG9 to transferrin-receptor positive recycling endosomes that have accumulated sphingomyelin. Importantly, bacterial sphingomyelinase reverts the autophagy defect in Neimann–Pick type A fibroblasts, whereas the main sphingomyelin metabolite C16-ceramide does not, suggesting that defective autophagy in Neimann–Pick type A disease is the result of sphingomyelin accumulation rather than lysosomal dysfunction or other secondary effects of SMPD1 deficiency (Corcelle-Termeau et al., 2016).

Mitochondrion-targeting ceramides, such as C6-pyridinium ceramide, selectively promote autophagosome formation (Hou et al., 2011). Interestingly, C18-pyridinium ceramide or the accumulation of endogenous C18-ceramide by ceramide synthase 1 (CerS1) stimulates the direct interaction of ceramide and LC3B-II for the induction of autophagic cell death independent of apoptosis (Fig. 4C) (Dany et al., 2016; Sentelle et al., 2012). Moreover, as depletion of CerS1 abrogates sodium selenite–induced mitophagy, the interaction of C18-ceramide and LC3B-II also appears to regulate canonical mitophagy. Sentelle et al. identified that the lipidation and Ile35/Phe52 residues of LC3B are important for ceramide binding; however, ceramide binding does not promote LC3 lipidation to support the notion that the interaction targets mitochondria to autophagosomal membranes. Interestingly, LC3B interacts with C18- and C16-ceramides as well as sphingosine but not C14-, C22-, C24-ceramide, or S1P to demonstrate selectivity in sphingolipid binding (Sentelle et al., 2012); however, the physiological relevance of this selectivity remains unclear.

It is intriguing to speculate a role for sphingolipids in the shaping, elongation, and/or closure of the phagophore. Inhibition of serine palmitoyltransferase (SPT) or inositol phosphorylceramide synthesis in yeast decreases autophagic activity independent of the Atg12–Atg5 and Atg8–PE conjugation systems or the formation of preautophagosomal structures (the yeast equivalent of the isolation membrane) to suggest that sphingolipid balance may play a critical role in the expansion and/or closure of autophagosomal membranes (Yamagata, Obara, & Kihara, 2011). Indeed, ceramide colocalizes with LC3 in autophagic cells (Sims et al., 2010), and the catabolism of sphingolipids is observed to increase in parallel with autophagic activity in the absence of sphingolipid biosynthesis (Ghidoni et al., 1996). Moreover, liver-specific deletion of the essential autophagy gene ATG7 accumulates sphingolipids, including ceramide (Alexaki et al., 2014), to support the hypothesis that sphingolipids are present within the autophagosomal membrane and/or cargo (Fig. 4C). Furthermore, sphingolipid-metabolizing enzymes, such as SphK1, associate with LC3 foci during autophagy in a cell-type specific manner (Moruno Manchon et al., 2015; Moruno Manchon, Uzor, Finkbeiner, & Tsvetkov, 2016). Moreover, S1P and S1PR2 as well as the negative regulator of de novo ceramide synthesis ORMDL1 are autophagic substrates to indicate that autophagy can modulate sphingolipid signaling (Fig. 4C) (Slattum, Gu, Sabbadini, & Rosenblatt, 2014; Wang, Robinet, Smith, & Gulshan, 2015).

2.5 Phagophore Closure

The closure or sealing of the phagophore generates the completed autophagosome and triggers the dissociation of associated ATG proteins and recruitment of essential factors for lysosomal fusion. While closure is required for the efficient degradation of autophagic cargo, the molecular mechanism of this process is poorly understood. The LC3 family member GABARAP, ER-resident protein VMP1, and autophagy proteins ATG2A and ATG2B have been implicated in phagophore closure, but the molecular mechanism by which the proteins modulate membrane dynamics is not clear (Itakura & Mizushima, 2010; Kishi-Itakura, Koyama-Honda, Itakura, & Mizushima, 2014; Velikkakath, Nishimura, Oita, Ishihara, & Mizushima, 2012; Weidberg et al., 2010). Interestingly, the sealing of the phagophore to form the double-membrane autophagosome is a membrane fission (rather than fusion) event to suggest that the process may involve the endosomal sorting complex required for transport (ESCRT) machinery, which regulate topologically similar membrane scission events in a number of intracellular processes (Manil-Segalén, Lefebvre, Culetto, & Legouis, 2012). Continued investigation is needed to define the molecular mechanism of phagophore closure and examine the role of sphingolipids in autophagosome completion.

2.6 Autophagosome–Lysosome Fusion

The final step in autophagy is the fusion of autophagosomes with late endosomes to form amphisomes (which go on to fuse with lysosomes for degradation) or lysosomes to form autolysosomes. Late endosomes and lysosomes represent the endpoint in autophagy, and thus, any disruption in endocytic membrane trafficking or lysosomal function will impair autophagic flux (Razi, Chan, & Tooze, 2009). Autophagosome–lysosome fusion is regulated by the cytoskeleton, Rab GTPases, membrane-tethering complexes, and soluble N-ethylmaleimide–sensitive factor attachment protein receptors (SNAREs) (Fig. 5) (Nakamura & Yoshimori, 2017). The dynein–dynactin motor complex associates with autophagosomes via the recruitment of Rab7 and its effector Rab-interacting lysosomal protein to transport autophagosomes along microtubule tracks toward perinuclear late endosomes and lysosomes (Kimura, Noda, & Yoshimori, 2008; Köchl, Hu, Chan, & Tooze, 2006; Ravikumar et al., 2005; Wijdeven et al., 2016). Rab7 can also direct autophagosomes to the cell periphery via FYVE and coiled-coil domain-containing 1 and kinesin-driven movement along microtubules (Pankiv et al., 2010).

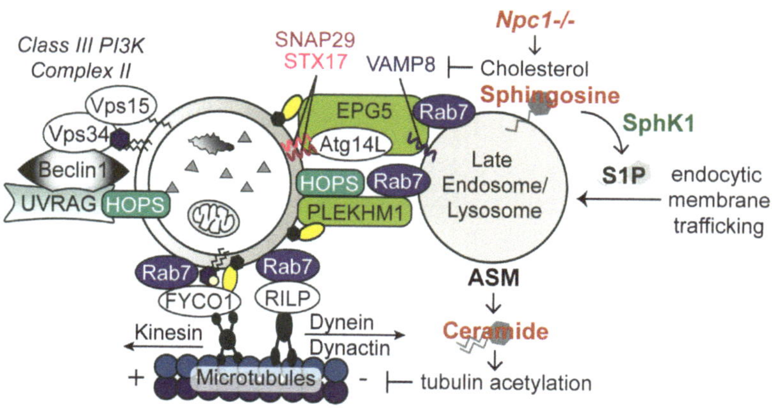

Figure 5 ***Autophagosome–lysosome fusion.*** Autophagosome maturation is regulated by microtubule-dependent transport, the Rab GTPase Rab7, membrane-tethering complexes (EPG5, ATG14, PLEKHM1/HOPS), and SNAREs (SNAP29/STX17 and VAMP8). Sphingolipids regulate autophagosome–lysosome fusion via direct effects on late endosome/lysosome function and trafficking, or indirectly by effects on endocytic membrane trafficking. ASM, acid sphingomyelinase. Autophagic regulation by sphingolipids in *red text* has been implicated with toxic autophagy, whereas *green text* represents survival autophagy. *See text for further details.*

SNAREs are membrane-anchored proteins that promote membrane fusion through the assembly of a trans-SNARE complex consisting of a vesicle (v)-SNARE and two or more target (t)-SNAREs on the acceptor membrane. Autophagosome–lysosome fusion is mediated by the t-SNAREs syntaxin 17 (STX17) and synaptosome-associated protein 29 (SNAP29) on autophagosomes and v-SNARE vesicle–associated membrane protein 8 (VAMP8) on late endosomes/lysosomes (Furuta, Fujita, Noda, Yoshimori, & Amano, 2010; Itakura, Kishi-Itakura, & Mizushima, 2012; Takáts et al., 2013). SNARE-mediated membrane fusion is assisted by the recruitment of membrane tethers that bridge the two membranes and promote specificity for membrane fusion. Ultraviolet irradiation resistance-associated gene (UVRAG), a member of the class III PI3K complex II, associates with autophagosomes and endosomes and interacts with the homotypic fusion and protein sorting (HOPS) complex subunit VPS16 to stimulate Rab7 GTPase activity (Liang et al., 2008). Rab7 recruits effectors such as PLEKHM1, which binds HOPS–SNARE complexes and LC3 proteins to tether autophagosome and endolysosomal membranes (McEwan et al., 2015). In addition, ectopic P granules protein 5 is recruited to late endosomes and lysosomes along with Rab7 and VAMP8 and interacts with LC3 (via an LIR motif) and STX17–SNAP29 binary complexes to bridge autophagosomal and endolysosomal membranes and promote trans-SNARE complex assembly (Wang et al., 2016). Similarly, oligomeric ATG14L stabilizes binary STX17 and SNAP29 t-SNARE complexes on autophagosomal membranes to promote interaction with VAMP8 on endolysosomes (Diao et al., 2015). Fusion of the autophagosome and lysosome results in the degradation of inner autophagosomal membrane and autophagic cargo by lysosomal proteases and hydrolases and the release of nutrients to the cytosol.

2.6.1 Sphingolipids in Autophagosome Maturation

Autophagosome maturation is tightly linked to endocytic membrane trafficking and requires a functional pool of late endosomes and/or lysosomes. As sphingolipids are trafficked, sorted, and metabolized along the endocytic pathway, sphingolipid metabolites can modulate autophagic flux by: (1) indirect effects on endocytic membrane trafficking or (2) direct effects on late endosome/lysosome function (Fig. 5). The cholesterol-mediated activation of ASM increases ceramide to promote tubulin acetylation on stabilized microtubules and impair autophagosome trafficking (Toops, Tan, Jiang, Radu, & Lakkaraju, 2015). Moreover, ASM is upregulated in Alzheimer's disease and impairs autophagic flux via defects in lysosome biogenesis (Lee et al., 2014). In contrast, enhanced ceramide metabolism by the overexpression of acid ceramidase increases lysosomal density and autophagic flux in prostate cancer cells (Turner et al., 2011), while the loss of ASAH1 due to chromosome 8p loss of heterozygosity in breast cancer enhances autophagic flux (Cai et al., 2016) to suggest that ceramide metabolism may have cell-type dependent effects.

Impaired autophagic degradation is a common mechanism that underlies many lysosomal storage diseases, including Niemann–Pick disease type A (NPA) and Niemann-Pick disease type B (Canonico et al., 2016; Gabandé-Rodríguez, Boya, Labrador, Dotti, & Ledesma, 2014; Seranova et al., 2017). The accumulation of autophagic structures due to loss of function ASM mutations in NPA has been suggested to arise from sphingomyelin-mediated lysosomal membrane permeabilization and inefficient autolysosomal clearance (Gabandé-Rodríguez et al., 2014). Others report that ASM deficiency suppresses lysosomal trafficking to impair autophagosome–lysosome fusion (Li et al., 2014). As discussed above, the loss of ASM in NPA mislocalizes ATG9 to sphingomyelin-enriched recycling endosomes to suppress autophagosome biogenesis (Corcelle-Termeau et al., 2016). Interestingly, inhibition of ASM impairs transferrin-receptor trafficking while accumulating cholesterol in late endosomes/lysosomes (Kuzu, Gowda, Noory, & Robertson, 2017). As sphingomyelin cycles between recycling endosomes and the plasma membrane to facilitate the trafficking of transferrin receptors (Gagescu et al., 2000; Shakor et al., 2011), impaired autophagy in ASM-deficient cells may due to the combination of defects in endocytic recycling (ATG9 trafficking) and autophagosome maturation (late endosome/lysosome function). Further study is needed to reconcile the mechanisms of impaired autophagy reported in ASM-deficient cells.

Cholesterol accumulation in endomembranes of lysosomal storage diseases impairs SNARE sorting and recycling that is required for endocytic membrane trafficking (Fraldi et al., 2010). Indeed, autophagic flux defects in Niemann–Pick disease type C1 (NPC1) result from defective amphisome formation due to failed recruitment of the SNARE protein VAMP8 to late endosomes (Sarkar et al., 2013). Notably, NPC1 mutant cells maintain functional endocytic cargo degradation and lysosomal proteolysis, and surprisingly, autophagic function can be restored by simply inducing autophagy, which stimulates the direct fusion of autophagosomes with lysosomes, independent of amphisome formation or alteration in late endosomal cholesterol (Sarkar et al., 2013). Moreover, NPC1 is a sphingosine storage disease containing decreased SphK1 activity and accumulated lysosomal sphingosine (Lloyd-Evans et al., 2008). As SphK1 enhances autophagic flux and S1P-metabolizing enzymes (SPP1, SGPL1) decrease autophagic flux (Moruno Manchon et al., 2015), the metabolism of sphingosine and/ or S1P may be important for autophagic degradation.

We and others have shown that SphK1 localizes to sphingosine-enriched endocytic vesicles and that inhibition of SphK1 by sphingosine-based SphK1 inhibitors (SK1-I or FTY720) or the potent small molecule SphK1 inhibitor PF-543 in the presence of exogenous sphingosine leads to the accumulation of enlarged late endosomes and amphisomes (Lima, Milstien, & Spiegel, 2017; Young et al., 2016). Notably, SK1-I initially impairs autophagic flux due to defects in late endosome and amphisome function, but flux later becomes restored in a manner dependent on lysosomal biogenesis and CerS activity (Young et al., 2016). While the molecular mechanism of restored autophagic flux requires further study, we hypothesize that lysosomal biogenesis may increase a pool of lysosomes that can fuse with defective late endosomes or directly with autophagosomes independent of amphisome formation, as described in NPC1 mutant cells (Sarkar et al., 2013).

Interestingly, knockdown of SphK1 impairs endocytic trafficking (H. Shen et al., 2014). SphK1 recruitment to early endocytic membranes occurs via curvature-sensitive interactions mediated by a hydrophobic patch and positively charged electrostatic domain on the surface of SphK1 (Pulkoski-Gross et al., 2018; H. Shen et al., 2014). We propose that an additional mechanism by which SphK1 regulates autophagic flux via indirect effects on endocytic membrane trafficking. We have shown that catalytically inactive SphK1 (G82D) promotes the fusion of late endosomes in the presence of sphingosine analogues or exogenous sphingosine to suggest that SphK1 can enhance membrane fusion independent of its catalytic activity (Young et al., 2016). Lima et al. reveal the colocalization of SphK1 and the membrane-remodeling BAR domain proteins Endophilin-A2 and Endophilin-B1 at late endosomal membranes of SK1-I—treated cells to suggest that SphK1 may aid in the recruitment of membrane-remodeling factors (Lima et al., 2017). In addition, sphingosine is a fusogenic lipid that enhances SNARE-mediated membrane fusion during neuronal exocytosis (Castro, Prieto, & Silva, 2014; Darios et al., 2003; Goni & Alonso, 2006; Williams, Vicôgne, Zaitseva, McLaughlin, & Pessin, 2009). As the core machinery of SNARE-mediated membrane fusion is conserved in neuronal and endosomal systems (Antonin et al., 2000) and autophagy requires SNARE-mediated membrane fusion, the role of sphingosine, S1P, and SphK1 in autophagosome—lysosome fusion is an area of future interest.

3. SPHINGOLIPID-MEDIATED AUTOPHAGY IN CANCER: DR. JEKYLL AND MR. HYDE

Autophagy is an evolutionarily conserved stress response that typically promotes survival and antagonizes apoptosis. However, excessive autophagy was found to induce cell death accompanied by the accumulation of autophagic structures but morphologically distinct from apoptosis or necrosis, thereby coining the term "autophagic cell death" or "type II cell death." Autophagy-dependent cell death is defined as a form of regulated cell death (RCD) that is dependent on the autophagic machinery (Galluzzi et al., 2018). In addition to cell killing by excessive cytoplasmic degradation, autophagy also initiates apoptosis. As many autophagy proteins have autophagy-independent functions and several pharmacological inhibitors of autophagy have off-target effects, two or more components of the autophagic machinery should be shown to be required prior to labeling cell death as autophagy dependent. Furthermore, the term should be avoided in instances where autophagy is activated alongside RCD or in cases where autophagy merely facilitates the execution of RCD (Galluzzi et al., 2018). Autophagy has multiple faces in cancer, as it can: (1) promote cancer cell survival, (2) induce an alternative mechanism of cell death, or (3) initiate the apoptotic cascade. Here, we will summarize how sphingolipid-mediated autophagy drives cancer cell survival as "Dr. Jekyll" but can be unleashed for cancer cell death as "Mr. Hyde."

3.1 Dr. Jekyll: A Driver of Cancer Cell Survival

Sphingolipids promote the induction of cytoprotective autophagy that fuels cancer survival, metastasis, and drug resistance. Loss of ASM or SphK1 in acute myeloid leukemia (AML) or breast cancer cell lines, respectively, impairs starvation-induced autophagy to suggest that sphingomyelin and/or sphingosine metabolism may drive tumorigenesis and tumor progression via autophagy (Lavieu et al., 2006; Taniguchi et al., 2012). Moreover, multiple cancer cell lines have been shown to elevate dihydroceramide at the expense of ceramide in response to hypoxia with levels returning to normal on reoxygenation (Devlin et al., 2011). As hypoxia is a well-established inducer of autophagy, it will be interesting to investigate whether increases in dihydroceramide may promote cancer cell survival in the hypoxic tumor microenvironment by limiting the accumulation of prodeath ceramide and activating autophagic signaling. Exogenous short-chain ceramide analogues also trigger the rapid induction of prosurvival autophagy by the activation of AMPK (Guenther et al., 2008) or ERK1/2 signaling (Fan et al., 2017). Interestingly, inhibitors of autophagic flux synergistically enhance C6-ceramide—induced apoptosis in hepatocarcinoma and colorectal cancer cells to suggest that targeting autophagy can increase the therapeutic efficacy of short-chain ceramide analogues (Adiseshaiah et al., 2013).

Sphingolipid-induced autophagy also promotes tumor progression and metastasis. Notably, the loss of ASAH1 increases autophagic flux to enhance the invasiveness and growth of breast cancer (Cai et al., 2016). Moreover, SphK1 drives epithelial-to-mesenchymal transition in hepatocellular carcinoma by generating S1P for the TRAF2-mediated activation of Beclin 1 and autophagic degradation of E-cadherin (Liu et al., 2017). While most epithelial cells extrude apically to be eliminated by the lumen, epithelial cells containing oncogenic K-Ras predominantly extrude basally to result in increased invasion into underlying tissues and migration to secondary sites. Interestingly, oncogenic K-Ras(V12) suppresses the S1P signaling required for apical extrusion by inducing autophagy and the autophagy-dependent degradation of S1P and S1PR2 (Slattum et al., 2014). Sphingolipid-mediated autophagy also enhances drug resistance, as the accumulation of dihydroceramide in gastric and colon cancer cell lines treated with tetrahydrocannabinol (THC) or dihydroceramide desaturase inhibitors induces cell cycle arrest and activates cytoprotective autophagy that suppresses the chemotherapeutic efficacy of etoposide (Hernández-Tiedra et al., 2016). Similarly, overexpression of acid ceramidase in prostate cancer enhances autophagic flux to promote resistance to C6-ceramide—induced cell death (Turner et al., 2011). Many chemotherapeutics generate proapoptotic ceramide while also stimulating autophagy as mechanism to evade cell death; thus, establishing a connection between ceramide metabolism and autophagy may provide rational for combination with sphingolipid-based therapeutics as a means to synergistically enhance the antitumor efficacy of established therapeutics.

3.2 Mr. Hyde: Inducer of Autophagy-Dependent Cancer Cell Death

Like Mr. Hyde, sphingolipid modulation can unleash a switch between prosurvival and lethal autophagy. Interestingly, different sphingolipid species appear to associate with distinct modes of autophagy-dependent cell death, including lethal mitophagy, noncanonical autophagic cell death, autophagy-dependent lysosomal membrane permeabilization, and autophagy-dependent apoptosis.

3.2.1 Lethal Mitophagy

Mitochondrial-targeted C18-pyridinium ceramide or the CerS1-mediated generation of endogenous C18-ceramide induce lethal mitophagy in head and neck squamous carcinoma cells. Notably, cell death is dependent on ATG3, ATG7, or ATG5 but independent of BAX/BAK, caspase-3/7, or RIP kinase activity to indicate autophagy-dependent cell death. In the same study, inducible expression of CerS1 was shown to suppress tumor growth in UM-SCC-22A xenografts in a manner dependent on LC3B to suggest that the activation of CerS1 may be an attractive therapeutic strategy to induce lethal autophagy in cancer. A similar result was demonstrated in AML in response to FLT3 inhibition (Dany et al., 2016). Here, pharmacological inhibition of mutant FLT3 triggers the reactivation of CerS1/C18-ceramide synthesis and accumulation of mitochondrial ceramide for the recruitment of lipidated LC3 and induction of lethal mitophagy. Cell death was suppressed by knockdown of Beclin 1, ATG5, or LC3 or on the expression of LC3 lacking the ceramide-interacting residues Ile35 or Phe52 but was not altered by inhibitors of apoptosis or necroptosis. Importantly, CerS1 enhances the in vivo efficacy of FLT3-targeted therapy in AML (Dany et al., 2016). While CerS1 is highly expressed in the central nervous system, human glioma tissues have significantly reduced C18-ceramide compared to control (Wang et al., 2017). Interestingly, the overexpression of CerS1 or exogenous C18-ceramide induces lethal autophagy in this model and sensitizes cells to chemotherapy (Wang et al., 2017) to suggest that increased C18-ceramide can unleash autophagic cell death. Likewise, exogenous C16-ceramide triggers autophagic cell death in colon cancer cells (Deroyer et al., 2014). As C16-ceramide is also reported to directly interact with

LC3B (Sentelle et al., 2012), further characterization of lethal mitophagy induced by C18- and/or C16-ceramides may expose a target for the induction of nonapoptotic cancer cell death.

3.2.2 Autophagic Cell Death by Noncanonical Autophagy

Resveratrol induces autophagic cell death in lung cancer cell lines that is accompanied by the accumulation of multiple sphingolipids, including ceramide, sphingosine, S1P, and sphingomyelin (Dasari et al., 2017). While independent of apoptotic or necroptotic signaling, resveratrol-induced cell death requires the LC3-lipidation machinery (i.e., ATG7, ATG12, ATG4B, MAP1LC3B) but is independent of ULK1 or Beclin 1. Noncanonical autophagy occurs in the absence of some components of the autophagy machinery (Codogno, Mehrpour, & Proikas-Cezanne, 2011); however, the function of such autophagy in cancer is unclear. Dasari et al. propose that resveratrol-induced cell death occurs by hyperactivation of noncanonical autophagy that is initiated by LC3 lipidation at the Golgi (Dasari et al., 2017; Gao et al., 2016). In the same study, glucocerebrosidase (GBA1) was identified as an essential gene for LC3 lipidation and resveratrol-induced toxicity. In contrast, GBA1 is not required for prosurvival autophagy in response to nutrient starvation to suggest that GBA1 and/or sphingolipid metabolism serve different roles during canonical versus noncanonical autophagy (Dasari et al., 2017).

3.2.3 Autophagy-Dependent Lysosomal Membrane Permeabilization

Lethal autophagy by Δ^9-THC occurs by the autophagy-dependent delivery of dihydroceramide to the lysosome to cause membrane destabilization, lysosomal membrane permeabilization, and the induction of apoptosis (Hernández-Tiedra et al., 2016). In contrast to starvation-induced autophagy, THC-induced autophagy requires the SPT generation of sphingolipids and increase in dihydroceramide-to-ceramide ratio. Similarly, inhibition of dihydroceramide desaturase (DEGS1) was sufficient to accumulate dihydroceramide, induce autophagy, and stimulate cathepsin release from lysosomes (Hernández-Tiedra et al., 2016). This result is supported by other studies demonstrating autophagy-dependent apoptosis in response to THC in melanoma and glioma cell lines (Armstrong et al., 2015; Velasco et al., 2009). While just scratching the surface, these studies suggest that an accumulation or imbalance of dihydroceramide within the autophagosomal membrane or autophagic cargo can trigger cell death on delivery to the lysosome.

3.2.4 Autophagy-Dependent Noncanonical Caspase-8 Activation

We and others have established that autophagosomal membranes can serve as platforms for the noncanonical activation of caspase-8 via the assembly of an intracellular death-inducing signaling complex (iDISC) (Bell et al., 2008; Deegan et al., 2014; Huang, Okamoto, Yu, & Sinicrope, 2013; Jiang et al., 2011; Laussmann et al., 2011; Villar et al., 2017; Young et al., 2012; Zhang, Gong, Xing, Zheng, & Ding, 2013). While iDISC-mediated apoptosis has since been shown to occur in response to a variety of stimuli, we initially characterized iDISC-mediated cell death in response to the pan SphK1/2 inhibitor SKI-I, suggesting that sphingolipid dysregulation by SphK1/2 may be an approach to switch cytoprotective autophagy to apoptosis (Young et al., 2012). SKI-I triggers the recruitment of pro—caspase-8 to LC3- and ATG5-containing autophagic membranes for iDISC assembly via two arms: (1) LC3 ~ p62 ~ caspase-8, and (2) ATG5 ~ FADD ~ caspase-8. The concentration of pro—caspase-8 at the autophagosomal membrane promotes caspase-8 self-oligomerization and activation to initiate apoptosis. Importantly, iDISC-mediated apoptosis is autophagy dependent and requires the lipidation of LC3, as loss of ATG3 or ATG5 suppresses caspase activation and SKI-I—mediated apoptosis. Subsequent work by our group has established that a defect in autophagosome completion on the loss of ATG2A/B accumulates immature autophagosomal membranes for the iDISC assembly and the induction of noncanonical apoptosis (Tang et al., 2017). Thus, it is tempting to speculate that changes in sphingolipids on the inhibition of SphK1/2 may impair phagophore closure to trigger iDISC-mediated cell death. Interestingly, cell death induced by the sphingosine-based SphK1—selective inhibitor SK1-I is Beclin 1 and ATG5-dependent and accompanied by caspase activation; however, the mechanism appears distinct from iDISC-mediated apoptosis as caspase activation only partially contributes to cell death (Lima et al., 2018). Thus, more work is needed to carefully dissect the cross talk between sphingolipid metabolism, phagophore closure, and iDISC-mediated apoptosis.

3.2.5 Autophagy-Dependent Apoptosis via the Calpain-Mediated Cleavage of ATG5

In addition to serving as platforms for caspase activation, several autophagy proteins can initiate apoptosis, including Beclin 1, ATG5, ATG3, ATG4D, AMBRA1 (Young & Wang, 2013). Notably, the calpain-mediated cleavage of ATG5 impairs autophagosome biogenesis and generates an N-terminal fragment that translocates to the mitochondria and interacts with Bcl-xL to convert autophagy to apoptosis (Yousefi et al., 2006). Loss of SPP1 accumulates S1P to drive the induction of cytoprotective autophagy (Lépine, Allegood, Park et al., 2011). Interestingly, doxorubicin increases de novo ceramide synthesis in SPP1-depleted breast cancer cells to stimulate the calpain-mediated cleavage of ATG5 and switch the high levels of cytoprotective autophagy to apoptosis (Lépine, Allegood, Edmonds et al., 2011). Similarly, melanoma differentiation–associated gene 7/interleukin-24 stimulates de novo ceramide synthesis in prostate cancer to induce autophagy that is switched to apoptosis by the calpain-dependent generation of a proapoptotic ATG5 fragment (Bhutia et al., 2010). Collectively, these studies highlight the potential roles of ceramide and S1P metabolism in regulating the switch between cytoprotective to lethal autophagy for cancer therapy.

4. CONCLUSION AND FUTURE DIRECTIONS

In this chapter, we aimed to provide a comprehensive review of sphingolipids in the regulation of autophagosome biogenesis, autophagic flux, and autophagic outcome (survival or death) in cancer. As a natural caveat to this approach, the data were collected using a range of cell lines and experimental conditions. This is noteworthy because cell line–dependent protein expression may drive autophagic response to sphingolipids. For example, extracellular S1P inhibits autophagy in S1PR3-expressing HL-60 leukemia cells but promotes autophagy in S1PR5-expressing PC3 prostate cancer cells (Chang et al., 2009; Taniguchi et al., 2012). Moreover, different cancer cell lines have varying degrees of autophagy dependence; for example, K-Ras mutations are associated with high autophagy dependence (Guo et al., 2011; J. H. Kim et al., 2011; M. J. Kim et al., 2011). Despite these limitations, we can begin to appreciate the importance of sphingolipids in autophagy, as sphingolipid metabolism regulates each step of autophagosome biogenesis and maturation. Moreover, as the same sphingolipid influences autophagy at multiple steps, other factors such as intracellular location, local metabolism, and/or species-dependent effects (i.e., acyl chain length, saturation) are likely to be involved, as is suggested by the role of mitochondrial localized C18-ceramide in lethal mitophagy (Sentelle et al., 2012). Continued investigation can begin to dissect some intriguing, but unanswered questions, including:

- What is the sphingolipid composition of the ER during autophagy initiation and phagophore nucleation?
- Are sphingolipids essential components of the autophagosomal membrane?
- Does the sphingolipid composition of the autophagosomal membrane change at each step of autophagy?
- Can autophagy regulate sphingolipid homeostasis via the transport and delivery of sphingolipids for lysosomal catabolism?
- What determines whether sphingolipid-induced autophagy is cytoprotective or lethal? Intracellular signaling? Intracellular location? Fatty acyl chain length? Biophysical properties?

Sphingolipids and autophagy are both dysregulated in cancer, and thus establishing connections between sphingolipid metabolism and autophagy can significantly advance our understanding of cancer biology and therapy. Defining the molecular mechanism of sphingolipids in stress-induced autophagy can reveal insight into the mechanisms driving tumorigenesis and tumor progression, whereas sphingolipid modulation can provide a novel therapeutic approach via the induction of autophagy-dependent cell death. As many cancers show some degree of autophagy dependence, the ability to hijack the autophagy machinery for the induction of lethal autophagy by excessive degradation or for the induction of noncanonical apoptosis would be therapeutically beneficial. Collectively, although we have just begun to scratch the surface, the regulation of autophagy by sphingolipids is evident and warrants continued investigation.

REFERENCES

Adiseshaiah, P. P., Clogston, J. D., McLeland, C. B., Rodriguez, J., Potter, T. M., Neun, B. W., et al. (2013). Synergistic combination therapy with nanoliposomal C6-ceramide and vinblastine is associated with autophagy dysfunction in hepatocarcinoma and colorectal cancer models. *Cancer Letters, 337*(2), 254–265. https://doi.org/10.1016/j.canlet.2013.04.034.

Alexaki, A., Gupta, S. D., Majumder, S., Kono, M., Tuymetova, G., Harmon, J. M., et al. (2014). Autophagy regulates sphingolipid levels in the liver. *Journal of Lipid Research, 55*(12), 2521–2531. https://doi.org/10.1194/jlr.M051862.

Antonin, W., Holroyd, C., Fasshauer, D., Pabst, S., Von Mollard, G. F., & Jahn, R. (2000). A SNARE complex mediating fusion of late endosomes defines conserved properties of SNARE structure and function. *The EMBO Journal, 19*(23), 6453–6464. https://doi.org/10.1093/emboj/19.23.6453.

Armstrong, J. L., Hill, D. S., McKee, C. S., Hernandez-Tiedra, S., Lorente, M., Lopez-Valero, I., et al. (2015). Exploiting cannabinoid-induced cytotoxic autophagy to drive melanoma cell death. *The Journal of Investigative Dermatology, 135*(6), 1629–1637. https://doi.org/10.1038/jid.2015.45.

Bell, B. D., Leverrier, S., Weist, B. M., Newton, R. H., Arechiga, A. F., Luhrs, K. A., et al. (2008). FADD and caspase-8 control the outcome of autophagic signaling in proliferating T cells. *Proceedings of the National Academy of Sciences, 105*(43), 16677–16682. https://doi.org/10.1073/pnas.0808597105.

Bhutia, S. K., Dash, R., Das, S. K., Azab, B., Su, Z., Lee, S.-G., et al. (2010). Mechanism of autophagy to apoptosis switch triggered in prostate cancer cells by antitumor cytokine melanoma differentiation-associated gene 7/Interleukin-24. *Cancer Research, 70*(9), 3667–3676. https://doi.org/10.1158/0008-5472.CAN-09-3647.

Bjørkøy, G., Lamark, T., Brech, A., Outzen, H., Perander, M., Øvervatn, A., et al. (2005). p62/SQSTM1 forms protein aggregates degraded by autophagy and has a protective effect on huntingtin-induced cell death. *The Journal of Cell Biology, 171*(4), 603–614. https://doi.org/10.1083/jcb.200507002.

Cai, Y., Crowther, J., Pastor, T., Abbasi Asbagh, L., Baietti, M. F., De Troyer, M., et al. (2016). Loss of chromosome 8p governs tumor progression and drug response by altering lipid metabolism. *Cancer Cell, 29*(5), 751–766. https://doi.org/10.1016/j.ccell.2016.04.003.

Canonico, B., Cesarini, E., Salucci, S., Luchetti, F., Falcieri, E., Di Sario, G., et al. (2016). Defective autophagy, mitochondrial clearance and lipophagy in niemann-pick type B lymphocytes. *PLoS One, 11*(10), e0165780. https://doi.org/10.1371/journal.pone.0165780.

Castro, B. M., Prieto, M., & Silva, L. C. (2014). Ceramide: A simple sphingolipid with unique biophysical properties. *Progress in Lipid Research, 54*, 53–67. https://doi.org/10.1016/j.plipres.2014.01.004.

Chang, C.-L., Ho, M.-C., Lee, P.-H., Hsu, C.-Y., Huang, W.-P., & Lee, H. (2009). S1P5 is required for sphingosine 1-phosphate-induced autophagy in human prostate cancer PC-3 cells. *American Journal of Physiology – Cell Physiology, 297*(2), C451–C458. https://doi.org/10.1152/ajpcell.00586.2008.

Codogno, P., Mehrpour, M., & Proikas-Cezanne, T. (2011). Canonical and non-canonical autophagy: Variations on a common theme of self-eating? *Nature Reviews Molecular Cell Biology, 13*(1), 7–12. https://doi.org/10.1038/nrm3249.

Copetti, T., Bertoli, C., Dalla, E., Demarchi, F., & Schneider, C. (2009). p65/RelA modulates BECN1 transcription and autophagy. *Molecular and Cellular Biology, 29*(10), 2594–2608. https://doi.org/10.1128/MCB.01396-08.

Corcelle-Termeau, E., Vindeløv, S. D., Hämälistö, S., Mograbi, B., Keldsbo, A., Bräsen, J. H., et al. (2016). Excess sphingomyelin disturbs ATG9A trafficking and autophagosome closure. *Autophagy, 12*(5), 833–849. https://doi.org/10.1080/15548627.2016.1159378.

Dany, M., Gencer, S., Nganga, R., Thomas, R. J., Oleinik, N., Baron, K. D., et al. (2016). Targeting FLT3-ITD signaling mediates ceramide-dependent mitophagy and attenuates drug resistance in AML. *Blood, 128*(15), 1944–1958. https://doi.org/10.1182/blood-2016-04-708750.

Darios, F., Corti, O., Lucking, C. B., Hampe, C., Muriel, M. P., Abbas, N., et al. (2003). Parkin prevents mitochondrial swelling and cytochrome c release in mitochondria-dependent cell death. *Human Molecular Genetics, 12*(5), 517–526.

Dasari, S. K., Bialik, S., Levin-Zaidman, S., Levin-Salomon, V., Merrill, A. H., Futerman, A. H., et al. (2017). Signalome-wide RNAi screen identifies GBA1 as a positive mediator of autophagic cell death. *Cell Death and Differentiation, 24*(7), 1288–1302. https://doi.org/10.1038/cdd.2017.80.

Deegan, S., Saveljeva, S., Logue, S. E., Pakos-Zebrucka, K., Gupta, S., Vandenabeele, P., et al. (2014). Deficiency in the mitochondrial apoptotic pathway reveals the toxic potential of autophagy under ER stress conditions. *Autophagy, 10*(11), 1921–1936. https://doi.org/10.4161/15548627.2014.981790.

Deroyer, C., Rénert, A.-F., Merville, M.-P., & Fillet, M. (2014). New role for EMD (emerin), a key inner nuclear membrane protein, as an enhancer of autophagosome formation in the C16-ceramide autophagy pathway. *Autophagy, 10*(7), 1229–1240. https://doi.org/10.4161/auto.28777.

Devlin, C. M., Lahm, T., Hubbard, W. C., Van Demark, M., Wang, K. C., Wu, X., et al. (2011). A dihydroceramide-based response to hypoxia. *Journal of Biological Chemistry, 286*(44), 38069–38078. https://doi.org/10.1074/jbc.M111.297994.

Diao, J., Liu, R., Rong, Y., Zhao, M., Zhang, J., Lai, Y., et al. (2015). ATG14 promotes membrane tethering and fusion of autophagosomes to endolysosomes. *Nature, 520*(7548), 563–566. https://doi.org/10.1038/nature14147.

Egan, D. F., Shackelford, D. B., Mihaylova, M. M., Gelino, S., Kohnz, R. A., Mair, W., et al. (2011). Phosphorylation of ULK1 (hATG1) by AMP-activated protein kinase connects energy sensing to mitophagy. *Science, 331*(6016), 456–461. https://doi.org/10.1126/science.1196371.

Fan, C., Liu, Y., Zhao, M., Zhan, R., Cui, W., Jin, H., et al. (2017). Autophagy inhibits C2-ceramide-mediated cell death by decreasing the reactive oxygen species levels in SH-SY5Y cells. *Neuroscience Letters, 651*, 198–206. https://doi.org/10.1016/j.neulet.2017.03.006.

Feng, Y., He, D., Yao, Z., & Klionsky, D. J. (2014). The machinery of macroautophagy. *Cell Research, 24*(1), 24–41. https://doi.org/10.1038/cr.2013.168.

Fimia, G. M., Stoykova, A., Romagnoli, A., Giunta, L., Di Bartolomeo, S., Nardacci, R., et al. (2007). Ambra1 regulates autophagy and development of the nervous system. *Nature, 447*(7148), 1121–1125. https://doi.org/10.1038/nature05925.

Fraldi, A., Annunziata, F., Lombardi, A., Kaiser, H.-J., Medina, D. L., Spampanato, C., et al. (2010). Lysosomal fusion and SNARE function are impaired by cholesterol accumulation in lysosomal storage disorders. *The EMBO Journal, 29*(21), 3607–3620. https://doi.org/10.1038/emboj.2010.237.

Fujita, N., Itoh, T., Omori, H., Fukuda, M., Noda, T., & Yoshimori, T. (2008). The Atg16L complex specifies the site of LC3 lipidation for membrane biogenesis in autophagy. *Molecular Biology of the Cell, 19*(5), 2092–2100. https://doi.org/10.1091/mbc.E07-12-1257.

Furuta, N., Fujita, N., Noda, T., Yoshimori, T., & Amano, A. (2010). Combinational soluble N-ethylmaleimide-sensitive factor attachment protein receptor proteins VAMP8 and Vti1b mediate fusion of antimicrobial and canonical autophagosomes with lysosomes. *Molecular Biology of the Cell, 21*(6), 1001–1010. https://doi.org/10.1091/mbc.E09-08-0693.

Gabandé-Rodríguez, E., Boya, P., Labrador, V., Dotti, C. G., & Ledesma, M. D. (2014). High sphingomyelin levels induce lysosomal damage and autophagy dysfunction in Niemann Pick disease type A. *Cell Death and Differentiation, 21*(6), 864–875. https://doi.org/10.1038/cdd.2014.4.

Gagescu, R., Demaurex, N., Parton, R. G., Hunziker, W., Huber, L. A., & Gruenberg, J. (2000). The recycling endosome of madin-darby canine kidney cells is a mildly acidic compartment rich in raft components. *Molecular Biology of the Cell, 11*(8), 2775–2791. https://doi.org/10.1091/mbc.11.8.2775.

Galluzzi, L., Baehrecke, E. H., Ballabio, A., Boya, P., Bravo-San Pedro Pedro, J. M., Cecconi, F., et al. (2017). Molecular definitions of autophagy and related processes. *The EMBO Journal, 36*(13), 1811–1836. https://doi.org/10.15252/embj.201796697.

Galluzzi, L., Pietrocola, F., Bravo-San Pedro, J. M., Amaravadi, R. K., Baehrecke, E. H., et al. (2015). Autophagy in malignant transformation and cancer progression. *The EMBO Journal, 34*(7), 856–880. https://doi.org/10.15252/embj.201490784.

Galluzzi, L., Vitale, I., Aaronson, S. A., Abrams, J. M., Adam, D., Agostinis, P., et al. (2018). Molecular mechanisms of cell death: Recommendations of the Nomenclature Committee on cell death 2018. *Cell Death and Differentiation, 1*. https://doi.org/10.1038/s41418-017-0012-4.

Gao, Y., Liu, Y., Hong, L., Yang, Z., Cai, X., Chen, X., et al. (2016). Golgi-associated LC3 lipidation requires V-ATPase in noncanonical autophagy. *Cell Death and Disease,* 7(8), e2330. https://doi.org/10.1038/cddis.2016.236.

Garofalo, T., Matarrese, P., Manganelli, V., Marconi, M., Tinari, A., Gambardella, L., et al. (2016). Evidence for the involvement of lipid rafts localized at the ER-mitochondria associated membranes in autophagosome formation. *Autophagy, 12*(6), 917–935. https://doi.org/10.1080/15548627.2016.1160971.

Geng, J. F., & Klionsky, D. J. (2008). The Atg8 and Atg12 ubiquitin-like conjugation systems in macroautophagy. *EMBO Reports, 9*(9), 859–864. https://doi.org/10.1038/embor.2008.163.

Ghidoni, R., Houri, J.-J., Giuliani, A., Ogier-Denis, E., Parolari, E., Botti, S., et al. (1996). The metabolism of Sphingo(glyco)lipids is correlated with the differentiation-dependent autophagic pathway in HT-29 cells. *European Journal of Biochemistry, 237*(2), 454–459. https://doi.org/10.1111/j.1432-1033.1996.0454k.x.

Goni, F. M., & Alonso, A. (2006). Biophysics of sphingolipids I. Membrane properties of sphingosine, ceramides and other simple sphingolipids. *Biochimica et Biophysica Acta, 1758*(12), 1902–1921. https://doi.org/10.1016/j.bbamem.2006.09.011.

Guenther, G. G., Peralta, E. R., Rosales, K. R., Wong, S. Y., Siskind, L. J., & Edinger, A. L. (2008). Ceramide starves cells to death by downregulating nutrient transporter proteins. *Proceedings of the National Academy of Sciences, 105*(45), 17402–17407. https://doi.org/10.1073/pnas.0802781105.

Guo, J. Y., Chen, H. Y., Mathew, R., Fan, J., Strohecker, A. M., Karsli-Uzunbas, G., et al. (2011). Activated Ras requires autophagy to maintain oxidative metabolism and tumorigenesis. *Genes and Development, 25*(5), 460–470. https://doi.org/10.1101/gad.2016311.

Hamasaki, M., Furuta, N., Matsuda, A., Nezu, A., Yamamoto, A., Fujita, N., et al. (2013). Autophagosomes form at ER-mitochondria contact sites. *Nature, 495*(7441), 389–393. https://doi.org/10.1038/nature11910.

Hanahan, D., & Weinberg, R. A. (2011). Hallmarks of cancer: The next generation. *Cell, 144*(5), 646–674. https://doi.org/10.1016/j.cell.2011.02.013.

Hara, T., Takamura, A., Kishi, C., Iemura, S.-I., Natsume, T., Guan, J.-L., et al. (2008). FIP200, a ULK-interacting protein, is required for autophagosome formation in mammalian cells. *The Journal of Cell Biology, 181*(3), 497–510. https://doi.org/10.1083/jcb.200712064.

Hernández-Tiedra, S., Fabriàs, G., Dávila, D., Salanueva, Í. J., Casas, J., Montes, L. R., et al. (2016). Dihydroceramide accumulation mediates cytotoxic autophagy of cancer cells via autolysosome destabilization. *Autophagy, 12*(11), 2213–2229. https://doi.org/10.1080/15548627.2016.1213927.

Hinkovska-Galcheva, V. T., Boxer, L. A., Mansfield, P. J., Harsh, D., Blackwood, A., & Shayman, J. A. (1998). The formation of ceramide-1-phosphate during neutrophil phagocytosis and its role in liposome fusion. *The Journal of Biological Chemistry, 273*(50), 33203–33209.

Hosokawa, N., Hara, T., Kaizuka, T., Kishi, C., Takamura, A., Miura, Y., et al. (2009). Nutrient-dependent mTORC1 association with the ULK1-Atg13-FIP200 complex required for autophagy. *Molecular Biology of the Cell, 20*(7), 1981–1991. https://doi.org/10.1091/mbc.E08-12-1248.

Hosokawa, N., Sasaki, T., Iemura, S., Natsume, T., Hara, T., & Mizushima, N. (2009). Atg101, a novel mammalian autophagy protein interacting with Atg13. *Autophagy, 5*(7), 973–979.

Hou, Q., Jin, J., Zhou, H., Novgorodov, S. A., Bielawska, A., Szulc, Z. M., et al. (2011). Mitochondrially targeted ceramides preferentially promote autophagy, retard cell growth, and induce apoptosis. *Journal of Lipid Research, 52*(2), 278–288. https://doi.org/10.1194/jlr.M012161.

Huang, S., Okamoto, K., Yu, C., & Sinicrope, F. A. (2013). p62/Sequestosome-1 Up-regulation promotes ABT-263-induced caspase-8 aggregation/activation on the autophagosome. *Journal of Biological Chemistry, 288*(47), 33654–33666. https://doi.org/10.1074/jbc.M113.518134.

Itakura, E., Kishi, C., Inoue, K., & Mizushima, N. (2008). Beclin 1 forms two distinct phosphatidylinositol 3-kinase complexes with mammalian Atg14 and UVRAG. *Molecular Biology of the Cell, 19*(12), 5360–5372. https://doi.org/10.1091/mbc.E08-01-0080.

Itakura, E., Kishi-Itakura, C., & Mizushima, N. (2012). The hairpin-type tail-anchored SNARE syntaxin 17 targets to autophagosomes for fusion with endosomes/lysosomes. *Cell, 151*(6), 1256–1269. https://doi.org/10.1016/j.cell.2012.11.001.

Itakura, E., & Mizushima, N. (2010). Characterization of autophagosome formation site by a hierarchical analysis of mammalian Atg proteins. *Autophagy, 6*(6), 764–776. https://doi.org/10.4161/auto.6.6.12709.

Jiang, H., White, E. J., Ríos-Vicil, C. I., Xu, J., Gomez-Manzano, C., & Fueyo, J. (2011). Human adenovirus type 5 induces cell lysis through autophagy and autophagy-triggered caspase activity. *Journal of Virology, 85*(10), 4720–4729. https://doi.org/10.1128/JVI.02032-10.

Justice, M. J., Bronova, I., Schweitzer, K. S., Poirier, C., Blum, J. S., Berdyshev, E. V., et al. (2018). Inhibition of acid sphingomyelinase disrupts LYNUS signaling and triggers autophagy. *Journal of Lipid Research.* https://doi.org/10.1194/jlr.M080242.

Karanasios, E., Walker, S. A., Okkenhaug, H., Manifava, M., Hummel, E., Zimmermann, H., et al. (2016). Autophagy initiation by ULK complex assembly on ER tubulovesicular regions marked by ATG9 vesicles. *Nature Communications,* 7(12420). https://doi.org/10.1038/ncomms12420.

Kim, J., Kundu, M., Viollet, B., & Guan, K.-L. (2011). AMPK and mTOR regulate autophagy through direct phosphorylation of Ulk1. *Nature Cell Biology, 13*(2), 132–141. https://doi.org/10.1038/ncb2152.

Kim, J. H., Kim, H. Y., Lee, Y. K., Yoon, Y. S., Xu, W. G., Yoon, J. K., et al. (2011). Involvement of mitophagy in oncogenic K-Ras-induced transformation: Overcoming a cellular energy deficit from glucose deficiency. *Autophagy,* 7(10), 1187–1198. https://doi.org/10.4161/auto.7.10.16643.

Kimura, S., Noda, T., & Yoshimori, T. (2008). Dynein-dependent movement of autophagosomes mediates efficient encounters with lysosomes. *Cell Structure and Function, 33*(1), 109–122. https://doi.org/10.1247/csf.08005.

Kim, M. J., Woo, S. J., Yoon, C. H., Lee, J. S., An, S., Choi, Y. H., et al. (2011). Involvement of autophagy in oncogenic K-Ras-induced malignant cell transformation. *Journal of Biological Chemistry, 286*(15), 12924–12932. https://doi.org/10.1074/jbc.M110.138958.

Kishi-Itakura, C., Koyama-Honda, I., Itakura, E., & Mizushima, N. (2014). Ultrastructural analysis of autophagosome organization using mammalian autophagy-deficient cells. *Journal of Cell Science, 127*(18), 4089–4102. https://doi.org/10.1242/jcs.156034.

Klionsky, D. J., Abdelmohsen, K., Abe, A., Abedin, M. J., Abeliovich, H., et al. (2016). Guidelines for the use and interpretation of assays for monitoring autophagy (3rd ed.) *Autophagy, 12*(1), 1–222. https://doi.org/10.1080/15548627.2015.1100356.

Köchl, R., Hu, X. W., Chan, E. Y. W., & Tooze, S. A. (2006). Microtubules facilitate autophagosome formation and fusion of autophagosomes with endosomes. *Traffic,* 7(2), 129–145. https://doi.org/10.1111/j.1600-0854.2005.00368.x.

Kuzu, O. F., Gowda, R., Noory, M. A., & Robertson, G. P. (2017). Modulating cancer cell survival by targeting intracellular cholesterol transport. *British Journal of Cancer, 117*(4), 513–524. https://doi.org/10.1038/bjc.2017.200.

Lamb, C. A., Yoshimori, T., & Tooze, S. A. (2013). The autophagosome: Origins unknown, biogenesis complex. *Nature Reviews Molecular Cell Biology, 14*(12), 759–774. https://doi.org/10.1038/nrm3696.

Laussmann, M. A., Passante, E., Düssmann, H., Rauen, J. A., Würstle, M. L., Delgado, M. E., et al. (2011). Proteasome inhibition can induce an autophagy-dependent apical activation of caspase-8. *Cell Death and Differentiation, 18*(10), 1584–1597. https://doi.org/10.1038/cdd.2011.27.

Lavieu, G., Scarlatti, F., Sala, G., Carpentier, S., Levade, T., Ghidoni, R., et al. (2006). Regulation of autophagy by sphingosine kinase 1 and its role in cell survival during nutrient starvation. *Journal of Biological Chemistry, 281*(13), 8518–8527. https://doi.org/10.1074/jbc.M506182200.

Lazova, R., Camp, R. L., Klump, V., Siddiqui, S. F., Amaravadi, R. K., & Pawelek, J. M. (2012). Punctate LC3B expression is a common feature of solid tumors and associated with proliferation, metastasis, and poor outcome. *Clinical Cancer Research, 18*(2), 370–379. https://doi.org/10.1158/1078-0432.CCR-11-1282.

Lee, J. K., Jin, H. K., Park, M. H., Kim, B., Lee, P. H., Nakauchi, H., et al. (2014). Acid sphingomyelinase modulates the autophagic process by controlling lysosomal biogenesis in Alzheimer's disease. *The Journal of Experimental Medicine, 211*(8), 1551–1570. https://doi.org/10.1084/jem.20132451.

Lee, J. W., Park, S., Takahashi, Y., & Wang, H. G. (2010). The association of AMPK with ULK1 regulates autophagy. *PLoS One, 5*(11), e15394. https://doi.org/10.1371/journal.pone.0015394.

Lépine, S., Allegood, J. C., Edmonds, Y., Milstien, S., & Spiegel, S. (2011). Autophagy induced by deficiency of sphingosine-1-phosphate phosphohydrolase 1 is switched to apoptosis by calpain-mediated autophagy-related gene 5 (Atg5) cleavage. *The Journal of Biological Chemistry, 286*(52), 44380–44390. https://doi.org/10.1074/jbc.M111.257519.

Lépine, S., Allegood, J. C., Park, M., Dent, P., Milstien, S., & Spiegel, S. (2011). Sphingosine-1-phosphate phosphohydrolase-1 regulates ER stress-induced autophagy. *Cell Death and Differentiation, 18*(2), 350–361. https://doi.org/10.1038/cdd.2010.104.

Levy, J. M. M., Towers, C. G., & Thorburn, A. (2017). Targeting autophagy in cancer. *Nature Reviews Cancer, 17*(9), 528–542. https://doi.org/10.1038/nrc.2017.53.

Li, X., Xu, M., Pitzer, A. L., Xia, M., Boini, K. M., Li, P.-L., et al. (2014). Control of autophagy maturation by acid sphingomyelinase in mouse coronary arterial smooth muscle cells: Protective role in atherosclerosis. *Journal of Molecular Medicine, 92*(5), 473–485. https://doi.org/10.1007/s00109-014-1120-y.

Liang, C., Lee, J., Inn, K.-S., Gack, M. U., Li, Q., Roberts, E. A., et al. (2008). Beclin1-binding UVRAG targets the class C Vps complex to coordinate autophagosome maturation and endocytic trafficking. *Nature Cell Biology, 10*(7), 776–787. https://doi.org/10.1038/ncb1740.

Lima, S., Milstien, S., & Spiegel, S. (2017). Sphingosine and sphingosine kinase 1 involvement in endocytic membrane trafficking. *The Journal of Biological Chemistry, 292*(8), 3074–3088. https://doi.org/10.1074/jbc.M116.762377.

Lima, S., Takabe, K., Newton, J., Saurabh, K., Young, M. M., Leopoldino, A. M., et al. (2018). TP53 is required for BECN1- and ATG5-dependent cell death induced by sphingosine kinase 1 inhibition. *Autophagy*, 1–50. https://doi.org/10.1080/15548627.2018.1429875.

Lindqvist, L. M., Heinlein, M., Huang, D. C. S., & Vaux, D. L. (2014). Prosurvival Bcl-2 family members affect autophagy only indirectly, by inhibiting Bax and Bak. *Proceedings of the National Academy of Sciences of the United States of America, 111*(23), 8512–8517. https://doi.org/10.1073/pnas.1406425111.

Liu, H., Ma, Y., He, H.-W., Zhao, W.-L., & Shao, R.-G. (2017). SPHK1 (sphingosine kinase 1) induces epithelial-mesenchymal transition by promoting the autophagy-linked lysosomal degradation of CDH1/E-cadherin in hepatoma cells. *Autophagy, 13*(5), 900–913. https://doi.org/10.1080/15548627.2017.1291479.

Lloyd-Evans, E., Morgan, A. J., He, X., Smith, D. A., Elliot-Smith, E., Sillence, D. J., et al. (2008). Niemann-Pick disease type C1 is a sphingosine storage disease that causes deregulation of lysosomal calcium. *Nature Medicine, 14*(11), 1247–1255. https://doi.org/10.1038/nm.1876.

Lu, J., He, L., Behrends, C., Araki, M., Araki, K., Wang, Q. J., et al. (2014). NRBF2 regulates autophagy and prevents liver injury by modulating Atg14L-linked phosphatidylinositol-3 kinase III activity. *Nature Communications, 5*(3920). https://doi.org/10.1038/ncomms4920.

Manil-Segalén, M., Lefebvre, C., Culetto, E., & Legouis, R. (2012). Need an ESCRT for autophagosomal maturation? *Communicative and Integrative Biology, 5*(6), 566–571. https://doi.org/10.4161/cib.21522.

Mari, M., Griffith, J., Rieter, E., Krishnappa, L., Klionsky, D. J., & Reggiori, F. (2010). An Atg9-containing compartment that functions in the early steps of autophagosome biogenesis. *The Journal of Cell Biology, 190*(6), 1005–1022. https://doi.org/10.1083/jcb.200912089.

Matarrese, P., Garofalo, T., Manganelli, V., Gambardella, L., Marconi, M., Grasso, M., et al. (2014). Evidence for the involvement of GD3 ganglioside in autophagosome formation and maturation. *Autophagy, 10*(5), 750–765. https://doi.org/10.4161/auto.27959.

McEwan, D. G., Popovic, D., Gubas, A., Terawaki, S., Suzuki, H., Stadel, D., et al. (2015). PLEKHM1 regulates autophagosome-lysosome fusion through HOPS complex and LC3/GABARAP proteins. *Molecular Cell, 57*(1), 39–54. https://doi.org/10.1016/j.molcel.2014.11.006.

Mercer, C. A., Kaliappan, A., & Dennis, P. B. (2009). A novel, human Atg13 binding protein, Atg101, interacts with ULK1 and is essential for macroautophagy. *Autophagy, 5*(5), 649–662.

Mercer, T. J., Gubas, A., & Tooze, S. A. (2018). A molecular perspective of mammalian autophagosome biogenesis. *The Journal of Biological Chemistry*. https://doi.org/10.1074/jbc.R117.810366.

Mikhaylova, O., Stratton, Y., Hall, D., Kellner, E., Ehmer, B., Drew, A. F., et al. (2012). VHL-regulated MiR-204 suppresses tumor growth through inhibition of LC3B-mediated autophagy in renal clear cell carcinoma. *Cancer Cell, 21*(4), 532–546. https://doi.org/10.1016/j.ccr.2012.02.019.

Mishra, S. K., Gao, Y.-G., Deng, Y., Chalfant, C. E., Hinchcliffe, E. H., & Brown, R. E. (2017). CPTP: A sphingolipid transfer protein that regulates autophagy and inflammasome activation. *Autophagy*, 1–46. https://doi.org/10.1080/15548627.2017.1393129.

Mitroi, D. N., Karunakaran, I., Gräler, M., Saba, J. D., Ehninger, D., Ledesma, M. D., et al. (2017). SGPL1 (sphingosine phosphate lyase 1) modulates neuronal autophagy via phosphatidylethanolamine production. *Autophagy, 13*(5), 885–899. https://doi.org/10.1080/15548627.2017.1291471.

Mizushima, N. (2010). The role of the Atg1/ULK1 complex in autophagy regulation. *Current Opinion in Cell Biology, 22*(2), 132–139. https://doi.org/10.1016/j.ceb.2009.12.004.

Molejon, M. I., Ropolo, A., Re, A. L., Boggio, V., & Vaccaro, M. I. (2013). The VMP1-Beclin 1 interaction regulates autophagy induction. *Scientific Reports, 3*. https://doi.org/10.1038/srep01055.

Moruno Manchon, J. F., Uzor, N.-E., Dabaghian, Y., Furr-Stimming, E. E., Finkbeiner, S., & Tsvetkov, A. S. (2015). Cytoplasmic sphingosine-1-phosphate pathway modulates neuronal autophagy. *Scientific Reports, 5*(15213). https://doi.org/10.1038/srep15213.

Moruno Manchon, J. F., Uzor, N.-E., Finkbeiner, S., & Tsvetkov, A. S. (2016). SPHK1/sphingosine kinase 1-mediated autophagy differs between neurons and SH-SY5Y neuroblastoma cells. *Autophagy, 12*(8), 1418–1424. https://doi.org/10.1080/15548627.2016.1183082.

Nakamura, S., & Yoshimori, T. (2017). New insights into autophagosome-lysosome fusion. *Journal of Cell Science, 130*(7), 1209–1216. https://doi.org/10.1242/jcs.196352.

Nguyen, T. N., Padman, B. S., Usher, J., Oorschot, V., Ramm, G., & Lazarou, M. (2016). Atg8 family LC3/GABARAP proteins are crucial for autophagosome–lysosome fusion but not autophagosome formation during PINK1/Parkin mitophagy and starvation. *The Journal of Cell Biology, 215*(6), 857–874. https://doi.org/10.1083/jcb.201607039.

Orsi, A., Razi, M., Dooley, H. C., Robinson, D., Weston, A. E., Collinson, L. M., et al. (2012). Dynamic and transient interactions of Atg9 with autophagosomes, but not membrane integration, are required for autophagy. *Molecular Biology of the Cell, 23*(10), 1860–1873. https://doi.org/10.1091/mbc.E11-09-0746.

Pankiv, S., Alemu, E. A., Brech, A., Bruun, J. A., Lamark, T., Overvatn, A., et al. (2010). FYCO1 is a Rab7 effector that binds to LC3 and PI3P to mediate microtubule plus end-directed vesicle transport. *The Journal of Cell Biology, 188*(2), 253–269. https://doi.org/10.1083/jcb.200907015.

Pankiv, S., Clausen, T. H., Lamark, T., Brech, A., Bruun, J.-A., Outzen, H., et al. (2007). p62/SQSTM1 binds directly to Atg8/LC3 to facilitate degradation of ubiquitinated protein aggregates by autophagy. *Journal of Biological Chemistry, 282*(33), 24131–24145. https://doi.org/10.1074/jbc.M702824200.

Park, J.-M., Jung, C. H., Seo, M., Otto, N. M., Grunwald, D., Kim, K. H., et al. (2016). The ULK1 complex mediates MTORC1 signaling to the autophagy initiation machinery via binding and phosphorylating ATG14. *Autophagy, 12*(3), 547–564. https://doi.org/10.1080/15548627.2016.1140293.

Park, J.-M., Seo, M., Jung, C. H., Grunwald, D., Stone, M., Otto, N. M., et al. (2018). ULK1 phosphorylates Ser30 of BECN1 in association with ATG14 to stimulate autophagy induction. *Autophagy*, 1–28. https://doi.org/10.1080/15548627.2017.1422851.

Pattingre, S., Tassa, A., Qu, X., Garuti, R., Liang, X. H., Mizushima, N., et al. (2005). Bcl-2 antiapoptotic proteins inhibit Beclin 1-dependent autophagy. *Cell, 122*(6), 927–939. https://doi.org/10.1016/j.cell.2005.07.002.

Proikas-Cezanne, T., Takacs, Z., Dönnes, P., & Kohlbacher, O. (2015). WIPI proteins: Essential PtdIns3P effectors at the nascent autophagosome. *Journal of Cell Science, 128*(2), 207–217. https://doi.org/10.1242/jcs.146258.

Pulkoski-Gross, M. J., Jenkins, M. L., Truman, J.-P., Salama, M. F., Clarke, C. J., Burke, J. E., et al. (2018). An intrinsic lipid-binding interface controls sphingosine kinase 1 function. *Journal of Lipid Research*. https://doi.org/10.1194/jlr.M081307.

Ravikumar, B., Acevedo-Arozena, A., Imarisio, S., Berger, Z., Vacher, C., O'Kane, C. J., et al. (2005). Dynein mutations impair autophagic clearance of aggregate-prone proteins. *Nature Genetics, 37*(7), 771–776. https://doi.org/10.1038/ng1591.

Razi, M., Chan, E. Y. W., & Tooze, S. A. (2009). Early endosomes and endosomal coatomer are required for autophagy. *The Journal of Cell Biology, 185*(2), 305–321. https://doi.org/10.1083/jcb.200810098.

Rebecca, V. W., & Amaravadi, R. K. (2015). Emerging strategies to effectively target autophagy in cancer. *Oncogene*. https://doi.org/10.1038/onc.2015.99.

Russell, R. C., Tian, Y., Yuan, H., Park, H. W., Chang, Y.-Y., Kim, J., et al. (2013). ULK1 induces autophagy by phosphorylating Beclin-1 and activating VPS34 lipid kinase. *Nature Cell Biology, 15*(7), 741–750. https://doi.org/10.1038/ncb2757.

Sarkar, S., Carroll, B., Buganim, Y., Maetzel, D., Ng, A. H. M., Cassady, J. P., et al. (2013). Impaired autophagy in the lipid-storage disorder Niemann-Pick Type C1 disease. *Cell Reports, 5*(5), 1302–1315. https://doi.org/10.1016/j.celrep.2013.10.042.

Scarlatti, F., Bauvy, C., Ventruti, A., Sala, G., Cluzeaud, F., Vandewalle, A., et al. (2004). Ceramide-mediated macroautophagy involves inhibition of protein kinase B and up-regulation of Beclin 1. *Journal of Biological Chemistry, 279*(18), 18384–18391. https://doi.org/10.1074/jbc.M313561200.

Sentelle, R. D., Senkal, C. E., Jiang, W., Ponnusamy, S., Gencer, S., Panneer Selvam, S., et al. (2012). Ceramide targets autophagosomes to mitochondria and induces lethal mitophagy. *Nature Chemical Biology, 8*(10), 831–838. https://doi.org/10.1038/nchembio.1059.

Seranova, E., Connolly, K. J., Zatyka, M., Rosenstock, T. R., Barrett, T., Tuxworth, R. I., et al. (2017). Dysregulation of autophagy as a common mechanism in lysosomal storage diseases. *Essays in Biochemistry, 61*(6), 733–749. https://doi.org/10.1042/EBC20170055.

Shakor, A. B. A., Taniguchi, M., Kitatani, K., Hashimoto, M., Asano, S., Hayashi, A., et al. (2011). Sphingomyelin synthase 1-generated sphingomyelin plays an important role in transferrin trafficking and cell proliferation. *Journal of Biological Chemistry, 286*(41), 36053–36062. https://doi.org/10.1074/jbc.M111.228593.

Shen, H., Giordano, F., Wu, Y., Chan, J., Zhu, C., Milosevic, I., et al. (2014). Coupling between endocytosis and sphingosine kinase 1 recruitment. *Nature Cell Biology, 16*(7), 652–662. https://doi.org/10.1038/ncb2987.

Shen, W., Henry, A. G., Paumier, K. L., Li, L., Mou, K., Dunlop, J., et al. (2014). Inhibition of glucosylceramide synthase stimulates autophagy flux in neurons. *Journal of Neurochemistry, 129*(5), 884–894. https://doi.org/10.1111/jnc.12672.

Shibutani, S. T., & Yoshimori, T. (2014). A current perspective of autophagosome biogenesis. *Cell Research, 24*(1), 58–68. https://doi.org/10.1038/cr.2013.159.

Simanshu, D. K., Kamlekar, R. K., Wijesinghe, D. S., Zou, X., Zhai, X., Mishra, S. K., et al. (2013). Non-vesicular trafficking by a ceramide-1-phosphate transfer protein regulates eicosanoids. *Nature, 500*(7463), 463–467. https://doi.org/10.1038/nature12332.

Sims, K., Haynes, C. A., Kelly, S., Allegood, J. C., Wang, E., Momin, A., et al. (2010). Kdo2-lipid A, a TLR4-specific agonist, induces de novo sphingolipid biosynthesis in RAW264.7 macrophages, which is essential for induction of autophagy. *The Journal of Biological Chemistry, 285*(49), 38568–38579. https://doi.org/10.1074/jbc.M110.170621.

Slattum, G., Gu, Y., Sabbadini, R., & Rosenblatt, J. (2014). Autophagy in oncogenic K-Ras promotes basal extrusion of epithelial cells by degrading S1P. *Current Biology, 24*(1), 19–28. https://doi.org/10.1016/j.cub.2013.11.029.

Song, D.-D., Zhang, T.-T., Chen, J.-L., Xia, Y.-F., Qin, Z.-H., Waeber, C., et al. (2017). Sphingosine kinase 2 activates autophagy and protects neurons against ischemic injury through interaction with Bcl-2 via its putative BH3 domain. *Cell Death and Disease, 8*(7), e2912. https://doi.org/10.1038/cddis.2017.289.

Sou, Y., Waguri, S., Iwata, J., Ueno, T., Fujimura, T., Hara, T., et al. (2008). The Atg8 conjugation system is indispensable for proper development of autophagic isolation membranes in mice. *Molecular Biology of the Cell, 19*(11), 4762–4775. https://doi.org/10.1091/mbc.E08-03-0309.

Sun, T., Li, D., Wang, L., Xia, L., Ma, J., Guan, Z., et al. (2011). c-Jun NH2-terminal kinase activation is essential for up-regulation of LC3 during ceramide-induced autophagy in human nasopharyngeal carcinoma cells. *Journal of Translational Medicine, 9*(161). https://doi.org/10.1186/1479-5876-9-161.

Takahashi, Y., Meyerkord, C. L., Hori, T., Runkle, K., Fox, T. E., Kester, M., et al. (2011). Bif-1 regulates Atg9 trafficking by mediating the fission of Golgi membranes during autophagy. *Autophagy,* 7(1), 61–73. https://doi.org/10.4161/auto.7.1.14015.

Takahashi, Y., Tsotakos, N., Liu, Y., Young, M. M., Serfass, J., Tang, Z., et al. (2016). The Bif-1-Dynamin 2 membrane fission machinery regulates Atg9-containing vesicle generation at the Rab11-positive reservoirs. *Oncotarget,* 7(15), 20855–20868. https://doi.org/10.18632/oncotarget.8028.

Takáts, S., Nagy, P., Varga, Á., Pircs, K., Kárpáti, M., Varga, K., et al. (2013). Autophagosomal Syntaxin17-dependent lysosomal degradation maintains neuronal function in Drosophila. *The Journal of Cell Biology, 201*(4), 531–539. https://doi.org/10.1083/jcb.201211160.

Tang, Z., Takahashi, Y., Chen, C., Liu, Y., He, H., Tsotakos, N., et al. (2017). Atg2A/B deficiency switches cytoprotective autophagy to non-canonical caspase-8 activation and apoptosis. *Cell Death and Differentiation, 24*(12), 2127–2138. https://doi.org/10.1038/cdd.2017.133.

Taniguchi, M., Kitatani, K., Kondo, T., Hashimoto-Nishimura, M., Asano, S., Hayashi, A., et al. (2012). Regulation of autophagy and its associated cell death by "Sphingolipid Rheostat" reciprocal role of ceramide and sphingosine 1-phosphate in the mammalian target of rapamycin pathway. *Journal of Biological Chemistry, 287*(47), 39898–39910. https://doi.org/10.1074/jbc.M112.416552.

Toops, K. A., Tan, L. X., Jiang, Z., Radu, R. A., & Lakkaraju, A. (2015). Cholesterol-mediated activation of acid sphingomyelinase disrupts autophagy in the retinal pigment epithelium. *Molecular Biology of the Cell, 26*(1), 1–14. https://doi.org/10.1091/mbc.E14-05-1028.

Tooze, S. A., Dooley, H. C., Jefferies, H. B. J., Joachim, J., Judith, D., Lamb, C. A., et al. (2015). Assessing mammalian autophagy. *Methods in Molecular Biology, 1270,* 155–165. https://doi.org/10.1007/978-1-4939-2309-0_12.

Tsuboyama, K., Koyama-Honda, I., Sakamaki, Y., Koike, M., Morishita, H., & Mizushima, N. (2016). The ATG conjugation systems are important for degradation of the inner autophagosomal membrane. *Science (New York, N.Y.), 354*(6315), 1036–1041. https://doi.org/10.1126/science.aaf6136.

Turner, L. S., Cheng, J. C., Beckham, T. H., Keane, T. E., Norris, J. S., & Liu, X. (2011). Autophagy is increased in prostate cancer cells overexpressing acid ceramidase and enhances resistance to C6 ceramide. *Prostate Cancer and Prostatic Diseases, 14*(1), 30–37. https://doi.org/10.1038/pcan.2010.47.

Velasco, G., Salazar, M., Carracedo, A., Salanueva, I. J., Hernandez-Tiedra, S., Lorente, M., et al. (2009). Cannabinoid action induces autophagy-mediated cell death through stimulation of ER stress in human glioma cells. *Journal of Clinical Investigation, 119*(5), 1359–1372. https://doi.org/10.1172/JCI37948.

Velikkakath, A. K. G., Nishimura, T., Oita, E., Ishihara, N., & Mizushima, N. (2012). Mammalian Atg2 proteins are essential for autophagosome formation and important for regulation of size and distribution of lipid droplets. *Molecular Biology of the Cell, 23*(5), 896–909. https://doi.org/10.1091/mbc.E11-09-0785.

Villar, V. H., Nguyen, T. L., Delcroix, V., Terés, S., Bouchecareilh, M., Salin, B., et al. (2017). mTORC1 inhibition in cancer cells protects from glutaminolysis-mediated apoptosis during nutrient limitation. *Nature Communications, 8*(14124). https://doi.org/10.1038/ncomms14124.

Wang, S., Robinet, P., Smith, J. D., & Gulshan, K. (2015). ORMDL orosomucoid-like proteins are degraded by free-cholesterol-loading-induced autophagy. *Proceedings of the National Academy of Sciences of the United States of America, 112*(12), 3728–3733. https://doi.org/10.1073/pnas.1422455112.

Wang, Z., Miao, G., Xue, X., Guo, X., Yuan, C., Wang, Z., et al. (2016). The vici syndrome protein EPG5 is a Rab7 effector that determines the fusion specificity of autophagosomes with late endosomes/lysosomes. *Molecular Cell, 63*(5), 781–795. https://doi.org/10.1016/j.molcel.2016.08.021.

Wang, Z., Wen, L., Zhu, F., Wang, Y., Xie, Q., Chen, Z., et al. (2017). Overexpression of ceramide synthase 1 increases C18-ceramide and leads to lethal autophagy in human glioma. *Oncotarget, 8*(61), 104022–104036. https://doi.org/10.18632/oncotarget.21955.

Weidberg, H., Shvets, E., Shpilka, T., Shimron, F., Shinder, V., & Elazar, Z. (2010). LC3 and GATE-16/GABARAP subfamilies are both essential yet act differently in autophagosome biogenesis. *The EMBO Journal, 29*(11), 1792–1802. https://doi.org/10.1038/emboj.2010.74.

Wijdeven, R. H., Janssen, H., Nahidiazar, L., Janssen, L., Jalink, K., Berlin, I., et al. (2016). Cholesterol and ORP1L-mediated ER contact sites control autophagosome transport and fusion with the endocytic pathway. *Nature Communications, 7*(11808). https://doi.org/10.1038/ncomms11808.

Williams, D., Vicôgne, J., Zaitseva, I., McLaughlin, S., & Pessin, J. E. (2009). Evidence that electrostatic interactions between vesicle-associated membrane protein 2 and acidic phospholipids may modulate the fusion of transport vesicles with the plasma membrane. *Molecular Biology of the Cell, 20*(23), 4910–4919. https://doi.org/10.1091/mbc.E09-04-0284.

Wong, P.-M., Feng, Y., Wang, J., Shi, R., & Jiang, X. (2015). Regulation of autophagy by coordinated action of mTORC1 and protein phosphatase 2A. *Nature Communications, 6*(8048). https://doi.org/10.1038/ncomms9048.

Yamagata, M., Obara, K., & Kihara, A. (2011). Sphingolipid synthesis is involved in autophagy in *Saccharomyces cerevisiae. Biochemical and Biophysical Research Communications, 410*(4), 786–791. https://doi.org/10.1016/j.bbrc.2011.06.061.

Yang, S., Wang, X., Contino, G., Liesa, M., Sahin, E., Ying, H., et al. (2011). Pancreatic cancers require autophagy for tumor growth. *Genes and Development, 25*(7), 717–729. https://doi.org/10.1101/gad.2016111.

Young, A. R. J., Chan, E. Y. W., Hu, X. W., Köchl, R., Crawshaw, S. G., High, S., et al. (2006). Starvation and ULK1-dependent cycling of mammalian Atg9 between the TGN and endosomes. *Journal of Cell Science, 119*(18), 3888–3900. https://doi.org/10.1242/jcs.03172.

Young, M. M., Takahashi, Y., Fox, T. E., Yun, J. K., Kester, M., & Wang, H.-G. (2016). Sphingosine kinase 1 cooperates with autophagy to maintain endocytic membrane trafficking. *Cell Reports, 17*(6), 1532–1545. https://doi.org/10.1016/j.celrep.2016.10.019.

Young, M. M., Takahashi, Y., Khan, O., Park, S., Hori, T., Yun, J., et al. (2012). Autophagosomal membrane serves as platform for intracellular death-inducing signaling complex (iDISC)-mediated caspase-8 activation and apoptosis. *Journal of Biological Chemistry, 287*(15), 12455–12468. https://doi.org/10.1074/jbc.M111.309104.

Young, M., & Wang, H. (2013). The cross talk between apoptosis and autophagy. In H. G. Wang (Ed.), *Autophagy and cancer* (Vol. 8, pp. 205–224). New York, NY: Springer-Verlag.
Yousefi, S., Perozzo, R., Schmid, I., Ziemiecki, A., Schaffner, T., Scapozza, L., et al. (2006). Calpain-mediated cleavage of Atg5 switches autophagy to apoptosis. *Nature Cell Biology, 8*(10), 1124–1132. https://doi.org/10.1038/Ncb1482.
Zhang, Y.-B., Gong, J.-L., Xing, T.-Y., Zheng, S.-P., & Ding, W. (2013). Autophagy protein p62/SQSTM1 is involved in HAMLET-induced cell death by modulating apoptosis in U87MG cells. *Cell Death and Disease, 4*, e550. https://doi.org/10.1038/cddis.2013.77.
Zhi, X., & Zhong, Q. (2015). Autophagy in cancer. *F1000prime Reports,* 7(18). https://doi.org/10.12703/P7-18.

CHAPTER THREE

Role and Function of Sphingomyelin Biosynthesis in the Development of Cancer

Giovanni D'Angelo*, Sitapriya Moorthi[§] and Chiara Luberto[§,1]
*Institute of Protein Biochemistry, National Research Council of Italy, Naples, Italy
[§]Department of Physiology and Biophysics, Stony Brook University, Stony Brook, NY, United States
[1]Corresponding author: E-mail: chiara.luberto@stonybrook.edu

Contents

Advances in Cancer Research, Volume 140
ISSN 0065-230X
https://doi.org/10.1016/bs.acr.2018.04.009

Abstract

Sphingomyelin (SM) biosynthesis represents a complex, finely regulated process, mostly occurring in vertebrates. It is intimately linked to lipid transport and it is ultimately carried out by two enzymes, SM synthase 1 and 2, selectively localized in the Golgi and plasma membrane. In the course of the SM biosynthetic reaction, various lipids are metabolized. Because these lipids have both structural and signaling functions, the SM biosynthetic process has the potential to affect diverse important cellular processes (such as cell proliferation, cell survival, and migration). Thus defects in SM biosynthesis might directly or indirectly impact the normal physiology of the cell and eventually of the organism. In this chapter, we will focus on evidence supporting a role for SM biosynthesis in specific cellular functions and how its dysregulation can affect neoplastic transformation.

1. STRUCTURE AND INTRACELLULAR DISTRIBUTION OF SPHINGOMYELIN

Sphingomyelin (SM) is a phosphosphingolipid, mostly present in vertebrates representing 5%–10% of total phospholipids (for a review see Hanada, 2014). It is an essential component of mammalian cells, as cells devoid of SM synthesis do not proliferate without exogenous addition of SM (Shakor et al., 2011; Yamaoka, Miyaji, Kitano, Umehara, & Okazaki, 2004) and mice lacking all enzymes responsible for SM synthesis do not progress through embryonic development.

SM was the first sphingolipid identified in 1884 by the German biochemist Thudichum (1884), and it is characterized by a sphingoid backbone to which a fatty acid is linked with an amide bond and a phosphorylcholine head group (Fig. 1). The nature of the sphingoid backbone and of the fatty acid confers heterogeneity to SM molecules; indeed the fatty acid moiety, mostly saturated, varies in length from C14 to C26 (with C16:0,

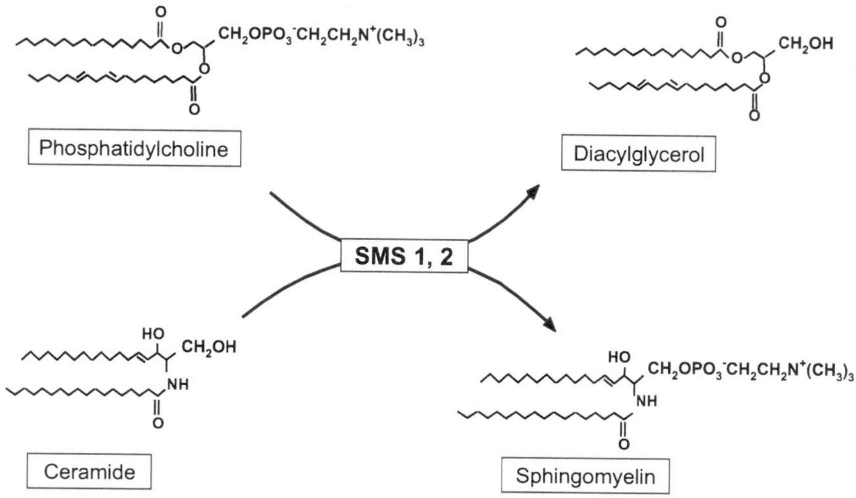

Figure 1 ***Schematic representation of the sphingomyelin (SM) synthase reaction.*** Represented in the figure are the substrates, ceramide and phosphatidylcholine and products, SM, and diacylglycerol (DAG) of the biochemical reaction catalyzed by SM synthase 1 and 2 (SMS1 and 2). The reaction can also proceed in the reverse orientation.

C24:0, and C24:1 being the most represented), whereas the sphingoid backbone can be either saturated (dihydrosphinganine) or unsaturated in position C4 (sphingosine) or hydroxylated (4-hydroxysphinganine) (for a review see Slotte, 2013). The role and function of the different molecular species of SM within cells is not currently well understood, with very few exceptions (Contreras et al., 2012; Sakamoto et al., 2017). However, imaging of SM using matrix-assisted laser desorption (MALDI) is beginning to point to specific patterns of distribution of SM species within different organs (i.e., lungs, kidneys, and brain) and in pathological conditions (i.e., fungal infections or cancer) (Jones et al., 2014; Miyamoto et al., 2016; Qureshi et al., 2010; Sugimoto et al., 2015, 2016). As more studies in this direction will be performed, our understanding of the functional meaning of these observations will be clarified and will certainly represent one of the next frontiers in sphingolipid biology.

In cells, the localization of SM has been established using a variety of approaches including measurements of SM mass levels in subcellular fractions, analysis of the cellular distribution of SM-binding toxins, and indirectly through the targeted expression of bacterial sphingomyelinase in different subcellular compartments (Bakrac et al., 2010; Barlic et al., 2004; Birbes, El Bawab, Hannun, & Obeid, 2001; Dai et al., 2004;

Deevska et al., 2017; Devaux & Morris, 2004; Ishitsuka, Yamaji-Hasegawa, Makino, Hirabayashi, & Kobayashi, 2004; Makino et al., 2015; van Meer, Voelker, & Feigenson, 2008; Schon et al., 2008; Stoffel et al., 2016; Yachi et al., 2012). Altogether, these studies have shown that SM is present in most intracellular compartments including the Golgi where the bulk of SM is produced and the plasma membrane (PM) where it is highly enriched in the outer leaflet.

2. THE SPHINGOMYELIN SYNTHASE ENZYMES

The class of enzymes directly responsible for SM synthesis is called phosphatidylcholine:ceramide cholinephosphotransferase or sphingomyelin synthase (SMS). The reaction proceeds with the transfer of a phosphocholine moiety from phosphatidylcholine (PC) onto the primary hydroxyl group of ceramide forming SM and diacylglycerol (DAG) (Fig. 1) (Diringer, Marggraf, Koch, & Anderer, 1972; Ullman & Radin, 1974) (for an historical perspective of the characterization of the SMS reaction see Holthuis & Luberto, 2010). In 2004, two genes encoding for proteins with SMS activity were identified: *SGMS1*, located on chromosome 10, and *SGMS2*, located on chromosome 4 (Huitema, van den Dikkenberg, Brouwers, & Holthuis, 2004; Yamaoka et al., 2004).

The SMS proteins share ~60% identity, with the biggest structural difference being the presence in SMS1 of a sterile alpha motif at the N-terminus, whose function remains unknown (Huitema et al., 2004). SMS1 and SMS2 are integral membrane proteins with six transmembrane domains and both amino and carboxy termini oriented toward the cytosol. A recent report indicates that both SMS1 and SMS2 exist as homodimers formed through interactions primarily between the C-termini (Hayashi et al., 2017). The authors also showed that the homodimerization is important for the export of SMSs from the endoplasmic reticulum (ER) en route to the Golgi apparatus, where both SMSs reside. Importantly, SMS2 is also localized at the PM by virtue of the palmitoylation of four cysteines at the C-terminus (Tani & Kuge, 2009) and, at the PM, is the only enzyme able to synthesize SM (Tafesse et al., 2007). Similar to lipid phosphate phosphatases, the SMSs active site comprises of two conserved histidines and an aspartate, all oriented toward the lumen of the Golgi or, for SMS2 at the PM, oriented toward the extracellular space (Holthuis & Luberto, 2010). Both enzymes can also catalyze the reverse reaction in vitro (Huitema et al., 2004), corroborating a previous report that described the conversion of DAG and SM into ceramide and PC on the basolateral membranes of Madin–Darby Canine Kidney (MDCK) cells (van Helvoort, van't Hof, Ritsema, Sandra, & van Meer, 1994). Finally, both SMSs are also able to utilize phosphoethanolamine instead of PC thus forming ceramide phosphoethanolamine (CPE) instead of SM (Ding et al., 2015). This is the same reaction specifically catalyzed by the SMS-related protein (SMSr) (Huitema et al., 2004); however, while SMSr is thought to be a sensor for ceramide levels in the ER (Vacaru et al., 2009), the biological significance of the CPE synthase activity for SMSs is not known.

In cells, SMS1 is thought to play a fundamental role in the synthesis of the bulk of SM, whereas SMS2 at the PM is thought to regulate the levels of ceramide and SM in response to local acute perturbations. On the other hand, the contribution of SMS2 in the Golgi in the regulation of de novo SM synthesis and thus the mass of SM at the PM is still a point of debate (Asano et al., 2012; Ding et al., 2008; Li et al., 2007; Mitsutake et al., 2011; Shakor et al., 2011; Tafesse et al., 2007, 2013; Villani et al., 2008; Yamaoka et al., 2004) and it seems to depend on the cell type, perhaps pointing to specialized functions of SMS1 versus SMS2 in different cell. Some observations support this possibility. For instance, while *SMS1* expression is ubiquitous among cell types, *SMS2* expression varies among cells, being generally expressed in adherent cells but present in very low or undetectable levels in suspension cells (Holthuis & Luberto, 2010; Shakor et al., 2011). Moreover, the distinct localization pattern of SMS1 and 2 in neuronal subtypes (Kidani et al., 2012) and the different phenotypes of SMS1 versus SMS2 knockout mice also point to distinct functions of these enzymes (Taniguchi & Okazaki, 2014).

3. FUNCTIONS OF THE LIPIDS REGULATED BY THE SPHINGOMYELIN SYNTHASE REACTION

The lipids regulated by the SMS reaction exert specific functions, some of which could participate in promoting tumorigenesis. Below we discuss some of these lipids' functions that could affect cell signaling (Fig. 2).

3.1 Sphingomyelin

Several functions for SM have been proposed, and they are mostly linked to its presence in the PM, in the Golgi or in the nucleus.

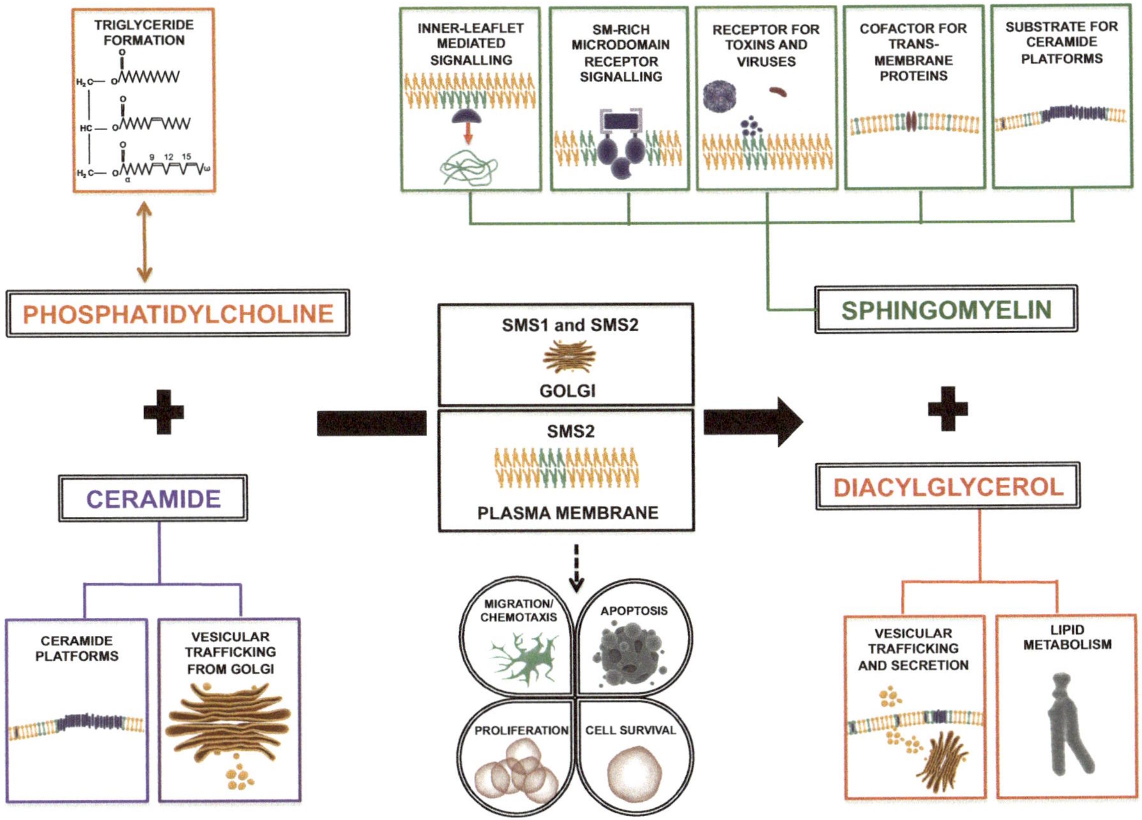

Figure 2 ***Cellular functions affected by the sphingomyelin synthase (SMS) reaction.*** As discussed in the text, the lipids directly regulated by SMSs can exert specific cellular functions. In the rectangles are reported the cellular functions directly linked to each of these lipids in response to modulation of SMSs/SM synthesis. In the cloverleaf are reported cellular functions affected by modulation of SMSs for which the specific mechanism is not known. In the *orange rectangles* are functions linked to phosphatidylcholine; in the *green rectangles* are functions mediated by sphingomyelin; in the *blue rectangles* are functions exerted by ceramide; and in the *red rectangles* are functions mediated by diacylglycerol.

3.1.1 Regulation of Lipid Microdomains in the Outer Leaflet of the Plasma Membrane and Impact on Receptor-Mediated Signaling

SM is enriched in the outer leaflet of the PM where together with cholesterol is responsible for the formation of SM-rich microdomains (Slotte, 2013). Indeed the saturation of the acyl chains of SM (as compared with the more unsaturated glycerophospholipids), the ability of establishing hydrogen bonds between SM molecules, and the ability of the polar phosphocholine head group to mask the hydrophobicity of cholesterol (umbrella model) (Ikonen, 2008) are all chemical features thought to favor the solubilization of cholesterol in the membrane and therefore the formation of highly packed microdomains (for a comprehensive discussion see Slotte, 2013). Furthermore, the fact that SM and cholesterol biosynthesis are finely coordinated (Ridgway, 2000) strengthens the biological relevance of their structural association. As SM-rich microdomains are thought to be a platform for selective recruitment and activation of surface receptors, it is not surprising that changes in SM biosynthesis, which lead to alteration of SM levels at the PM, have been implicated in the regulation of several membrane receptors. Modulation of SMS1 has been shown to regulate CD95 activation and Fas signaling in murine lymphoid cells (Miyaji et al., 2005), TCR in Jurkat cells (Jin et al., 2008), and murine CD41 T-cells (Dong et al., 2012), whereas modulation of SMS2 provided a link to regulation of TLR4 in macrophages (Hailemariam et al., 2008) and of TNFα receptor 1 in human embryonic kidney 293 cells (Li et al., 2007).

Modulation of SM on the PM through SMS2 was also shown to affect the fatty acid transporter CD36, with important implications for the homeostasis of lipid droplets (Mitsutake et al., 2011). This regulation occurs in the caveolae, discrete invaginations of PM, characterized by the presence of the scaffolding protein, caveolin 1 and containing SM-rich microdomains. In these structures of the PM, SMS2 associates with caveolin 1 and the CD36 transporter.

In addition to forming SM-rich microdomains, hydrolysis of SM may also promote the formation of ceramide-enriched platforms, PM structures larger than microdomains, which are considered signaling hubs (for more details see the paragraph below in Section 3.3.1) (Zhang, Li, Becker, & Gulbins, 2009).

SM can also control clathrin-dependent endocytosis of specific targets (Shakor et al., 2011). Indeed, SM-deficient lymphoid cells were unable to internalize the transferrin/transferrin receptor 1 complexes once they form in the PM, while restoration of SM biosynthesis, through expression of SMS1 (but not SMS2), restored the process. On the other hand, this regulatory effect of SM was not observed for low density lipoproteins (LDL) internalization, which also proceeds in a clathrin-dependent manner; while these data indicate that this is a transferrin-specific effect, the mechanism for such specificity remains unknown.

3.1.2 Signaling Functions at the Plasma Membrane Independent From Receptor Activation

SM present in microdomains has been found to allow the accumulation of phosphatidylinositol-4,5-bisphosphate [PtdIns(4,5)P_2] in the inner leaflet of the PM, at the cleavage furrow during cytokinesis (Abe et al., 2012). The accumulation of PtdIns(4,5)P_2 is essential for enabling cytokinesis, and, even though the precise mechanism leading to PtdIns(4,5)P_2 accumulation is not clear, it has been proposed that the long fatty acid chains of SM in the exoleaflet, by intercalating into the endoleaflet, could restrict the diffusion of PtdIns(4,5)P_2 in the membrane (Slotte, 2013).

Another example of signaling functions of SM at the PM independent from receptor activation is the regulation of the levels of phosphatidylserine (PS) in the inner leaflet following modulation of SM, with important consequences in regulation of KRAS oncogene-mediated signaling (van der Hoeven et al., 2017).

3.1.3 Specific Interactions With Membrane Proteins and Nucleic Acids

It has been long known that SM acts as a receptor for toxins produced by certain bacteria, viruses, fungi, and animals which are either secreted or present on the surface of these microorganisms. Some of these toxins, on binding to SM, oligomerize and form pores allowing the uncontrolled exchange of solutes between the cells and the extracellular environment, ultimately leading to cell lysis. Lysenin and EqtII are examples of such toxins (Schon et al., 2008; Yilmaz, Yamaji-Hasegawa, Hullin-Matsuda, & Kobayashi, 2018). Others, like *Pasteurella multocida* toxin, bind to SM to penetrate into the host cells, causing pleiotropic effects (Brothers et al., 2011). The vacuolating cytotoxin of the gastric pathogen *Helicobacter pylori* binds to SM and exploits the machinery present in lipid microdomains to enter epithelial cells, ultimately resulting in cellular vacuolation (Gupta et al., 2008).

Other studies proposed an inhibitory function for SM in the switch of voltage-activated potassium channels; interestingly, the regulation was specific for the Kv2.1 channel subtype, it was dependent on the phospho-head group of SM and independent from the presence of cholesterol thus pointing to a specific function of SM and not of microdomains (Ramu, Xu, & Lu, 2006; Xu, Ramu, & Lu, 2008).

Importantly, it has also been demonstrated that SM may act as cofactor for transmembrane proteins involved in regulating retrograde trafficking from the Golgi to the ER via COPI vesicles (Contreras et al., 2012). While it was shown that the amount of SM and cholesterol is very low in COPI vesicles compared with the Golgi, the specific presence of 18:0 SM regulated the activity of the transmembrane protein p24, a key modulator of vesicle biogenesis. A conserved domain in the transmembrane portion of p24 was found to be responsible for this interaction and it has been identified also in several other transmembrane proteins, mostly localized in the PM, suggesting the potential role of SM in mediating other cell functions through regulation of these groups of transmembrane proteins.

In the nucleus, SM has been proposed both to form platforms with cholesterol in the inner nuclear membrane, serving as an anchor for transcriptionally active chromatin (Scassellati et al., 2010), and also to protect double-strand RNA from RNAse degradation (Micheli, Albi, Leray, & Magni, 1998).

3.2 Diacylglycerol

DAG has been long recognized a key lipid signaling molecule, associated with a plethora of cellular processes such proliferation and/or survival, cytoskeletal reorganization, membrane trafficking, exocytosis, synaptic formation, phagocytosis, and a key intermediate of several lipid metabolic pathways such as synthesis of phospholipids and triglycerides (TGs). The heterogeneity of DAG's signaling functions is a consequence of the vast number of proteins it interacts with through a high affinity C1 domain present in these targets, like classical and novel protein kinase Cs (PKCs), protein kinase D (PKD), lipid kinases to name a few (for a comprehensive review see Almena & Merida, 2011). Evidence for regulation of DAG by SMSs has been more elusive than changes in SM and only in very few cell systems, modulation of either SMS produced detectable and significant changes of DAG levels in whole cells (Burns et al., 2013; Ding et al., 2008; Li et al., 2007). This may indicate a tight regulation of such pool and/or very localized changes within the cell, which would render the identification of specific functions for SMS-derived DAG challenging.

3.2.1 Signaling Through Protein Kinase D at the Golgi

In Hela cells, downregulation of either SMS did not change the amount of total cellular DAG (Tafesse et al., 2007; Villani et al., 2008); on the other hand, loss of SMS activity significantly impaired the levels of DAG at the Golgi. Indeed by using a fluorescent DAG-binding reporter (YFP-DBD) comprising of the C1 domain of novel PKCβ, it was shown that downregulation of either SMS1 or SMS2 greatly reduced the localization of the YFP-DBD at the Golgi. Furthermore, in addition to DAG, also the localization at the Golgi of the DAG-binding PKD was affected, indicating that SMSs are able to regulate the levels of DAG in this organelle. In line with the well-established role of Golgi PKD in regulating vesicular trafficking to the PM, inhibition or downregulation of either SMS also impaired this process and insulin secretion (Subathra, Qureshi, & Luberto, 2011).

3.2.2 Effect on Protein Kinase Cα

In bone marrow-derived macrophages, stimulation with the lipid mediator, ceramide-1-phosphate led to SMS activation followed by induction of PKCα and proliferation (Gangoiti, Granado, Arana, Ouro, & Gomez-Munoz, 2010). However, in this study, the molecular mechanism linking SMS and PKCα was not investigated.

3.2.3 Lipid Metabolism

Some evidence exists suggesting that DAG produced through SMS can be channeled to either synthesis of TGs or back to form PC.

In Baby Hamster Kidney cells, activation of SM synthesis following bacterial sphingomyelinase treatment failed to accumulate DAG while an increase of TGs was measured, suggesting a metabolic link between the two (Allan & Quinn, 1988). However, these results seem to contradict a recent report in which overexpression of SMS1 in hepatocytes caused a decrease of TGs (Deevska et al., 2017). In the study, it was shown that a common pool of DAG is used for the synthesis of either TGs or PC destined to SM synthesis. Upregulation of SMS1, by consuming PC, shunted more DAG toward de novo synthesis of PC causing a drop of TGs (Sarri, Sicart, Lazaro-Dieguez, & Egea, 2011). Additionally, DAG produced by SMS could also be utilized to resynthesize PC thus participating in a positive feedback that results in the overall increase of PC levels.

In addition to the link with TGs, DAG produced by the SMS reaction could be converted back into PC via the SMS reverse reaction. Such reaction has been documented in vitro for both SMS1 and SMS2 and in cells (Huitema et al., 2004) on the basolateral membrane of MDCK cells (van Helvoort et al., 1994).

3.3 Ceramide

Ceramide is one of the substrates of the SMS reaction and, as such, modulation of the SMS enzymes could affect its levels. The role and functions of ceramide have been extensively discussed in several comprehensive reviews (Bartke & Hannun, 2009; Hannun & Obeid, 2011; Hernandez-Corbacho, Salama, Canals, Senkal, & Obeid, 2017; Ogretmen, 2018) and include growth arrest, apoptosis, differentiation, inflammation, and autophagy. Here, we will only briefly summarize the main evidence reporting ceramide functions associated with the PM and Golgi complex, which would be the most proximal sites impacted by alteration in SM synthesis.

3.3.1 Ceramide-Enriched Platforms at the Plasma Membrane

It has been shown that ceramide accumulation generated through hydrolysis of SM in the outer leaflet of the PM contributes to transforming small rafts into larger signaling hubs, the so-called "ceramide platforms," which promote clustering of receptor molecules for enhanced signaling (Babiychuk, Monastyrskaya, & Draeger, 2008; Zhang et al., 2009). Hence these membrane structures are larger than SM-rich microdomains, they are generally formed in response to acid sphingomyelinase activation and they have been visualized with the use of ceramide antibodies. The formation of these platforms has been shown in response to CD95/Fas activation by Fas ligand (FasL) (Grassme et al., 2001), and on infections with *Pseudomonas aeruginosa* (Grassme et al., 2003) or the ebola virus (Miller, Adhikary, Kolokoltsov, & Davey, 2012). In particular, activation of Fas by FasL generates a "cap" on the surface of the cells initiated by hydrolysis of SM and accumulation of ceramide, resulting in the oligomerization of Fas and recruitment of different proteins to form the DISC, death-inducing signaling complex. In murine lymphoid cells engineered to express Fas and lacking SM synthesis (and with no detectable SM on the PM), stimulation with FasL generated poor capping and impaired the apoptotic response (Miyaji et al., 2005). Restoration of SM synthesis with overexpression of SMS1 increased the formation of ceramide in the detergent-resistant fractions of the membranes (where caps would form), doubled the amount of capping on cell surface, and enhanced apoptosis (Miyaji et al., 2005).

These results suggest that the generation of ceramide on the PM depends on the expression of SMS1 to produce critical levels of SM.

3.3.2 Signaling at the Plasma Membrane Independent From Ceramide-Enriched Platforms

SMS2 at the PM has its active site oriented toward the extracellular space, thus, when present, it may directly influence the levels of ceramide and SM in the outer leaflet. On the other hand, it cannot be excluded that ceramide in the outer leaflet may flip to the inner leaflet where it could interact with cytosolic targets. Alternatively, ceramide could be generated directly from hydrolysis of SM present in the inner leaflet. Indeed, an older study provided evidence to support the latter (Linardic & Hannun, 1994). In this study, treatment of HL-60 cells with vitamin D_3 or TNFα generated ceramide at the PM from hydrolysis of a pool of SM, which was resistant to treatment with exogenously added bacterial sphingomyelinase and thus thought to be present in the inner leaflet. Several proteins such as protein phosphatase 1 and protein phosphatase 2A (PP2A), the inhibitor 2 of PP2A, cathepsin D, and kinase suppressor of Ras (Ogretmen, 2018) have been shown to directly interact with ceramide, representing potential signaling targets for ceramide facing the cytosol.

3.3.3 Ceramide Functions at the Golgi

Not much is known about ceramide functions at the Golgi. Most investigations have addressed this point by treating cells with ceramide analogues carrying short acyl chains, such as C2- or C6-ceramide, as they have been shown to easily enter the cells and accumulate in the Golgi where they are metabolized into short-chain SM and glucosylceramide. The addition of these analogues has been shown to inhibit protein transport from the *trans*-Golgi network and to impair constitutive and regulated secretion (Rosenwald & Pagano, 1993; Shinkai, Takeyama, Ueda, Hori, & Yamamoto, 1997). A recent study proposed that it is the conversion of short-chain ceramide into short-chain SM that ultimately inhibits trafficking by altering lipid order in the Golgi membrane (Duran et al., 2012), whereas another one suggested that this effect is due to sphingosine generated from hydrolysis of ceramide (Hu et al., 2005) and another to the existence of a negative feedback loop that would inhibit SM synthesis, DAG production, and PKD activation (Capasso et al., 2017). The assessment of the physiological significance of these observations is the next step in this area of investigation. In fact the high concentrations of ceramide analogues used in some of the studies and their unnatural structure could complicate the interpretation of the results. One study showed that the effects of these short-chain ceramide analogues in combination with the Golgi disrupting drug Brefeldin A are opposite to the effects observed when natural ceramide was added (Fukunaga et al., 2000). While short-chain analogues inhibited the disruption of the Golgi induced by Brefeldin A, the addition of natural ceramide or endogenous accumulation of ceramide augmented Brefeldin A-mediated effects, perhaps underscoring the functional difference between ceramide analogues and natural ceramide.

4. REGULATION OF SPHINGOMYELIN BIOSYNTHESIS

4.1 Sphingomyelin Synthesis via the Ceramide Transfer Protein

As mentioned earlier, the bulk of SM synthesis occurs in the Golgi where both SMS1 and, in some cell types, SMS2 reside. Ceramide, the substrate of the SMS reaction, is synthesized in the ER, and it has to make its way to the Golgi for SM synthesis. Thus the regulation of ceramide availability for SMS in the Golgi will have a significant impact for SM biosynthesis. In most cells, ceramide destined to SM synthesis reaches the Golgi through protein-mediated transport via the Ceramide Transfer Protein, CERT (Fig. 3A) (Hanada et al., 2003).

CERT is a 68 KDa cytosolic protein carrying different functional domains or motifs each regulating different activities of the protein (Fig. 3B). An N-terminal pleckstrin homology (PH) domain allows the binding to the *trans*-Golgi-enriched phosphoinositide phosphatidylinositol-4-phosphate [PtdIns(4)*P*] pool and targets CERT to the Golgi complex. The C-terminal steroidogenic acute regulatory protein-related lipid transfer (START) domain is responsible for binding and transferring ceramide between membranes (Hanada et al., 2003). The START domain is able to accommodate different ceramide species ranging from C6 to C20 Cer with no extra space at the ceramide head for the binding/transfer of ceramide derivatives bearing bulky headgroups (Kudo et al., 2008; Kumagai et al., 2005). Finally, an FFAT (two phenylalanines in an acidic tract) motif present in the intervening region between the PH and START domains allows CERT interaction with the ER integral proteins

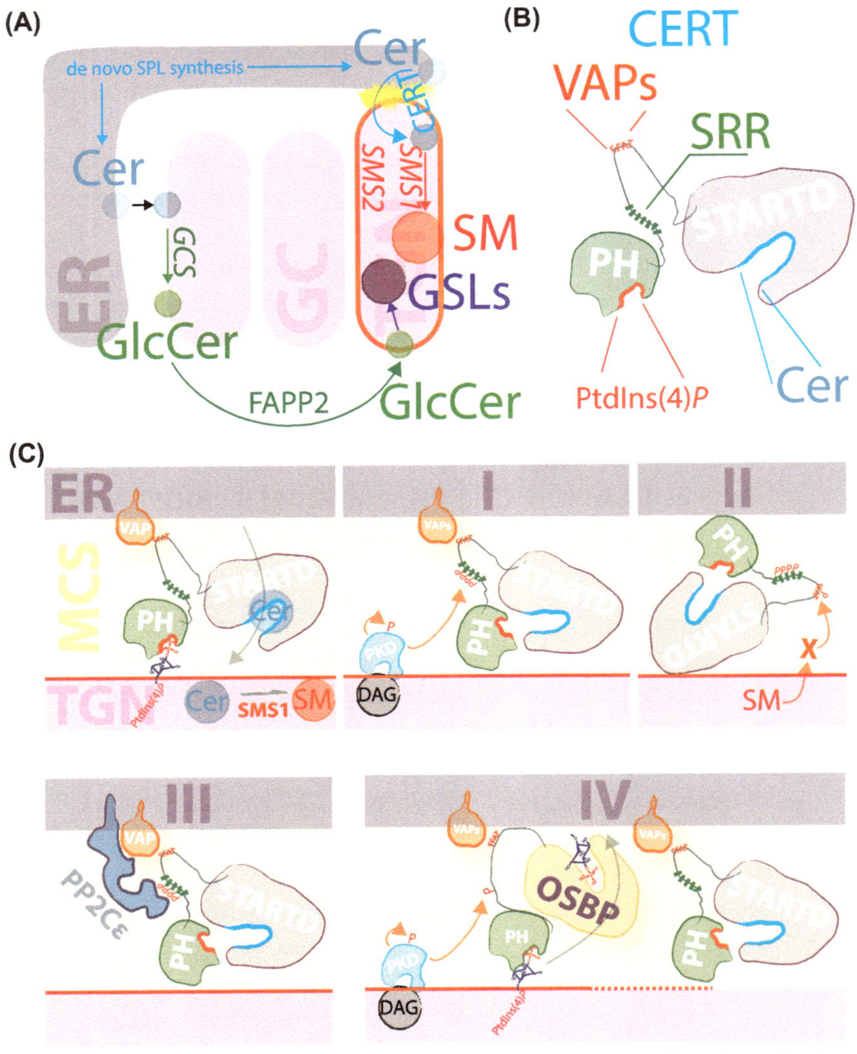

Figure 3 ***Regulation of ceramide transfer protein (CERT)-mediated synthesis of sphingomyelin at the endoplasmic reticulum (ER)-Golgi interface.*** (A) Schematic representation of SPL metabolism and transport at the ER–Golgi complex interface. *Cer,* ceramide; *GCS,* GlcCer synthase; *GlcCer,* glucosylceramide; *SM,* sphingomyelin; *SMS1 and 2,* sphingomyelin synthases 1 and 2. (B) Schematic representation of CERT molecular organization. *FFAT,* "two phenylalanines in an acidic tract" motif; *PH,* pleckstrin homology domain (in red is highlighted the PtdIns(4)*P* binding site); *SRR,* serine-rich region; *STARTD,* steroidogenic acute regulatory protein-related lipid transfer domain (in cyan is highlighted the Cer binding pocket); *VAPs,* vesicle-associated membrane protein (VAMP)-associated proteins. (C) CERT regulation. CERT contacts Trans Golgi Network (TGN) membranes by the binding of its PH domain to PtdIns(4)*P* and ER membranes by the interaction of its FFAT motif with VAPs. At the ER-TGN membrane contact

VAMP-associated Proteins A and B (VAP-A and VAP-B) (Kawano, Kumagai, Nishijima, & Hanada, 2006).

By bridging the ER and the *trans*-Golgi membranes, CERT is thought to dynamically reside at ER–Golgi membrane contact sites, where it extracts ceramide from the ER membranes and delivers it in a nonvesicular fashion to the *trans* cisternae of the Golgi complex for its conversion into SM (Hanada et al., 2003).

CERT activity is subject to a tight regulation (Fig. 3C). Phosphorylation of CERT at different sites in a serine (Ser)-rich region (SRR) at the C-terminal end of its PH domain triggers a conformational change in the protein that impedes PtdIns(4)*P* binding and thus hampers ceramide transfer to the Golgi (Kumagai, Kawano, Shinkai-Ouchi, Nishijima, & Hanada, 2007; Prashek et al., 2017).

Interestingly, the kinase initiating the phosphorylation of CERT at SRR is PKD (Fugmann et al., 2007). As discussed, PKD is activated at the Golgi by the local formation of DAG produced by the SMS reaction (Villani et al., 2008). Thus CERT activity is inhibited by phosphorylation in response to SM production, leading to the concept that a negative feedback exists linking SM synthesis to ceramide provision to the primary SM-synthesizing compartment (i.e., the Golgi complex) (Weber, Hornjik, Olayioye, Hausser, & Radde, 2015).

CERT dephosphorylation at SRR is mediated by the protein phosphatase 2Cε (PP2Cε), an integral ER protein, which interacts with VAP-A and has its catalytic site facing the cytosol (Saito et al., 2008). Following the PP2Cε-mediated dephosphorylation, CERT is recruited to the Golgi complex and SM production is stimulated (Saito et al., 2008). Also the CERT-VAPs interaction is subjected to regulation. Phosphorylation of CERT at a specific Ser located in proximity of the FFAT motif (Ser315)

sites (MCS) CERT transfers Cer from ER to TGN for SMS operated SM synthesis. (I) Diacylglycerol (DAG) produced at the TGN by SMS in the same reaction that leads to SM synthesis locally activates protein kinase D (PKD). PKD phosphorylates CERT at its SRR inducing a conformational change in CERT that result in its inactivation. (II) Plasma membrane SM levels are sensed by a so far nonidentified factor to trigger a signaling reaction leading to CERT phosphorylation in proximity of its FFAT motif and to the inhibition of CERT–VAPs interaction. (III) VAPs interact with PP2Cε that dephosphorylates CERT SRR leading to its activation. (IV) DAG produced in SM synthesis triggers a PKD-dependent signaling reaction leading to OSBP1-mediated relocation of PtdIns(4)*P* to ER membranes and consequent inactivation of proteins using PtdIns(4)*P* as a TGN targeting determinant (including CERT).

enhances CERT-VAPs interaction and promotes ceramide transfer and SM synthesis (Kumagai, Kawano-Kawada, & Hanada, 2014). Interestingly, phosphorylation at Ser315 is enhanced under conditions where cellular SM content is reduced suggesting that CERT activity is tuned according to the instant cellular need for SM synthesis.

Along similar lines a further regulatory loop exists to link CERT-mediated lipid transfer, SM synthesis, and PtdIns(4)*P* turnover at the ER–Golgi membrane contact sites. In fact CERT-mediated ceramide transfer and SM synthesis strictly depend on the PtdIns(4)*P* levels at the Golgi complex, and inhibition of PtdIns(4)*P* production leads to interruption of SM synthesis (D'Angelo et al., 2013; Toth et al., 2006). Conversely, CERT-mediated SM production has been recently shown to trigger an additional negative feedback loop involving PKD-mediated activation of the cholesterol/PtdIns(4)*P* transfer protein, OSBP1 at the ER–Golgi membrane contact sites (Mesmin et al., 2013), which leads to PtdIns(4)*P* relocation to the ER for ER-localized conversion into PtdIns (Capasso et al., 2017).

Altogether, work in the last two decades led to the discovery of a homeostatically regulated nonvesicular lipid trafficking pathway between ER and Golgi complex accounting for a substantial fraction of SM production where CERT is the central molecular player. Importantly, given the primary function of CERT for SM synthesis, misregulation of any of CERT's regulatory hubs would have significant impact on SM synthesis. Furthermore, given the importance of the points of contact between the ER and *trans*-Golgi for CERT activity, it is also likely that any condition in which the Golgi morphology is altered with loss of these contact points will negatively affect SM synthesis (Chandran & Machamer, 2008; Spessott, Uliana, & Maccioni, 2010).

While in many cell types CERT is the main route for transport of ceramide from the ER to the Golgi for SM synthesis, in some cell types like C6 glioma, SM synthesis also depends on ceramide transported through a nitric oxide-dependent vesicular trafficking, adding another layer of regulation to SM production (Giussani et al., 2008).

Transport of ceramide to SMS through CERT and vesicular trafficking regulates SM synthesis in the Golgi through SMS1 and possibly Golgi-localized SMS2. On the other hand, not much is known about regulation of SMS2 activity at the PM.

4.2 Other Mechanisms of Regulation

While not much is known about transcriptional and posttranscriptional regulation of human *SMS2*, very interesting details are emerging about the regulation of *SMS1*'s transcription and translation. The human *SMS1* locus spans over 320 kb, counting 11 annotated exons. Its canonical 5′untranslated region (UTR) is very long (over 900 nt) compared with the average 5′UTR of human genes (between 100 and 200 nt) and it was recently shown to harbor regulatory features that limit the translation efficiency of the *SMS1* mRNA (Moorthi, Burns, Yu, & Luberto, 2018; Vladychenskaya, Dergunova, Dmitrieva, & Limborska, 2004). Indeed SMS1 protein expression may be regulated by shifts in promoter utilization to modulate the length of the 5′UTR with significant consequences on translation efficiency of full length SMS1 (Moorthi et al., 2018). Additional evidence pointing to the existence of different alternative exons 1 has been provided for both human and mouse *SMS1* with possible consequences for the regulation of alternative splicing, as well (Sudarkina, Filippenkov, Brodsky, Limborska, & Dergunova, 2015; Yang, Jean-Baptiste, Khoury, & Greenwood, 2005). Moreover, the long 5′UTR of *SMS1* may contribute to the formation of circular RNAs, which, by acting as sponges, are thought to counteract the effects of microRNAs targeting that same portion of the *SMS1* gene (Filippenkov, Sudarkina, Limborska, & Dergunova, 2015). Finally, it was also reported that the 3′UTR of human *SMS1* mRNA is extraordinarily long and contains features that could greatly affect its stability (Dergunova, Rozhkova, Sudarkina, & Limborska, 2013).

When considering regulation of the SMS enzymatic function in cells, different conditions have been shown to affect SMS activity, in addition to the already mentioned oligomerization of SMS1 or SMS2 necessary for them to reach the Golgi from the ER (Hayashi et al., 2017). Recent work in chondrocytes has highlighted how the presence of neutral sphingomyelinase 2 (N-SMase 2) in the Golgi is important to maintain the localization of SMS1 in the detergent-insoluble fraction of the Golgi membranes allowing optimal in situ metabolic homeostasis among levels of ceramide, SM, and DAG to ensure proper vesicular transport (Stoffel et al., 2016).

Anecdotal evidence also suggested that expression of the ABCA8 transporter in oligodendrocytes increases expression and activity of SMS1 (Kim et al., 2013); whereas the activation of peroxisome proliferator-activated receptor delta stimulates SMS2 activity in macrophage-derived foam cells (Mou, Yang, Qu, Chen, & Zhang, 2016). Moreover, by tracing DAG formation in the luminal leaflet of the Golgi using fluorescence resonance energy transfer (FRET)-based probes, it was shown that increasing calcium levels with thapsigargin stimulates DAG production at sites of SM synthesis (Ueda et al., 2015).

Finally, it was reported that the overexpression of constitutively active PI-3 kinase increased SMS activity in natural killer (NK) cells (Taguchi et al., 2004), whereas overexpression of PKCδ increased nuclear SMS activity in melanoma cells (Albi, La Porta, Cataldi, & Magni, 2005).

5. SPHINGOMYELIN SYNTHESIS AND THE REGULATION OF CELLULAR FUNCTIONS THAT CAN AFFECT CANCER BIOLOGY

Alteration of SMS activity and modulation of the expression of SMS1 or SMS2 have been associated with changes of cellular functions that could contribute to cancer development. Below we discuss evidence for each of these cellular processes.

5.1 Proliferation

Early studies have hinted at a proproliferative function of SMS by showing a positive association between proliferation and stimulation of SM synthesis. This was observed in regenerating rat liver (Miro-Obradors, Osada, Aylagas, Sanchez-Vegazo, & Palacios-Alaiz, 1993) and in quiescent astrocytes stimulated with basic fibroblast growth factor, where the observed increase of SM synthesis aligned with a decrease of ceramide (Riboni, Viani, Bassi, Giussani, & Tettamanti, 2001).

More recently, possible mechanisms that could mediate some proproliferative effects of SMSs have been proposed; however, these events may be specific to certain cell types and growth conditions. In one study, for instance, SMS1 expression in lymphoid cells allowed for the optimal clathrin-dependent internalization of the transferrin–transferrin receptor complexes and the release in cells of iron, a crucial cofactor for energy metabolism and DNA synthesis (Shakor et al., 2011). Another study using murine embryonic fibroblasts (MEFs) showed that expression of SMS1 or SMS2 enhanced the CD98-mediated mTor response to essential amino acids, which promotes proliferation in cooperation with growth factors (Zama, Mitsutake, Okazaki, & Igarashi, 2018). Both of these proproliferative effects are thought to occur via modulation of SM-rich microdomains by SMSs.

In HeLa cells, downregulation of either SMS1 or SMS2 caused accumulation of ceramide, reduced levels of SM, and impaired cell growth. Different from the previously cited studies, cell growth was not restored on addition of exogenous SM, suggesting that the effects were triggered by changes in the bioactive lipids, ceramide, and/or DAG (Tafesse et al., 2007). In addition to affecting the rate of proliferation, SMS activity was also shown to associate with regulation of cell cycle. Pharmacological inhibition of SMS with D609 during induction of stroke in rats blocked the cell cycle via upregulation of p21 and dephosphorylation of Rb; these effects were elicited possibly by the observed accumulation of ceramide (Adibhatla & Hatcher, 2010). Similar cellular effects were reported in the neuronal Neuro-2a cells on downregulation of SMS1. In these conditions, a G0/G1 cell cycle arrest and reduced proliferation was proposed to occur through the observed upregulation of p27 and decreased levels of cyclin D1 and phospho-Akt (Wesley, Hatcher, & Dempsey, 2015).

Finally, the already mentioned regulatory effects of PM SM on cytokinesis via modulation of PtdIns(4,5)P_2 at the cleavage furrow during cytokinesis could represent yet another mechanism by which SMS activity could affect progression through cell cycle (Slotte, 2013).

5.2 Apoptosis and Cell Survival

Evidence for a potential prosurvival function of SMS was first presented when inhibition of SMS activity was observed following treatment of Kym-1 cells with TNFα, and it was proposed that the SMS inhibition contributed to the cytotoxic accumulation of ceramide (Bourteele et al., 1998). More recently, similar observations were reported following cell death of NK cells and curcumin-induced cell death in multidrug-resistant HL-60 cells (Shakor et al., 2014; Taguchi et al., 2004). In case of NK cells, addition of IL-2 rescued cell viability in part by increasing SMS activity in a PI3kinase-dependent manner. For curcumin-induced death, SMS inhibition seemed to be downstream of activation of N-SMase 2.

Inhibition of SMS1 activity was also reported following Fas treatment of Jurkat and Hela cells, and it was proposed that it contributed to accumulation of ceramide and induction of cell death (Lafont et al., 2010). Very interestingly, the inhibition occurred through a caspase-mediated cleavage of the protein that caused the relocalization of SMS1 from the Golgi into the cytoplasm. Furthermore, downregulation of SMS1 was also shown to enhance Fas cytotoxicity, whereas overexpression protected in part from cell death. It is important to point out that these results may seem to contradict the observations supporting the critical role of SM on the PM inactivation of the death receptor, discussed in the earlier section. On the other hand, those effects were reported on massive loss of SM levels in the PM, which prevented proper receptor activation, while in this later study the effect of

SMS1 downregulation on the levels of SM on the PM was limited and still enabling the full engagement of the death receptor.

Further in line with a prosurvival function of SMS, downregulation of SMS1 exacerbated the cytotoxic effect of photodamage in Jurkat cells and enhanced the accumulation of ceramide (Separovic et al., 2007), whereas in yeast, SMS1 was among murine genes from a T-cell library that protected cells from death induced by different cytotoxic stresses such as Bax expression, hydrogen peroxide, osmotic stress, and high temperatures (Yang, Khoury, Jean-Baptiste, & Greenwood, 2006).

The importance of SMSs in the neurotoxic response was also proposed. In Neuro-2a cells, upregulation of SMS1 in response to hydrogen peroxide (H_2O_2) counteracted the cytotoxic effect of H_2O_2, possibly by keeping the levels of ceramide in check (Tu, Yang, & Hu, 2016). In a mouse model that tested the mechanisms of neurodamage after prenatal alcohol exposure, the absence of SMS2 exacerbated alcohol-induced apoptosis in mossy cells of hippocampus in newborns, likely by enhancing accumulation of ceramide (Wang et al., 2015).

Along similar lines, *cert*$^{-/-}$ mice embryonically die at day E11.5 due to growth arrest, defective organogenesis, and mitochondrial toxicity induced by intracellularly accumulated ceramide (Wang et al., 2009). Moreover, fibroblasts derived from *cert*$^{-/-}$ embryos display compromised viability, ER, and oxidative stress and premature senescence (Rao et al., 2014). Finally, defects in CERT activity were also shown to be responsible for inhibition of SM synthesis in response to UVB treatment of keratinocytes (Charruyer et al., 2008). Indeed, UVB induced the formation of nonfunctional CERT homotrimers and the consequent block of SM synthesis partly responsible for the cytotoxic increase of ceramide.

5.3 Differentiation

Little is known about the effect of SMSs in cell differentiation, but the few reported observations are in line with a positive association between SM synthesis and the differentiation process. For instance, during neuronal development, the increase of SM is essential to acquire full maturation (Ledesma, Brugger, Bunning, Wieland, & Dotti, 1999), and during treatment of the pheochromocytoma cell line PC12 with the differentiating agent nerve growth factor, a fourfold increase in SM biosynthesis was observed (Piccinotti et al., 2000). More recently, the ability of SMS1 to maintain the homeostasis of PM microdomains was invoked as the mechanism by which the enzyme regulates epithelial differentiation of MDCK cells in response to hypertonic conditions. In fact, downregulation of SMS1 impaired the formation of mature adherens junction complexes, a critical event to acquire the differentiated phenotype and known to require detergent-resistant microdomains, hence PM SM (Favale, Santacreu, Pescio, Marquez, & Sterin-Speziale, 2015).

5.4 Chemotaxis and Migration

A limited number of studies have addressed the role of SMSs in chemotaxis and migration. In Neuro-2a cells, downregulation of SMS1 impaired their migration by decreasing the production of metalloproteinases (Wesley et al., 2015), whereas in SMS2-deficient brains, a substantial decrease of ezrin and β-actin expression has been reported (Zhang, Dong, Zhu, Wang, & Yang, 2011). As active ezrin regulates migration and adhesion (and positively correlates with metastasis), these results seem to point to a positive regulation of migration by SMS activity. On the other hand, opposite to neuronal cells, downregulation of SMS1, SMS2, or CERT in breast cancer cells with high CERT expression increased migration by promoting lateral diffusion and activation of the epidermal growth factor (EGF) receptor and consequently stimulating EGF-mediated signaling (Heering et al., 2012). Furthermore, loss of SMSs' expression also enhanced migration of MEFs by promoting their response to chemokines (Asano et al., 2012). In particular, it was observed that loss of SMSs facilitated the dimerization and activation of the chemokine receptor CXCR4 in response to the CXCL12 chemokine. Addition of SM or overexpression of SMS1 or SMS2 impaired the receptor activation thus confirming the negative effect of SM.

5.5 Exosomes

Exosomes have been identified as mediators of cell-to-cell communication as they transfer bioactive molecules such as nucleic acids, proteins, and lipids into recipient cells. Cancer-derived exosomes affect the tumor microenvironment by regulating immunity, angiogenesis, and metastasis and their release not only influence the invasive potentials of proximal cells but also affect distantly located tissues. Until now there is no evidence linking SMS to cancer-derived exosomes; however, it has been reported that downregulation of SMS2 in neuronal cells enhanced the secretion of exosomes (Yuyama, Sun, Mitsutake, & Igarashi, 2012), thus it is tempting to speculate the possibility that SMS2 might regulate also secretion of cancer-derived exosomes.

6. SPHINGOMYELIN SYNTHESIS AND CANCER

6.1 Sphingomyelin Levels in Transformed Conditions

Altered levels of SM have been reported when comparing normal and transformed tissues, but the specific changes seem to depend on the type of cancer. For instance, in rat hepatomas, SM levels were elevated in nuclei, mitochondria, and PM compared with normal liver. Moreover, the fatty acyl composition of SM in hepatomas did not show organelle-specific compositions typical of normal livers and SM rich in oleate was overrepresented (Hostetler, Zenner, & Morris, 1976; Wood, Upreti, & deAntueno, 1986). Similarly, elevated SM accompanied by decreased ceramide was measured by MALDI-MS in human specimens of hepatocellular carcinoma compared with normal paracarcinoma tissue (Li et al., 2017). SM enrichment was also reported in microsomes isolated from human brain tumors compared with gray matter (Sun & Leung, 1974). In line with these lipid changes, increased SMS activity has been reported in rat hepatocellular carcinoma compared with normal liver and in SV-40 transformed human lung fibroblasts as compared with the normal counterpart (Luberto & Hannun, 1998). Thus, in these conditions, increased SMS and higher SM associate with the transformed phenotype.

On the other hand, SM levels seem to be inversely correlated with the aggressiveness of some colorectal and breast cancer cell lines. A metastatic colorectal cell line showed lower C16-SM levels compared with the isogenic nonmetastatic parental cells (Fhaner, Liu, Ji, Simpson, & Reid, 2012), whereas untargeted membrane lipid profiling in different breast cancer cell lines relative to normal control cells (MCF10A) showed that the levels of SM 34:0 were the lowest in the most invasive cell lines, MDA-MB-231 and MDA-MB-157 (He, Guo, & Li, 2015; Wang et al., 2016). In agreement with downregulated SM synthesis in aggressive breast cancer, human triple negative and basal breast cancer specimens and cell lines showed decreased expression of CERT (Heering et al., 2012; Lee et al., 2012). Finally, decreased levels of SM were also reported in prostate cancer as compared with portions of benign hypertrophic tissue, in nonsmall lung cancer compared with matched nonmalignant lung tissue and in esophageal tumor compared with normal tissue (Marien et al., 2015; Merchant, de Graaf, Minsky, Obertop, & Glonek, 1993; Narayan & Dahiya, 1991).

Thus overall it seems that different transformed conditions are characterized by specific SM/SMS fingerprints.

6.2 Regulation of Sphingomyelin Synthesis in Transformed Phenotypes: A Double-Edged Sword

6.2.1 Protumorigenic Effects of Sphingomyelin Biosynthesis

SM biosynthesis has been reported to be both a downstream target of oncogenes or to potentially alter oncogenic activity. Indeed, the oncogene Bcr-abl, responsible for the insurgence of chronic myelogenous leukemia (CML), has been shown to upregulate SMS1 expression and activity in CML cell lines (Burns et al., 2013). Pharmacological and siRNA-mediated inhibition of SMS1 in these cells slowed down cell growth significantly and induced accumulation of ceramide together with a dramatic drop of DAG levels. As both lipids have been shown to modulate cell proliferation in other cell types, it is possible that the effect on cell growth following manipulations of SMS1 could be due to the impact on either or both of these signaling lipids.

On the other hand, regulation of oncogenic KRAS by SM has also been proposed. Indeed, the maintenance of the correct distribution and/or level of SM at the PM is crucial to preserve the localization of oncogenic KRAS at the inner leaflet where it carries out its protumorigenic signaling functions (van der Hoeven et al., 2017). In this context, it was also shown that SM-mediated regulation of KRAS occurs indirectly via regulation of PS levels in the cytosolic leaflet of the PM. Using a functional reporter assay for KRAS in *Caenorhabditis elegans*, the authors of the study observed that knock down of *SMS1* (as well as other genes able to modulate SM levels) prevented in part the multivulva phenotype, typical of the activity of oncogenic KRAS in this organism. Considering that lipid levels were not determined to confirm the expected effect on SM after downregulation of the different genes and considering also that in *C. elegans* there are four additional SMS homologues, the full protective effect of downregulation of SM synthesis via SMS on oncogenic KRAS functions could be underestimated. Nevertheless these results imply that alteration of SM distribution at the PM could represent a novel approach to block the transforming activity of oncogenic KRAS. Thus, in the cases of Bcr-abl and KRAS, SM synthesis seems to exert a protumorigenic function.

In line with protumorigenic functions of SMSs, an early report indicated that samples of patients with chemoresistant leukemias showed elevated SMS activity, together with the elevated synthesis of glucosylated sphingolipids and lower ceramide levels (Itoh et al., 2003). More recent studies seem also to confirm a potential link between SM synthesis and drug resistance.

In fact, loss of SMS2 in mouse brain reduced the expression and activity of the multidrug resistance gene Mdr1/P-glycoprotein (Zhang et al., 2011). Similar results were also observed in human intestinal epithelial Caco-2 cells where siRNA-mediated downregulation of SMS1 or SMS2 reduced the expression of the multidrug resistance proteins Mdr1 and MRP2 (Jin et al., 2016). These observations thus indicate that overexpression or increased activity of SMSs in tumors may induce drug resistance via modulation of Mdr proteins. Importantly, indirect substantiation for a role of SM in regulating sensitivity to chemotherapeutics comes from evidence showing that silencing of CERT sensitized different cancer cell lines to multiple cytotoxic agents, including paclitaxel (Swanton et al., 2007). Downregulation of CERT was shown to synergistically augment paclitaxel-induced ER stress likely due to enhanced accumulation of ceramide in the ER. Moreover, CERT expression was found to be increased in drug-resistant cell lines and in residual ovarian cancers following chemotherapy with paclitaxel. Finally, increased expression of CERT was also found in HER2+ versus HER2− breast cancer specimens and in primary breast cancers of patients receiving adjuvant chemotherapy, where high expression was associated with poor prognosis (Lee et al., 2012). Importantly, inhibition of CERT sensitized HER2+ cells to chemotherapy by promoting autophagy. These observations thus suggest that CERT expression may have a prognostic value for predicting treatment outcome in specific groups of cancer patients.

6.2.2 Antitumorigenic Effects of Sphingomyelin Biosynthesis

Differently from what is discussed above, other evidence point to prodifferentiation/antitumor effects promoted by SMSs or SM in different cell contexts. Inhibition of SMS1 in primary renal papillary collecting duct cells caused a discontinuous presence of α-catenin at the PM, suggesting that loss of SMS1 may alter adherens junctions (Brandan et al., 2018). Pharmacological inhibition of SMS with D609 showed similar effects, and these were reverted by supplementation with exogenous SM. Although altered localization was also shown for β-catenin, no changes were reported for talin. Because talin in these cells is present only in focal adhesions, it seems that inhibition of SMS1 affected specifically adherens junctions. The changes in adherens junctions versus focal adhesions were also accompanied by separation between cells and no significant change in attachment to the extracellular matrix supported by intact focal adhesion and thus preserved viability of the cells. Interestingly, pharmacological inhibition of SMSs also increased the number of fibroblasts-resembling cells, which expressed high levels of α-smooth muscle actin and vimentin, two mesenchymal markers. Although the ability of exogenously added SM to revert the phenotype was not tested for the later results, the authors speculated that inhibition of SMS could induce an epithelial to mesenchymal transition. This possibility awaits further confirmation.

In addition, as mentioned earlier, low expression of CERT seems to promote EGF-mediated migration of aggressive triple negative breast cancer cells without affecting cell proliferation (Heering et al., 2012). However, overexpression of CERT in these cells had only a minimal effect on migration and invasion as tested in vitro. This mild effect thus raises questions on the potential therapeutic efficacy of solely boosting SM synthesis in this intrinsic cancer model.

On the other hand, massive activation of SM synthesis with 2-hydroxyoleic acid (2OHOA, Minerval) impaired cell viability in U118 glioma cells, which have very low levels of SM compared with normal MRC5 cells (Barcelo-Coblijn et al., 2011). The overall cytotoxic effect of 2OHOA is the result of the combination of several events such as cell cycle arrest, differentiation, ER stress, autophagy, and sphingolipidosis (Llado et al., 2014). Supporting the concept that low SM synthesis may be advantageous to brain tumors and elevation of SM synthesis could be unfavorable for the tumor, expression analysis of glioma samples from patients showed the association between low expression of *SMS1* and reduced patient survival (Llado et al., 2014). Treatment with 2OHOA has been molecularly linked to dephosphorylation of Rb, inactivation of E2F1, and reduction of DNA synthesis (Llado et al., 2014). In addition, displacement and inactivation of KRAS has been also reported possibly due to decreased levels of phosphatidylethanolamine (PE) and PS (Barcelo-Coblijn et al., 2011; van der Hoeven et al., 2017). However, the cytotoxic effect of 2OHOA is not entirely mediated by activation of SM synthesis as in the lung cancer SK-LU-Ad1 cells, cytotoxicity was observed without significant changes in SM levels (Luberto, unpublished observations). Indeed other lipid changes including decreased levels of PE and oleic acid have been reported on treatment with 2OHOA (Llado et al., 2014), thus the challenge is to predict which of these targets becomes relevant in each cancer model.

Finally, SM levels in the PM are critical to ascertain effectiveness of alkyl-lysophospholipids, a novel class of membrane-targeting chemotherapeutic agents that penetrate in the cells via lipid microdomain-dependent endocytosis and disturb lipid metabolism (van der Luit et al., 2007; Van der Luit et al., 2007). In fact, when added to cells, these unnatural phospholipids accumulate preferentially in detergent-resistant, SM-rich microdomains from where they are internalized through dynamin-dependent and clathrin-independent endocytosis. Given the stabilizing function of SM for these microdomains, its presence favors the internalization of these drugs and their cytotoxic effect.

7. CONCLUSIONS AND UNRESOLVED QUESTIONS

With the identification of the genes responsible for SMS activity, great progress has been made particularly in the study of the role of SM in the organization and functions of microdomains on the PM. On the other hand, the impact of physiological and pathophysiological modulation of SMS activity on these events, as opposed to complete knockout, still needs further clarification. This is exemplified by the difference between SMS1 knockout versus knock down cells in their response to FasL (Lafont et al., 2010; Miyaji et al., 2005). Moreover, we are lacking a clear understanding of the functions and molecular targets for ceramide and DAG regulated by SMSs, and whether some of the already identified targets for these lipids participate in the ability of SMSs to impact tumorigenesis. The relationship between DAG generated at the Golgi, PKD, and vesicular trafficking is the best characterized example of the signaling potential of these hubs but, as we start to define them, additional questions arise such as if other functions are associated with regulation of PKD by SMS-derived DAG, if other targets for SMS-generated DAG at the Golgi exist and what other functions may be affected, and if ceramide itself regulates Golgi specific targets. Even though, most of SMSs-dependent effects have been linked to their enzymatic activity, one report suggests the possibility of SMS2 functions regulated in a catalytically independent manner opening the door for unforeseen effects (Hayashi et al., 2014).

While remarkable progress has been made in the understanding of the negative and positive feedback mechanisms targeting CERT activity to locally regulate SM synthesis, currently we do not know whether SM synthesis is the only target for CERT. Other ceramide metabolizing enzymes may be localized at the Golgi, such as ceramide kinase (Boath et al., 2008; Lamour et al., 2007) and alkaline ceramidase 2 (Xu et al., 2006), thus it is possible that some of the phenotypes observed on alteration of CERT and ascribed to impaired SM synthesis may be due to other enzymes. This clarification becomes important when targeting CERT as a surrogate for Golgi SMSs.

Also significant progress is being made toward uncovering modes of regulation of SMS1, whereas nothing is known about regulation of SMS2.

As the impact of SM synthesis in tumors varies depending on the context, both inhibitors and activators of SMSs and CERT are needed for therapeutic applications. Moreover, together with drugs that augment SM biosynthesis, the potential benefit of supplementation of SM needs also further consideration. In some experimental mouse models, administration of SM, either in the diet or intravenously, was able to impair progression of colon cancer or potentiate the effect of the chemotherapeutic drug, 5-FU (Lemonnier et al., 2003; Modrak, Lew, Goldenberg, & Blumenthal, 2000). However, these results seem to depend on the animal model used, thus the understanding of the real potential benefits of such approach needs further investigation.

Finally, most observations regarding the functions of SM synthesis in cancer are intrinsic to cancer cell lines and there are no data addressing if and how SMSs in the tumor microenvironment, and therefore in the host, have any function that could affect the development of the tumor.

These are some of the questions that will need to be addressed moving forward with the investigations of the roles and functions of SM synthesis in tumorigenesis.

ACKNOWLEDGMENTS

This work was supported by NCI grant PO1 CA097132 to C.L. for project #4 and by the Stony Brook Scholars in Biomedical Sciences Award to S.M.

REFERENCES

Abe, M., Makino, A., Hullin-Matsuda, F., Kamijo, K., Ohno-Iwashita, Y., Hanada, K., et al. (2012). A role for sphingomyelin-rich lipid domains in the accumulation of phosphatidylinositol-4,5-bisphosphate to the cleavage furrow during cytokinesis. *Molecular and Cellular Biology, 32*(8), 1396–1407.

Adibhatla, R. M., & Hatcher, J. F. (2010). Protection by D609 through cell-cycle regulation after stroke. *Molecular Neurobiology, 41*(2–3), 206–217.

Albi, E., La Porta, C. A., Cataldi, S., & Magni, M. V. (2005). Nuclear sphingomyelin-synthase and protein kinase C delta in melanoma cells. *Archives of Biochemistry and Biophysics, 438*(2), 156–161.

Allan, D., & Quinn, P. (1988). Resynthesis of sphingomyelin from plasma-membrane phosphatidylcholine in BHK cells treated with *Staphylococcus aureus* sphingomyelinase. *Biochemical Journal, 254*(3), 765–771.

Almena, M., & Merida, I. (2011). Shaping up the membrane: Diacylglycerol coordinates spatial orientation of signaling. *Trends Biochemical Science, 36*(11), 593–603.

Asano, S., Kitatani, K., Taniguchi, M., Hashimoto, M., Zama, K., Mitsutake, S., et al. (2012). Regulation of cell migration by sphingomyelin synthases: Sphingomyelin in lipid rafts decreases responsiveness to signaling by the CXCL12/CXCR4 pathway. *Molecular and Cellular Biology, 32*(16), 3242–3252.

Babiychuk, E. B., Monastyrskaya, K., & Draeger, A. (2008). Fluorescent annexin A1 reveals dynamics of ceramide platforms in living cells. *Traffic, 9*(10), 1757–1775.

Bakrac, B., Kladnik, A., Macek, P., McHaffie, G., Werner, A., Lakey, J. H., et al. (2010). A toxin-based probe reveals cytoplasmic exposure of Golgi sphingomyelin. *Journal of Biological Chemistry, 285*(29), 22186–22195.

Barcelo-Coblijn, G., Martin, M. L., de Almeida, R. F., Noguera-Salva, M. A., Marcilla-Etxenike, A., Guardiola-Serrano, F., et al. (2011). Sphingomyelin and sphingomyelin synthase (SMS) in the malignant transformation of glioma cells and in 2-hydroxyoleic acid therapy. *Proceedings of the National Academy of Sciences of the United States of America, 108*(49), 19569–19574.

Barlic, A., Gutierrez-Aguirre, I., Caaveiro, J. M., Cruz, A., Ruiz-Arguello, M. B., Perez-Gil, J., et al. (2004). Lipid phase coexistence favors membrane insertion of equinatoxin-II, a pore-forming toxin from *Actinia equina*. *Journal of Biological Chemistry, 279*(33), 34209–34216.

Bartke, N., & Hannun, Y. A. (2009). Bioactive sphingolipids: Metabolism and function. *Journal of Lipid Research, 50*(Suppl), S91–S96.

Birbes, H., El Bawab, S., Hannun, Y. A., & Obeid, L. M. (2001). Selective hydrolysis of a mitochondrial pool of sphingomyelin induces apoptosis. *FASEB J, 15*(14), 2669–2679.

Boath, A., Graf, C., Lidome, E., Ullrich, T., Nussbaumer, P., & Bornancin, F. (2008). Regulation and traffic of ceramide 1-phosphate produced by ceramide kinase: Comparative analysis to glucosylceramide and sphingomyelin. *Journal of Biological Chemistry, 283*(13), 8517–8526.

Bourteele, S., Hausser, A., Doppler, H., Horn-Muller, J., Ropke, C., Schwarzmann, G., et al. (1998). Tumor necrosis factor induces ceramide oscillations and negatively controls sphingolipid synthases by caspases in apoptotic Kym-1 cells. *Journal of Biological Chemistry, 273*(47), 31245–31251.

Brandan, Y. R., Guaytima, E. D. V., Favale, N. O., Pescio, L. G., Sterin-Speziale, N. B., & Marquez, M. G. (2018). The inhibition of sphingomyelin synthase 1 activity induces collecting duct cells to lose their epithelial phenotype. *Biochimica et Biophysica Acta, 1865*(2), 309–322.

Brothers, M. C., Ho, M., Maharjan, R., Clemons, N. C., Bannai, Y., Waites, M. A., et al. (2011). Membrane interaction of *Pasteurella multocida* toxin involves sphingomyelin. *FEBS Journal, 278*(23), 4633–4648.

Burns, T. A., Subathra, M., Signorelli, P., Choi, Y., Yang, X., Wang, Y., et al. (2013). Sphingomyelin synthase 1 activity is regulated by the BCR-ABL oncogene. *Journal of Lipid Research, 54*(3), 794–805.

Capasso, S., Sticco, L., Rizzo, R., Pirozzi, M., Russo, D., Dathan, N. A., et al. (2017). Sphingolipid metabolic flow controls phosphoinositide turnover at the trans-Golgi network. *EMBO Journal, 36*(12), 1736–1754.

Chandran, S., & Machamer, C. E. (2008). Acute perturbations in Golgi organization impact de novo sphingomyelin synthesis. *Traffic, 9*(11), 1894–1904.

Charruyer, A., Bell, S. M., Kawano, M., Douangpanya, S., Yen, T. Y., Macher, B. A., et al. (2008). Decreased ceramide transport protein (CERT) function alters sphingomyelin production following UVB irradiation. *Journal of Biological Chemistry, 283*(24), 16682–16692.

Contreras, F. X., Ernst, A. M., Haberkant, P., Bjorkholm, P., Lindahl, E., Gonen, B., et al. (2012). Molecular recognition of a single sphingolipid species by a protein's transmembrane domain. *Nature, 481*(7382), 525–529.

D'Angelo, G., Uemura, T., Chuang, C. C., Polishchuk, E., Santoro, M., Ohvo-Rekila, H., et al. (2013). Vesicular and non-vesicular transport feed distinct glycosylation pathways in the Golgi. *Nature, 501*(7465), 116–120.

Dai, Q., Liu, J., Chen, J., Durrant, D., McIntyre, T. M., & Lee, R. M. (2004). Mitochondrial ceramide increases in UV-irradiated HeLa cells and is mainly derived from hydrolysis of sphingomyelin. *Oncogene, 23*(20), 3650–3658.

Deevska, G. M., Dotson, P. P., 2nd, Karakashian, A. A., Isaac, G., Wrona, M., Kelly, S. B., et al. (2017). Novel interconnections in lipid metabolism revealed by overexpression of sphingomyelin synthase-1. *Journal of Biological Chemistry, 292*(12), 5110–5122.

Dergunova, L. V., Rozhkova, A. V., Sudarkina, O. Y., & Limborska, S. A. (2013). The use of alternative polyadenylation in the tissue-specific regulation of human SMS1 gene expression. *Molecular Biology Reports, 40*(12), 6685–6690.

Devaux, P. F., & Morris, R. (2004). Transmembrane asymmetry and lateral domains in biological membranes. *Traffic, 5*(4), 241–246.

Ding, T., Kabir, I., Li, Y., Lou, C., Yazdanyar, A., Xu, J., et al. (2015). All members in the sphingomyelin synthase gene family have ceramide phosphoethanolamine synthase activity. *Journal of Lipid Research, 56*(3), 537–545.

Ding, T., Li, Z., Hailemariam, T., Mukherjee, S., Maxfield, F. R., Wu, M. P., et al. (2008). SMS overexpression and knockdown: Impact on cellular sphingomyelin and diacylglycerol metabolism, and cell apoptosis. *Journal of Lipid Research, 49*(2), 376–385.

Diringer, H., Marggraf, W. D., Koch, M. A., & Anderer, F. A. (1972). Evidence for a new biosynthetic pathway of sphingomyelin in SV 40 transformed mouse cells. *Biochemical and Biophysical Research Communications, 47*(6), 1345–1352.

Dong, L., Watanabe, K., Itoh, M., Huan, C. R., Tong, X. P., Nakamura, T., et al. (2012). CD4+ T-cell dysfunctions through the impaired lipid rafts ameliorate concanavalin A-induced hepatitis in sphingomyelin synthase 1-knockout mice. *International Immunology, 24*(5), 327–337.

Duran, J. M., Campelo, F., van Galen, J., Sachsenheimer, T., Sot, J., Egorov, M. V., et al. (2012). Sphingomyelin organization is required for vesicle biogenesis at the Golgi complex. *EMBO Journal, 31*(24), 4535–4546.

Favale, N. O., Santacreu, B. J., Pescio, L. G., Marquez, M. G., & Sterin-Speziale, N. B. (2015). Sphingomyelin metabolism is involved in the differentiation of MDCK cells induced by environmental hypertonicity. *Journal of Lipid Research, 56*(4), 786–800.

Fhaner, C. J., Liu, S., Ji, H., Simpson, R. J., & Reid, G. E. (2012). Comprehensive lipidome profiling of isogenic primary and metastatic colon adenocarcinoma cell lines. *Analytical Chemistry, 84*(21), 8917–8926.

Filippenkov, I. B., Sudarkina, O. Y., Limborska, S. A., & Dergunova, L. V. (2015). Circular RNA of the human sphingomyelin synthase 1 gene: Multiple splice variants, evolutionary conservatism and expression in different tissues. *RNA Biology, 12*(9), 1030–1042.

Fugmann, T., Hausser, A., Schoffler, P., Schmid, S., Pfizenmaier, K., & Olayioye, M. A. (2007). Regulation of secretory transport by protein kinase D-mediated phosphorylation of the ceramide transfer protein. *Journal of Cell Biology, 178*(1), 15–22.

Fukunaga, T., Nagahama, M., Hatsuzawa, K., Tani, K., Yamamoto, A., & Tagaya, M. (2000). Implication of sphingolipid metabolism in the stability of the Golgi apparatus. *Journal of Cell Science, 113*(Pt 18), 3299–3307.

Gangoiti, P., Granado, M. H., Arana, L., Ouro, A., & Gomez-Munoz, A. (2010). Activation of protein kinase C-alpha is essential for stimulation of cell proliferation by ceramide 1-phosphate. *FEBS Letters, 584*(3), 517–524.

Giussani, P., Colleoni, T., Brioschi, L., Bassi, R., Hanada, K., Tettamanti, G., et al. (2008). Ceramide traffic in C6 glioma cells: Evidence for CERT-dependent and independent transport from ER to the Golgi apparatus. *Biochimica et Biophysica Acta, 1781*(1–2), 40–51.

Grassme, H., Jekle, A., Riehle, A., Schwarz, H., Berger, J., Sandhoff, K., et al. (2001). CD95 signaling via ceramide-rich membrane rafts. *Journal of Biological Chemistry, 276*(23), 20589–20596.

Grassme, H., Jendrossek, V., Riehle, A., von Kurthy, G., Berger, J., Schwarz, H., et al. (2003). Host defense against *Pseudomonas aeruginosa* requires ceramide-rich membrane rafts. *Nature Medicine, 9*(3), 322–330.

Gupta, V. R., Patel, H. K., Kostolansky, S. S., Ballivian, R. A., Eichberg, J., & Blanke, S. R. (2008). Sphingomyelin functions as a novel receptor for *Helicobacter pylori* VacA. *PLoS Pathogens, 4*(5), e1000073.

Hailemariam, T. K., Huan, C., Liu, J., Li, Z., Roman, C., Kalbfeisch, M., et al. (2008). Sphingomyelin synthase 2 deficiency attenuates NFkappaB activation. *Arteriosclerosis, Thrombosis, and Vascular Biology, 28*(8), 1519–1526.

Hanada, K. (2014). Co-evolution of sphingomyelin and the ceramide transport protein CERT. *Biochimica et Biophysica Acta, 1841*(5), 704–719.

Hanada, K., Kumagai, K., Yasuda, S., Miura, Y., Kawano, M., Fukasawa, M., et al. (2003). Molecular machinery for non-vesicular trafficking of ceramide. *Nature, 426*(6968), 803–809.

Hannun, Y. A., & Obeid, L. M. (2011). Many ceramides. *Journal of Biological Chemistry, 286*(32), 27855–27862.

Hayashi, Y., Nemoto-Sasaki, Y., Matsumoto, N., Tanikawa, T., Oka, S., Tanaka, Y., et al. (2017). Carboxyl-terminal tail-mediated homodimerizations of sphingomyelin synthases are responsible for efficient export from the endoplasmic reticulum. *Journal of Biological Chemistry, 292*(3), 1122–1141.

Hayashi, Y., Nemoto-Sasaki, Y., Tanikawa, T., Oka, S., Tsuchiya, K., Zama, K., et al. (2014). Sphingomyelin synthase 2, but not sphingomyelin synthase 1, is involved in HIV-1 envelope-mediated membrane fusion. *Journal of Biological Chemistry, 289*(44), 30842–30856.

Heering, J., Weis, N., Holeiter, M., Neugart, F., Staebler, A., Fehm, T. N., et al. (2012). Loss of the ceramide transfer protein augments EGF receptor signaling in breast cancer. *Cancer Research, 72*(11), 2855–2866.

He, M., Guo, S., & Li, Z. (2015). In situ characterizing membrane lipid phenotype of breast cancer cells using mass spectrometry profiling. *Scientific Reports, 5*, 11298.

van Helvoort, A., van't Hof, W., Ritsema, T., Sandra, A., & van Meer, G. (1994). Conversion of diacylglycerol to phosphatidylcholine on the basolateral surface of epithelial (Madin-Darby canine kidney) cells. Evidence for the reverse action of a sphingomyelin synthase. *Journal of Biological Chemistry, 269*(3), 1763–1769.

Hernandez-Corbacho, M. J., Salama, M. F., Canals, D., Senkal, C. E., & Obeid, L. M. (2017). Sphingolipids in mitochondria. *Biochimica et Biophysica Acta, 1862*(1), 56–68.

van der Hoeven, D., Cho, K. J., Zhou, Y., Ma, X., Chen, W., Naji, A., et al. (2017). Sphingomyelin metabolism is a regulator of KRAS function. *Molecular and Cellular Biology*. https://doi.org/10.1128/MCB.00373-17.

Holthuis, J. C., & Luberto, C. (2010). Tales and mysteries of the enigmatic sphingomyelin synthase family. *Advances in Experimental Medicine and Biology, 688*, 72–85.

Hostetler, K. Y., Zenner, B. D., & Morris, H. P. (1976). Abnormal membrane phospholipid content in subcellular fractions from the Morris 7777 hepatoma. *Biochimica et Biophysica Acta, 441*(2), 231–238.

Huitema, K., van den Dikkenberg, J., Brouwers, J. F., & Holthuis, J. C. (2004). Identification of a family of animal sphingomyelin synthases. *EMBO Journal, 23*(1), 33–44.

Hu, W., Xu, R., Zhang, G., Jin, J., Szulc, Z. M., Bielawski, J., et al. (2005). Golgi fragmentation is associated with ceramide-induced cellular effects. *Molecular Biology of the Cell, 16*(3), 1555–1567.

Ikonen, E. (2008). Cellular cholesterol trafficking and compartmentalization. *Nature Reviews. Molecular Cell Biology, 9*(2), 125–138.

Ishitsuka, R., Yamaji-Hasegawa, A., Makino, A., Hirabayashi, Y., & Kobayashi, T. (2004). A lipid-specific toxin reveals heterogeneity of sphingomyelin-containing membranes. *Biophysical Journal, 86*(1 Pt 1), 296–307.

Itoh, M., Kitano, T., Watanabe, M., Kondo, T., Yabu, T., Taguchi, Y., et al. (2003). Possible role of ceramide as an indicator of chemoresistance: Decrease of the ceramide content via activation of glucosylceramide synthase and sphingomyelin synthase in chemoresistant leukemia. *Clinical Cancer Research, 9*(1), 415–423.

Jin, Z. X., Huang, C. R., Dong, L., Goda, S., Kawanami, T., Sawaki, T., et al. (2008). Impaired TCR signaling through dysfunction of lipid rafts in sphingomyelin synthase 1 (SMS1)-knockdown T cells. *International Immunology, 20*(11), 1427–1437.

Jin, G., Li, Y., Zhu, Y., Du, L., Yan, J., & Yang, Q. (2016). SMS regulates the expression and function of P-gp and MRP2 in Caco-2 cells. *Cell Biology and Toxicology, 32*(6), 483–497.

Jones, E. E., Dworski, S., Canals, D., Casas, J., Fabrias, G., Schoenling, D., et al. (2014). On-tissue localization of ceramides and other sphingolipids by MALDI mass spectrometry imaging. *Analytical Chemistry, 86*(16), 8303–8311.

Kawano, M., Kumagai, K., Nishijima, M., & Hanada, K. (2006). Efficient trafficking of ceramide from the endoplasmic reticulum to the Golgi apparatus requires a VAMP-associated protein-interacting FFAT motif of CERT. *Journal of Biological Chemistry, 281*(40), 30279–30288.

Kidani, Y., Ohshima, K., Sakai, H., Kohno, T., Baba, A., & Hattori, M. (2012). Differential localization of sphingomyelin synthase isoforms in neurons regulates sphingomyelin cluster formation. *Biochemical and Biophysical Research Communications, 417*(3), 1014–1017.

Kim, W. S., Hsiao, J. H., Bhatia, S., Glaros, E. N., Don, A. S., Tsuruoka, S., et al. (2013). ABCA8 stimulates sphingomyelin production in oligodendrocytes. *Biochemical Journal, 452*(3), 401–410.

Kudo, N., Kumagai, K., Tomishige, N., Yamaji, T., Wakatsuki, S., Nishijima, M., et al. (2008). Structural basis for specific lipid recognition by CERT responsible for nonvesicular trafficking of ceramide. *Proceedings of the National Academy of Sciences of the United States of America, 105*(2), 488–493.

Kumagai, K., Kawano-Kawada, M., & Hanada, K. (2014). Phosphoregulation of the ceramide transport protein CERT at serine 315 in the interaction with VAMP-associated protein (VAP) for inter-organelle trafficking of ceramide in mammalian cells. *Journal of Biological Chemistry, 289*(15), 10748–10760.

Kumagai, K., Kawano, M., Shinkai-Ouchi, F., Nishijima, M., & Hanada, K. (2007). Interorganelle trafficking of ceramide is regulated by phosphorylation-dependent cooperativity between the PH and START domains of CERT. *Journal of Biological Chemistry, 282*(24), 17758–17766.

Kumagai, K., Yasuda, S., Okemoto, K., Nishijima, M., Kobayashi, S., & Hanada, K. (2005). CERT mediates intermembrane transfer of various molecular species of ceramides. *Journal of Biological Chemistry, 280*(8), 6488–6495.

Lafont, E., Milhas, D., Carpentier, S., Garcia, V., Jin, Z. X., Umehara, H., et al. (2010). Caspase-mediated inhibition of sphingomyelin synthesis is involved in FasL-triggered cell death. *Cell Death and Differentiation, 17*(4), 642–654.

Lamour, N. F., Stahelin, R. V., Wijesinghe, D. S., Maceyka, M., Wang, E., Allegood, J. C., et al. (2007). Ceramide kinase uses ceramide provided by ceramide transport protein: Localization to organelles of eicosanoid synthesis. *Journal of Lipid Research, 48*(6), 1293–1304.

Ledesma, M. D., Brugger, B., Bunning, C., Wieland, F. T., & Dotti, C. G. (1999). Maturation of the axonal plasma membrane requires upregulation of sphingomyelin synthesis and formation of protein-lipid complexes. *EMBO Journal, 18*(7), 1761–1771.

Lee, A. J., Roylance, R., Sander, J., Gorman, P., Endesfelder, D., Kschischo, M., et al. (2012). CERT depletion predicts chemotherapy benefit and mediates cytotoxic and polyploid-specific cancer cell death through autophagy induction. *Journal of Pathology, 226*(3), 482–494.

Lemonnier, L. A., Dillehay, D. L., Vespremi, M. J., Abrams, J., Brody, E., & Schmelz, E. M. (2003). Sphingomyelin in the suppression of colon tumors: Prevention versus intervention. *Archives of Biochemistry and Biophysics, 419*(2), 129–138.

Li, Z., Guan, M., Lin, Y., Cui, X., Zhang, Y., Zhao, Z., et al. (2017). Aberrant lipid metabolism in hepatocellular carcinoma revealed by liver lipidomics. *International Journal of Molecular Sciences, 18*(12).

Li, Z., Hailemariam, T. K., Zhou, H., Li, Y., Duckworth, D. C., Peake, D. A., et al. (2007). Inhibition of sphingomyelin synthase (SMS) affects intracellular sphingomyelin accumulation and plasma membrane lipid organization. *Biochimica et Biophysica Acta, 1771*(9), 1186–1194.

Linardic, C. M., & Hannun, Y. A. (1994). Identification of a distinct pool of sphingomyelin involved in the sphingomyelin cycle. *Journal of Biological Chemistry, 269*(38), 23530–23537.

Llado, V., Lopez, D. J., Ibarguren, M., Alonso, M., Soriano, J. B., Escriba, P. V., et al. (2014). Regulation of the cancer cell membrane lipid composition by NaCHOleate: Effects on cell signaling and therapeutical relevance in glioma. *Biochimica et Biophysica Acta, 1838*(6), 1619–1627.

Luberto, C., & Hannun, Y. A. (1998). Sphingomyelin synthase, a potential regulator of intracellular levels of ceramide and diacylglycerol during SV40 transformation. Does sphingomyelin synthase account for the putative phosphatidylcholine-specific phospholipase C? *Journal of Biological Chemistry, 273*(23), 14550–14559.

van der Luit, A. H., Vink, S. R., Klarenbeek, J. B., Perrissoud, D., Solary, E., Verheij, M., et al. (2007). A new class of anticancer alkylphospholipids uses lipid rafts as membrane gateways to induce apoptosis in lymphoma cells. *Molecular Cancer Therapeutics, 6*(8), 2337–2345.

Makino, A., Abe, M., Murate, M., Inaba, T., Yilmaz, N., Hullin-Matsuda, F., et al. (2015). Visualization of the heterogeneous membrane distribution of sphingomyelin associated with cytokinesis, cell polarity, and sphingolipidosis. *FASEB J, 29*(2), 477–493.

Marien, E., Meister, M., Muley, T., Fieuws, S., Bordel, S., Derua, R., et al. (2015). Non-small cell lung cancer is characterized by dramatic changes in phospholipid profiles. *International Journal of Cancer, 137*(7), 1539–1548.

van Meer, G., Voelker, D. R., & Feigenson, G. W. (2008). Membrane lipids: Where they are and how they behave. *Nature Reviews. Molecular Cell Biology, 9*(2), 112–124.

Merchant, T. E., de Graaf, P. W., Minsky, B. D., Obertop, H., & Glonek, T. (1993). Esophageal cancer phospholipid characterization by 31P NMR. *NMR in Biomedicine, 6*(3), 187–193.

Mesmin, B., Bigay, J., Moser von Filseck, J., Lacas-Gervais, S., Drin, G., & Antonny, B. (2013). A four-step cycle driven by PI(4)P hydrolysis directs sterol/PI(4)P exchange by the ER-Golgi tether OSBP. *Cell, 155*(4), 830–843.

Micheli, M., Albi, E., Leray, C., & Magni, M. V. (1998). Nuclear sphingomyelin protects RNA from RNase action. *FEBS Letters, 431*(3), 443–447.

Miller, M. E., Adhikary, S., Kolokoltsov, A. A., & Davey, R. A. (2012). Ebolavirus requires acid sphingomyelinase activity and plasma membrane sphingomyelin for infection. *Journal of Virology, 86*(14), 7473–7483.

Miro-Obradors, M. J., Osada, J., Aylagas, H., Sanchez-Vegazo, I., & Palacios-Alaiz, E. (1993). Microsomal sphingomyelin accumulation in thioacetamide-injured regenerating rat liver: Involvement of sphingomyelin synthase activity. *Carcinogenesis, 14*(5), 941–946.

Mitsutake, S., Zama, K., Yokota, H., Yoshida, T., Tanaka, M., Mitsui, M., et al. (2011). Dynamic modification of sphingomyelin in lipid microdomains controls development of obesity, fatty liver, and type 2 diabetes. *Journal of Biological Chemistry, 286*(32), 28544–28555.

Miyaji, M., Jin, Z. X., Yamaoka, S., Amakawa, R., Fukuhara, S., Sato, S. B., et al. (2005). Role of membrane sphingomyelin and ceramide in platform formation for Fas-mediated apoptosis. *Journal of Experimental Medicine, 202*(2), 249–259.

Miyamoto, S., Hsu, C. C., Hamm, G., Darshi, M., Diamond-Stanic, M., Decleves, A. E., et al. (2016). Mass spectrometry imaging reveals elevated glomerular ATP/AMP in diabetes/obesity and identifies sphingomyelin as a possible mediator. *EBioMedicine, 7*, 121–134.

Modrak, D. E., Lew, W., Goldenberg, D. M., & Blumenthal, R. (2000). Sphingomyelin potentiates chemotherapy of human cancer xenografts. *Biochemical and Biophysical Research Communications, 268*(2), 603–606.

Moorthi, S., Burns, T. A., Yu, G.-Q., & Luberto, C. (2018). Bcr-abl regulation of sphingomyelin synthase 1 reveals a novel oncogenic-driven mechanism of protein up-regulation. *FASEB J.* Mar 13: fj201701016R. https://doi.org/10.1096/fj.201701016R [Epub ahead of print].

Mou, D., Yang, H., Qu, C., Chen, J., & Zhang, C. (2016). Pharmacological activation of peroxisome proliferator-activated receptor {delta} increases sphingomyelin synthase activity in THP-1 macrophage-derived foam cell. *Inflammation, 39*(4), 1538–1546.

Narayan, P., & Dahiya, R. (1991). Alterations in sphingomyelin and fatty acids in human benign prostatic hyperplasia and prostatic cancer. *Biomedica Biochimica Acta, 50*(9), 1099–1108.

Ogretmen, B. (2018). Sphingolipid metabolism in cancer signalling and therapy. *Nature Reviews Cancer, 18*(1), 33–50.

Piccinotti, A., Benaglia, G., Bresciani, R., Zizioli, D., Presta, M., Preti, A., et al. (2000). Nerve growth factor induces sphingomyelin accumulation in pheochromocytoma cells. *FEBS Letters, 472*(1), 143–147.

Prashek, J., Bouyain, S., Fu, M., Li, Y., Berkes, D., & Yao, X. (2017). Interaction between the PH and START domains of ceramide transfer protein competes with phosphatidylinositol 4-phosphate binding by the PH domain. *Journal of Biological Chemistry, 292*(34), 14217–14228.

Qureshi, A., Subathra, M., Grey, A., Schey, K., Del Poeta, M., & Luberto, C. (2010). Role of sphingomyelin synthase in controlling the antimicrobial activity of neutrophils against *Cryptococcus neoformans. PLoS One, 5*(12), e15587.

Ramu, Y., Xu, Y., & Lu, Z. (2006). Enzymatic activation of voltage-gated potassium channels. *Nature, 442*(7103), 696–699.

Rao, R. P., Scheffer, L., Srideshikan, S. M., Parthibane, V., Kosakowska-Cholody, T., Masood, M. A., et al. (2014). Ceramide transfer protein deficiency compromises organelle function and leads to senescence in primary cells. *PLoS One, 9*(3), e92142.

Riboni, L., Viani, P., Bassi, R., Giussani, P., & Tettamanti, G. (2001). Basic fibroblast growth factor-induced proliferation of primary astrocytes. evidence for the involvement of sphingomyelin biosynthesis. *Journal of Biological Chemistry, 276*(16), 12797–12804.

Ridgway, N. D. (2000). Interactions between metabolism and intracellular distribution of cholesterol and sphingomyelin. *Biochimica et Biophysica Acta, 1484*(2–3), 129–141.

Rosenwald, A. G., & Pagano, R. E. (1993). Inhibition of glycoprotein traffic through the secretory pathway by ceramide. *Journal of Biological Chemistry, 268*(7), 4577–4579.

Saito, S., Matsui, H., Kawano, M., Kumagai, K., Tomishige, N., Hanada, K., et al. (2008). Protein phosphatase 2Cepsilon is an endoplasmic reticulum integral membrane protein that dephosphorylates the ceramide transport protein CERT to enhance its association with organelle membranes. *Journal of Biological Chemistry, 283*(10), 6584–6593.

Sakamoto, H., Yoshida, T., Sanaki, T., Shigaki, S., Morita, H., Oyama, M., et al. (2017). Possible roles of long-chain sphingomyelines and sphingomyelin synthase 2 in mouse macrophage inflammatory response. *Biochemical and Biophysical Research Communications, 482*(2), 202–207.

Sarri, E., Sicart, A., Lazaro-Dieguez, F., & Egea, G. (2011). Phospholipid synthesis participates in the regulation of diacylglycerol required for membrane trafficking at the Golgi complex. *Journal of Biological Chemistry, 286*(32), 28632–28643.
Scassellati, C., Albi, E., Cmarko, D., Tiberi, C., Cmarkova, J., Bouchet-Marquis, C., et al. (2010). Intranuclear sphingomyelin is associated with transcriptionally active chromatin and plays a role in nuclear integrity. *Biology of the Cell, 102*(6), 361–375.
Schon, P., Garcia-Saez, A. J., Malovrh, P., Bacia, K., Anderluh, G., & Schwille, P. (2008). Equinatoxin II permeabilizing activity depends on the presence of sphingomyelin and lipid phase coexistence. *Biophysical Journal, 95*(2), 691–698.
Separovic, D., Hanada, K., Maitah, M. Y., Nagy, B., Hang, I., Tainsky, M. A., et al. (2007). Sphingomyelin synthase 1 suppresses ceramide production and apoptosis post-photodamage. *Biochemical and Biophysical Research Communications, 358*(1), 196–202.
Shakor, A. B., Atia, M., Ismail, I. A., Alshehri, A., El-Refaey, H., Kwiatkowska, K., et al. (2014). Curcumin induces apoptosis of multidrug-resistant human leukemia HL60 cells by complex pathways leading to ceramide accumulation. *Biochimica et Biophysica Acta, 1841*(12), 1672–1682.
Shakor, A. B., Taniguchi, M., Kitatani, K., Hashimoto, M., Asano, S., Hayashi, A., et al. (2011). Sphingomyelin synthase 1-generated sphingomyelin plays an important role in transferrin trafficking and cell proliferation. *Journal of Biological Chemistry, 286*(41), 36053–36062.
Shinkai, M., Takeyama, Y., Ueda, T., Hori, Y., & Yamamoto, M. (1997). Inhibitory action of sphingosine or ceramide on amylase secretion from isolated rat pancreatic acini. *Biochemical and Biophysical Research Communications, 235*(1), 197–200.
Slotte, J. P. (2013). Biological functions of sphingomyelins. *Progress in Lipid Research, 52*(4), 424–437.
Spessott, W., Uliana, A., & Maccioni, H. J. (2010). Cog2 null mutant CHO cells show defective sphingomyelin synthesis. *Journal of Biological Chemistry, 285*(53), 41472–41482.
Stoffel, W., Hammels, I., Jenke, B., Binczek, E., Schmidt-Soltau, I., Brodesser, S., et al. (2016). Neutral sphingomyelinase (SMPD3) deficiency disrupts the Golgi secretory pathway and causes growth inhibition. *Cell Death & Disease, 7*(11), e2488.
Subathra, M., Qureshi, A., & Luberto, C. (2011). Sphingomyelin synthases regulate protein trafficking and secretion. *PLoS One, 6*(9), e23644.
Sudarkina, O. Y., Filippenkov, I. B., Brodsky, I. B., Limborska, S. A., & Dergunova, L. V. (2015). Comparative analysis of sphingomyelin synthase 1 gene expression at the transcriptional and translational levels in human tissues. *Molecular and Cellular Biochemistry, 406*(1–2), 91–99.
Sugimoto, M., Shimizu, Y., Yoshioka, T., Wakabayashi, M., Tanaka, Y., Higashino, K., et al. (2015). Histological analyses by matrix-assisted laser desorption/ionization-imaging mass spectrometry reveal differential localization of sphingomyelin molecular species regulated by particular ceramide synthase in mouse brains. *Biochimica et Biophysica Acta, 1851*(12), 1554–1565.
Sugimoto, M., Wakabayashi, M., Shimizu, Y., Yoshioka, T., Higashino, K., Numata, Y., et al. (2016). Imaging mass spectrometry reveals acyl-chain- and region-specific sphingolipid metabolism in the kidneys of sphingomyelin synthase 2-deficient mice. *PLoS One, 11*(3), e0152191.
Sun, G. Y., & Leung, B. S. (1974). Phospholipids and acyl groups of subcellular membrane fractions from human intracranial tumors. *Journal of Lipid Research, 15*(4), 423–431.
Swanton, C., Marani, M., Pardo, O., Warne, P. H., Kelly, G., Sahai, E., et al. (2007). Regulators of mitotic arrest and ceramide metabolism are determinants of sensitivity to paclitaxel and other chemotherapeutic drugs. *Cancer Cell, 11*(6), 498–512.

Tafesse, F. G., Huitema, K., Hermansson, M., van der Poel, S., van den Dikkenberg, J., Uphoff, A., et al. (2007). Both sphingomyelin synthases SMS1 and SMS2 are required for sphingomyelin homeostasis and growth in human HeLa cells. *Journal of Biological Chemistry, 282*(24), 17537–17547.

Tafesse, F. G., Sanyal, S., Ashour, J., Guimaraes, C. P., Hermansson, M., Somerharju, P., et al. (2013). Intact sphingomyelin biosynthetic pathway is essential for intracellular transport of influenza virus glycoproteins. *Proceedings of the National Academy of Sciences of the United States of America, 110*(16), 6406–6411.

Taguchi, Y., Kondo, T., Watanabe, M., Miyaji, M., Umehara, H., Kozutsumi, Y., et al. (2004). Interleukin-2-induced survival of natural killer (NK) cells involving phosphatidylinositol-3 kinase-dependent reduction of ceramide through acid sphingomyelinase, sphingomyelin synthase, and glucosylceramide synthase. *Blood, 104*(10), 3285–3293.

Taniguchi, M., & Okazaki, T. (2014). The role of sphingomyelin and sphingomyelin synthases in cell death, proliferation and migration-from cell and animal models to human disorders. *Biochimica et Biophysica Acta, 1841*(5), 692–703.

Tani, M., & Kuge, O. (2009). Sphingomyelin synthase 2 is palmitoylated at the COOH-terminal tail, which is involved in its localization in plasma membranes. *Biochemical and Biophysical Research Communications, 381*(3), 328–332.

Thudichum, J. (1884). *A treatise on the chemical constitution of the brain*. London: Bailliere, Tindall and Cox. Republished in 1962 by Archon Books, Hamden, Connecticut.

Toth, B., Balla, A., Ma, H., Knight, Z. A., Shokat, K. M., & Balla, T. (2006). Phosphatidylinositol 4-kinase IIIbeta regulates the transport of ceramide between the endoplasmic reticulum and Golgi. *Journal of Biological Chemistry, 281*(47), 36369–36377.

Tu, R., Yang, W., & Hu, Z. (2016). Inhibition of sphingomyelin synthase 1 affects ceramide accumulation and hydrogen peroxide-induced apoptosis in Neuro-2a cells. *Neuroreport, 27*(13), 967–973.

Ueda, Y., Ogiso, H., Sato, M., Umezawa, Y., Okazaki, T., & Kobayashi, T. (2015). Asymmetrical diacylglycerol dynamics on the cytosolic and lumenal sides of a single endomembrane in living cells. *Scientific Reports, 5*, 12960.

Ullman, M. D., & Radin, N. S. (1974). The enzymatic formation of sphingomyelin from ceramide and lecithin in mouse liver. *Journal of Biological Chemistry, 249*(5), 1506–1512.

Vacaru, A. M., Tafesse, F. G., Ternes, P., Kondylis, V., Hermansson, M., Brouwers, J. F., et al. (2009). Sphingomyelin synthase-related protein SMSr controls ceramide homeostasis in the ER. *Journal of Cell Biology, 185*(6), 1013–1027.

Van der Luit, A. H., Budde, M., Zerp, S., Caan, W., Klarenbeek, J. B., Verheij, M., et al. (2007). Resistance to alkyl-lysophospholipid-induced apoptosis due to downregulated sphingomyelin synthase 1 expression with consequent sphingomyelin- and cholesterol-deficiency in lipid rafts. *Biochemical Journal, 401*(2), 541–549.

Villani, M., Subathra, M., Im, Y. B., Choi, Y., Signorelli, P., Del Poeta, M., et al. (2008). Sphingomyelin synthases regulate production of diacylglycerol at the Golgi. *Biochemical Journal, 414*(1), 31–41.

Vladychenskaya, I. P., Dergunova, L. V., Dmitrieva, V. G., & Limborska, S. A. (2004). Human gene MOB: Structure specification and aspects of transcriptional activity. *Gene, 338*(2), 257–265.

Wang, S., Chen, X., Luan, H., Gao, D., Lin, S., Cai, Z., et al. (2016). Matrix-assisted laser desorption/ionization mass spectrometry imaging of cell cultures for the lipidomic analysis of potential lipid markers in human breast cancer invasion. *Rapid Communications in Mass Spectrometry, 30*(4), 533–542.

Wang, X., Rao, R. P., Kosakowska-Cholody, T., Masood, M. A., Southon, E., Zhang, H., et al. (2009). Mitochondrial degeneration and not apoptosis is the primary cause of embryonic lethality in ceramide transfer protein mutant mice. *Journal of Cell Biology, 184*(1), 143–158.

Wang, L., Wu, L., Wang, X., Deng, J., Ma, Z., Fan, W., et al. (2015). Prenatal alcohol exposure inducing the apoptosis of mossy cells in hippocampus of SMS2−/− mice. *Environmental Toxicology and Pharmacology, 40*(3), 975–982.

Weber, P., Hornjik, M., Olayioye, M. A., Hausser, A., & Radde, N. E. (2015). A computational model of PKD and CERT interactions at the trans-Golgi network of mammalian cells. *BMC Systems Biology, 9*, 9.

Wesley, U. V., Hatcher, J. F., & Dempsey, R. J. (2015). Sphingomyelin synthase 1 regulates neuro-2a cell proliferation and cell cycle progression through modulation of p27 expression and Akt signaling. *Molecular Neurobiology, 51*(3), 1530–1541.

Wood, R., Upreti, G. C., & deAntueno, R. J. (1986). A comparison of lipids from liver and hepatoma subcellular membranes. *Lipids, 21*(4), 292–300.

Xu, R., Jin, J., Hu, W., Sun, W., Bielawski, J., Szulc, Z., et al. (2006). Golgi alkaline ceramidase regulates cell proliferation and survival by controlling levels of sphingosine and S1P. *FASEB J, 20*(11), 1813–1825.

Xu, Y., Ramu, Y., & Lu, Z. (2008). Removal of phospho-head groups of membrane lipids immobilizes voltage sensors of K+ channels. *Nature, 451*(7180), 826–829.

Yachi, R., Uchida, Y., Balakrishna, B. H., Anderluh, G., Kobayashi, T., Taguchi, T., et al. (2012). Subcellular localization of sphingomyelin revealed by two toxin-based probes in mammalian cells. *Genes to Cells, 17*(8), 720–727.

Yamaoka, S., Miyaji, M., Kitano, T., Umehara, H., & Okazaki, T. (2004). Expression cloning of a human cDNA restoring sphingomyelin synthesis and cell growth in sphingomyelin synthase-defective lymphoid cells. *Journal of Biological Chemistry, 279*(18), 18688–18693.

Yang, Z., Jean-Baptiste, G., Khoury, C., & Greenwood, M. T. (2005). The mouse sphingomyelin synthase 1 (SMS1) gene is alternatively spliced to yield multiple transcripts and proteins. *Gene, 363*, 123–132.

Yang, Z., Khoury, C., Jean-Baptiste, G., & Greenwood, M. T. (2006). Identification of mouse sphingomyelin synthase 1 as a suppressor of Bax-mediated cell death in yeast. *FEMS Yeast Research, 6*(5), 751–762.

Yilmaz, N., Yamaji-Hasegawa, A., Hullin-Matsuda, F., & Kobayashi, T. (2018). Molecular mechanisms of action of sphingomyelin-specific pore-forming toxin, lysenin. *Seminars in Cell & Developmental Biology, 73*, 188–198.

Yuyama, K., Sun, H., Mitsutake, S., & Igarashi, Y. (2012). Sphingolipid-modulated exosome secretion promotes clearance of amyloid-beta by microglia. *Journal of Biological Chemistry, 287*(14), 10977–10989.

Zama, K., Mitsutake, S., Okazaki, T., & Igarashi, Y. (2018). Sphingomyelin in microdomains of the plasma membrane regulates amino acid-stimulated mTOR signal activation. *Cell Biology International.*

Zhang, Y., Dong, J., Zhu, X., Wang, W., & Yang, Q. (2011). The effect of sphingomyelin synthase 2 (SMS2) deficiency on the expression of drug transporters in mouse brain. *Biochemical Pharmacology, 82*(3), 287–294.

Zhang, Y., Li, X., Becker, K. A., & Gulbins, E. (2009). Ceramide-enriched membrane domains—structure and function. *Biochimica et Biophysica Acta, 1788*(1), 178–183.

CHAPTER FOUR

Neutral Sphingomyelinases in Cancer: Friend or Foe?

Christopher J. Clarke[1]
Department of Medicine and Cancer Center, Stony Brook University, Stony Brook, NY, United States
[1]Corresponding author: E-mail: Christopher.clarke@stonybrookmedicine.edu

Contents

Abstract

For many years, neutral sphingomyelinases (N-SMases) were long thought to be anticancer enzymes owing to their roles as key producers of ceramide linked to apoptosis, growth arrest, and the chemotherapeutic response. However, in recent years, with the cloning of multiple isoforms and with new information on their cellular roles, particularly for nSMase2, a more complex picture is emerging suggesting that N-SMases have both pro- and anticancer roles. In this chapter, we will summarize current knowledge on N-SMase expression in cancer and the roles of N-SMase activity and specific isoforms in cancer-relevant biologies. We will also discuss what we see as the major challenges ahead for research into N-SMases in cancer.

Advances in Cancer Research, Volume 140
ISSN 0065-230X
https://doi.org/10.1016/bs.acr.2018.04.010

1. THE NEUTRAL SPHINGOMYELINASE FAMILY

Early studies into the signaling and metabolism of sphingolipids (SLs) established the hydrolysis of sphingomyelin (SM) as a major pathway of ceramide (Cer) production activated in response to external stimuli and focused much attention on the enzymes that mediate this reaction. The sphingomyelinases (SMases) are broadly divided into three major classes based on their pH optima of activity. Neutral sphingomyelinase (N-SMase) activity was first described from Niemann–Pick fibroblasts deficient in the previously described acid SMase (Schneider & Kennedy, 1967). However, it was some 30 years before the molecular identity of the first mammalian N-SMase termed nSMase1 (sphingomyelinase phosphodiesterase 2; SMPD2) was reported (Tomiuk, Hofmann, Nix, Zumbansen, & Stoffel, 1998). This was subsequently followed by cloning of nSMase2 (SMPD3) (Hofmann, Tomiuk, Wolff, & Stoffel, 2000), nSMase3 (SMPD4) (Krut, Wiegmann, Kashkar, Yazdanpanah, & Krönke, 2006), and mitochondria-associated nSMase (MA-nSMase; SMPD5) (Wu, Clarke, & Hannun, 2010; Yabu, Shimuzu, & Yamashita, 2009). Unlike their acid SMase and alkaline SMase counterparts—which exist as single gene products (Gorelik, Liu, Illes, & Nagar, 2017; Schuchman, Levran, Pereira, & Desnick, 1992) (although for acid SMase, this product can be alternately trafficked to the lysosome and secretory pathways)—the number and diverse compartmentalization of N-SMase family members stands out and suggests distinct and nonoverlapping functions for each isoform related to their cellular localizations.

2. NEUTRAL SPHINGOMYELINASES IN TUMORIGENESIS

Since their cloning, research on mammalian N-SMases has focused on defining specific mechanisms of regulation, their downstream-signaling effectors, and their biological roles both in the context of cellular functions and the pathologies and diseases where they are involved. Here, we will discuss studies of N-SMase expression and activity in cancer and those N-SMase regulated biologies that are particularly relevant for cancer initiation, progression, metastasis, and therapeutics. For more in-depth overviews on the roles and regulation of N-SMases, we recommend the reviews by Shamseddine, Airola, and Hannun (2015), Airola and Hannun (2013), and Wu, Rajagopalan, Roddy, Clarke, and Hannun (2010).

2.1 Neutral Sphingomyelinases Expression in Cancer

Deregulation of cellular SL levels has been reported in many cancers (reviewed in Ogretmen, 2018) leading to increased interest in targeting the SL network for cancer therapeutics. This also prompted efforts aimed at defining tumor-specific alterations in SL enzymes in comparison to normal tissue using microarray data, as well as in vitro studies comparing the SL network of cancer cells to nontransformed cells. Here, information on the expression of mammalian N-SMases is somewhat limited—most likely owing to the relatively recent cloning of nSMase3 and the as-yet unidentified human homologue of MA-nSMase. Despite being the first cloned N-SMase, there is also limited published information on nSMase1, although a study in breast cancer reveals no difference in nSMase1 between tumor and normal tissue (Schiffmann et al., 2009).

In contrast to this, there are several studies that suggest alterations in nSMase2 expression is associated with cancer development and progression. The first insight into this came from the determination that the nSMase2 gene was localized on chromosome 16q22.1—a common site of loss of heterozygosity in breast and prostate cancers—and was linked to early-stage breast carcinogenesis (Driouch et al., 1997). As such losses can be indicative of a tumor suppressor gene, this was suggestive that decreased nSMase2 levels may be associated with cancer. Consistent with this, a homozygous loss of nSMase2 was characterized in a mouse osteosarcoma cell line and, when nSMase2 was reconstituted, resulted in suppression of cell growth (Kim et al., 2008). At the epigenetic level, a number of studies have reported suppressive changes at the nSMase2 gene in a number of cancers. An initial report in 2009 found hypermethylation of the nSMase2 gene in aggressive breast cancer cells (Demircan et al., 2009). These findings were independently validated in a later study, demonstrating hypermethylation of the nSMase2 gene in basal MDA-MB-231 cells but not luminal MCF7 cells resulting in lower nSMase2 expression (Clarke et al., 2016). Importantly, such observations were not limited to in vitro cell lines with studies in clear cell renal carcinoma and hepatocellular carcinoma (HCC) tumors reporting increased hypermethylation of the nSMase2 gene when compared to normal tissue (Revill et al., 2013; Wang et al., 2015). In renal carcinoma, higher nSMase2 methylation was somewhat correlated with larger tumors and higher tumor grade (Wang et al., 2015), and in HCC, low levels of nSMase2 were associated with earlier recurrence (Revill et al., 2013).

Outside of gene expression, Haber et al. identified point mutations in nSMase2 in 5% of acute myeloid leukemias and 6% of acute lymphoid leukemias (Kim et al., 2008). Although the functional consequences of most remain unclear, two of the mutations resulted in decreased protein stability (D358G) and mislocalization (G248S) of nSMase2, respectively. Finally, a proteomic study of blood vessel cells found that nSMase2 was increased in breast tumor–associated cells compared with those of normal tissue (Bhati et al., 2008). Thus, it will be important in the future to distinguish levels in tumor cells from those of associated stroma, immune cells, and blood vessels.

2.2 Cancer-Relevant Biologies Regulated by Neutral Sphingomyelinases

2.2.1 Apoptosis

A major hallmark of cancer is the ability of tumor cells to overcome the normal apoptotic signaling pathways and evade cell death (Hanahan & Weinberg, 2011). Since the discovery of their signaling functions (Hannun, Loomis, Merrill, & Bell, 1986), Cer and other SLs such as sphingosine (Sph) were classically identified as mediators of apoptosis—stimulating interest in defining the specific SL enzymes and pathways involved. The activation of SM turnover by many stress-inducing stimuli implicated SMases and since then, increases in N-SMase activity were linked to TNF, Fas ligand, heat stress, ischemia/reperfusion, and, discussed below, chemotherapeutic agents (reviewed in Nikolova-Karakashian & Rozenova, 2010). Subsequently, in leukemic cells, serum withdrawal increased Reactive Oxygen Species (ROS)-dependent N-SMase activity (Fernández-Ayala et al., 2000) and in keratinocytes, ultraviolet-B induces N-SMase activity and apoptosis (Magnoni et al., 2002). Finally, N-SMase activation was also observed in response to ionizing radiation (IR) (Chmura et al., 1997). However, independent studies have described acid SMase-deficient cells as both IR-resistant (fibroblast) (Lozano et al., 2001) and sensitive (lymphoid) (Bezombes et al., 2001) suggesting potential cell type specific involvement of N-SMase activity.

Studies exploring N-SMase isoform roles in apoptosis placed nSMase2 in the response to TNF in MCF-7 and MG-63 cells (Luberto et al., 2002; Martin, Williams, & Chaterjee, 2006). Notably, in the former, this was attributed to generation of Cer, but in the latter, further metabolism of nSMase2-derived Cer to lactosylceramide was required for cell death. Consistent with this, both nSMase2 and nSMase3 were implicated in

TNF/cycloheximide-induced apoptosis (Chipuk et al., 2012) although, again, this appeared not to be through Cer generation but subsequent downstream metabolites. Despite these studies, there is also clear cell-type specificity for the role for nSMase2 as it was found to be dispensable for TNF and OxLDL-induced apoptosis of fibroblasts (Devillard et al., 2010). Moreover, more recent work has implicated nSMase1 in apoptosis induced by heat shock, UV, and hydrogen peroxide in Jurkat T cells (Yabu et al., 2015) and was also found to be activated by extracellular CCN proteins in human skin fibroblasts, synergizing with FasL and TNF to promote apoptosis (Juric, Chen, & Lau, 2012). The involvement of CCN here could account for why an earlier study found no effect of nSMase1 on FasL-induced apoptosis (Tepper, Ruurs, Borst, & van Blitterswijk, 2001).

Functionally, and where it is involved, N-SMase activity is important for commitment of the cells to apoptosis, with the inhibitor GW4869 preventing TNF-induced mitochondrial permeablization (MOMP), cytochrome C release, and caspase 9 activation in MCF7 cells (Luberto et al., 2002). Further evidence for this came from the study of Chipuk et al. (2012) who demonstrated cooperation of nSMase2 and nSMase3 activity with Bax and Bak to induce MOMP and cell death in HeLa cells treated with TNF/CHX. In contrast to these findings, while nSMase1 regulation of TNF/CCN and FasL/CCN-induced apoptosis was found to occur upstream of MOMP, this was through ROS generation (Juric et al., 2012) Finally, studies of factor associated with N-SMase, a TNF receptor—binding protein required for N-SMase activation placed it upstream of caspase 8 activation and lysosomal permeablization—although the N-SMase isoform involved here was not identified (Werneburg, Guicciardi, Yin, & Gores, 2004).

2.2.2 Chemotherapy-Induced Apoptosis

Regulation of SL metabolism has emerged as a central component of the cellular response to chemotherapies across diverse cancers. Indeed, accumulation of Cer (and other proapoptotic SLs) has been reported in response to anthracyclines, taxanes, etoposide, and 5-flurouracil (reviewed in Ogretmen, 2018) among others and, functionally, neutralizing the accumulation of such SLs has been shown to blunt cell death in many cases. Early efforts to identify the primary pathways of chemotherapy-induced Cer production demonstrated increased N-SMase activity in response to many drugs including daunorubicin-stimulated leukemia cells (Jaffrézou et al., 1996), doxorubicin-treated cardiomyocytes (Delpy et al., 1999), cisplatin-treated glioma cells (Noda et al., 2001), and in HCT116 colon cancer cells treated with oxaliplatin (Nemoto et al., 2009). Consistent with this, a study utilizing aSMase-deficient lymphoid cells suggested N-SMase was the primary SMase activated by anthracyclines (Bezombes et al., 2001). While collectively suggestive of a role for N-SMase in the chemotherapy response, however, the lack of pharmacological inhibitors and molecular tools made it difficult to attribute these responses to specific isoforms. Moreover, in addition to N-SMases, it was also clear that de novo SL synthesis, aSMases, and the salvage pathway played varying roles in the responses to some stimuli. More recent efforts into isoform specific roles of N-SMase have attributed anthracycline-induced N-SMase activity in MCF7 cells directly to transcriptional induction of nSMase2 (Ito et al., 2009; Shamseddine, Clarke, et al., 2015), and in medulloblastoma cells treated with cyclopamine (Meyers-Needham et al., 2012). In all cases, nSMase2 induction was required for the anticancer effects of the drugs, consistent with a separate study implicating a GW4869-sensitive N-SMase in Dox-induced death of MCF7 cells (Liu et al., 2013). However, there also appear to be cell-specific differences as a separate study reported Dox induction of nSMase3 in colon cancer cells and a role in cell death was inferred (Corcoran et al., 2008) although it should be noted that effects of Dox and nSMase3 knockdown on SL levels were not determined in this study.

2.2.3 Growth Arrest and Proliferation

In addition to avoiding apoptosis, the ability of tumor cells to avoid normal growth checkpoints and proliferate independently of growth factors represents two major hallmarks of cancer (Hanahan & Weinberg, 2011). Notably, aside from their well-established roles in mediating apoptosis and cell survival, various bioactive SLs have also been linked to regulation of cell proliferation and the cell cycle. Indeed, some of the earliest studies on SLs connected Cer accumulation to growth arrest in a variety of cell types and, crucially, found it was distinct from its apoptotic effects (Zhang et al., 1996). Subsequent studies extended these observations, demonstrating induction of G0/G1 growth arrest by both exogenous and endogenously generated Cer, and these were linked to dephosphorylation of retinoblastoma (Dbaibo et al., 1995; Venable, Lee, Smyth, Bielawska, & Obeid, 1995), modulation of cyclin-dependent kinases (Lee, Bielawska, & Obeid, 2000), and inhibition of telomerase (Ogretmen et al., 2001). The N-SMases first entered this discussion in the mid 1990s when serum withdrawal was shown to increase N-SMase activity and Cer levels prior to growth arrest in Molt4 leukemia cells (Jayadev et al., 1995). This was followed by studies showing activation of N-SMases prior to daunorubicin-induced growth arrest (Ogretmen et al., 2001) and as fibroblasts entered the senescent state (Venable et al., 1995). Isoform-specific roles of N-SMase in regulation of growth has primarily focused on nSMase2. Initial studies demonstrated upregulation of nSMase2 mRNA and activity occurred on cell confluence of MCF7, WI-38 HDF, and SK-NHS cells and, in the case of MCF7 cells, was required for subsequent G1 growth arrest through effects on p21 and Rb (Marchesini et al., 2004). Similar upregulation of nSMase2 was also reported in neuroblastoma cells plated at high density compared with those at low density (Spassieva et al., 2012). Outside of cell–cell contact, omega-3 fatty acids were reported to inhibit breast cancer cell growth through a GW4869-sensitive N-SMase (Wu et al., 2005). Moreover, transcriptional induction of nSMase2 through epigenetic regulation was identified as the primary retinoic acid all-trans retinoic acid (ATRA)-regulated node in the SL network of breast cancer cells and was required for ATRA-induced G1 growth arrest through modulation of ribosomal S6 kinase and translation (Clarke, Mediwala, et al., 2011). Finally, growth arrest of MCF7 cells induced by sublethal doses of Dox also required nSMase2 induction (Shamseddine, Clarke, et al., 2015). Consistent with all these observations, stable nSMase2 overexpression is sufficient to decrease proliferation of both MCF7 and HCC cells (Marchesini et al., 2004; Revill et al., 2013) increasing cells in both G1 and G2 phases of the cell cycle (Clarke, Mediwala, et al., 2011).

While the above studies are supportive of an antigrowth role for N-SMases and particularly nSMase2, there is also considerable evidence connecting nSMase2 with proliferation. Indeed, studies from the Negre-Salvayre and Salvayre groups have placed nSMase2 upstream of cell growth and ERK activation in response to various stimuli (TNF, OxLDL, uPA) and in a number of "normal" cell types (smooth muscle, fibroblasts, epithelial) (Devillard et al., 2010; Maupas-Schwalm et al., 2009; Tellier et al., 2007). A common theme of this work demonstrates acute and transient increases in nSMase2 activity directly coupled to activation of sphingosine kinase 1 (SphK1), leading to S1P generation that functions as the mitogenic signal. Considering these observations together, it seems broadly that the pro- and antigrowth roles of nSMase2 could be differentiated by time period of enzyme activation—early, transient activation seems more associated with pro-growth effects whereas more sustained activation, including transcriptional upregulation of nSMase2, is broadly (although not exclusively) associated with increased Cer levels and antigrowth phenotypes. Clearly, the coupling of N-SMase activity to downstream-metabolizing enzymes such as SphK1 or ceramide kinase (CERK) is also important and may be particularly relevant for those cancers with higher levels of such proteins.

2.2.4 Exosome Release

Growing appreciation that cancer cells are very adept at shaping their microenvironment to promote tumor growth and metastasis has increased interest in mechanisms of cell–cell communication. In this context, recent attention has focused on exosomes—30–100 nm vesicles secreted by cells that contain protein, lipids, nucleic acids, and RNA as cargo that can be transferred into target cells to modulate their signaling pathways (reviewed in Zhao et al., 2018). Evidence linking exosomes to various aspects of tumorigenesis has led to increased interest in therapeutically targeting this pathway.

The first identification of N-SMases as regulators of this pathway came from the seminal study of Trajkovic et al. (2008) who demonstrated that Cer can promote exosome formation in the multivesicular body and using loss of function siRNA approaches, specifically implicated nSMase2 in this process. Since then, multiple studies have consolidated and extended these findings. Results by Kosaka et al. (2010) suggested that nSMase2 regulated the amount of exosomes secreted rather than influencing their cargo per se and reported that secretion of miRNAs was primarily in exosomes regulated by the nSMase2–Cer axis. Similar effects of nSMase2 inhibition on miRNA secretion have since been reported in various cancer cells including colon and liver (Kubota, Chiba, Watanabe, Sakamoto, & Watanabe, 2015), esophageal (Liao, Liu, Shi, Yin, & Pu, 2016), duodenal (Ohshima et al., 2014), and breast (Kosaka et al., 2013; Singh, Pochampally, Watabe, Lu, & Mo, 2014). However, nSMase2 does not appear to influence exosome secretion from the PC-3 prostate cancer cell line (Phuyal, Hessvik, Skotland, Sandvig, & Llorente, 2014) suggesting this pathway is not a universal mechanism relevant to all cancers and cell types. Additionally, a more recent study added an extra wrinkle finding that while inhibition of nSMase2 prevented secretion of exosomes, it conversely resulted in enhanced secretion of the larger microvesicles (Menck et al., 2017). This study also found that nSMase1 may also be a player in this pathway, suggesting some degree of redundancy.

The relevance of the nSMase2–exosome pathway to cancer has also begun to be investigated. Following up their earlier work, Kosaka et al. (2013) connected nSMase2 regulation of exosomal miR-210 to increased in vivo metastasis in basal breast cancer by promoting local angiogenesis. Along these lines, an independent study demonstrated that nSMase2-regulated secretion of miR-10b from basal breast cancer cells promoted invasion of nontransformed mammary epithelial cells (Singh et al., 2014). While these studies point to pro-tumor functions for nSMase2 and exosomes, a study by Bieberich et al. suggested that nSMase2-derived exosomes may also have antitumor relevance (Kong et al., 2015). Here, activation of nSMase2 in the basal MDA-MB-231 cell line induced exosomal secretion of breast cancer resistance protein, a promoter of multidrug resistance, leading to increased Dox sensitivity. However, on the flip side, a study in oral squamous carcinoma cells found that exosomal secretion of miR-21 from drug resistant cells was nSMase2 dependent and that transfer into drug-sensitive cells promoted drug resistance (Liu et al., 2017). Taken together, while study of the nSMase2–exosome connection in cancer is still relatively in its infancy, the potential of this pathway to be of high relevance for cancer progression and by extension, a therapeutic target, is clear and certainly warrants further study. For a more in-depth discussion of this, the reader is referred to Side Effects in Cancer Therapy: Are Sphingolipids to Blame? by Patel and Spassieva (2018) in this book.

2.2.5 Inflammation

The inflammatory response is normally a beneficial process required to repair damaged tissue and destroy foreign agents but, if left unchecked, substantially increases the chances of malignant cell transformation. Indeed, connections between chronic inflammation and cancer have long been appreciated and it is now considered a major enabling characteristic of tumorigenesis (Hanahan & Weinberg, 2011). Increased local inflammation, and the resulting recruitment of immune cells to the microenvironment, provides a range of bioactive molecules such as growth factors, cytokines, and chemokines, which can drive cell proliferation, promote cell survival, increase angiogenesis, and lead to alterations in responsiveness to targeted and cytotoxic chemotherapies (reviewed in Yang & Lin, 2017).

SLs are well-established mediators of the inflammatory response (reviewed in Maceyeka & Spiegel, 2014). The first evidence for the role of SLs and, indeed, the SMases in the inflammatory response came from studies demonstrating activation of SM turnover by various cytokines including TNF-alpha, interleukin (IL)-1 beta, and IFN-gamma with early studies linking Cer generation to the production of prostaglandin E2 (PGE2) and IL-6 (Ballou, Chao, Holness, Barker, & Raghow, 1992; Laulederkind et al., 1995). Subsequent research suggested this could be through effects on arachidonic acid (AA) production (Klapisz, Masliah, Béréziat, Wolf, & Koumanov, 2000), as well as induction of COX-2 (Subbaramaiah, Chung, & Dannenberg, 1998). Further insight into both these pathways came in the early 2000s when it was suggested that Cer, per se, was not the bioactive SL of importance in these circumstances. In contrast, the metabolism of Cer to S1P through SphK1 was required for cytokine induction of Cox-2 (Pettus, Bielawski, Porcelli, et al., 2003) whereas phosphorylation to C1P by CERK was required for cPLA2 activation and AA generation (Pettus, Bielawska, Spiegel, et al., 2003).

Although these studies clarified the crucial SL mediators of PGE2 production, the upstream source of the Cer for both these pathways remained an outstanding question. In this context, early evidence for a role of N-SMases in this process was promising. Aside from the acute activation of N-SMases by cytokines, induction of COX-2 in TNF-stimulated H292 lung cancer cells could be inhibited by glutathione (GSH) pretreatment (Chen, Sun, Chen, & Chang, 2001), whereas acid SMase knockdown had no effect on TNF-stimulated PGE2 production in L929 fibroblasts (Zeidan et al., 2006). Further supporting evidence demonstrated that nSMase2 was required for TNF-stimulated S1P generation and endothelial nitric oxide synthase (eNOS) activation in HUVEC cells (De Palma, Meacci, Perrotta, Bruni, & Clementi, 2006) and identified a role for acute and transient activation of nSMase2 in regulation of the adhesion proteins VCAM and ICAM (Clarke, Truong, & Hannun, 2007). Finally, a GW4869-sensitive N-SMase was reported to be upstream of CERK and required for eicosanoid synthesis in neuroblastoma cells (Barth et al., 2012). Taken together with findings that nSMase2 is the primary TNF-activated N-SMase in multiple cell types (Clarke, Mediwala, et al., 2011), these studies suggest that by serving a source of Cer for downstream-metabolizing enzymes, N-SMases and specifically nSMase2 functions as an upstream regulator of proinflammatory cytokine signaling—and can contribute to upregulation of both prostaglandins, adhesion proteins, and immune cell recruitment.

In addition to TNF, the efforts of Nikolova-Karakashian et al. have conclusively linked N-SMase activation to the IL-1 beta–signaling cascade in hepatocytes. After first establishing nSMase2 as the primary

IL-1 beta—activated isoform, they found that nSMase2-regulated activation of JNK in the IL-1 beta response (Karakashian, Giltiay, Smith, & Nikolova-Karakashian, 2004). Follow-up studies revealed that this was required for IL-1—mediated translocation of the transcription factor Fox01 into the nucleus and that effects of nSMase2 were mediated through PP2A regulation of IRAK-1, preventing its degradation and resulting in potentiated signaling through the IL-1 receptor (Dobierzewska, Giltiay, Sabapathi, Karakashian, & Nikolova-Karakashian, 2011). Physiologically, this was particularly relevant in the context of aging, with hepatocytes from older animals displaying hyper-responsiveness to IL-1 beta as a direct consequence of redox activation of nSMase2 (Rutkute, Karakashian, Giltiay, Dobierzewska, & Nikolova-Karakashian, 2007). While the SL involved was not specifically identified, the involvement of PP2A implies that Cer is the mediator of this process. Taken together with the TNF studies above, this demonstrates that N-SMases and nSMase2 can function as central mediators of proinflammatory signaling through regulating the levels of multiple bioactive SLs.

3. THE CHALLENGES AHEAD

In the last two decades, knowledge of the mammalian N-SMases has increased considerably, implicating them in many cellular processes and suggesting specific isoforms as targets for therapeutic intervention in various pathologies. Yet despite this progress, there remain many gaps in our understanding of N-SMases, particularly when considering the context of cancer. Here, we will briefly discuss what we see as some of the issues pertaining to N-SMases in general and in relation to tumorigenesis.

3.1 The Known Unknowns

At present, the clear majority of studies investigating downstream biological roles of the mammalian N-SMases have focused primarily on nSMase2. The reasons for this are twofold. For some isoforms i.e., MA-nSMase (SMPD5), this can be attributed to their relatively recent discovery and, as human homologues have yet to be identified, their roles in human cancers cannot be explored as yet. For others, this is likely due to uncertainty as to whether the enzymes in question function as bona fide N-SMases. The best example of this is nSMase1 that possesses the N-SMase family catalytic core and displays in vitro N-SMase activity, yet does not affect cellular SL levels when overexpressed (Sawai, Domae, Nagan, & Hannun, 1999). Similarly, the nSMase3 isoform—which lacks the conserved "catalytic core" residues and bears little resemblance to the broader N-SMase family—did not seem to possess in vitro or in vivo N-SMase activity when overexpressed (Clarke, Cloessner, Roddy, & Hannun, 2011). In contrast to these findings, more recent studies have reported that both these isoforms can modulate cellular SL levels in some circumstances (Moylan et al., 2014; Yabu et al., 2015). While some progress has been made with nSMase1, further investigation of the biological function of both enzymes is needed. Moreover, in studies of nSMase2 functions, both should be utilized as additional controls for overexpression or knockdown, particularly pertaining to lipid levels and biological outcomes. This is not only important for demonstrating specificity of responses but also for establishing those biologies where there is redundancy in isoforms or overlap in cellular functions. Such knowledge is crucial when considering N-SMase inhibitors as therapeutics as isoform-specific inhibitors would be less than ideal if other N-SMases isoforms are able to compensate.

3.2 Shaping the Environment: Thinking Outside the Cell

In recent years, the importance of the microenvironment as a key component of cancer development and progression has become clear. The use of xenograft models in immunodeficient mice has given the scientific field a wealth of information about the cell intrinsic effects of SL metabolism in tumor progression and have been valuable tools for proof of principle testing of therapeutics. However, such models cannot fully recapitulate the complexity of the tumor microenvironment, particularly with regards to tumor-associated immune cells. This raises the strong possibility that functions of SLs in shaping the tumor microenvironment may have been overlooked.

The connection of N-SMases (and nSMase2) to exosome secretion provides a direct mechanistic link between alterations of tumor cell SL metabolism and modulating tumor-associated stromal and immune cells. While initial studies in this area have shown promise, linking nSMase2 to exosomal miRNA and demonstrating effects on nontransformed epithelial and endothelial cells, there remain significant areas of exploration for future work. A major unanswered question pertains to the involvement of nSMase2 in exosome secretion—namely, is it a regulated ("activated") process in cancer cells and tumors or is it constitutive? If the latter, does this mean that there are functions for nSMase2 and exosomes in nontransformed cells? Understanding such functions would be particularly important to ensure that development of the nSMase2–exosome axis as a therapeutic target does not have adverse effects detrimental to noncancer cells. Conversely, if nSMase2-driven exosome secretion is an activated process in cancer, then understanding the "triggers"—Are there specific oncogenes? Are there specific factors in the microenvironment?—would provide insight into tumorigenesis and provide biomarkers for those cancers that would benefit the most from targeting this pathway. Finally, while many studies on nSMase2 exosomes in cancer have focused on tumor cells and subsequent effects on the microenvironment, it is also important to consider the reverse process of stromal cells influencing tumor cell biology and function. Indeed, accumulating evidence has linked stromal cell–derived exosomes to cancer migration, invasion, chemotherapy resistance, and metastasis (reviewed in Zhao et al., 2018). Determining if such effects are through N-SMase regulation of exosome secretion in the stroma would bolster the case of therapeutically targeting this pathway. This would also be relevant for those cancers where the tumor cells secrete exosomes independently of N-SMase activity and nSMase2 (Phuyal et al., 2014).

While exosomes can phenotypically modulate the tumor stroma, there is also potential for N-SMases to influence the size, composition, and function of the tumor-immune microenvironment. This could be effected through various proinflammatory signaling pathways discussed above such as upregulation of adhesion proteins or synthesis of eicosanoids. Alternatively, this could also occur through direct regulation of the levels of S1P and C1P as both these lipids are well known to promote migration of immune cells. Indeed, S1P can induce chemotaxis of macrophages, $CD4^+$ T-cells, $CD8^+$ T-cells, and dendritic cells (Graeler, Shankar, & Goetzl, 2002; Gude et al., 2008; Maeda et al., 2007), whereas C1P can promote both immune cell migration and proliferation (Arana et al., 2013; Katz et al., 2016). As it is now accepted that variations in the tumor immune infiltrate are linked to increased tumor growth, metastasis, and more recently to response to therapy including chemotherapy, the pathological implications of such regulation are clear. Moreover, the relative expression of N-SMases or other downstream SL enzymes in tumor cells may be surrogate biomarkers for therapeutically beneficial tumor immune microenvironments. Notably, a recent study detected high levels of S1P in the tumor interstitial fluid and suggested that they were correlated with breast tumor growth (Nagahashi et al., 2016). Thus, it would be of interest to determine if variations in such levels are associated with differences in tumor immune cell infiltration and if targeting SphK1 or the N-SMases upstream are able to modulate their levels.

3.3 Inhibitor Development

A major hindrance to the early study of N-SMases in cell signaling and biology was the lack of molecular tools as well as N-SMase inhibitors. The cloning of the various N-SMase isoforms enabled gain of function approaches and the subsequent use of siRNA technology for loss of function experiments. However, this led to the new problem of lack of specific pharmacological inhibitors, both for dissecting the roles of individual isoforms complementary to genetic approaches, for use in various animal models, and as starting points for potential therapeutic development. Many different compounds have been reported as early inhibitors of N-SMase activity. Scyphostatin, a chemical isolated from *Discomycetes* was shown to inhibit N-SMase activity and cell death in neurons (Brann, Tcherpakov, Williams, Futerman, & Fainzilber, 2002). Similarly, C11AG selectively suppressed N-SMase activity in T-cells and macrophages (Amtmann & Zoller, 2005) and manumycin A, although a known inhibitor of Ras farnesyltransferase, also decreased TNF-stimulated N-SMase activity in HUVEC cells (De Palma et al., 2006). The SM analogue 3-O-methylsphingomyelin was reported to have effects on glial cell signaling whereas inhibitors of acid SMase or de novo Cer synthesis did not (Won, Im, Khan, Singh, & Singh, 2004). While the inhibitor GW4869 was identified through a high-throughput screen and reported to function as a reversible noncompetitive inhibitor of N-SMase, reducing SM hydrolysis and Cer production in TNF-α—stimulated cell death in MCF7 cells (Luberto et al., 2002). However, aside from potential issues with off-target effects (manumycin A) and membrane permeability of these compounds (3-O-methylsphingomyelin), solubility is also a problem (GW4869). More recent efforts have addressed this issue somewhat. For example, the study by Lee et al. (2011) utilized partially purified nSMase2 and identified d-*lyxo*phytosphingosine and d-*arabino*-phytosphingosine as inhibitors, albeit not particularly potent—with IC50s of around 100 μM. Similarly, a high-throughput study using recombinant human nSMase2 identified cambinol as an uncompetitive inhibitor with an in vitro IC50 of around 5 μM and defined the chemical elements necessary for inhibitory effects. Crucially, cambinol treatment (10 μM) inhibited TNF-induced Cer generation and cell death in neurons, a process previously linked to nSMase2 (Figuera-Losada et al., 2015) However, while these inhibitors are promising, an ongoing problem is that the specificity regarding the cloned N-SMase isoforms has not clearly been tested. Given the conservation of the core catalytic residues across the N-SMase family (except for nSMase3), this is particularly crucial as, for example, while GW4869 was previously reported to inhibit nSMase2 (Marchesini, Luberto, & Hannun, 2003), subsequent research has suggested that it may also inhibit nSMase1 (Menck et al., 2017). An alternative strategy, moving forward, could design inhibitors based on their regulation at the molecular level as, for example, nSMase2 requires anionic phospholipids (Marchesini et al., 2003) for full activity whereas nSMase1 does not (Tomiuk et al., 1998). Thus, compounds aimed at disrupting this action in nSMase2 may afford isoform specificity. Now, with the recent cloning and solving of the crystal structure of nSMase2 (Airola et al., 2017), the resulting insight into enzyme activation will hopefully provide valuable information for the development of potent, isoform-specific inhibitors. These will be critical both as research tools as well as for therapeutic development.

3.4 Are Neutral Sphingomyelinases Viable Targets for Cancer Therapy?

While there is a paucity of potent and isoform specific N-SMase inhibitors, an additional point that must be considered and is worthy of discussion, particularly in the context of cancer, is whether N-SMases should be targeted at all. Indeed, based on current knowledge, it is clear that there are both pros and cons. As an example, there is accumulating evidence that targeting exosomes may be of benefit to prevent cancer metastasis, lending weight to nSMase2 as a viable target to accomplish this. However, considering the cancer cell-intrinsic functions of nSMase2 in mediating confluence-induced growth arrest, is it possible that targeting nSMase2 could lead to enhanced primary tumor growth? Could they have adverse effects on nontransformed cells leading to hyperplasia? Along these lines, it is important to note that current cancer therapies are rarely given in isolation and nSMase2 has been strongly connected to the chemotherapeutic response. Thus, it is plausible that targeting nSMase2 to prevent exosome secretion could interfere with the anticancer efficacy of cytotoxic drugs? Similarly, as nSMase2 seems to be an important part of the apoptosis machinery in response to other extracellular stresses, would N-SMase inhibitors enable cancer cells in nutrient or oxygen-poor tumor regions to survive? Taking this further, the roles of nSMase2 in other tissues must also be considered. For example, loss of nSMase2 can adversely affect bone mineralization and chondrocyte differentiation (Khavandgar et al., 2011). Thus, is it possible that nSMase2 inhibitors may adversely affect bone density or the skeleton? These questions are not intended to discourage the pursuit of N-SMase inhibitors as cancer therapeutics but more to encourage researchers to think of the bigger picture, particularly when it pertains to translation of research findings from the bench to the bedside.

4. CONCLUDING REMARKS

The early connections of N-SMase activation to cellular stress and chemotherapy responses, growth arrest, and apoptosis suggested a relatively straightforward paradigm of an anticancer enzyme producing a putative tumor suppressor lipid and potential for manipulating this pathway to enhance drug efficacy and promote cell death. Now, after two decades of research and the identification of multiple isoforms, the role of N-SMases in cancer is a more complicated picture with both pro- and antitumor roles emerging (Fig. 1). Hopefully, the next 20 years will shed more light on these

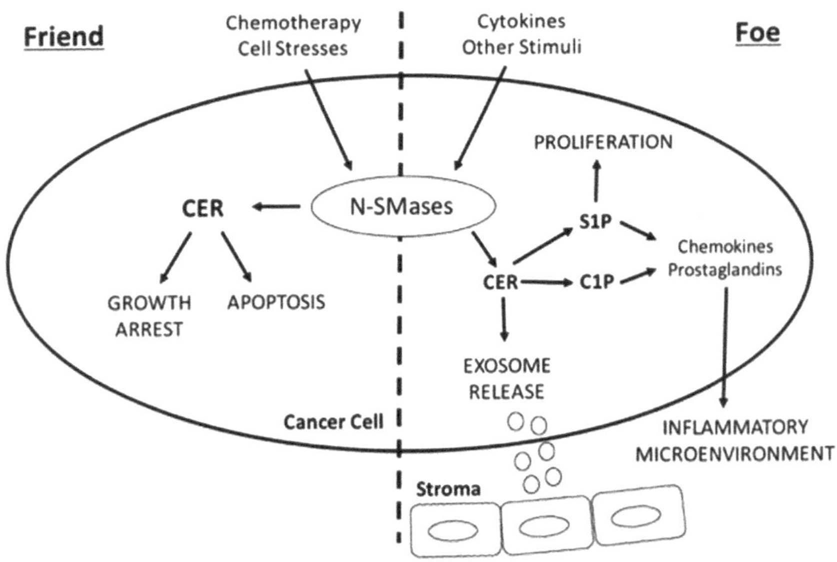

Figure 1 ***The two faces of N-SMases in cancer.*** Current knowledge of the mammalian N-SMases in cancer has suggested they can have both anticancer (friend) and procancer (foe) roles. Increased N-SMase activity and Cer generation has been linked to growth arrest and apoptosis of cancer cells, particularly in the response to chemotherapies. Conversely, N-SMase–mediated exosome release has been shown to promote tumorigenesis through effects on the phenotypes of tumor stromal cells and can also transfer drug resistance to sensitive cancer cells. In addition, cytokine activation of N-SMases could generate proinflammatory lipids and mediators that could adversely affect the tumor microenvironment or promote proliferation through production of S1P. *N-SMases*, neutral sphingomyelinases.

complex roles, the upstream mechanisms of N-SMase regulation and the downstream targets that mediate their functions. Such a wider understanding will be essential to answer the questions of if and when to target the N-SMases for effective cancer therapy.

ACKNOWLEDGMENTS

The author wishes to thank Drs. Michael Airola and Yusuf Hannun for critical reading and review of the manuscript, would like to acknowledge all those scientists who have contributed to the N-SMase field, and to apologize for those whose work he could not mention owing to space limitations.

REFERENCES

Airola, M. V., & Hannun, Y. A. (2013). Sphingolipid metabolism and neutral sphingomyelinases. *Handbook of Experimental Pharmacology, 215*, 57–76.

Airola, M. V., Shanbhogue, P., Shamseddine, A. A., Guja, K. E., Senkal, C. E., Maini, R., et al. (2017). Structure of human nSMase2 reveals an interdomain allosteric activation mechanism for ceramide generation. *Proceedings of the National Academy of Sciences of the United States of America, 114*(28), E5549–E5558.

Amtmann, E., & Zoller, M. (2005). Stimulation of CD95-induced apoptosis in T-cells by a subtype specific neutral sphingomyelinase inhibitor. *Biochemical Pharmacology, 69*, 1141–1148.

Arana, L., Ordoñez, M., Ouro, A., Rivera, I. G., Gangoiti, P., Trueba, M., et al. (2013). Ceramide 1-phosphate induces macrophage chemoattractant protein-1 release: Involvement in ceramide 1-phosphate-stimulated cell migration. *American Journal of Physiology. Endocrinology and Metabolism, 304*(11), E1213–E1226.

Ballou, L. R., Chao, C. P., Holness, M. A., Barker, S. C., & Raghow, R. (1992). Interleukin-1-mediated PGE2 production and sphingomyelin metabolism. Evidence for the regulation of cyclooxygenase gene expression by sphingosine and ceramide. *Journal of Biological Chemistry, 267*(28), 20044–20050.

Barth, B. M., Gustafson, S. J., Hankins, J. L., Kaiser, J. M., Haakenson, J. K., Kester, M., & Kuhn, T. B. (2012). Ceramide kinase regulates TNFα-stimulated NADPH oxidase activity and eicosanoid biosynthesis in neuroblastoma cells. *Cell Signalling, 24*(6), 1126–1133.

Bezombes, C., Ségui, B., Cuvillier, O., Bruno, A. P., Uro-Coste, E., Gouazé, V., et al. (2001). Lysosomal sphingomyelinase is not solicited for apoptosis signaling. *FASEB Journal, 15*(2), 297–299.

Bhati, R., Patterson, C., Livasy, C. A., Fan, C., Ketelsen, D., Hu, Z., et al. (2008). Molecular characterization of human breast tumor vascular cells. *American Journal of Pathology, 172*(5), 1381–1390.

Brann, A. B., Tcherpakov, M., Williams, I. M., Futerman, A. H., & Fainzilber, M. (2002). Nerve growth factor-induced p75-mediated death of cultured hippocampal neurons is age-dependent and transduced through ceramide generated by neutral sphingomyelinase. *Journal of Biological Chemistry, 277*, 9812–9818.

Chen, C. C., Sun, Y. T., Chen, J. J., & Chang, Y. J. (2001). Tumor necrosis factor-alpha-induced cyclooxygenase-2 expression via sequential activation of ceramide-dependent mitogen-activated protein kinases, and IkappaB kinase 1/2 in human alveolar epithelial cells. *Molecular Pharmacology, 59*(3), 493–500.

Chipuk, J. E., McStay, G. P., Bharti, A., Kuwana, T., Clarke, C. J., Siskind, L. J., et al. (2012). Sphingolipid metabolism cooperates with BAK and BAX to promote the mitochondrial pathway of apoptosis. *Cell, 148*(5), 988–1000.

Chmura, S. J., Nodzenski, E., Beckett, M. A., Kufe, D. W., Quintans, J., & Weichselbaum, R. R. (1997). Loss of ceramide production confers resistance to radiation-induced apoptosis. *Cancer Research, 57*(7), 1270–1275.

Clarke, C. J., Cloessner, E. A., Roddy, P. L., & Hannun, Y. A. (2011). Neutral sphingomyelinase-2 is the primary neutral sphingomyelinase activated by tumor necrosis factor-α in MCF-7 cells. *Biochemical Journal, 435*, 381–390.

Clarke, C. J., Mediwala, K., Jenkins, R. W., Sutton, C. A., Tholanikunnel, B., & Hannun, Y. A. (2011). Neutral sphingomyelinase-2 mediates growth arrest by retinoic acid through modulation of S6 kinase. *Journal of Biological Chemistry, 286*(24), 21565–21576.

Clarke, C. J., Shamseddine, A. A., Jacob, J. J., Khalife, G., Burns, T., & Hannun, Y. A. (2016). All trans retinoic acid (ATRA) transcriptionally induces nSMase2 through CBP/p300-mediated histone acetylation. *Journal of Lipid Research, 57*, 868–881.

Clarke, C. J., Truong, T.-G., & Hannun, Y. A. (2007). Role for neutral sphingomyelinase-2 in tumor necrosis factor alpha-stimulated expression of vascular cell adhesion molecule-1 (VCAM1) and intercellular adhesion molecule-1 (ICAM1) in lung epithelial cells: p38 MAPK is an upstream regulator of nSMase2. *Journal of Biological Chemistry, 282*, 1384–1396.

Corcoran, C. A., He, Q., Ponnusamy, S., Ogretmen, B., Huang, Y., & Sheikh, M. S. (2008). Neutral sphingomyelinase-3 is a DNA damage and nongenotoxic stress-regulated gene that is deregulated in human malignancies. *Molecular Cancer Research, 6*(5), 795–807.

Dbaibo, G. S., Pushkareva, M. Y., Jayadev, S., Schwarz, J. K., Horowitz, J. M., Obeid, L. M., et al. (1995). Retinoblastoma gene product as a downstream target for a ceramide-dependent pathway of growth arrest. *Proceedings of the National Academy of Sciences of the United States of America, 92*(5), 1347–1351.

De Palma, C., Meacci, E., Perrotta, C., Bruni, P., & Clementi, E. (2006). Endothelial nitric oxide synthase activation by tumor necrosis factor alpha through neutral sphingomyelinase 2, sphingosine kinase 1, and sphingosine 1 phosphate receptors: A novel pathway relevant to the pathophysiology of endothelium. *Arteriosclerosis Thrombosis and Vascular Biology, 26*, 99–105.

Delpy, E., Hatem, S. N., Andrieu, N., de Vaumas, C., Henaff, M., Rucker-Martin, C., et al. (1999). Doxorubicin induces slow ceramide accumulation and late apoptosis in cultured adult rat ventricular myocytes. *Cardiovascular Research, 43*(2), 398–407.

Demircan, B., Dyer, L. M., Gerace, M., Lobenhofer, E. K., Robertson, K. D., & Brown, K. D. (2009). Comparative epigenomics of human and mouse mammary tumors. *Genes Chromosomes & Cancer, 48*, 83–97.

Devillard, R., Galvani, S., Thiers, J. C., Guenet, J. L., Hannun, Y., Bielawski, J., et al. (2010). Stress-induced sphingolipid signaling: Role of type-2 neutral sphingomyelinase in murine cell apoptosis and proliferation. *PLoS One, 5*(3), e2896.

Dobierzewska, A., Giltiay, N. V., Sabapathi, S., Karakashian, A. A., & Nikolova-Karakashian, M. N. (2011). Protein phosphatase 2A and neutral sphingomyelinase 2 regulate IRAK-1 protein ubiquitination and degradation in response to interleukin-1beta. *Journal of Biological Chemistry, 286*(37), 32064–32073.

Driouch, K., Dorion-Bonnet, F., Briffod, M., Champeme, M. H., Longy, M., & Lidereau, R. (1997). Loss of heterozygosity on chromosome arm 16q in breast cancer metastases. *Genes Chromosomes & Cancer, 19*(3), 185–191.

Fernández-Ayala, D. J., Martín, S. F., Barroso, M. P., Gómez-Díaz, C., Villalba, J. M., Rodríguez-Aguilera, J. C., et al. (2000). Coenzyme Q protects cells against serum withdrawal-induced apoptosis by inhibition of ceramide release and caspase-3 activation. *Antioxidants & Redox Signaling, 2*(2), 263–275.

Figuera-Losada, M., Stathis, M., Dorskind, J. M., Thomas, A. G., Bandaru, V. V., Yoo, S. W., et al. (2015). Cambinol, a novel inhibitor of neutral sphingomyelinase 2 shows neuroprotective properties. *PLoS One, 10*(5), e0124481.

Gorelik, A., Liu, F., Illes, K., & Nagar, B. (2017). Crystal structure of the human alkaline sphingomyelinase provides insights into substrate recognition. *Journal of Biological Chemistry, 292*(17), 7087–7094.

Graeler, M., Shankar, G., & Goetzl, E. J. (2002). Cutting edge: suppression of T cell chemotaxis by sphingosine 1-phosphate. *Journal of Immunology, 169*(8), 4084–4087.

Gude, D. R., Alvarez, S. E., Paugh, S. W., Mitra, P., Yu, J., Griffiths, R., et al. (2008). Apoptosis induces expression of sphingosine kinase 1 to release sphingosine-1-phosphate as a "come-and-get-me" signal. *FASEB Journal, 22*(8), 2629–2638.

Hanahan, D., & Weinberg, R. A. (2011). Hallmarks of cancer: The next generation. *Cell, 144*(5), 646–674.

Hannun, Y. A., Loomis, C. R., Merrill, A. H., Jr., & Bell, R. M. (1986). Sphingosine inhibition of protein kinase C activity and of phorbol dibutyrate binding in vitro and in human platelets. *Journal of Biological Chemistry, 261*(27), 12604–12609.

Hofmann, K., Tomiuk, S., Wolff, G., & Stoffel, W. (2000). Cloning and characterization of the mammalian brain-specific, Mg2+-dependent neutral sphingomyelinase. *Proceedings of the National Academy of Sciences of the United States of America, 97*(11), 5895–5900.

Ito, H., Murakami, M., Furuhata, A., Gao, S., Yoshida, K., Sobue, S., et al. (2009). Transcriptional regulation of neutral sphingomyelinase 2 gene expression of a human breast cancer cell line, MCF-7, induced by the anti-cancer drug, daunorubicin. *Biochimica et Biophysica Acta, 1789*, 681–690.

Jaffrézou, J. P., Levade, T., Bettaïeb, A., Andrieu, N., Bezombes, C., Maestre, N., et al. (1996). Daunorubicin-induced apoptosis: Triggering of ceramide generation through sphingomyelin hydrolysis. *EMBO Journal, 15*(10), 2417–2424.

Jayadev, S., Liu, B., Bielawska, A. E., Lee, J. Y., Nazaire, F., Pushkareva, M., et al. (1995). Role for ceramide in cell cycle arrest. *Journal of Biological Chemistry, 270*, 2047–2052.

Juric, V., Chen, C. C., & Lau, L. F. (2012). TNFα-induced apoptosis enabled by CCN1/CYR61: Pathways of reactive oxygen species generation and cytochrome c release. *PLoS One, 7*(2), e31303.

Karakashian, A. A., Giltiay, N. V., Smith, G. M., & Nikolova-Karakashian, M. N. (2004). Expression of neutral sphingomyelinase-2 (NSMase-2) in primary rat hepatocytes modulates IL-beta-induced JNK activation. *FASEB Journal, 18*, 968–970.

Katz, S., Ernst, O., Avni, D., Athamna, M., Philosoph, A., Arana, L., et al. (2016). Exogenous ceramide-1-phosphate (C1P) and phospho-ceramide analogue-1 (PCERA-1) regulate key macrophage activities via distinct receptors. *Immunology Letters, 169*, 73–81.

Khavandgar, Z., Poirier, C., Clarke, C. J., Li, J. J., Wang, N., McKee, M. D., et al. (2011). A cell autonomous role for sphingomyelin phosphodiesterase 3 in bone mineralization. *Journal of Cell Biology, 194*(2), 277–289.

Kim, W. J., Okimoto, R. A., Purton, L. E., Goodwin, M., Haserlat, S. M., Dayyani, F., et al. (2008). Mutations in the neutral sphingomyelinase gene SMPD3 implicate the ceramide pathway in human leukemias. *Blood, 111*(9), 4716–4722.

Klapisz, E., Masliah, J., Béréziat, G., Wolf, C., & Koumanov, K. S. (2000). Sphingolipids and cholesterol modulate membrane susceptibility to cytosolic phospholipase A(2). *Journal of Lipid Research, 41*(10), 1680–1688.

Kong, J. N., He, Q., Wang, G., Dasgupta, S., Dinkins, M. B., Zhu, G., et al. (2015). Guggulsterone and bexarotene induce secretion of exosome-associated breast cancer resistance protein and reduce doxorubicin resistance in MDA-MB-231 cells. *International Journal of Cancer, 137*(7), 1610–1620.

Kosaka, N., Iguchi, H., Hagiwara, K., Yoshioka, Y., Takeshita, F., & Ochiya, T. (2013). Neutral sphingomyelinase 2 (nSMase2)-dependent exosomal transfer of angiogenic microRNAs regulate cancer cell metastasis. *Journal of Biological Chemistry, 288*(15), 10849–10859.

Kosaka, N., Iguchi, H., Yoshioka, Y., Takeshita, F., Matsuki, Y., & Ochiya, T. (2010). Secretory mechanisms and intercellular transfer of microRNAs in living cells. *Journal of Biological Chemistry, 285*(23), 17442–17452.

Krut, O., Wiegmann, K., Kashkar, H., Yazdanpanah, B., & Krönke, M. (2006). Novel tumor necrosis factor-responsive mammalian neutral sphingomyelinase-3 is a C-tail-anchored protein. *Journal of Biological Chemistry, 281*(19), 13784–13793.

Kubota, S., Chiba, M., Watanabe, M., Sakamoto, M., & Watanabe, N. (2015). Secretion of small/microRNAs including miR-638 into extracellular spaces by sphingomyelin phosphodiesterase 3. *Oncology Reports, 33*(1), 67–73.

Laulederkind, S. J., Bielawska, A., Raghow, R., Hannun, Y. A., & Ballou, L. R. (1995). Ceramide induces interleukin 6 gene expression in human fibroblasts. *Journal of Experimental Medicine, 182*(2), 599–604.

Lee, J. Y., Bielawska, A. E., & Obeid, L. M. (2000). Regulation of cyclin-dependent kinase 2 activity by ceramide. *Experimental Cell Research, 261*(2), 303–311.

Lee, D. H., Kim, S. H., Ahn, K. H., Kim, S. K., Choi, J. M., Ji, J. E., et al. (2011). Identification and evaluation of neutral sphingomyelinase 2 inhibitors. *Archives of Pharmacal Research, 34*(2), 229–236.

Liao, J., Liu, R., Shi, Y. J., Yin, L. H., & Pu, Y. P. (2016). Exosome-shuttling microRNA-21 promotes cell migration and invasion-targeting PDCD4 in esophageal cancer. *International Journal of Oncology, 48*(6), 2567–2579.

Liu, T., Chen, G., Sun, D., Lei, M., Li, Y., Zhou, C., et al. (2017). Exosomes containing miR-21 transfer the characteristic of cisplatin resistance by targeting PTEN and PDCD4 in oral squamous cell carcinoma. *Acta Biochimica Biophysica Sinica (Shanghai), 49*(9), 808–816.

Liu, H., Wu, X., Dong, Z., Luo, Z., Zhao, Z., Xu, Y., et al. (2013). Fatty acid synthase causes drug resistance by inhibiting TNF-α and ceramide production. *Journal of Lipid Research, 54*(3), 776–785.

Lozano, J., Menendez, S., Morales, A., Ehleiter, D., Liao, W. C., Wagman, R., et al. (2001). Cell autonomous apoptosis defects in acid sphingomyelinase knockout fibroblasts. *Journal of Biological Chemistry, 276*(1), 442–448.

Luberto, C., Hassler, D. F., Signorelli, P., Okamoto, Y., Sawai, H., Boros, E., et al. (2002). Inhibition of tumor necrosis factor-induced cell death in MCF7 by a novel inhibitor of neutral sphingomyelinase. *Journal of Biological Chemistry, 277*(43), 41128–41139.

Maceyeka, M., & Spiegel, S. (2014). Sphingolipid metabolites in inflammatory disease. *Nature, 510*, 58–67.

Maeda, Y., Matsuyuki, H., Shimano, K., Kataoka, H., Sugahara, K., & Chiba, K. (2007). Migration of CD4 T cells and dendritic cells toward sphingosine 1-phosphate (S1P) is mediated by different receptor subtypes: S1P regulates the functions of murine mature dendritic cells via S1P receptor type 3. *Journal of Immunology, 178*(6), 3437–3446.

Magnoni, C., Euclidi, E., Benassi, L., Bertazzoni, G., Cossarizza, A., Seidenari, S., et al. (2002). Ultraviolet B radiation induces activation of neutral and acidic sphingomyelinases and ceramide generation in cultured normal human keratinocytes. *Toxicology In Vitro, 16*(4), 349–355.

Marchesini, N., Luberto, C., & Hannun, Y. A. (2003). Biochemical properties of mammalian neutral sphingomyelinase 2 and its role in sphingolipid metabolism. *Journal of Biological Chemistry, 278*, 13775–13783.

Marchesini, N., Osta, W., Bielawski, J., Luberto, C., Obeid, L. M., & Hannun, Y. A. (2004). Role for mammalian neutral sphingomyelinase 2 in confluence-induced growth arrest of MCF7 cells. *Journal of Biological Chemistry, 279*, 25101—25111.
Martin, S. F., Williams, N., & Chaterjee, S. (2006). Lactosylceramide is required in apoptosis induced by N-SMase. *Glycoconjugate Journal, 23*(3—4), 147—157.
Maupas-Schwalm, F., Bedel, A., Augé, N., Grazide, M. H., Mucher, E., Thiers, J. C., et al. (2009). Integrin alpha(v) beta(3), metalloproteinases, and sphingomyelinase-2 mediate urokinase mitogenic effect. *Cell Signalling, 21*, 1925—1934.
Menck, K., Sönmezer, C., Worst, T. S., Schulz, M., Dihazi, G. H., Streit, F., et al. (2017). Neutral sphingomyelinases control extracellular vesicles budding from the plasma membrane. *Journal of Extracellular Vesicles, 6*(1), 1378056.
Meyers-Needham, M., Lewis, J., Gencer, S., Sentelle, R. D., Clarke, C. J., Hannun, Y. A., et al. (2012). Off-target function of the Sonic hedgehog inhibitor cyclopamine in mediating apoptosis via nitric oxide-dependent neutral sphingomyelinase 2/ceramide induction. *Molecular Cancer Therapeutics, 11*, 1092—1102.
Moylan, J. S., Smith, J. D., Wolf Horrell, E. M., McLean, J. B., Deevska, G. M., Bonnell, M. R., et al. (2014). Neutral sphingomyelinase-3 mediates TNF-stimulated oxidant activity in skeletal muscle. *Redox Biology, 2*, 910—920.
Nagahashi, M., Yamada, A., Miyazaki, H., Allegood, J. C., Tsuchida, J., Aoyagi, T., et al. (2016). Interstitial fluid sphingosine-1-phosphate in murine mammary gland and cancer and human breast tissue and cancer determined by novel methods. *Journal of Mammary Gland Biology and Neoplasia, 21*, 9—17.
Nemoto, S., Nakamura, M., Osawa, Y., Kono, S., Itoh, Y., Okano, Y., et al. (2009). Sphingosine kinase isoforms regulate oxaliplatin sensitivity of human colon cancer cells through ceramide accumulation and Akt activation. *Journal of Biological Chemistry, 284*(16), 10422—10432.
Nikolova-Karakashian, M. N., & Rozenova, K. (2010). Ceramide in stress response. *Advances in Experimental Medicine and Biology, 688*, 86—108.
Noda, S., Yoshimura, S., Sawada, M., Naganawa, T., Iwama, T., Nakashima, S., et al. (2001). Role of ceramide during cisplatin-induced apoptosis in C6 glioma cells. *Journal of Neuro-oncology, 52*(1), 11—21.
Ogretmen, B. (2018). Sphingolipid metabolism in cancer signalling and therapy. *Nature Reviews Cancer, 18*(1), 33—50.
Ogretmen, B., Kraveka, J. M., Schady, D., Usta, J., Hannun, Y. A., & Obeid, L. M. (2001). Role of ceramide in mediating the inhibition of telomerase activity in A549 human lung adenocarcinoma cells. *Journal of Biological Chemistry, 276*, 32506—32514.
Ohshima, K., Kanto, K., Hatakeyama, K., Ide, T., Wakabayashi-Nakao, K., Watanabe, Y., et al. (2014). Exosome-mediated extracellular release of polyadenylate-binding protein 1 in human metastatic duodenal cancer cells. *Proteomics, 14*(20), 2297—2306.
Patel, F., & Spassieva, S. D. (2018). Side effects in cancer therapy: are sphingolipids to blame?. In C. E. Chalfant, & P. B. Fisher (Eds.), *Advances in Cancer Research* (Vol. 140) (in this volume).
Pettus, B. J., Bielawska, A., Spiegel, S., Roddy, P., Hannun, Y. A., & Chalfant, C. E. (2003). Ceramide kinase mediates cytokine- and calcium ionophore-induced arachidonic acid release. *Journal of Biological Chemistry, 278*(40), 38206—38213.
Pettus, B. J., Bielawski, J., Porcelli, A. M., Reames, D. L., Johnson, K. R., Morrow, J., et al. (2003). The sphingosine kinase 1/sphingosine-1-phosphate pathway mediates COX-2 induction and PGE2 production in response to TNF-alpha. *FASEB Journal, 17*(11), 1411—1421.
Phuyal, S., Hessvik, N. P., Skotland, T., Sandvig, K., & Llorente, A. (2014). Regulation of exosome release by glycosphingolipids and flotillins. *FEBS Letters, 281*(9), 2214—2217.

Revill, K., Wang, T., Lachenmayer, A., Kojima, K., Harrington, A., Li, J., et al. (2013). Genome-wide methylation analysis and epigenetic unmasking identify tumor suppressor genes in hepatocellular carcinoma. *Gastroenterology, 145*, 1424–1435.

Rutkute, K., Karakashian, A. A., Giltiay, N. V., Dobierzewska, A., & Nikolova-Karakashian, M. N. (2007). Aging in rat causes hepatic hyperresposiveness to interleukin-1beta which is mediated by neutral sphingomyelinase-2. *Hepatology, 46*, 1166–1176.

Sawai, H., Domae, N., Nagan, N., & Hannun, Y. A. (1999). Function of the cloned putative neutral sphingomyelinase as lyso-platelet activating factor-phospholipase C. *Journal of Biological Chemistry, 274*(53), 38131–38139.

Schiffmann, S., Sandner, J., Birod, K., Wobst, I., Angioni, C., Ruckhäberle, E., et al. (2009). Ceramide synthases and ceramide levels are increased in breast cancer tissue. *Carcinogenesis, 30*(5), 745–752.

Schneider, P. B., & Kennedy, E. P. (1967). Sphingomyelinase in normal human spleens and in spleens from subjects with Niemann-Pick disease. *Journal of Lipid Research, 8*, 202–209.

Schuchman, E. H., Levran, O., Pereira, L. V., & Desnick, R. J. (1992). Structural organization and complete nucleotide sequence of the gene encoding human acid sphingomyelinase (SMPD1). *Genomics, 12*(2), 197–205.

Shamseddine, A. A., Airola, M. V., & Hannun, Y. A. (2015). Roles and regulation of neutral sphingomyelinase-2 in cellular and pathological process. *Advances in Biological Regulation, 57*, 24–41.

Shamseddine, A. A., Clarke, C. J., Carroll, B., Airola, M. V., Mohammed, S., Rella, A., et al. (2015). Neutral sphingomylinase-2 (nSMase2) is the primary sphingomyelinase activated by doxorubicin via a p53-dependent mechanism. *Cell Death & Disease, 6*, e1947.

Singh, R., Pochampally, R., Watabe, K., Lu, Z., & Mo, Y. Y. (2014). Exosome-mediated transfer of miR-10b promotes cell invasion in breast cancer. *Molecular Cancer, 13*, 256.

Spassieva, S. D., Rahmaniyan, M., Bielawski, J., Clarke, C. J., Kraveka, J. M., & Obeid, L. M. (2012). Cell density-dependent inhibition of diyhydroceramide desaturase in neuroblastoma cells. *Journal of Lipid Research, 53*(5), 918–928.

Subbaramaiah, K., Chung, W. J., & Dannenberg, A. J. (1998). Ceramide regulates the transcription of cyclooxygenase-2. Evidence for involvement of extracellular signal-regulated kinase/c-Jun N-terminal kinase and p38 mitogen-activated protein kinase pathways. *Journal of Biological Chemistry, 273*(49), 32943–32949.

Tellier, E., Nègre-Salvayre, A., Bocquet, B., Itohara, S., Hannun, Y. A., Salvayre, R., et al. (2007). Role for furin in tumor necrosis factor alpha-induced activation of the matrix metalloproteinase/sphingolipid mitogenic pathway. *Molecular and Cellular Biology, 27*(8), 2997–3007.

Tepper, A. D., Ruurs, P., Borst, J., & van Blitterswijk, W. J. (2001). Effect of overexpression of a neutral sphingomyelinase on CD95-induced ceramide production and apoptosis. *Biochemical and Biophysical Research, 280*, 634–639.

Tomiuk, S., Hofmann, K., Nix, M., Zumbansen, M., & Stoffel, W. (1998). Cloned mammalian neutral sphingomyelinase: Functions in sphingolipid signaling? *Proceedings of the National Academy of Sciences of the United States of America, 95*(7), 3638–3643.

Trajkovic, K., Hsu, C., Chiantia, S., Rajendran, L., Wenzel, D., Wieland, F., et al. (2008). Ceramide triggers budding of exosome vesicles into multivesicular endosomes. *Science, 319*, 1244–1247.

Venable, M. E., Lee, J. Y., Smyth, M. J., Bielawska, A., & Obeid, L. M. (1995). Role of ceramide in cellular senescence. *Journal of Biological Chemistry, 270*, 30701–30708.

Wang, J., Li, J., Gu, J., Yu, J., Guo, S., Zhu, Y., & Ye, D. (2015). Abnormal methylation status of FBXW10 and **SMPD3**, and associations with clinical characteristics in clear cell renal cell carcinoma. *Oncol. Lett, 10*(5), 3073–3080.

Werneburg, N., Guicciardi, M. E., Yin, X. M., & Gores, G. J. (2004). TNF-alpha-mediated lysosomal permeabilization is FAN and caspase 8/Bid dependent. *American Journal of Physiology — Gastrointestinal and Liver Physiology, 287*(2), G436–G443.

Won, J. S., Im, Y. B., Khan, M., Singh, A. K., & Singh, I. (2004). The role of neutral sphingomyelinase produced ceramide in lipopolysaccharide-mediated expression of inducible nitric oxide synthase. *Journal of Neurochemsitry, 88*, 583–593.

Wu, B. X., Clarke, C. J., & Hannun, Y. A. (2010). Neutral sphingomyelinases: Regulation and roles in cell signaling responses. *NeuroMolecular Medicine, 12*(4), 320–330.

Wu, M., Harvey, K. A., Ruzmetov, N., Welch, Z. R., Sech, L., Jackson, K., et al. (2005). Omega-3 polyunsaturated fatty acids attenuate breast cancer growth through activation of a neutral sphingomyelinase-mediated pathway. *International Journal of Cancer, 117*(3), 340–348.

Wu, B. X., Rajagopalan, V., Roddy, P. L., Clarke, C. J., & Hannun, Y. A. (2010b). Identification and characterization of murine mitochondrial-associated sphingomyelinase (MA-nSMase), the mammalian sphingomyelin phosphodiesterase 5. *Journal of Biological Chemistry, 285*(23), 17993–18002.

Yabu, T., Shiba, H., Shibasaki, Y., Nakanishi, T., Imamura, S., Touhata, K., et al. (2015). Stress-induced ceramide generation and apoptosis via the phosphorylation and activation of nSMase1 by JNK signaling. *Cell Death & Differentiation, 22*(2), 258–273.

Yabu, T., Shimuzu, A., & Yamashita, M. (2009). A novel mitochondrial sphingomyelinase in zebrafish cells. *Journal of Biological Chemistry, 284*(30), 29971–29982.

Yang, L., & Lin, P. C. (2017). Mechanisms that drive inflammatory tumor microenvironment, tumor heterogeneity, and metastatic progression. *Seminal Cancer Biology, 47*, 185–195.

Zeidan, Y. H., Pettus, B. J., Elojeimy, S., Taha, T., Obeid, L. M., Kawamori, T., et al. (2006). Acid ceramidase but not acid sphingomyelinase is required for tumor necrosis factor-{alpha}-induced PGE2 production. *Journal of Biological Chemistry, 281*(34), 24695–24703.

Zhang, J., Alter, N., Reed, J. C., Borner, C., Obeid, L. M., & Hannun, Y. A. (1996). Bcl-2 interrupts the ceramide-mediated pathway of cell death. *Proceedings of the National Academy of Sciences of the United States of America, 93*, 5325–5328.

Zhao, H., Achreja, A., Lessi, E., Logozzi, M., Mizzoni, D., Di Raimo, R., et al. (2018). The key role of extracellular vesicles in the metastatic process. *Biochimica et Biophysica Acta, 1869*(1), 64–77.

CHAPTER FIVE

Ceramide and Exosomes: A Novel Target in Cancer Biology and Therapy

Ahmed Elsherbini and Erhard Bieberich[1]
Department of Physiology, University of Kentucky, Lexington, KY, United States
[1]Corresponding author: E-mail: erhard.bieberich@uky.edu

Contents

Abstract

Exosomes are secreted extracellular vesicles (EVs) that carry micro RNAs and other factors to reprogram cancer cells and tissues affected by cancer. Exosomes are exchanged between cancer cells and other tissues, often to prepare a premetastatic niche, escape immune surveillance, or spread multidrug resistance. Only a few studies investigated the function of lipids in exosomes although their lipid composition is different from that of the secreting cells. Ceramide is one of the lipids critical for exosome formation, and it is also enriched in these EVs. New research suggests that lipids in the exosomal membrane may organize and transmit "mobile rafts" that turn exosomes into extracellular signalosomes spreading activation of cell signaling pathways in oncogenesis and metastasis. Ceramide may modulate the function of mobile rafts and their effect on these cell signaling pathways. The critical role of lipids and, in particular, ceramide for formation, secretion, and function of exosomes may lead to a radically new understanding of cancer biology and therapy.

Advances in Cancer Research, Volume 140
ISSN 0065-230X
https://doi.org/10.1016/bs.acr.2018.05.004

1. GOT EXOSOMES: WHAT'S (REALLY) IN YOUR PREP?

Exosomes are a type of extracellular vesicles (EVs) that are formed as intraluminal vesicles in multivesicular endosomes (MVEs). The MVEs fuse with the plasma membrane and release these vesicles to the extracellular space (Fig. 1A). It is assumed that there are two different pathways, leading to exosome formation, endosomal sorting complexes required for transport (ESCRT)-dependent and independent (Colombo et al., 2013; Colombo, Raposo, & Thery, 2014; Hurley, 2015; Hurley & Odorizzi, 2012; Juan & Furthauer, 2017; Kowal, Tkach, & Thery, 2014; Marsh & van Meer, 2008; Stoorvogel, 2015; Trajkovic et al., 2008; Villarroya-Beltri, Baixauli, Gutierrez-Vazquez, Sanchez-Madrid, & Mittelbrunn, 2014). ESCRT proteins, such as Alix1 and TSG101, form a neck-shaped complex with

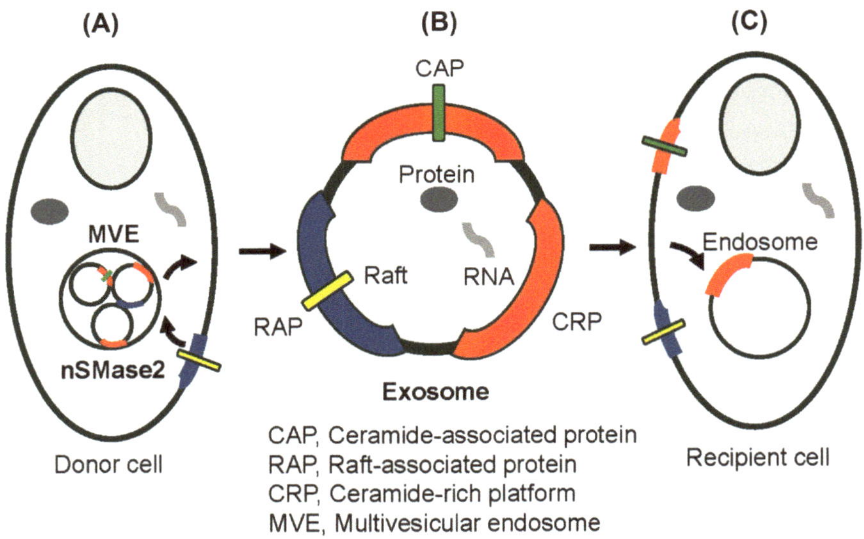

Figure 1 Ceramide in exosome formation and the concept of mobile rafts. (A) nSMase2 generates ceramide that induces formation of ceramide-enriched intraluminal vesicles in MVEs. These vesicles can contain endocytosed lipid rafts and are secreted as exosomes after fusion of MVEs with the plasma membrane of the donor cell. In addition to lipids, RNA and protein is packaged into exosomes. (B) Exosomes can contain mobile rafts with raft-associated protein (RAP) and ceramide-rich platforms (CRPs) with ceramide-associated protein (CAP). (C) Exosomes are taken up by recipient cells via endocytosis, pinocytosis, or fusion with the plasma membrane. Uptake leads to incorporation of lipid rafts or CRPs into the plasma membrane or the endosome. RNA and proteins inside of exosomes are released to the cytosol. *MVE*, multivesicular endosome; *nSMase2*, neutral sphingomyelinase 2.

other proteins such as vacuolar sorting-associated protein 4, initiating budding of exosomes at the MVE membrane. ESCRT-independent exosome formation relies on ceramide generation by neutral sphingomyelinase 2 (nSMase2), a key cell signaling enzyme (Fig. 1A). However, this distinction may not be as strict as previously thought, and the same cell type can secrete different types of exosomes. Exosomes are distinct from microvesicles, another type of EVs that are released by blebbing off the plasma membrane. These two types of EVs are different not only in size, function, and cargo but also in the intrinsic signals leading to their secretion (for comprehensive reviews on EV formation and secretion, see Desrochers, Antonyak, & Cerione, 2016; Edgar, 2016; Hyenne, Lefebvre, & Goetz, 2017; Kowal et al., 2014; van Niel, D'Angelo, & Raposo, 2018; Raposo & Stoorvogel, 2013; Stoorvogel, 2015; Tkach, Kowal, & Thery, 2018; Tkach & Thery, 2016). Exosomes carry RNA (micro RNA [miRNA], long noncoding RNA, messenger RNA), mitochondrial DNA, signaling proteins, enzymes, metabolites, and lipids from the donor to a recipient cell (Choi, Kim, Kim, & Gho, 2013; Kai, Dittmar, & Sen, 2017; Kinoshita, Yip, Spence, & Liu, 2017; Record, Carayon, Poirot, & Silvente-Poirot, 2014; Salehi & Sharifi, 2018; Zhang & Grizzle, 2014; Fig. 1). The recipient cell takes up exosomes by endocytosis, pinocytosis, or simply fusing the exosomal with the plasma membrane (Abels & Breakefield, 2016; Costa Verdera, Gitz-Francois, Schiffelers, & Vader, 2017; Horibe, Tanahashi, Kawauchi, Murakami, & Rikitake, 2018; Jiang, Li, Li, & Xia, 2015; McKelvey, Powell, Ashton, Morris, & McCracken, 2015; Tian et al., 2013; Tian et al., 2014). A multitude of different effects of exosomes was reported. In cancer, exosomes reprogram tissue to allow growth of metastases, act as decoys to escape the immune system, or spread factors conferring resistance to drugs used in chemotherapy (Andaloussi, Mager, Breakefield, & Wood, 2013; Chiarugi & Cirri, 2016; Desrochers et al., 2016; Dorsam, Reiners, & von Strandmann, 2018; Dreyer & Baur, 2016; Kahlert & Kalluri, 2013; Kalluri, 2016; Ruivo, Adem, Silva, & Melo, 2017; Silva & Melo, 2015; Steinbichler, Dudas, Riechelmann, & Skvortsova, 2017; Weidle, Birzele, Kollmorgen, & Ruger, 2017; Zhang & Grizzle, 2014; Zhao et al., 2017). Exosomes secreted by tumor cells can be found in the blood and urine, which makes analysis of exosomal content attractive for early cancer diagnosis in "liquid biopsy" (Kai et al., 2017; Minciacchi, Freeman, & Di Vizio, 2015; Minciacchi, Zijlstra, Rubin, & Di Vizio, 2017; Sharma, Khatun, & Shiras, 2016; Yokoi, Yoshioka, & Ochiya, 2015; Yoshioka et al., 2014; Zhang et al., 2017). Since exosomes are the physiological equivalent to liposomes, they can be used as vector to deliver miRNAs and drugs or for cell-free vaccination in cancer therapy (Pitt et al., 2014; Schorey & Bhatnagar, 2008; Viaud et al., 2010). Our own research showed that stimulating exosome secretion from breast cancer stem-like cells can break their resistance to chemotherapy (Kong, He, et al., 2015). Others have reported that exosomes from stem cells can be used for "cell-free stem cell therapy" (Phinney & Pittenger, 2017; Rani, Ryan, Griffin, & Ritter, 2015). However, most of the in vitro studies rely on the preparation of exosomes from medium of a (large) donor cell culture to incubate a (smaller) recipient culture. Despite the exploding number of publications on exosomes, only a few studies strived to answer a simple but fundamental biochemical question: what is the stoichiometry between the effector molecule and the effect? In other words, is the number of exosomes in a given tissue really sufficient to recapitulate in vivo the effects seen in vitro? The simple answer to this question is "no one really knows since the in vivo function of exosomes is largely unclear and difficult to test."

The difficulty in testing the function of exosomes stems from the dirty secret of their preparation; exosomes are intrinsically inhomogeneous. Most exosome preparations rely on two methods based on centrifugation: differential ultracentrifugation and low-speed centrifugation using a polymer matrix, such as polyethylene glycol (for reviews on exosome preparation, see Brown & Yin, 2017; Carpintero-Fernandez, Fafian-Labora, & O'Loghlen, 2017; Coumans et al., 2017; Helwa et al., 2017; P. Li, Kaslan, Lee, Yao, & Gao, 2017; Taylor & Shah, 2015; Xu, Greening, Zhu, Takahashi, & Simpson, 2016; Zeringer, Barta, Li, & Vlassov, 2015). These methods do not completely remove other vesicular and particulate contaminants, such as microvesicles, lipoproteins, or simply cell debris. This is not as much of a problem for in vitro studies when harvesting exosomes secreted by cultured cells into the medium supernatant. Provided that exosome- and lipoprotein-free serum is used for cultivation of cells; main contaminants are cell debris (including apoptotic bodies) and microvesicles. The bulk of these contaminants can be removed by precentrifugation at 20,000× g or filtration of the cell culture medium at 0.2 μm prior to ultracentrifugation at 110,000× g. Microvesicles are still a concern since their size varies broadly around 1 μm, while exosomes constitute a population of vesicles with a narrow size range from 50 to 150 nm. Density-gradient ultracentrifugation improves exosome purity; however, densities of lipoproteins and different vesicle population are too close to being efficiently separated. Exosome purity is also a concern when using polymers for dehydration leading to precipitation of exosomes at low-speed centrifugation since these methods do not distinguish between different vesicle populations. Therefore, exosomes and microvesicles are often combined under the term of EVs, with some of the functions such as delivery of effector molecules being shared between these two vesicle populations (for reviews comparing exosomes with microvesicles, see Lee, El Andaloussi, & Wood, 2012; P. Li et al., 2017; Minciacchi et al., 2015; Raposo & Stoorvogel, 2013). The problem of purity is even more serious when it comes to preparation of exosomes from tissues and bodily fluids, such as blood, urine, or cerebrospinal fluid. Tissue dissociation using proteases and gentle mechanical disruption can introduce artifacts by releasing intracellular vesicles into the preparation or contaminating the sample with endogenous and exogenously added proteases that associate with exosomes. While exosome preparation from serum is still easier than that from solid tissue, vesicles can be contaminated with lipoprotein particles, which limit interpretability of "liquid biopsies." Immunoprecipitation using antibodies against exosome surface proteins can alleviate this problem and even separate exosomes derived from different cell types. However, the yield is usually lower than with centrifugation techniques and often compromises a quantitative analysis of exosome function (Brett et al., 2017; Momen-Heravi et al., 2013; Nakai et al., 2016; Szatanek, Baran, Siedlar, & Baj-Krzyworzeka, 2015). Recently, great strides were made in the field of exosome isolation, introducing new techniques such as ultrafiltration and microfluidics-based separation. Although each newer technique has its own advantage, they also pose distinct limitation, keeping the isolation of pure and homogenous exosomes a challenging task (He & Zeng, 2016; Momen-Heravi et al., 2013; Pietrowska et al., 2017; Sharma, Scholz-Romero, Rice, & Salomon, 2018; Szatanek et al., 2015; Xu, Simpson, & Greening, 2017). To date, there is no reliable method for efficient purification or sorting of exosomes from tissues, regardless of being soft or more solid, tumor, or healthy tissues.

Methods used in our lab and that of other groups largely rely on a hit-or-miss approach. The purity of a particular exosome preparation is not predictable but has to be rigorously tested after the preparation was accomplished. Apart from mistakenly analyzing exosomes that originate from the tissue's blood supply, main contaminants in exosome preparations from tissues, including tumors, are vesicles released from damaged cells. The postpreparation quality control is critical and encompasses an array of physical tests, detection of exosomal marker proteins, and loss-of-function assays. Must-have physical tests include laser light scattering microscopy, electron microscopy, or tunable resistive pulse sensing to visualize and quantify exosomes (Akers et al., 2016; Coscia et al., 2016; Koritzinsky, Street, Star, & Yuen, 2017; Maas, Broekman, & de Vrij, 2017; Vogel et al., 2016). We use nanoparticle tracking analysis based on laser light scattering microscopy to reliably quantify the number and size distribution of exosomes and other vesicles in our preparation (Coumans et al., 2017; Dinkins, Enasko, et al., 2016; Helwa et al., 2017; Kong, He, et al., 2015; Koritzinsky et al., 2017; Szatanek et al., 2017; Tkach et al., 2018). For a preparation with high purity, the number of exosomes should match up with the quantity of marker proteins such as Alix1, TSG101, Flotillin, or CD63 as detectable in immunoblots. Finally, loss-of-function assays will help exclude other sources of bioactive molecules in the exosomes preparation. These assays include ultraviolet irradiation (to destroy RNA in exosomes) or mild detergent treatment, which either obliterates the effect of exosomes (if based on RNA) or exosomes themselves (Coumans et al., 2017; Kinoshita et al., 2017; Osteikoetxea et al., 2015). Our group has specialized on prepurification loss-of-function assays by preventing formation or secretion of exosomes in vitro and in vivo (Dinkins, Dasgupta, Wang, Zhu, & Bieberich, 2014; Dinkins, Enasko, et al., 2016; Kong, Hardin, et al., 2015; Wang et al., 2012). To date, inhibition or genetic deficiency of nSMase2 is a method widely used to prevent exosome secretion in various cell types and tissues (Chairoungdua, Smith, Pochard, Hull, & Caplan, 2010; Dinkins et al., 2014; Dinkins, Enasko, et al., 2016; Dinkins, Wang, & Bieberich, 2016; Goldkorn, Chung, & Filosto, 2013; Guo, Bellingham, & Hill, 2015; Kong, He, et al., 2015; Kosaka et al., 2010; Marsh & van Meer, 2008; Menck et al., 2017; Shamseddine, Airola, & Hannun, 2015; Tan et al., 2013; Trajkovic et al., 2008; Yuyama, Sun, Mitsutake, & Igarashi, 2012). nSMase2 converts sphingomyelin into ceramide, a reaction shown to be instrumental for exosome secretion in vitro (Trajkovic et al., 2008) (Fig. 1A). Our laboratory showed that inhibition and genetic deficiency of nSMase2 also suppress exosome secretion in vivo (Dinkins et al., 2014; Dinkins, Enasko, et al., 2016; Dinkins, Wang, et al., 2016). These discoveries and the critical role of ceramide for the function of exosomes, particularly in cancer, will be discussed in the next sections of this review.

2. WHY CERAMIDE IN EXOSOMES? IT'S ALL IN THE NUMBERS

The answer to the initial question of just how much of an effector molecule needs to be in exosomes to actually affect the recipient cell has remained controversial. Exosomes are not the only form of transferring bioactive molecules. Prior to the massive upswing of exosome research, most of the intercellular communication was thought to be managed by exchange of humoral factors such growth factors, cytokines, hormones, and other secreted smaller and larger molecules. The importance of these factors does not vanish with the growing importance of exosomes. One has to realize that exosomes simply add to the secretome by stabilizing and combining certain signaling molecules such as miRNAs for long-distance communication and targeted delivery (for review on secretome, including EVs, see Sinha et al., 2018; van der Pol, Boing, Harrison, Sturk, & Nieuwland, 2012; Vizoso, Eiro, Cid, Schneider, & Perez-Fernandez, 2017). Enrichment of specific miRNAs in exosomes may be suitable for early diagnosis in cancer, but it still needs to be tested if the amount of miRNA transferred in exosomes is really sufficient to reprogram tissue for metastasis or fulfill other functions of exosomes found in vitro. If looking for signaling molecules that are transferred in functionally significant quantities one may want to shift focus away from RNA and include other signaling molecules transported by exosomes. But first, one will need to do the numbers to assess the contribution of RNA and other factors to cell signaling in cancer.

Despite the excitement emerging from the function of RNA in reprogramming tissue, their content in exosomes appears to be very low. According to a study from 2014 the abundance of miRNA is less than one molecule per 100 exosomes (Chevillet et al., 2014). In a more recent study by our collaborators and us, very low abundance (copy number) of two specific miRNAs (miRNA-16 and 451) was confirmed using different methods for exosome preparation (Helwa et al., 2017). However, these numbers were based on the assumption that miRNAs were evenly distributed over the entire exosome population. Our study showed that based on an average miRNA length of 22 nucleotides (double-stranded), an average molecular mass of 640 g/mol base pair and recovery of about 25 ng miRNA/5 $\times$ 10^{11} exosomes isolated from 5 mL serum, there is about one-to-two molecules of miRNA in a single exosome (Helwa et al., 2017). With respect to miRNA the concentration of a specific miRNA per cell needs to reach a threshold to efficiently silence its target gene expression (Brown et al., 2007; Mukherji et al., 2011). Titration experiments show that the typical concentrations are in the range of several hundred to more than 1000 copies of a specific miRNA/cell, which is similar to the threshold levels for gene silencing (Bissels et al., 2009). Provided that a specific miRNA copy is less than one in 100 exosomes, one will need at least 10^4 to 10^5 exosomes taken up by a single cell to reach a concentration comparable to endogenous miRNAs. At an average size of 100 nm per exosome and 10 μm per cell (surface ratio 1:10^4), a single cell will have to take up exosomes equivalent to 1–10 times of its own cell surface to reach an efficient copy number of a specific miRNA. The 2014 study provided a potential solution to this dilemma. Since the copy number of miRNA was quantified as an average over the entire exosome population, "low occupancy/high abundance" exosomes may be rare (≤1/100) but could deliver ≥10–100 copies of a specific miRNA/exosome (Chevillet et al., 2014). In this case, taking up the equivalent of ≤1/10 of the cell surface could be sufficient to reach the efficient copy number/cell. This does not appear unreasonable since some cells such as macrophages can pinocytose an equivalent of the cell surface in 33 min (Chevillet et al., 2014). However, this calculation implies two additional predicaments: (1) The concentration of a specific miRNA is massively enriched in exosomes (≥10- to 100-fold), requiring an efficient sorting mechanism. This assumption is supported by a recent study that identified a set of exosomal miRNAs abundant in exosomes derived from human neural stem cells with at least 10 copies per exosome (Stevanato, Thanabalasundaram, Vysokov, & Sinden, 2016). Upon function transfer analysis the group showed that exosomal miRNA can reach the molecular machinery of miRNA-suppressed gene repression at a physiologically relevant level. (2) The recipient cell is specifically targeted by "low occupancy/high abundance" exosomes. Cell culture supernatants of 10^7 donor cells (equivalent to a confluent 100 mm dish with 10 mL medium) can yield about 10^{11} exosomes after overnight incubation (equivalent to number of exosomes in 0.5–1 mL serum (Helwa et al., 2017)). At 1% abundance of a specific miRNA copy, one will have to collect 10^{12} exosomes to achieve 1000 copies miRNA/cell transferred to 10^7 recipient cells, which corresponds to conditioned medium from about 10 dishes (or exosomes from 5 to 10 mL serum). However, if many identical copies are packaged into few exosomes, the "harvest" of 10^7 cells may be enough to transfer a sufficient number of miRNA copies to 10^6 cells. Therefore, one of the main reasons for exosomal miRNA being found to reprogram cells may simply rely on using a large amount of donor cells (or serum exosomes) to incubate a much smaller number of recipient cells. Another reason may result from mistaking the effect of miRNA with that of other more abundant cell signaling molecules transferred by exosomes.

In addition to miRNA, proteins transferred by exosomes were discussed to affect cancer cells (for reviews and exemplary studies on proteomics and protein function in exosomes see Azmi, Bao, & Sarkar, 2013; Choi et al., 2013; Greening, Xu, Gopal, Rai, & Simpson, 2017; Intasqui, Bertolla, & Sadi, 2018; Jakobsen et al., 2015; A. Li, Zhang, Zheng, Liu, & Chen, 2017; van Niel et al., 2018; Pietrowska et al., 2017; Pocsfalvi et al., 2016; Sandfeld-Paulsen, Aggerholm-Pedersen, et al., 2016; Sandfeld-Paulsen, Jakobsen, et al., 2016; Sinha et al., 2018; Tanase et al., 2017; Taylor, Zacharias, & Gercel-Taylor, 2011; Weidle et al., 2017; Fig. 1). For example, exosomal fibronectin can bind to integrin and enhance cell adhesion and prepare a premetastatic niche for fibrosarcoma cells (Sung, Ketova, Hoshino, Zijlstra, & Weaver, 2015; Tkach & Thery, 2016). The effect of exosomes on preparing a premetastic niche will be discussed in more detail in the next section of this review. Among exosomal proteins, enzymes are particularly attractive since it only takes a few molecules to amplify their activity in the target cells. Matrix metalloproteinases (MMPs) are transferred by exosomes and facilitate cancer cell invasion (Hakulinen, Sankkila, Sugiyama, Lehti, & Keski-Oja, 2008; Sanderson, Bandari, & Vlodavsky, 2017; Shay, Lynch, & Fingleton, 2015). However, while stoichiometry is most likely the greatest barrier in understanding the effect of exosomal miRNAs, topology is the problem with most exosomal proteins. Unless they are bound to the exosomal surface such as MT1-MMP, the majority of proteins discussed to promote neoplasia and metastasis are encaged inside of exosomes (Fig. 1). Hence, it is difficult to reconcile the activity of proteins modifying the extracellular matrix (ECM) with topological enclosure inside of exosomes. Solutions such as slow or induced disintegration of the exosomal membrane for controlled release and activation of exosomal enzymes are conceivable, but membrane disintegration is in stark contrast to the presumed stability of exosomes when discussing other functions of exosomes, such as miRNA transfer (Kumeda et al., 2017). Perhaps, there are functionally specialized exosomes that may be prone to release of their contents, while others remain stable and are meant for long-distance transport of bioactive molecules. It was suggested that sphingomyelin is one of the lipids stabilizing the exosomal membrane (Record et al., 2014). Secreted sphingomyelinases may convert sphingomyelin into ceramide and potentially destabilize the exosomal membrane, thereby releasing proteins and other factors enclosed in exosomes. This speculation brings us the next class of molecules that is far less explored, albeit it may offer several explanations for yet unresolved obstacles in understanding exosomes—lipids.

The few in-depth analyses of exosomal lipids have shown that the lipid composition of the exosomal membrane is distinct from that of most cellular membranes in the donor cells (Record et al., 2014). For starters, a simple geometrical correlation between surface and volume of a vesicle implies that the smaller the vesicle, the larger the surface in comparison to volume. Hence, it is not surprising that many lipids appear to be enriched in the exosomal membrane when normalizing lipid to protein content, the latter being mainly correlated with volume. Apparent lipid enrichment due to this geometrical correlation may thus not ignite much excitement, but one should also realize that it implies a major function of exosomes as vector for autocrine and paracrine membrane and lipid transport and exchange between cells. Independent of geometry, there is specific enrichment when comparing the molar proportions of lipids in the exosomal to other cellular membranes. This difference is one of the most important features that distinguishes exosomes from microvesicles, the lipid composition of which is very similar to that of the parental plasma membrane (Record et al., 2014). A recent lipidomics analysis using liquid chromatography—tandem mass spectrometry showed that the molar proportions of cholesterol, glycosphingolipids, and sphingomyelin are enriched by up to 3-fold compared with lipids abundant in other membranes such as phosphatidylcholine, the proportion of which is reduced in exosomes (Record et al., 2014). Lysobisphosphatidic acid, an endosomal membrane lipid also known as bis-monoacylglycerophosphate is exclusively contained in exosomes and was shown to partake in ESCRT-dependent exosome formation and exosomal membrane stability (Record et al., 2014; Tkach & Thery, 2016). However, in contrast to the proposed functions of exosomal miRNA and proteins in cancer, the role of most exosomal lipids has remained unclear.

3. THE INCREDIBLE JOURNEY OF A LIPID: RAFTING THE EXTRACELLULAR SPACE ON EXOSOMES

As discussed in the previous section, ceramide is a sphingolipid that is critical for exosome formation and/or secretion. While it is not understood how ceramide may contribute to the inward budding of vesicles at the MVE, facilitating or inducing membrane curvature may be one of its tasks (for biophysical properties of ceramide in membrane curvature, see Burgert et al., 2017; Goni & Alonso, 2006; Goni, Contreras, Montes, Sot, & Alonso, 2005; Pinto, Silva, Futerman, & Prieto, 2011; Silva et al., 2012). Our studies showed that exosomes contain mainly two ceramide species: C18:0 (about

70% of total ceramide) and C24:1 ceramide, while isolated intracellular vesicles also contain C16:0 ceramide (He et al., 2014; G. Wang et al., 2012). In synthetic vesicles, mixtures of C18:0 and C24:1 ceramide lead to in-plane phase separation at $\geq$1% ceramide, and upregulation of medium-chain ceramides (such as C16:0 and C18:0 ceramide) favors negative curvature (inward budding) of the membrane (Carrer & Maggio, 1999; Carrer, Hartel, Monaco, & Maggio, 2003; Castro, Prieto, & Silva, 2014; Goni & Alonso, 2006; Goni et al., 2005; Pinto et al., 2011). Interaction of very long chain sphingolipids in the outer leaflet with phosphatidylserine in the inner leaflet was suggested to facilitate the enrichment of cholesterol in the exosomal membrane, probably to increase its stability and/or prevent phagocytosis through binding to phosphatidylserine exposed to Annexin V type receptors on macrophages (Skotland, Sandvig, & Llorente, 2017). Anisotropic demixing of membrane lipids is a necessary condition for curvature, suggesting that formation of ceramide microdomains and their inward budding may critically contribute to the intraluminal formation of exosomes in MVEs. Using an anticeramide antibody developed in our laboratory and TEM, we showed that enrichment of ceramide is common to several intraluminal membranes in MVEs, including exosomes and multilamellar vesicles (Fig. 2). Therefore, it appears reasonable to presume that whatever the function of ceramide in exosomes, it is the composition of lipids that is critical. Hence, instead of looking at enrichment of individual lipids, one should rather look at ratios. Lipidomics analyses showed that the molar proportion of ceramide was enriched by 1.3- to 3-fold when compared to that of membranes in the donor cells, which was also found in our studies on exosomes secreted by astrocytes (Record et al., 2014; G. Wang et al., 2012). However, when comparing the proportion of ceramide to cholesterol, we and others found that ceramide was >3-fold more enriched than cholesterol (Record et al., 2014). Based on this observation, the function of ceramide in exosomes may not just rely on its enrichment, but its composition and proportion in comparison to other lipids such as cholesterol.

There are only a few studies showing that uptake of exosomes actually increases the lipid content of the recipient cell (Record et al., 2014). More likely is the localized alteration of the membrane lipid proportion by exosomal lipids (e.g., after fusion with the cell membrane or endocytotic uptake). As discussed in the previous section, membrane lipids are not homogeneously distributed, but they are anisotropically organized in suprastructures, such as lipid rafts, microdomains, or platforms. These terms are often used interchangeably, and the reader interested in further details is

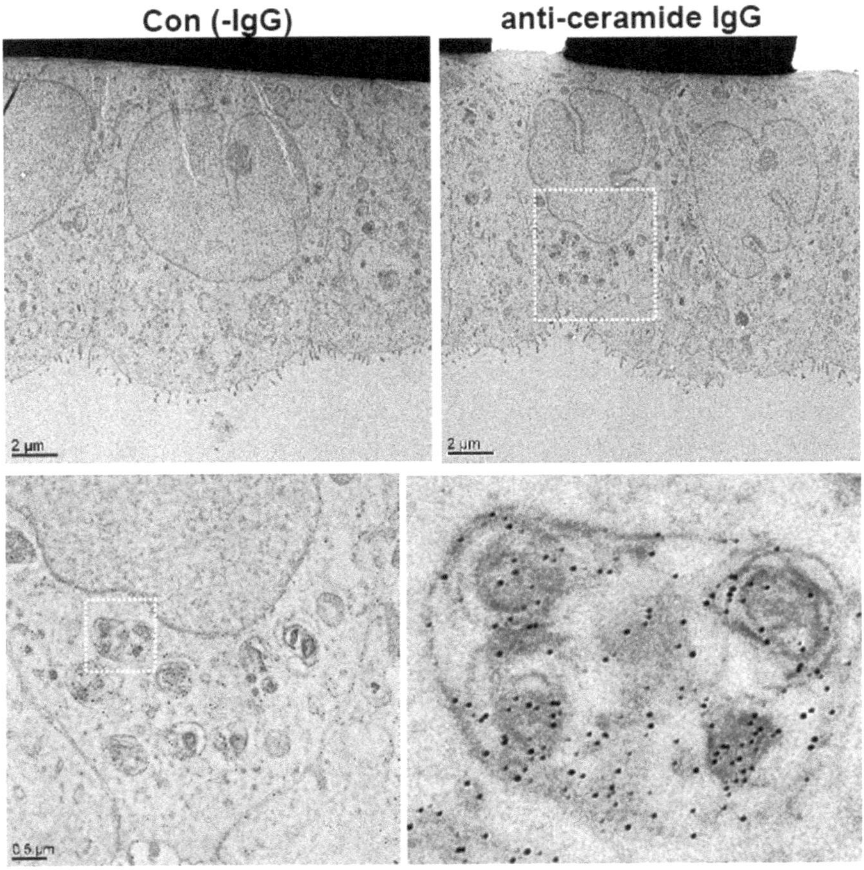

Figure 2 TEM of MVEs in MDCK cells. Anticeramide IgG was used for immunogold labeling. MVEs contain many ceramide-enriched multilamellar and unilamellar luminal vesicles. Unilammellar vesicles are presumed to be secreted as exosomes. *MDCK*, madin-darby canine kidney; *MVE*, multivesicular endosome; *TEM*, transmission electron microscopy.

referred to excellent reviews on this topic (Bieberich, 2008; Gulbins & Kolesnick, 2003; Lingwood & Simons, 2010; Lopez, 2015; Ma, Hinde, & Gaus, 2015; Mollinedo & Gajate, 2015; Owen, Magenau, Williamson, & Gaus, 2012; Simons & Gerl, 2010; Simons & Ikonen, 1997; Simons & Sampaio, 2011; Staubach & Hanisch, 2011; Zhang, Li, Becker, & Gulbins, 2009). In brief, the "classical" raft is about 50–200 nm in diameter, stabilized by complexes between cholesterol and sphingomyelin and enriched in glycosphingolipids such as ganglioside GM1 ("Raft" in Fig. 1B). The main function of suprastructures is to regulate proteins associated with rafts ("raft-associated protein" or RAP in Fig. 1B), such as receptors or caveolins, often modified by attachment to membrane anchors, such as glycosylphosphatidylinositol or palmitoyl residues (Busija, Patel, & Insel, 2017; Lajoie & Nabi, 2010; Simons & Toomre, 2000). As the result, lipid rafts can activate cell signaling pathways or induce endocytosis. In exosome biology, evidence was provided that parts or the entire exosomal membrane is derived from endocytosed lipid rafts (Skotland et al., 2017; Staubach & Hanisch, 2011; Tan et al., 2013; Fig. 1A). Not yet proven, but very likely is enrichment of exosomal lipids in endosomes of the recipient cell after endocytic or pinocytotic uptake (Fig. 1C). Also exciting is the idea that exosomes may actually transfer lipid rafts from the donor to the recipient cell (Fig. 1B). Most recently, research in our and our collaborators' laboratories has shown that ceramide-rich platforms (CRPs) are 50−200 nm in size and contain ≥20 ceramide molecules (Burgert et al., 2017). CRPs are formed when sphingomyelinases convert sphingomyelin into ceramide (Fig. 1A). As discussed, ceramide is ≥3-fold enriched over cholesterol in exosomes, suggesting that exosomes could serve as mobile ceramide-enriched lipid rafts or CRPs (Fig. 1B). The surface ratio between a sphere and a circle of equal diameter is 2, suggesting that the surface of a single 100 nm-sized exosome covers a plasma membrane (approximated as a plane) area of about 200 nm diameter, quite within a size range of larger CRPs. Therefore, it does not appear unreasonable to suggest that one of the functions of exosomes is to exchange lipid rafts between cells.

The main concept underlying these "mobile rafts" is that lipid rafts formed in donor cells are transported via exosomes to the recipient cell, thereby inducing the same cell signaling pathway as activated in the donor cells (Fig. 1). This mechanism is analogous to signalosomes in neurons, vesicles that form after receptor activation in synapse and transport the activated receptor to the cell body (Cosker & Segal, 2014; Cosker, Courchesne, & Segal, 2008). Hence, the function of mobile lipid rafts in regulating cancer cell signaling pathways is broadcasted to distant cells in a tumor or even to metastases. Experimental evidence for mobile rafts is provided by testing if a particular receptor or cell signaling pathway activation in recipient cells is similar to donor cells when exposed to donor-derived exosomes. Even though the overall lipid composition of the recipient cells due to uptake or fusion with exosomes does not change, the suprastructure of the plasma membrane or other cellular membranes may adopt the lipid raft organization of the donor cells (Fig. 1A and C). There are several areas in cancer biology that may benefit from broadcasting cell signaling events via mobile rafts in exosomes: (1) metabolic or morphogenetic reprogramming in a tumor in response to scarcity of resources such as oxygen or glucose; (2) preparation of a premetastatic niche; (3) communication with the immune system to escape surveillance; and (4) acquisition and spreading of resistance to drugs or other tumor suppressive signals. In all these instances, donor and recipient cells may be of the same or different cell types (e.g., cancer to cancer cell, cancer to tissue cell, tissue to cancer cell, or cancer to immune cell and vice versa), and the raft-activated cell signaling pathway may induce the same or a different response in donor and recipient cells. In principle, the range of possible combinations in donor and recipient cells and responses to mobile rafts is similar to the potential of miRNA-mediated programming through exosomes. However, since lipids in the exosomal membrane are abundant and instrumental to the activity of mobile rafts, the effect of exosomal lipids on recipient cells may outperform or at least complement that of miRNAs for reasons discussed in the previous section.

4. EXOSOMES IN CANCER: TARGET TISSUE BIOHACKING AND HIJACKING OF THE IMMUNE SYSTEM

In metabolic and morphogenetic reprogramming, cancer cells are often found to switch from mitochondrial to cytosolic generation of adenosine triphosphate (ATP) by limiting glucose consumption to aerobic glycolysis without invoking oxidative phosphorylation in mitochondria (Gatenby & Gillies, 2004; Liberti & Locasale, 2016). This so-called Warburg effect appears reasonable since oxygen becomes a scarce resource in a growing tumor. In a recent study, it was shown that patient-derived, cancer-associated fibroblasts secrete exosomes that suppress oxygen consumption in pancreatic and prostate cell culture by 80% within 24 h (Rabinowitz & Coller, 2016; Zhao et al., 2016). While the authors of the study hypothesize that miRNA-mediated silencing of oxidative metabolism genes and nutrients delivered by exosomes are accountable for this rapid effect, it was noted that the amount of exosomes and timing do not reconcile with this hypothesis (Rabinowitz & Coller, 2016). However, since lactic acid production is a consequence of aerobic glycolysis and low/acidic pH stimulates exosomes secretion (Parolini et al., 2009), it is possible that a positive feedback rapidly increases the amount of exosomes in the metastatic niche. As an alternative (or complement) to miRNA, mobile rafts transported by exosomes could regulate metabolism in cancer cells. The possibility of reprogramming cancer cell metabolism by mobile rafts has not been discussed yet, but it is known that cell signaling pathways promoting glucose uptake and metabolism, particularly the PI3K/Akt axis, are affected by lipid rafts and ceramide (Mollinedo & Gajate, 2015; Powell, Hajduch, Kular, & Hundal, 2003; Zhou, Summers, Birnbaum, & Pittman, 1998). For example, insulin and insulin-like growth factor 1 (IGF-1) receptors are embedded into lipid rafts and can activate PI3K/Akt to stimulate glucose uptake and fuel glycolysis (Mollinedo & Gajate, 2015). Exosomes from prostate cancer cells were found to spread RAPs (Fig. 1B), such as IGF-1 receptors (DeRita et al., 2017). We showed that the PI3K/Akt activator atypical protein kinase C is a ceramide-associated protein transported by exosomes, probably when bound to CRPs in mobile rafts (Fig. 1B; Wang et al., 2012). On the other hand, ceramide can inhibit PI3K/Akt cell signaling and, when delivered in nanoliposomes, counteract the Warburg effect in cancer (Jiang et al., 2011; Ryland et al., 2013; Tagaram et al., 2011). It is reasonable to speculate that different populations of exosomes, for example, depending on their ceramide content, may reprogram metabolism of cancer cells in distinct ways. Likewise, exosomes and EVs can reprogram tissues to allow tumor growth, particularly by inducing morphogenetic changes in endothelial cells and stem cells. Exosome-induced angiogenesis and capillary morphogenesis contribute to blood supply to tumors, which is an established research field addressing EV cancer biology (for reviews, see Grange et al., 2011; Mostefai, Andriantsitohaina, & Martinez, 2008). As with metabolic reprogramming, almost all the studies focus on effects mediated by miRNAs. However, it is well known that endothelial growth factor receptors are activated by lipid rafts and transported by EVs (Al-Nedawi, Meehan, Kerbel, Allison, & Rak, 2009; Pike, 2005). Secretion of sphingosine-1-phosphate (S1P), a ceramide derivative, is instrumental in promoting angiogenesis during tumor growth (Mizugishi et al., 2005; Nagahashi et al., 2012; Ogretmen, 2018; Pyne, El Buri, Adams, & Pyne, 2017; Spiegel & Kolesnick, 2002; Spiegel & Milstien, 2003; Takabe & Spiegel, 2014; Takabe, Paugh, Milstien, & Spiegel, 2008). It was found that stable complexes of S1P with G-protein-coupled receptors (GPCRs) are critical for formation of exosomes in MVEs through constitutive activation of Rho GTPases and stabilization of F-actin (Kajimoto et al., 2018). However, it has not been determined if these stable complexes are formed on exosomes that could then function as extracellular signalosomes. In colon cancer, *enterobacteria* trigger secretion of mucosa-derived exosomes that contain S1P and prostaglandin E2 (PGE2; Deng et al., 2015), an eicosanoid promoting cancer cell proliferation and angiogenesis. PGE2 in mucosa-derived exosomes has also been found to suppress liver-derived natural killer T (NKT) cells (Deng et al., 2013), a part of cancer exosome immune cell crosstalk discussed in the next paragraph. While it is known that the PGE2-EP4 receptor complex is activated in lipid rafts (Lin et al., 2017), it is currently unknown if this or other ligand-receptor complexes are embedded in mobile rafts and secreted via exosomes that function as extracellular signalosomes. The reason for this lack of knowledge is the technical difficulty to combine exosome preparation with the isolation and analysis of lipid rafts. The very few studies on this topic have shown that detergent-insoluble membranes isolated from exosomes—the next best equivalent to lipid rafts—contain a multitude of RAPs, including proteins regulating hepatocyte growth factor and EGF receptor cell signaling pathways in cancer cells (Ji et al., 2013).

By far, most of the studies on the effect of exosomes in cancer focus on their role in reprogramming tissue to generate a premetastatic niche and the cancer-to-immune cell crosstalk. As with other research testing the function of exosomes, most of these studies invoke miRNA and do not explore the function of exosomal lipids or mobile lipid rafts (Kai et al., 2017; Kinoshita et al., 2017; Salehi & Sharifi, 2018). miRNAs exchanged by exosomes between cancer cells promote cell growth (miRNA-155) modulate epithelial–mesenchymal transition (miRNA-23a, miRNA-191), contribute to breakdown of ECM (miRNA-29 family), and induce migration (miRNA-21). Hence the function of cancer exosomes is comparable to biohacking by genetic and epigenetic reprogramming of target tissue for metastasis. Up to recently, analysis of lipids in cancer metastasis was mainly focused on autocrine and paracrine signals mediated by S1P or lysophosphatidic acid, two moderately soluble lipids promoting cell growth and migration by targeting GPCRs and downstream PI3K/Akt and mammalian target of rapamycin (mTOR) cell signaling pathways (Radeff-Huang, Seasholtz, Matteo, & Brown, 2004; Ye, Ishii, Kingsbury, & Chun, 2002). The function of lipid rafts was invoked in facilitating GPCR activation by these lipids. However, only a few studies have investigated if EVs or exosomes exchange GPCRs between cells (Isola & Chen, 2016; Ye et al., 2002). Recently, ectosomes, EVs budding off cilia, were reported to accumulate GPCRs, but a role of ectosome-bound GPCRs in cancer has not been described yet (Isola & Chen, 2016; Nager et al., 2017; Soetedjo & Jin, 2014). One study found that exosomes can transfer the A2A adenosine receptor (A2AR) from A2AR expressing to nonexpressing cells with functional recovery of this GPCR, which suggests that exosomal transfer of receptors endows the recipient cell with the ability to respond to the same signals as the donor cell (Clayton, Al-Taei, Webber, Mason, & Tables lesi and 2011; Isola & Chen, 2016; Fig. 1).

Apart from spurring cell growth and migration, exosomes were invoked in preparing a premetastatic niche (for reviews, see Nogues, Benito-Martin, Hergueta-Redondo, & Peinado, 2017; Zhao et al., 2017). Initial mechanisms activated by exosomes are similar to those facilitating evasion from the primary tumor, particularly ECM breakdown and migration, but in reverse sequence. In the previous paragraph, we discussed MMPs transported by exosomes to degrade the ECM for cancer cell invasion. In addition to opening a route for invasion, cancer cells need to stop migrating and settle into the premetastatic niche. A study on human breast cancer MDA-MB-231 cells settling into a premetastatic niche in the liver showed that exosomal communication can be bidirectional. Upon priming of HepN liver cells with cancer cell–derived exosomes, liver cells secrete exosomes that turn down migratory gene expression and induce mesenchymal-to-epithelial reverse transition (MET) in cancer cells (Dioufa, Clark, Ma, Beckwitt, & Wells, 2017). As the result of this bidirectional communication, HepN-derived exosomes enhance cancer cell seeding but also suppress cell outgrowth, which renders cancer cells temporarily dormant once settled into their niche. The authors show that the behavior of the two cell types was due to exosome-induced changes in the miRNA composition although it was not clear how many exosomes from both cell types were required to induce this behavior. Interestingly the study also reported upregulation of E-cadherin and ZO-1 in the dormant cancer cells, two junctional cell adherence proteins associated with lipid rafts and promoting MET (Bruewer et al., 2003; Nusrat et al., 2000). CD44 is another exosome-associated transmembrane receptor glycoprotein the interaction of which with the ECM is regulated by lipid rafts (Murai, 2015). CD44 is upregulated in cancer stem cells, regulates adhesion and metastasis, and it is a well-known metastatic marker in exosomes (Senbanjo & Chellaiah, 2017). It was found to be delivered from ovarian cancer cells to mesothelial cells where it upregulates degradation of the ECM by inducing gene expression of MMPs to promote cancer cell invasion (Nakamura et al., 2017). A specific involvement of changes in the miRNA profile, either induced by CD44 or due to RNA transported by exosomes, was not noted. Since CD44 and tetraspanins are RAPs that are abundant in exosomes, it remains to be elucidated if they are also functional in mobile rafts (Rappa, Mercapide, Anzanello, Pope, & Lorico, 2013; Ronquist, Ronquist, Larsson, & Carlsson, 2010).

Crosstalk between cancer and immune cells is an area of exosome biology with great translational potential. As with vaccinations, immunizations, and other fields of immune biology involving the generation of antibodies, exosomes can be engineered to utilize the physiological potential of (re)programming the immune system to fight cancer cells. Therefore we will first discuss mechanisms by which cancer cells use exosomes to escape immune surveillance and then touch on the potential to hijack these mechanisms and restore vigilance of immune cells toward cancer. Not surprisingly the majority of these studies has focused on miRNA-mediated and immunoinhibitory protein—mediated reprogramming of immune cells, which is comprehensively reviewed in the following articles (Czernek & Duchler, 2017; Eichmuller, Osen, Mandelboim, & Seliger, 2017; Whiteside, 2017a, 2017b). However, it is known that antigen recognition and activation of immune cells critically relies on membrane lipids, which provides the unique opportunity to utilize lipids and mobile rafts in exosomes to manipulate the immune system. In the immune response to antigens, dendritic cells, B-cells, or macrophages act as antigen-presenting cells (APC) that process antigens and present them to lymphocytes, such as T-cells and NKTs (Balato, Unutmaz, & Gaspari, 2009; Chaplin, 2010). This presentation activates T-cells and leads to the production of antibodies against the antigen or phagocytosis. The activation relies on a complex between antigen peptide fragments bound to major histocompatibility complex (MHC) class I and II proteins on the APC surface and T-cell receptors (TCRs) on lymphocytes. This complex is also known as "immunological synapse" or supramolecular activation cluster, and it is critically dependent on glycolipid-associated and lipid raft—associated receptors in the APC and T-cell membrane (Dustin, 2014; Dustin & Baldari, 2017; Huang & Sauer, 2010). Macrophages (Daudi cells) were found to sort MHC proteins into exosomes via lipid rafts although it was not investigated what the function of these exosomes is once taken up by other cells (de Gassart, Geminard, Fevrier, Raposo, & Vidal, 2003). Lipid rafts were also shown to facilitate uptake of oncogenic viruses such as Epstein Barr Virus, and its association with exosomes to escape proteolytic degradation prior to exosomal spreading of the virus by B-lymphocytes (Verweij et al., 2011). It is unknown, however, if specific lipids in mobile rafts contribute to the immunosuppressive function of cancer cell—derived exosomes. In nanomedicine, a promising field is the use of raft-associated lipids in engineered exosomes or nanovesicles to activate T-cells (Fais et al., 2016). Dendritic cells are among the cells with the highest output of exosomes and can be genetically manipulated to produce exosomes with specific RNA content. Recently, tumor-derived exosomes were laced with alpha-galactosylceramide, a glycolipid activating the immune synapse, to enhance T-cell activation by dendritic cells, a promising strategy to use engineered exosomes for vaccination against cancer cells (Liu et al., 2017). Alpha-galactosylceramide binds to Cd1b and Cd1d, a lipid raft—associated MHC-like receptor on APCs to trigger TCR-mediated activation of T-cells (Lang, Maltsev, Besra, & Lang, 2004). It is not known if mobile rafts in exosomes can be used to enhance this process or exosomes secreted by tumor cells interrupt T-cell activation by interfering with activation of the immune synapse by lipid rafts.

Another area in exosome biology with great translational potential is understanding and utilizing exosome-regulated multidrug resistance (MDR) of cancer cells. There are several mechanisms by which exosomes or more generalized EVs modulate MDR: (1) spreading of MDR from resistant to nonresistant cancer cells; (2) loss of factors increasing MDR; (3) loss of factors suppressing MDR; and (4) acquisition of factors suppressing MDR. One may speculate that mechanisms (1) and (2), as well as (3) and (4), are complementary in that the secretion of a particular factor may be a loss for the donor and gain for the recipient cell. As a shared consequence of these distinct mechanisms, circulating EVs contain miRNAs or proteins critical for MDR and can be analyzed for cancer diagnosis and targeted chemotherapy in personalized nanomedicine (for reviews, see Batrakova & Kim, 2016; Bell, Kirk, Hiltbrunner, Gabrielsson, & Bultema, 2016; Fais et al., 2016; Mirzaei, Sahebkar, Jaafari, Goodarzi, & Mirzaei, 2017; Sharma, 2017; Sharma et al., 2016; Wang, Zheng, & Zhao, 2016). EV-mediated spreading of MDR is mostly attributed to transport of RNAs or proteins mediating export or inactivation of chemotherapeutic drugs, particularly ABC transport proteins and MDR-1/P-glycoprotein (Lopes-Rodrigues et al., 2013, 2016; Torreggiani, Roncuzzi, Perut, Zini, & Baldini, 2016). Our laboratory has found that induction of exosome secretion by activating ceramide generation through drug-mediated interference with sterol metabolism increases sensitivity of breast cancer stem-like cells to doxorubicin, a widely used drug in breast cancer therapy (Kong, He, et al., 2015; Spassieva & Bieberich, 2011). Increase of drug sensitivity was achieved by treating human breast cancer MDA-MB-231 cells with the farnesoid X receptor antagonist guggulsterone and the retinoid X receptor agonist bexarotene, and it was directly related to ceramide levels and secretion of breast cancer resistance protein/ABCG2 associated with exosomes. While this or similar mechanism of restoring drug sensitivity via EV-mediated secretion of MDR proteins (MRPs) has only been reported in a few studies, it provides the opportunity to break MDR by stimulating ceramide generation. On the other hand, restoring drug sensitivity and spreading of MRPs to other cancer cells may be prone to a delicate tradeoff, and it remains to be elucidated whether this mechanism can be utilized in cancer therapy.

5. CONCLUSIONS

While miRNA was targeted in exosome biology, the limited amount of RNA in exosomes makes it challenging to understand or assess their proposed impact. Instead, other signaling factors such as proteins and lipids abundant in exosomes may be better suited to explain many functions of exosomes in cancer. In particular, we have focused on lipid rafts in exosomes (mobile rafts) and the role of ceramide in cell signaling pathways activated in donor and recipient cells. Ceramide is unique in that its generation is required for exosome formation and secretion, and that it is a component of lipid rafts regulating cell signaling pathways for reprogramming of cancer cells and their microenvironment. This provides the intriguing opportunity of broadcasting and manipulating cell signaling events via mobile rafts in extracellular signalosomes that are regulated by ceramide. While the idea of mobile rafts and exosomes as vectors is novel and not yet explored, it is likely to generate a new target for cancer diagnostics and therapeutics.

ABBREVIATIONS

APC	Antigen-presenting cell
ATP	Adenosine triphosphate
CRP	Ceramide-rich platform
ECM	Extracellular matrix
EGF	Endothelial growth factor
EMT	Epithelial-to-mesenchymal transition
ESCRT	Endosomal sorting complexes required for transport
EV	Extracellular vesicle
GPI	Glycosylphosphatidylinositol
GPCR	G-protein coupled receptor
LBPA	Lysobisphosphatidic acid
LC-MS/MS	Liquid chromatography tandem mass spectrometry
lnc RNA	Long non-coding RNA
LPA	Lysophosphatidic acid
MDR	Multidrug resistance
MET	Mesenchymal-to-epithelial reverse transition

MHC	Major histocompatibility complex
miRNA	Micro RNA
MMP	Matrix metalloproteinases
MVE	Multivesicular endosome;
nSMase2	Neutral sphingomyelinase 2
NKT	Natural killer T cell
NTA	Nanoparticle tracking analysis
PGE2	Prostaglandin E2
PI3K	Phosphatidyl inositol 3-kinase
RAP	Raft-associated protein
S1P	Sphingosine-1-phosphate
TCR	T-cell receptor
TEM	Transmission electron microscopy.

ACKNOWLEDGMENTS

This work was the supported by the grants R01NS046835, R01AG034389, R01NS095215, and NSF 1615874.

REFERENCES

Abels, E. R., & Breakefield, X. O. (2016). Introduction to extracellular vesicles: Biogenesis, RNA cargo selection, content, release, and uptake. *Cellular and Molecular Neurobiology, 36*(3), 301–312. https://doi.org/10.1007/s10571-016-0366-z.

Akers, J. C., Ramakrishnan, V., Nolan, J. P., Duggan, E., Fu, C. C., Hochberg, F. H., et al. (2016). Comparative analysis of technologies for quantifying extracellular vesicles (EVs) in clinical cerebrospinal fluids (CSF). *PLoS One, 11*(2), e0149866. https://doi.org/10.1371/journal.pone.0149866.

Al-Nedawi, K., Meehan, B., Kerbel, R. S., Allison, A. C., & Rak, J. (2009). Endothelial expression of autocrine VEGF upon the uptake of tumor-derived microvesicles containing oncogenic EGFR. *Proceedings of the National Academy of Sciences of the United States of America, 106*(10), 3794–3799. https://doi.org/10.1073/pnas.0804543106.

Andaloussi S, E. L., Mager, I., Breakefield, X. O., & Wood, M. J. (2013). Extracellular vesicles: Biology and emerging therapeutic opportunities. *Nature Reviews. Drug Discovery, 12*(5), 347–357. https://doi.org/10.1038/nrd3978. pii:nrd3978.

Azmi, A. S., Bao, B., & Sarkar, F. H. (2013). Exosomes in cancer development, metastasis, and drug resistance: A comprehensive review. *Cancer Metastasis Reviews, 32*(3–4), 623–642. https://doi.org/10.1007/s10555-013-9441-9.

Balato, A., Unutmaz, D., & Gaspari, A. A. (2009). Natural killer T cells: An unconventional T-cell subset with diverse effector and regulatory functions. *The Journal of Investigative Dermatology, 129*(7), 1628–1642. https://doi.org/10.1038/jid.2009.30.

Batrakova, E. V., & Kim, M. S. (2016). Development and regulation of exosome-based therapy products. *Wiley Interdisciplinary Reviews: Nanomedicine and Nanobiotechnology, 8*(5), 744–757. https://doi.org/10.1002/wnan.1395.

Bell, B. M., Kirk, I. D., Hiltbrunner, S., Gabrielsson, S., & Bultema, J. J. (2016). Designer exosomes as next-generation cancer immunotherapy. *Nanomedicine, 12*(1), 163–169. https://doi.org/10.1016/j.nano.2015.09.011.

Bieberich, E. (2008). Ceramide signaling in cancer and stem cells. *Future Lipidology, 3*(3), 273–300. https://doi.org/10.2217/17460875.3.3.273.

Bissels, U., Wild, S., Tomiuk, S., Holste, A., Hafner, M., Tuschl, T., et al. (2009). Absolute quantification of microRNAs by using a universal reference. *RNA, 15*(12), 2375–2384. https://doi.org/10.1261/rna.1754109.

Brett, S. I., Lucien, F., Guo, C., Williams, K. C., Kim, Y., Durfee, P. N., et al. (2017). Immunoaffinity based methods are superior to kits for purification of prostate derived extracellular vesicles from plasma samples. *The Prostate*, 77(13), 1335–1343. https://doi.org/10.1002/pros.23393.

Brown, B. D., Gentner, B., Cantore, A., Colleoni, S., Amendola, M., Zingale, A., et al. (2007). Endogenous microRNA can be broadly exploited to regulate transgene expression according to tissue, lineage and differentiation state. *Nature Biotechnology*, *25*(12), 1457–1467. https://doi.org/10.1038/nbt1372.

Brown, P. N., & Yin, H. (2017). Polymer-based purification of extracellular vesicles. *Methods in Molecular Biology*, *1660*, 91–103. https://doi.org/10.1007/978-1-4939-7253-1_8.

Bruewer, M., Luegering, A., Kucharzik, T., Parkos, C. A., Madara, J. L., Hopkins, A. M., et al. (2003). Proinflammatory cytokines disrupt epithelial barrier function by apoptosis-independent mechanisms. *The Journal of Immunology: Official Journal of the American Association of Immunologists*, *171*(11), 6164–6172.

Burgert, A., Schlegel, J., Becam, J., Doose, S., Bieberich, E., Schubert-Unkmeir, A., et al. (2017). Characterization of plasma membrane ceramides by super-resolution microscopy. *Angewandte Chemie*. https://doi.org/10.1002/anie.201700570.

Busija, A. R., Patel, H. H., & Insel, P. A. (2017). Caveolins and cavins in the trafficking, maturation, and degradation of caveolae: Implications for cell physiology. *American Journal of Physiology. Cell Physiology*, *312*(4), C459–C477. https://doi.org/10.1152/ajpcell.00355.2016.

Carpintero-Fernandez, P., Fafian-Labora, J., & O'Loghlen, A. (2017). Technical advances to study extracellular vesicles. *Frontiers in Molecular Biosciences*, *4*, 79. https://doi.org/10.3389/fmolb.2017.00079.

Carrer, D. C., Hartel, S., Monaco, H. L., & Maggio, B. (2003). Ceramide modulates the lipid membrane organization at molecular and supramolecular levels. *Chemistry and Physics of Lipids*, *122*(1–2), 147–152.

Carrer, D. C., & Maggio, B. (1999). Phase behavior and molecular interactions in mixtures of ceramide with dipalmitoylphosphatidylcholine. *The Journal of Lipid Research*, *40*(11), 1978–1989.

Castro, B. M., Prieto, M., & Silva, L. C. (2014). Ceramide: A simple sphingolipid with unique biophysical properties. *Progress in Lipid Research*, *54*, 53–67. https://doi.org/10.1016/j.plipres.2014.01.004. pii:S0163-7827(14)00013-7pii.

Chairoungdua, A., Smith, D. L., Pochard, P., Hull, M., & Caplan, M. J. (2010). Exosome release of beta-catenin: A novel mechanism that antagonizes Wnt signaling. *The Journal of Cell Biology*, *190*(6), 1079–1091. https://doi.org/10.1083/jcb.201002049.

Chaplin, D. D. (2010). Overview of the immune response. *Journal of Allergy and Clinical Immunology*, *125*(2 Suppl. 2), S3–S23. https://doi.org/10.1016/j.jaci.2009.12.980.

Chevillet, J. R., Kang, Q., Ruf, I. K., Briggs, H. A., Vojtech, L. N., Hughes, S. M., et al. (2014). Quantitative and stoichiometric analysis of the microRNA content of exosomes. *Proceedings of the National Academy of Sciences of the United States of America*, *111*(41), 14888–14893. https://doi.org/10.1073/pnas.1408301111.

Chiarugi, P., & Cirri, P. (2016). Metabolic exchanges within tumor microenvironment. *Cancer Letters*, *380*(1), 272–280. https://doi.org/10.1016/j.canlet.2015.10.027.

Choi, D. S., Kim, D. K., Kim, Y. K., & Gho, Y. S. (2013). Proteomics, transcriptomics and lipidomics of exosomes and ectosomes. *Proteomics*, *13*(10–11), 1554–1571. https://doi.org/10.1002/pmic.201200329.

Clayton, A., Al-Taei, S., Webber, J., Mason, M. D., & Tabi, Z. (2011). Cancer exosomes express CD39 and CD73, which suppress T cells through adenosine production. *The Journal of Immunology: Official Journal of the American Association of Immunologists*, *187*(2), 676–683. https://doi.org/10.4049/jimmunol.1003884.

Colombo, M., Moita, C., van Niel, G., Kowal, J., Vigneron, J., Benaroch, P., et al. (2013). Analysis of ESCRT functions in exosome biogenesis, composition and secretion highlights the heterogeneity of extracellular vesicles. *Journal of Cell Science, 126*(Pt 24), 5553–5565. https://doi.org/10.1242/jcs.128868.

Colombo, M., Raposo, G., & Thery, C. (2014). Biogenesis, secretion, and intercellular interactions of exosomes and other extracellular vesicles. *Annual Review of Cell and Developmental Biology, 30*, 255–289. https://doi.org/10.1146/annurev-cellbio-101512-122326.

Coscia, C., Parolini, I., Sanchez, M., Biffoni, M., Boussadia, Z., Zanetti, C., et al. (2016). Generation, quantification, and tracing of metabolically labeled fluorescent exosomes. *Methods in Molecular Biology, 1448*, 217–235. https://doi.org/10.1007/978-1-4939-3753-0_16.

Cosker, K. E., Courchesne, S. L., & Segal, R. A. (2008). Action in the axon: Generation and transport of signaling endosomes. *Current Opinion in Neurobiology, 18*(3), 270–275. https://doi.org/10.1016/j.conb.2008.08.005.

Cosker, K. E., & Segal, R. A. (2014). Neuronal signaling through endocytosis. *Cold Spring Harbor Perspectives in Biology, 6*(2). https://doi.org/10.1101/cshperspect.a020669.

Costa Verdera, H., Gitz-Francois, J. J., Schiffelers, R. M., & Vader, P. (2017). Cellular uptake of extracellular vesicles is mediated by clathrin-independent endocytosis and macropinocytosis. *Journal of Controlled Release, 266*, 100–108. https://doi.org/10.1016/j.jconrel.2017.09.019.

Coumans, F. A. W., Brisson, A. R., Buzas, E. I., Dignat-George, F., Drees, E. E. E., El-Andaloussi, S., et al. (2017). Methodological guidelines to study extracellular vesicles. *Circulation Research, 120*(10), 1632–1648. https://doi.org/10.1161/CIRCRESAHA.117.309417.

Czernek, L., & Duchler, M. (2017). Functions of cancer-derived extracellular vesicles in immunosuppression. *Archivum Immunologiae et Therapiae Experimentalis, 65*(4), 311–323. https://doi.org/10.1007/s00005-016-0453-3.

Deng, Z., Mu, J., Tseng, M., Wattenberg, B., Zhuang, X., Egilmez, N. K., et al. (2015). Enterobacteria-secreted particles induce production of exosome-like S1P-containing particles by intestinal epithelium to drive Th17-mediated tumorigenesis. *Nature Communications, 6*, 6956. https://doi.org/10.1038/ncomms7956.

Deng, Z. B., Zhuang, X., Ju, S., Xiang, X., Mu, J., Liu, Y., et al. (2013). Exosome-like nanoparticles from intestinal mucosal cells carry prostaglandin E2 and suppress activation of liver NKT cells. *The Journal of Immunology: Official Journal of the American Association of Immunologists, 190*(7), 3579–3589. https://doi.org/10.4049/jimmunol.1203170.

DeRita, R. M., Zerlanko, B., Singh, A., Lu, H., Iozzo, R. V., Benovic, J. L., et al. (2017). c-Src, insulin-like growth factor I receptor, g-protein-coupled receptor kinases and focal adhesion kinase are enriched into prostate cancer cell exosomes. *Journal of Cellular Biochemistry, 118*(1), 66–73. https://doi.org/10.1002/jcb.25611.

Desrochers, L. M., Antonyak, M. A., & Cerione, R. A. (2016). Extracellular vesicles: Satellites of information transfer in cancer and stem cell biology. *Developmental Cell, 37*(4), 301–309. https://doi.org/10.1016/j.devcel.2016.04.019.

Dinkins, M. B., Dasgupta, S., Wang, G., Zhu, G., & Bieberich, E. (2014). Exosome reduction in vivo is associated with lower amyloid plaque load in the 5XFAD mouse model of Alzheimer's disease. *Neurobiology of Aging*. https://doi.org/10.1016/j.neurobiolaging.2014.02.012.

Dinkins, M. B., Enasko, J., Hernandez, C., Wang, G., Kong, J., Helwa, I., et al. (2016a). Neutral Sphingomyelinase-2 deficiency ameliorates Alzheimer's disease pathology and improves cognition in the 5XFAD mouse. *The Journal of Neuroscience: the Official Journal of the Society for Neuroscience, 36*(33), 8653–8667. https://doi.org/10.1523/JNEUROSCI.1429-16.2016.

Dinkins, M. B., Wang, G., & Bieberich, E. (2016b). Sphingolipid-enriched extracellular vesicles and Alzheimer's disease: A decade of research. *Journal of Alzheimer's Disease*. https://doi.org/10.3233/JAD-160567.

Dioufa, N., Clark, A. M., Ma, B., Beckwitt, C. H., & Wells, A. (2017). Bi-directional exosome-driven intercommunication between the hepatic niche and cancer cells. *Molecular Cancer, 16*(1), 172. https://doi.org/10.1186/s12943-017-0740-6.

Dorsam, B., Reiners, K. S., & von Strandmann, E. P. (2018). Cancer-derived extracellular vesicles: Friend and foe of tumour immunosurveillance. *Philosophical Transactions of the Royal Society of London B Biological Sciences, 373*(1737). https://doi.org/10.1098/rstb.2016.0481.

Dreyer, F., & Baur, A. (2016). Biogenesis and functions of exosomes and extracellular vesicles. *Methods in Molecular Biology, 1448*, 201–216. https://doi.org/10.1007/978-1-4939-3753-0_15.

Dustin, M. L. (2014). The immunological synapse. *Cancer Immunology Research, 2*(11), 1023–1033. https://doi.org/10.1158/2326-6066.CIR-14-0161.

Dustin, M. L., & Baldari, C. T. (2017). The immune synapse: Past, present, and future. *Methods in Molecular Biology, 1584*, 1–5. https://doi.org/10.1007/978-1-4939-6881-7_1.

Edgar, J. R. (2016). Q&A: What are exosomes, exactly? *BMC Biology, 14*, 46. https://doi.org/10.1186/s12915-016-0268-z.

Eichmuller, S. B., Osen, W., Mandelboim, O., & Seliger, B. (2017). Immune modulatory microRNAs involved in tumor attack and tumor immune escape. *Journal of the National Cancer Institute, 109*(10). https://doi.org/10.1093/jnci/djx034.

Fais, S., O'Driscoll, L., Borras, F. E., Buzas, E., Camussi, G., Cappello, F., et al. (2016). Evidence-based clinical use of nanoscale extracellular vesicles in nanomedicine. *ACS Nano, 10*(4), 3886–3899. https://doi.org/10.1021/acsnano.5b08015.

de Gassart, A., Geminard, C., Fevrier, B., Raposo, G., & Vidal, M. (2003). Lipid raft-associated protein sorting in exosomes. *Blood, 102*(13), 4336–4344. https://doi.org/10.1182/blood-2003-03-0871.

Gatenby, R. A., & Gillies, R. J. (2004). Why do cancers have high aerobic glycolysis? *Nature Reviews Cancer, 4*(11), 891–899. https://doi.org/10.1038/nrc1478.

Goldkorn, T., Chung, S., & Filosto, S. (2013). Lung cancer and lung injury: The dual role of ceramide. *Handbook of Experimental Pharmacology, 216*, 93–113. https://doi.org/10.1007/978-3-7091-1511-4_5.

Goni, F. M., & Alonso, A. (2006). Biophysics of sphingolipids I. Membrane properties of sphingosine, ceramides and other simple sphingolipids. *Biochimica et Biophysica Acta, 1758*(12), 1902–1921. https://doi.org/10.1016/j.bbamem.2006.09.011.

Goni, F. M., Contreras, F. X., Montes, L. R., Sot, J., & Alonso, A. (2005). Biophysics (and sociology) of ceramides. *Biochemical Society Symposium, 72*, 177–188.

Grange, C., Tapparo, M., Collino, F., Vitillo, L., Damasco, C., Deregibus, M. C., et al. (2011). Microvesicles released from human renal cancer stem cells stimulate angiogenesis and formation of lung premetastatic niche. *Cancer Research, 71*(15), 5346–5356. https://doi.org/10.1158/0008-5472.CAN-11-0241.

Greening, D. W., Xu, R., Gopal, S. K., Rai, A., & Simpson, R. J. (2017). Proteomic insights into extracellular vesicle biology - defining exosomes and shed microvesicles. *Expert Review of Proteomics, 14*(1), 69–95. https://doi.org/10.1080/14789450.2017.1260450.

Gulbins, E., & Kolesnick, R. (2003). Raft ceramide in molecular medicine. *Oncogene, 22*(45), 7070–7077. https://doi.org/10.1038/sj.onc.12071461207146 [pii].

Guo, B. B., Bellingham, S. A., & Hill, A. F. (2015). The neutral sphingomyelinase pathway regulates packaging of the prion protein into exosomes. *Journal of Biological Chemistry, 290*(6), 3455–3467. https://doi.org/10.1074/jbc.M114.605253.

Hakulinen, J., Sankkila, L., Sugiyama, N., Lehti, K., & Keski-Oja, J. (2008). Secretion of active membrane type 1 matrix metalloproteinase (MMP-14) into extracellular space in microvesicular exosomes. *Journal of Cellular Biochemistry, 105*(5), 1211–1218. https://doi.org/10.1002/jcb.21923.

Helwa, I., Cai, J., Drewry, M. D., Zimmerman, A., Dinkins, M. B., Khaled, M. L., et al. (2017). A comparative study of serum exosome isolation using differential ultracentrifugation and three commercial reagents. *PLoS One, 12*(1), e0170628. https://doi.org/10.1371/journal.pone.0170628.

He, M., & Zeng, Y. (2016). Microfluidic exosome analysis toward liquid biopsy for cancer. *Journal of Laboratory Automation, 21*(4), 599–608. https://doi.org/10.1177/2211068216651035.

He, Q., Wang, G., Wakade, S., Dasgupta, S., Dinkins, M., Kong, J. N., et al. (2014). Primary cilia in stem cells and neural progenitors are regulated by neutral sphingomyelinase 2 and ceramide. *Molecular Biology of the Cell, 25*(11), 1715–1729. doi:10/mbc.E13-12-0730, pii:mbc.E13-12-0730.

Horibe, S., Tanahashi, T., Kawauchi, S., Murakami, Y., & Rikitake, Y. (2018). Mechanism of recipient cell-dependent differences in exosome uptake. *BMC Cancer, 18*(1), 47. https://doi.org/10.1186/s12885-017-3958-1.

Huang, Y. H., & Sauer, K. (2010). Lipid signaling in T-cell development and function. *Cold Spring Harbor Perspectives in Biology, 2*(11), a002428. https://doi.org/10.1101/cshperspect.a002428.

Hurley, J. H. (2015). ESCRTs are everywhere. *The EMBO Journal, 34*(19), 2398–2407. https://doi.org/10.15252/embj.201592484.

Hurley, J. H., & Odorizzi, G. (2012). Get on the exosome bus with ALIX. *Nature Cell Biology, 14*(7), 654–655. https://doi.org/10.1038/ncb2530.

Hyenne, V., Lefebvre, O., & Goetz, J. G. (2017). Going live with tumor exosomes and microvesicles. *Cell Adhesion & Migration, 11*(2), 173–186. https://doi.org/10.1080/19336918.2016.1276694.

Intasqui, P., Bertolla, R. P., & Sadi, M. V. (2018). Prostate cancer proteomics: Clinically useful protein biomarkers and future perspectives. *Expert Review of Proteomics, 15*(1), 65–79. https://doi.org/10.1080/14789450.2018.1417846.

Isola, A. L., & Chen, S. (2016). Exosomes: The link between GPCR activation and metastatic potential? *Frontiers in Genetics,* 7, 56. https://doi.org/10.3389/fgene.2016.00056.

Jakobsen, K. R., Paulsen, B. S., Baek, R., Varming, K., Sorensen, B. S., & Jorgensen, M. M. (2015). Exosomal proteins as potential diagnostic markers in advanced non-small cell lung carcinoma. *Journal of Extracellular Vesicles, 4,* 26659. https://doi.org/10.3402/jev.v4.26659.

Ji, H., Greening, D. W., Barnes, T. W., Lim, J. W., Tauro, B. J., Rai, A., et al. (2013). Proteome profiling of exosomes derived from human primary and metastatic colorectal cancer cells reveal differential expression of key metastatic factors and signal transduction components. *Proteomics, 13*(10–11), 1672–1686. https://doi.org/10.1002/pmic.201200562.

Jiang, H., Li, Z., Li, X., & Xia, J. (2015). Intercellular transfer of messenger RNAs in multiorgan tumorigenesis by tumor cell-derived exosomes. *Molecular Medicine Reports, 11*(6), 4657–4663. https://doi.org/10.3892/mmr.2015.3312.

Jiang, Y., DiVittore, N. A., Kaiser, J. M., Shanmugavelandy, S. S., Fritz, J. L., Heakal, Y., et al. (2011). Combinatorial therapies improve the therapeutic efficacy of nanoliposomal ceramide for pancreatic cancer. *Cancer Biology & Therapy, 12*(7), 574–585.

Juan, T., & Furthauer, M. (2017). Biogenesis and function of ESCRT-dependent extracellular vesicles. *Seminars in Cell & Developmental Biology.* https://doi.org/10.1016/j.semcdb.2017.08.022.

Kahlert, C., & Kalluri, R. (2013). Exosomes in tumor microenvironment influence cancer progression and metastasis. *Journal of Molecular Medicine (Berlin), 91*(4), 431–437. https://doi.org/10.1007/s00109-013-1020-6.

Kai, K., Dittmar, R. L., & Sen, S. (2017). Secretory microRNAs as biomarkers of cancer. *Seminars in Cell & Developmental Biology*. https://doi.org/10.1016/j.semcdb.2017.12.011.

Kajimoto, T., Mohamed, N. N. I., Badawy, S. M. M., Matovelo, S. A., Hirase, M., Nakamura, S., et al. (2018). Involvement of Gbetagamma subunits of Gi protein coupled with S1P receptor on multivesicular endosomes in F-actin formation and cargo sorting into exosomes. *Journal of Biological Chemistry, 293*(1), 245–253. https://doi.org/10.1074/jbc.M117.808733.

Kalluri, R. (2016). The biology and function of exosomes in cancer. *Journal of Clinical Investigation, 126*(4), 1208–1215. https://doi.org/10.1172/JCI81135.

Kinoshita, T., Yip, K. W., Spence, T., & Liu, F. F. (2017). MicroRNAs in extracellular vesicles: Potential cancer biomarkers. *Journal of Human Genetics, 62*(1), 67–74. https://doi.org/10.1038/jhg.2016.87.

Kong, J. N., Hardin, K., Dinkins, M., Wang, G., He, Q., Mujadzic, T., et al. (2015). Regulation of Chlamydomonas flagella and ependymal cell motile cilia by ceramide-mediated translocation of GSK3. *Molecular Biology of the Cell, 26*(24), 4451–4465. https://doi.org/10.1091/mbc.E15-06-0371.

Kong, J. N., He, Q., Wang, G., Dasgupta, S., Dinkins, M. B., Zhu, G., et al. (2015). Guggulsterone and bexarotene induce secretion of exosome-associated breast cancer resistance protein and reduce doxorubicin resistance in MDA-MB-231 cells. *International Journal of Cancer, 137*(7), 1610–1620. https://doi.org/10.1002/ijc.29542.

Koritzinsky, E. H., Street, J. M., Star, R. A., & Yuen, P. S. (2017). Quantification of exosomes. *Journal of Cellular Physiology, 232*(7), 1587–1590. https://doi.org/10.1002/jcp.25387.

Kosaka, N., Iguchi, H., Yoshioka, Y., Takeshita, F., Matsuki, Y., & Ochiya, T. (2010). Secretory mechanisms and intercellular transfer of microRNAs in living cells. *Journal of Biological Chemistry, 285*(23), 17442–17452. https://doi.org/10.1074/jbc.M110.107821.

Kowal, J., Tkach, M., & Thery, C. (2014). Biogenesis and secretion of exosomes. *Current Opinion in Cell Biology, 29C*, 116–125. https://doi.org/10.1016/j.ceb.2014.05.004. pii:S0955-0674(14)00057-X.

Kumeda, N., Ogawa, Y., Akimoto, Y., Kawakami, H., Tsujimoto, M., & Yanoshita, R. (2017). Characterization of membrane integrity and morphological stability of human salivary exosomes. *Biological and Pharmaceutical Bulletin, 40*(8), 1183–1191. https://doi.org/10.1248/bpb.b16-00891.

Lajoie, P., & Nabi, I. R. (2010). Lipid rafts, caveolae, and their endocytosis. *Int Rev Cell Mol Biol, 282*, 135–163. https://doi.org/10.1016/S1937-6448(10) 82003–82009. pii: S1937-6448(10)82003-9.

Lang, G. A., Maltsev, S. D., Besra, G. S., & Lang, M. L. (2004). Presentation of alpha-galactosylceramide by murine CD1d to natural killer T cells is facilitated by plasma membrane glycolipid rafts. *Nature Reviews. Immunology, 112*(3), 386–396. https://doi.org/10.1111/j.1365-2567.2004.01896.x.

Lee, Y., El Andaloussi, S., & Wood, M. J. (2012). Exosomes and microvesicles: Extracellular vesicles for genetic information transfer and gene therapy. *Human Molecular Genetics, 21*(R1), R125–R134. https://doi.org/10.1093/hmg/dds317.

Li, A., Zhang, T., Zheng, M., Liu, Y., & Chen, Z. (2017). Exosomal proteins as potential markers of tumor diagnosis. *Journal of Hematology & Oncology, 10*(1), 175. https://doi.org/10.1186/s13045-017-0542-8.

Li, P., Kaslan, M., Lee, S. H., Yao, J., & Gao, Z. (2017). Progress in exosome isolation techniques. *Theranostics, 7*(3), 789–804. https://doi.org/10.7150/thno.18133.

Liberti, M. V., & Locasale, J. W. (2016). The Warburg effect: How does it benefit cancer cells? *Trends in Biochemical Sciences, 41*(3), 211–218. https://doi.org/10.1016/j.tibs.2015.12.001.

Lin, M. C., Chen, S. Y., Tsai, H. M., He, P. L., Lin, Y. C., Herschman, H., et al. (2017). PGE2/EP4 signaling controls the transfer of the mammary stem cell state by lipid rafts in extracellular vesicles. *Stem Cells, 35*(2), 425–444. https://doi.org/10.1002/stem.2476.

Lingwood, D., & Simons, K. (2010). Lipid rafts as a membrane-organizing principle. *Science, 327*(5961), 46–50. https://doi.org/10.1126/science.1174621. pii:327/5961/46.

Liu, H., Chen, L., Liu, J., Meng, H., Zhang, R., Ma, L., et al. (2017). Co-delivery of tumor-derived exosomes with alpha-galactosylceramide on dendritic cell-based immunotherapy for glioblastoma. *Cancer Letters, 411*, 182–190. https://doi.org/10.1016/j.canlet.2017.09.022.

Lopes-Rodrigues, V., Di Luca, A., Sousa, D., Seca, H., Meleady, P., Henry, M., et al. (2016). Multidrug resistant tumour cells shed more microvesicle-like EVs and less exosomes than their drug-sensitive counterpart cells. *Biochimica et Biophysica Acta, 1860*(3), 618–627. https://doi.org/10.1016/j.bbagen.2015.12.011.

Lopes-Rodrigues, V., Seca, H., Sousa, D., Sousa, E., Lima, R. T., & Vasconcelos, M. H. (2013). The network of P-glycoprotein and microRNAs interactions. *International Journal of Cancer*. https://doi.org/10.1002/ijc.28500.

Lopez, D. (2015). Molecular composition of functional microdomains in bacterial membranes. *Chemistry and Physics of Lipids, 192*, 3–11. https://doi.org/10.1016/j.chemphyslip.2015.08.015. pii:S0009-3084(15)30046-3.

Ma, Y., Hinde, E., & Gaus, K. (2015). Nanodomains in biological membranes. *Essays in Biochemistry, 57*, 93–107. https://doi.org/10.1042/bse0570093.

Maas, S. L., Broekman, M. L., & de Vrij, J. (2017). Tunable resistive pulse sensing for the characterization of extracellular vesicles. *Methods in Molecular Biology, 1545*, 21–33. https://doi.org/10.1007/978-1-4939-6728-5_2.

Marsh, M., & van Meer, G. (2008). Cell biology. No ESCRTs for exosomes. *Science, 319*(5867), 1191–1192. https://doi.org/10.1126/science.1155750.

McKelvey, K. J., Powell, K. L., Ashton, A. W., Morris, J. M., & McCracken, S. A. (2015). Exosomes: Mechanisms of uptake. *Journal of Circulating biomarkers, 4*, 7. https://doi.org/10.5772/61186.

Menck, K., Sonmezer, C., Worst, T. S., Schulz, M., Dihazi, G. H., Streit, F., et al. (2017). Neutral sphingomyelinases control extracellular vesicles budding from the plasma membrane. *Journal of Extracellular Vesicles, 6*(1), 1378056. https://doi.org/10.1080/20013078.2017.1378056.

Minciacchi, V. R., Freeman, M. R., & Di Vizio, D. (2015). Extracellular vesicles in cancer: Exosomes, microvesicles and the emerging role of large oncosomes. *Seminars in Cell & Developmental Biology, 40*, 41–51. https://doi.org/10.1016/j.semcdb.2015.02.010.

Minciacchi, V. R., Zijlstra, A., Rubin, M. A., & Di Vizio, D. (2017). Extracellular vesicles for liquid biopsy in prostate cancer: Where are we and where are we headed? *Prostate Cancer and Prostatic Diseases, 20*(3), 251–258. https://doi.org/10.1038/pcan.2017.7.

Mirzaei, H., Sahebkar, A., Jaafari, M. R., Goodarzi, M., & Mirzaei, H. R. (2017). Diagnostic and therapeutic potential of exosomes in cancer: The beginning of a new tale? *Journal of Cellular Physiology, 232*(12), 3251–3260. https://doi.org/10.1002/jcp.25739.

Mizugishi, K., Yamashita, T., Olivera, A., Miller, G. F., Spiegel, S., & Proia, R. L. (2005). Essential role for sphingosine kinases in neural and vascular development. *Molecular and Cellular Biology, 25*(24), 11113–11121.

Mollinedo, F., & Gajate, C. (2015). Lipid rafts as major platforms for signaling regulation in cancer. *Advances in Biological Regulation, 57*, 130–146. https://doi.org/10.1016/j.jbior.2014.10.003.

Momen-Heravi, F., Balaj, L., Alian, S., Mantel, P. Y., Halleck, A. E., Trachtenberg, A. J., et al. (2013). Current methods for the isolation of extracellular vesicles. *Biological Chemistry, 394*(10), 1253–1262. https://doi.org/10.1515/hsz-2013-0141.

Mostefai, H. A., Andriantsitohaina, R., & Martinez, M. C. (2008). Plasma membrane microparticles in angiogenesis: Role in ischemic diseases and in cancer. *Physiological Research, 57*(3), 311–320.

Mukherji, S., Ebert, M. S., Zheng, G. X., Tsang, J. S., Sharp, P. A., & van Oudenaarden, A. (2011). MicroRNAs can generate thresholds in target gene expression. *Nature Genetics, 43*(9), 854–859. https://doi.org/10.1038/ng.905.

Murai, T. (2015). Lipid raft-mediated regulation of hyaluronan-CD44 interactions in inflammation and cancer. *Frontiers in Immunology, 6*, 420. https://doi.org/10.3389/fimmu.2015.00420.

Nagahashi, M., Ramachandran, S., Kim, E. Y., Allegood, J. C., Rashid, O. M., Yamada, A., et al. (2012). Sphingosine-1-phosphate produced by sphingosine kinase 1 promotes breast cancer progression by stimulating angiogenesis and lymphangiogenesis. *Cancer Research, 72*(3), 726–735. https://doi.org/10.1158/0008-5472.CAN-11-2167.

Nager, A. R., Goldstein, J. S., Herranz-Perez, V., Portran, D., Ye, F., Garcia-Verdugo, J. M., et al. (2017). An actin network dispatches ciliary GPCRs into extracellular vesicles to modulate signaling. *Cell, 168*(1–2), 252–263 e214. https://doi.org/10.1016/j.cell.2016.11.036.

Nakai, W., Yoshida, T., Diez, D., Miyatake, Y., Nishibu, T., Imawaka, N., et al. (2016). A novel affinity-based method for the isolation of highly purified extracellular vesicles. *Scientific Reports, 6*, 33935. https://doi.org/10.1038/srep33935.

Nakamura, K., Sawada, K., Kinose, Y., Yoshimura, A., Toda, A., Nakatsuka, E., et al. (2017). Exosomes promote ovarian cancer cell invasion through transfer of CD44 to peritoneal mesothelial cells. *Molecular Cancer Research, 15*(1), 78–92. https://doi.org/10.1158/1541-7786.MCR-16-0191.

van Niel, G., D'Angelo, G., & Raposo, G. (2018). Shedding light on the cell biology of extracellular vesicles. *Nature Reviews. Molecular Cell Biology*. https://doi.org/10.1038/nrm.2017.125.

Nogues, L., Benito-Martin, A., Hergueta-Redondo, M., & Peinado, H. (2017). The influence of tumour-derived extracellular vesicles on local and distal metastatic dissemination. *Molecular Aspects of Medicine*. https://doi.org/10.1016/j.mam.2017.11.012.

Nusrat, A., Parkos, C. A., Verkade, P., Foley, C. S., Liang, T. W., Innis-Whitehouse, W., et al. (2000). Tight junctions are membrane microdomains. *Journal of Cell Science, 113*(Pt 10), 1771–1781.

Ogretmen, B. (2018). Sphingolipid metabolism in cancer signalling and therapy. *Nature Reviews Cancer, 18*(1), 33–50. https://doi.org/10.1038/nrc.2017.96.

Osteikoetxea, X., Sodar, B., Nemeth, A., Szabo-Taylor, K., Paloczi, K., Vukman, K. V., et al. (2015). Differential detergent sensitivity of extracellular vesicle subpopulations. *Organic & Biomolecular Chemistry, 13*(38), 9775–9782. https://doi.org/10.1039/c5ob01451d.

Owen, D. M., Magenau, A., Williamson, D., & Gaus, K. (2012). The lipid raft hypothesis revisited—new insights on raft composition and function from super-resolution fluorescence microscopy. *Bioessays, 34*(9), 739–747. https://doi.org/10.1002/bies.201200044.

Parolini, I., Federici, C., Raggi, C., Lugini, L., Palleschi, S., De Milito, A., et al. (2009). Microenvironmental pH is a key factor for exosome traffic in tumor cells. *Journal of Biological Chemistry, 284*(49), 34211–34222. https://doi.org/10.1074/jbc.M109.041152.

Phinney, D. G., & Pittenger, M. F. (2017). Concise review: Msc-derived exosomes for cell-free therapy. *Stem Cells, 35*(4), 851–858. https://doi.org/10.1002/stem.2575.

Pietrowska, M., Funk, S., Gawin, M., Marczak, L., Abramowicz, A., Widlak, P., et al. (2017). Isolation of exosomes for the purpose of protein cargo analysis with the use of mass spectrometry. *Methods in Molecular Biology, 1654*, 291–307. https://doi.org/10.1007/978-1-4939-7231-9_22.

Pike, L. J. (2005). Growth factor receptors, lipid rafts and caveolae: An evolving story. *Biochimica et Biophysica Acta, 1746*(3), 260–273. https://doi.org/10.1016/j.bbamcr.2005.05.005. pii:S0167-4889(05)00088-1.

Pinto, S. N., Silva, L. C., Futerman, A. H., & Prieto, M. (2011). Effect of ceramide structure on membrane biophysical properties: The role of acyl chain length and unsaturation. *Biochimica et Biophysica Acta, 1808*(11), 2753–2760. https://doi.org/10.1016/j.bbamem.2011.07.023. pii:S0005-2736(11)00228-8.

Pitt, J. M., Charrier, M., Viaud, S., Andre, F., Besse, B., Chaput, N., et al. (2014). Dendritic cell-derived exosomes as immunotherapies in the fight against cancer. *The Journal of Immunology: Official Journal of the American Association of Immunologists, 193*(3), 1006–1011. https://doi.org/10.4049/jimmunol.1400703.

Pocsfalvi, G., Stanly, C., Vilasi, A., Fiume, I., Capasso, G., Turiak, L., et al. (2016). Mass spectrometry of extracellular vesicles. *Mass Spectrometry Reviews, 35*(1), 3–21. https://doi.org/10.1002/mas.21457.

van der Pol, E., Boing, A. N., Harrison, P., Sturk, A., & Nieuwland, R. (2012). Classification, functions, and clinical relevance of extracellular vesicles. *Pharmacological Reviews, 64*(3), 676–705. https://doi.org/10.1124/pr.112.005983.

Powell, D. J., Hajduch, E., Kular, G., & Hundal, H. S. (2003). Ceramide disables 3-phosphoinositide binding to the pleckstrin homology domain of protein kinase B (PKB)/Akt by a PKCzeta-dependent mechanism. *Molecular and Cellular Biology, 23*(21), 7794–7808.

Pyne, N. J., El Buri, A., Adams, D. R., & Pyne, S. (2017). Sphingosine 1-phosphate and cancer. *Advances in Biological Regulation*. https://doi.org/10.1016/j.jbior.2017.09.006.

Rabinowitz, J. D., & Coller, H. A. (2016). Partners in the Warburg effect. *Elife, 5*, e15938. https://doi.org/10.7554/eLife.15938.

Radeff-Huang, J., Seasholtz, T. M., Matteo, R. G., & Brown, J. H. (2004). G protein mediated signaling pathways in lysophospholipid induced cell proliferation and survival. *Journal of Cellular Biochemistry, 92*(5), 949–966. https://doi.org/10.1002/jcb.20094.

Rani, S., Ryan, A. E., Griffin, M. D., & Ritter, T. (2015). Mesenchymal stem cell-derived extracellular vesicles: Toward cell-free therapeutic applications. *Molecular Therapy: the Journal of the American Society of Gene Therapy, 23*(5), 812–823. https://doi.org/10.1038/mt.2015.44.

Raposo, G., & Stoorvogel, W. (2013). Extracellular vesicles: Exosomes, microvesicles, and friends. *The Journal of Cell Biology, 200*(4), 373–383. https://doi.org/10.1083/jcb.201211138.

Rappa, G., Mercapide, J., Anzanello, F., Pope, R. M., & Lorico, A. (2013). Biochemical and biological characterization of exosomes containing prominin-1/CD133. *Molecular Cancer, 12*, 62. https://doi.org/10.1186/1476-4598-12-62.

Record, M., Carayon, K., Poirot, M., & Silvente-Poirot, S. (2014). Exosomes as new vesicular lipid transporters involved in cell-cell communication and various pathophysiologies. *Biochimica et Biophysica Acta, 1841*(1), 108–120. https://doi.org/10.1016/j.bbalip.2013.10.004.

Ronquist, K. G., Ronquist, G., Larsson, A., & Carlsson, L. (2010). Proteomic analysis of prostate cancer metastasis-derived prostasomes. *Anticancer Research, 30*(2), 285–290.

Ruivo, C. F., Adem, B., Silva, M., & Melo, S. A. (2017). The biology of cancer Exosomes: Insights and new perspectives. *Cancer Research,* 77(23), 6480–6488. https://doi.org/10.1158/0008-5472.CAN-17-0994.

Ryland, L. K., Doshi, U. A., Shanmugavelandy, S. S., Fox, T. E., Aliaga, C., Broeg, K., et al. (2013). C6-ceramide nanoliposomes target the Warburg effect in chronic lymphocytic leukemia. *PLoS One, 8*(12), e84648. https://doi.org/10.1371/journal.pone.0084648.

Salehi, M., & Sharifi, M. (2018). Exosomal miRNAs as novel cancer biomarkers: Challenges and opportunities. *Journal of Cellular Physiology*. https://doi.org/10.1002/jcp.26481.

Sanderson, R. D., Bandari, S. K., & Vlodavsky, I. (2017). Proteases and glycosidases on the surface of exosomes: Newly discovered mechanisms for extracellular remodeling. *Matrix Biology: Journal of the International Society for Matrix Biology*. https://doi.org/10.1016/j.matbio.2017.10.007.

Sandfeld-Paulsen, B., Aggerholm-Pedersen, N., Baek, R., Jakobsen, K. R., Meldgaard, P., Folkersen, B. H., et al. (2016). Exosomal proteins as prognostic biomarkers in non-small cell lung cancer. *Molecular Oncology, 10*(10), 1595–1602. https://doi.org/10.1016/j.molonc.2016.10.003.

Sandfeld-Paulsen, B., Jakobsen, K. R., Baek, R., Folkersen, B. H., Rasmussen, T. R., Meldgaard, P., et al. (2016). Exosomal proteins as diagnostic biomarkers in lung cancer. *Journal of Thoracic Oncology: Official Publication of the International Association for the Study of Lung Cancer, 11*(10), 1701–1710. https://doi.org/10.1016/j.jtho.2016.05.034.

Schorey, J. S., & Bhatnagar, S. (2008). Exosome function: From tumor immunology to pathogen biology. *Traffic, 9*(6), 871–881. https://doi.org/10.1111/j.1600-0854.2008.00734.x.

Senbanjo, L. T., & Chellaiah, M. A. (2017). CD44: A multifunctional cell surface adhesion receptor is a regulator of progression and metastasis of cancer cells. *Frontiers in Cell and Developmental Biology, 5*, 18. https://doi.org/10.3389/fcell.2017.00018.

Shamseddine, A. A., Airola, M. V., & Hannun, Y. A. (2015). Roles and regulation of neutral sphingomyelinase-2 in cellular and pathological processes. *Advances in Biological Regulation, 57*, 24–41. https://doi.org/10.1016/j.jbior.2014.10.002. pii:S2212-4926(14)00057-8.

Sharma, A. (2017). Chemoresistance in cancer cells: Exosomes as potential regulators of therapeutic tumor heterogeneity. *Nanomedicine, 12*(17), 2137–2148. https://doi.org/10.2217/nnm-2017-0184.

Sharma, A., Khatun, Z., & Shiras, A. (2016). Tumor exosomes: Cellular postmen of cancer diagnosis and personalized therapy. *Nanomedicine, 11*(4), 421–437. https://doi.org/10.2217/nnm.15.210.

Sharma, S., Scholz-Romero, K., Rice, G. E., & Salomon, C. (2018). Methods to enrich exosomes from conditioned media and biological fluids. *Methods in Molecular Biology, 1710*, 103–115. https://doi.org/10.1007/978-1-4939-7498-6_8.

Shay, G., Lynch, C. C., & Fingleton, B. (2015). Moving targets: Emerging roles for MMPs in cancer progression and metastasis. *Matrix Biology: Journal of the International Society for Matrix Biology, 44–46*, 200–206. https://doi.org/10.1016/j.matbio.2015.01.019.

Silva, L. C., Ben David, O., Pewzner-Jung, Y., Laviad, E. L., Stiban, J., Bandyopadhyay, S., et al. (2012). Ablation of ceramide synthase 2 strongly affects biophysical properties of membranes. *The Journal of Lipid Research, 53*(3), 430–436. https://doi.org/10.1194/jlr.M022715.

Silva, M., & Melo, S. A. (2015). Non-coding RNAs in Exosomes: New players in cancer biology. *Current Genomics, 16*(5), 295–303. https://doi.org/10.2174/1389202916666150707154719.

Simons, K., & Gerl, M. J. (2010). Revitalizing membrane rafts: New tools and insights. *Nature Reviews. Molecular Cell Biology, 11*(10), 688–699. https://doi.org/10.1038/nrm2977. pii:nrm2977.

Simons, K., & Ikonen, E. (1997). Functional rafts in cell membranes. *Nature, 387*(6633), 569–572. https://doi.org/10.1038/42408.
Simons, K., & Sampaio, J. L. (2011). Membrane organization and lipid rafts. *Cold Spring Harbor Perspectives in Biology, 3*(10), a004697. https://doi.org/10.1101/cshperspect.a004697.
Simons, K., & Toomre, D. (2000). Lipid rafts and signal transduction. *Nature Reviews. Molecular Cell Biology, 1*(1), 31–39. https://doi.org/10.1038/35036052.
Sinha, A., Principe, S., Alfaro, J., Ignatchenko, A., Ignatchenko, V., & Kislinger, T. (2018). Proteomic profiling of secreted proteins, exosomes, and microvesicles in cell culture conditioned media. *Methods in Molecular Biology, 1722*, 91–102. https://doi.org/10.1007/978-1-4939-7553-2_6.
Skotland, T., Sandvig, K., & Llorente, A. (2017). Lipids in exosomes: Current knowledge and the way forward. *Progress in Lipid Research, 66*, 30–41. https://doi.org/10.1016/j.plipres.2017.03.001.
Soetedjo, L., & Jin, H. (2014). Agonist-induced GPCR shedding from the ciliary surface is dependent on ESCRT-III and VPS4. *Current Biology: CB, 24*(5), 509–518. https://doi.org/10.1016/j.cub.2014.01.010.
Spassieva, S., & Bieberich, E. (2011). The gut-to-breast connection - interdependence of sterols and sphingolipids in multidrug resistance and breast cancer therapy. *Anticancer Agents Med Chem, 11*(9), 882–890. pii:BSP/ACAMC/E-Pub/00214.
Spiegel, S., & Kolesnick, R. (2002). Sphingosine 1-phosphate as a therapeutic agent. *Leukemia, 16*(9), 1596–1602. https://doi.org/10.1038/sj.leu.2402611.
Spiegel, S., & Milstien, S. (2003). Sphingosine-1-phosphate: An enigmatic signalling lipid. *Nature Reviews. Molecular Cell Biology, 4*(5), 397–407.
Staubach, S., & Hanisch, F. G. (2011). Lipid rafts: Signaling and sorting platforms of cells and their roles in cancer. *Expert Review of Proteomics, 8*(2), 263–277. https://doi.org/10.1586/epr.11.2.
Steinbichler, T. B., Dudas, J., Riechelmann, H., & Skvortsova, , II. (2017). The role of exosomes in cancer metastasis. *Seminars in Cancer Biology, 44*, 170–181. https://doi.org/10.1016/j.semcancer.2017.02.006.
Stevanato, L., Thanabalasundaram, L., Vysokov, N., & Sinden, J. D. (2016). Investigation of content, stoichiometry and transfer of miRNA from human neural stem cell line derived exosomes. *PLoS One, 11*(1), e0146353. https://doi.org/10.1371/journal.pone.0146353.
Stoorvogel, W. (2015). Resolving sorting mechanisms into exosomes. *Cell Research, 25*(5), 531–532. https://doi.org/10.1038/cr.2015.39.
Sung, B. H., Ketova, T., Hoshino, D., Zijlstra, A., & Weaver, A. M. (2015). Directional cell movement through tissues is controlled by exosome secretion. *Nature Communications, 6*, 7164. https://doi.org/10.1038/ncomms8164.
Szatanek, R., Baj-Krzyworzeka, M., Zimoch, J., Lekka, M., Siedlar, M., & Baran, J. (2017). The methods of choice for extracellular vesicles (EVs) characterization. *International Journal of Molecular Sciences, 18*(6). https://doi.org/10.3390/ijms18061153.
Szatanek, R., Baran, J., Siedlar, M., & Baj-Krzyworzeka, M. (2015). Isolation of extracellular vesicles: Determining the correct approach (Review). *International Journal of Molecular Medicine, 36*(1), 11–17. https://doi.org/10.3892/ijmm.2015.2194.
Tagaram, H. R., Divittore, N. A., Barth, B. M., Kaiser, J. M., Avella, D., Kimchi, E. T., et al. (2011). Nanoliposomal ceramide prevents in vivo growth of hepatocellular carcinoma. *Gut, 60*(5), 695–701. https://doi.org/10.1136/gut.2010.216671.
Takabe, K., Paugh, S. W., Milstien, S., & Spiegel, S. (2008). "Inside-out" signaling of sphingosine-1-phosphate: Therapeutic targets. *Pharmacological Reviews, 60*(2), 181–195. https://doi.org/10.1124/pr.107.07113. pii:pr.107.07113 [pii].
Takabe, K., & Spiegel, S. (2014). Export of sphingosine-1-phosphate and cancer progression. *The Journal of Lipid Research, 55*(9), 1839–1846. https://doi.org/10.1194/jlr.R046656.

Tanase, C. P., Codrici, E., Popescu, I. D., Mihai, S., Enciu, A. M., Necula, L. G., et al. (2017). Prostate cancer proteomics: Current trends and future perspectives for biomarker discovery. *Oncotarget, 8*(11), 18497–18512. https://doi.org/10.18632/oncotarget.14501.

Tan, S. S., Yin, Y., Lee, T., Lai, R. C., Yeo, R. W., Zhang, B., et al. (2013). Therapeutic MSC exosomes are derived from lipid raft microdomains in the plasma membrane. *Journal of Extracellular Vesicles, 2*. https://doi.org/10.3402/jev.v2i0.22614.

Taylor, D. D., & Shah, S. (2015). Methods of isolating extracellular vesicles impact downstream analyses of their cargoes. *Methods, 87*, 3–10. https://doi.org/10.1016/j.ymeth.2015.02.019.

Taylor, D. D., Zacharias, W., & Gercel-Taylor, C. (2011). Exosome isolation for proteomic analyses and RNA profiling. *Methods in Molecular Biology, 728*, 235–246. https://doi.org/10.1007/978-1-61779-068-3_15.

Tian, T., Zhu, Y. L., Hu, F. H., Wang, Y. Y., Huang, N. P., & Xiao, Z. D. (2013). Dynamics of exosome internalization and trafficking. *Journal of Cellular Physiology, 228*(7), 1487–1495. https://doi.org/10.1002/jcp.24304.

Tian, T., Zhu, Y. L., Zhou, Y. Y., Liang, G. F., Wang, Y. Y., Hu, F. H., et al. (2014). Exosome uptake through clathrin-mediated endocytosis and macropinocytosis and mediating miR-21 delivery. *Journal of Biological Chemistry, 289*(32), 22258–22267. https://doi.org/10.1074/jbc.M114.588046.

Tkach, M., Kowal, J., & Thery, C. (2018). Why the need and how to approach the functional diversity of extracellular vesicles. *Philosophical Transactions of the Royal Society of London B Biological Sciences, 373*(1737). https://doi.org/10.1098/rstb.2016.0479.

Tkach, M., & Thery, C. (2016). Communication by extracellular vesicles: Where we are and where we need to go. *Cell, 164*(6), 1226–1232. https://doi.org/10.1016/j.cell.2016.01.043.

Torreggiani, E., Roncuzzi, L., Perut, F., Zini, N., & Baldini, N. (2016). Multimodal transfer of MDR by exosomes in human osteosarcoma. *International Journal of Oncology, 49*(1), 189–196. https://doi.org/10.3892/ijo.2016.3509.

Trajkovic, K., Hsu, C., Chiantia, S., Rajendran, L., Wenzel, D., Wieland, F., et al. (2008). Ceramide triggers budding of exosome vesicles into multivesicular endosomes. *Science, 319*(5867), 1244–1247. https://doi.org/10.1126/science.1153124. pii:319/5867/1244.

Verweij, F. J., van Eijndhoven, M. A., Hopmans, E. S., Vendrig, T., Wurdinger, T., Cahir-McFarland, E., et al. (2011). LMP1 association with CD63 in endosomes and secretion via exosomes limits constitutive NF-kappaB activation. *The EMBO Journal, 30*(11), 2115–2129. https://doi.org/10.1038/emboj.2011.123.

Viaud, S., Thery, C., Ploix, S., Tursz, T., Lapierre, V., Lantz, O., et al. (2010). Dendritic cell-derived exosomes for cancer immunotherapy: what's next? *Cancer Research, 70*(4), 1281–1285. https://doi.org/10.1158/0008-5472.CAN-09-3276.

Villarroya-Beltri, C., Baixauli, F., Gutierrez-Vazquez, C., Sanchez-Madrid, F., & Mittelbrunn, M. (2014). Sorting it out: Regulation of exosome loading. *Seminars in Cancer Biology, 28C*, 3–13. https://doi.org/10.1016/j.semcancer.2014.04.009. pii: S1044-579X(14)00057-1.

Vizoso, F. J., Eiro, N., Cid, S., Schneider, J., & Perez-Fernandez, R. (2017). Mesenchymal stem cell secretome: Toward cell-free therapeutic strategies in regenerative medicine. *International Journal of Molecular Sciences, 18*(9). https://doi.org/10.3390/ijms18091852.

Vogel, R., Coumans, F. A., Maltesen, R. G., Boing, A. N., Bonnington, K. E., Broekman, M. L., et al. (2016). A standardized method to determine the concentration of extracellular vesicles using tunable resistive pulse sensing. *Journal of Extracellular Vesicles, 5*, 31242. https://doi.org/10.3402/jev.v5.31242.

Wang, G., Dinkins, M., He, Q., Zhu, G., Poirier, C., Campbell, A., et al. (2012). Astrocytes secrete exosomes enriched with proapoptotic ceramide and prostate apoptosis response 4 (PAR-4): Potential mechanism of apoptosis induction in Alzheimer disease (AD). *Journal of Biological Chemistry, 287*(25), 21384–21395. https://doi.org/10.1074/jbc.M112.340513. pii:M112.340513.

Wang, J., Zheng, Y., & Zhao, M. (2016). Exosome-based cancer therapy: Implication for targeting cancer stem cells. *Frontiers in Pharmacology, 7*, 533. https://doi.org/10.3389/fphar.2016.00533.

Weidle, U. H., Birzele, F., Kollmorgen, G., & Ruger, R. (2017). The multiple roles of exosomes in metastasis. *Cancer Genomics And Proteomics, 14*(1), 1–15. https://doi.org/10.21873/cgp.20015.

Whiteside, T. L. (2017a). The effect of tumor-derived exosomes on immune regulation and cancer immunotherapy. *Future Oncology, 13*(28), 2583–2592. https://doi.org/10.2217/fon-2017-0343.

Whiteside, T. L. (2017b). Exosomes carrying immunoinhibitory proteins and their role in cancer. *Clinical and Experimental Immunology, 189*(3), 259–267. https://doi.org/10.1111/cei.12974.

Xu, R., Greening, D. W., Zhu, H. J., Takahashi, N., & Simpson, R. J. (2016). Extracellular vesicle isolation and characterization: Toward clinical application. *Journal of Clinical Investigation, 126*(4), 1152–1162. https://doi.org/10.1172/JCI81129.

Xu, R., Simpson, R. J., & Greening, D. W. (2017). A protocol for isolation and proteomic characterization of distinct extracellular vesicle subtypes by sequential centrifugal ultrafiltration. *Methods in Molecular Biology, 1545*, 91–116. https://doi.org/10.1007/978-1-4939-6728-5_7.

Ye, X., Ishii, I., Kingsbury, M. A., & Chun, J. (2002). Lysophosphatidic acid as a novel cell survival/apoptotic factor. *Biochimica et Biophysica Acta, 1585*(2–3), 108–113. pii:S138819810200330X.

Yokoi, A., Yoshioka, Y., & Ochiya, T. (2015). Towards the realization of clinical extracellular vesicle diagnostics: Challenges and opportunities. *Expert Review of Molecular Diagnostics, 15*(12), 1555–1566. https://doi.org/10.1586/14737159.2015.1104249.

Yoshioka, Y., Kosaka, N., Konishi, Y., Ohta, H., Okamoto, H., Sonoda, H., et al. (2014). Ultra-sensitive liquid biopsy of circulating extracellular vesicles using ExoScreen. *Nature Communications, 5*, 3591. https://doi.org/10.1038/ncomms4591.

Yuyama, K., Sun, H., Mitsutake, S., & Igarashi, Y. (2012). Sphingolipid-modulated exosome secretion promotes the clearance of amyloid-beta by microglia. *Journal of Biological Chemistry*. https://doi.org/10.1074/jbc.M111.324616. pii:M111.324616.

Zeringer, E., Barta, T., Li, M., & Vlassov, A. V. (2015). Strategies for isolation of exosomes. *Cold Spring Harbor Protocols, 2015*(4), 319–323. https://doi.org/10.1101/pdb.top074476.

Zhang, H. G., & Grizzle, W. E. (2014). Exosomes: A novel pathway of local and distant intercellular communication that facilitates the growth and metastasis of neoplastic lesions. *The American Journal of Pathology, 184*(1), 28–41. https://doi.org/10.1016/j.ajpath.2013.09.027. pii:S0002-9440(13)00738-4.

Zhang, Y., Li, X., Becker, K. A., & Gulbins, E. (2009). Ceramide-enriched membrane domains—structure and function. *Biochimica et Biophysica Acta, 1788*(1), 178–183. https://doi.org/10.1016/j.bbamem.2008.07.030. pii:S0005-2736(08)00242-3.

Zhang, W., Xia, W., Lv, Z., Ni, C., Xin, Y., & Yang, L. (2017). Liquid biopsy for Cancer: Circulating tumor cells, circulating free DNA or exosomes? *Cellular Physiology and Biochemistry: International Journal of Experimental Cellular Physiology, Biochemistry, and Pharmacology, 41*(2), 755–768. https://doi.org/10.1159/000458736.

Zhao, H., Achreja, A., Iessi, E., Logozzi, M., Mizzoni, D., Di Raimo, R., et al. (2017). The key role of extracellular vesicles in the metastatic process. *Biochimica et Biophysica Acta, 1869*(1), 64–77. https://doi.org/10.1016/j.bbcan.2017.11.005.

Zhao, H., Yang, L., Baddour, J., Achreja, A., Bernard, V., Moss, T., et al. (2016). Tumor microenvironment derived exosomes pleiotropically modulate cancer cell metabolism. *Elife, 5*, e10250. https://doi.org/10.7554/eLife.10250.

Zhou, H., Summers, S. A., Birnbaum, M. J., & Pittman, R. N. (1998). Inhibition of Akt kinase by cell-permeable ceramide and its implications for ceramide-induced apoptosis. *Journal of Biological Chemistry, 273*(26), 16568–16575.

CHAPTER SIX

Sphingolipids at the Crossroads of NAFLD and Senescence

Mariana Nikolova-Karakashian
University of Kentucky College of Medicine, Department of Physiology, Lexington, KY, United States

Contents

Advances in Cancer Research, Volume 140
ISSN 0065-230X
https://doi.org/10.1016/bs.acr.2018.05.002

Abstract

Nonalcoholic fatty liver disease (NAFLD) is a group of liver disorders encompassing simple hepatic steatosis and its more aggressive forms of nonalcoholic steatohepatitis and cirrhosis. It is a rapidly growing health concern and the major cause for the increasing incidence of primary liver tumors. Unequivocal evidence shows that sphingolipid metabolism is altered in the course of the disease and these changes might contribute to NAFLD progression. Recent data provide solid support to the notion that deregulated ceramide and sphingosine-1-phosphate metabolism are present at all stages of NAFLD, i.e., steatosis, nonalcoholic steatohepatitis, advanced fibrosis, and hepatocellular carcinoma (HCC). Insulin sensitivity, de novo lipogenesis, and the resulting lipotoxicity, fibrosis, and angiogenesis are all seemingly regulated in a manner that involves either ceramide and/or sphingosine-1-phosphate. Sphingolipids might also participate in the onset of hepatocellular senescence. The latter has been shown to contribute to the advancement of cirrhosis to HCC in the classical cases of end-stage liver disease, i.e., viral- or alcohol-induced; however, emerging evidence suggests that senescence is also involved in the pathogenicity of NAFLD possibly via changes in ceramide metabolism.

1. INTRODUCTION

Primary liver cancer is the third largest contributor to cancer mortality in the world and seventh in the United States (Altekruse et al., 2014; Global Burden of Disease Liver Cancer Collaboration, 2017). Hepatocellular carcinoma (HCC) accounts for approximately 90% of all primary liver cancers, whereas intrahepatic cholangiocarcinoma and extrahepatic bile duct carcinoma account for the remaining cases. Viral infection, aflatoxin, alcohol abuse, and nonalcoholic fatty liver disease (NAFLD) are responsible for the vast majority of HCC cases. NAFLD is a rapidly growing concern and is the major cause for the increasing incidence of primary liver tumors. The etiology of NAFLD-related HCC is more diverse and less understood that of more traditional cases, such as HCC due to viral infection and alcohol abuse-associated cirrhosis.

In contrast to most other malignancies, HCC develops due to underlying chronic inflammation, typically, the result of an injury insult that failed to regenerate properly (Ringelhan et al., 2018). The induction of compensatory proliferation of hepatocytes, in an attempt to regenerate the lost or damaged tissue, is also associated with the onset of cellular senescence, which plays a complex role in HCC. On one hand, the onset of senescence in the regenerating hepatocytes limits their regenerative capacity, and at the same time promotes inflammation via the inflammasome-mediated induction of senescence-associated secretory proteins (SASPs). The onset of senescence is suggested to determine whether a hepatic injury will be contained or will progress to end-stage liver disease. Recently, senescent hepatocytes were also observed in pathologies where proliferation is not prominent, i.e., nonalcoholic steatohepatitis (NASH) and alcoholic fatty liver in mice (Yang et al., 2004) and humans (Aravinthan et al., 2013a, 2013b), indicating that the interplay between senescence and inflammation may be important in the progression of NAFLD to HCC.

Sphingolipids are a class of signaling molecules involved in the regulation of many biological processes, including senescence, tumorigenesis, and inflammation. Recent studies have revealed novel functions of sphingolipids in the metabolic syndrome and NAFLD. The focus of this chapter is the role sphingolipids play in NAFLD and in cellular senescence, and their potential role in the development of NAFLD-associated HCC.

2. HEPATOCARCINOGENESIS: AN OUTLINE

Primary liver tumors are derived from single cells, albeit which cells are responsible is still a question of debate (Alison, 2005). Hepatocytes themselves, oval/hepatic progenitor cells, and periductular cells have all been implicated as the lineage that responds to cell damaging insults and are also likely founder cells for the HCC tumors that subsequently develop. In contrast to most other malignancies, HCC develops due to underlying chronic inflammation, typically, the result of an injury insult that failed to regenerate properly (Ringelhan et al., 2018). This injury-associated necroinflammation and sustained cell death engage the local immune cell response, leading to activation of nonparenchymal cells, and the promotion of fibrosis and tumorigenesis.

Liver has a considerable capacity for regeneration. Although only 0.01% of adult liver hepatocytes are proliferating at any one time, in response to parenchyma cell loss the liver mass is restored by self-replication of the hepatocytes. Based on serial transplantation studies, up to 69 hepatocyte doublings can occur in rodent models of injured liver, confirming the clonogenic potential of hepatocytes (Overturf et al., 1997). The induction of such compensatory proliferation of hepatocytes following injury insult contributes to the deregulation of the immune network, DNA damage, epigenetic changes, and the onset chromosomal aberrations. Moreover, the surveillance of aberrant cells becomes deregulated during necroinflammation, thus impeding the capacity of the cells to control for aberrant cell propagation and accumulation.

3. NONALCOHOLIC FATTY LIVER DISEASE AS A RISK FACTOR FOR PRIMARY LIVER CANCER

Viral infections, alcohol abuse, and aflatoxin are well-recognized risk factors for the development of primary liver cancer (Duan et al., 2014). Between 15% and 50% of HCCs, however, develop in the absence of these etiological factors. NAFLD and the metabolic syndrome are considered the main culprits behind hepatic cancers of unknown origin (Bugianesi, 2007) and are also responsible for the recent rise in the incidence of liver cancer in Western countries (Torre et al., 2015). NAFLD is a group of liver disorders encompassing simple hepatic steatosis and its more aggressive forms of NASH and cirrhosis. Simple hepatic steatosis is a very mild form of NAFLD, which is characterized by excessive intrahepatocyte lipid accumulation. It follows a relatively benign clinical course and up to a third of the patients shows spontaneous histological improvement. In some cases, however, simple hepatic steatosis may progress to NASH, a state characterized with cell death and the resulting onset of "sterile" chronic inflammation, in addition to the presence of fat. On average, between one-third and one-half of these NASH patients develop progressive liver fibrosis within 5 years (Duan et al., 2014), whereas 20% will progress to advanced fibrosis (Matteoni et al., 1999), and between 3% and 12% of the patients will get cirrhosis (Calzadilla Bertot & Adams, 2016). Patients with NASH-related cirrhosis progress to HCC at a rate of between 4% and 27%. The longest time interval between NAFLD and HCC has been reported to be around 26 years, whereas the average length of NASH-related cirrhosis, before the diagnosis of HCC is 10–16 years. This is an indication as to the more oncogenic nature of NASH. More than half of the NAFLD-associated HCC occurs in men. NAFLD-associated HCC is usually manifested 4–6 years later than HCC in patients with cirrhosis of other etiology.

Recent reports have challenged the concept of linear progression from steatosis to NASH via chronic inflammation as 44% of NASH patients were without histological inflammation at baseline (McPherson et al., 2015). Hence, some authors consider steatosis and NASH to be separate entities. Similarly, although cirrhosis has long been regarded as an important risk factor for HCC, it is not a requisite stage for the development of HCC in patients with NAFLD. Only about half of the patients with NAFLD- and NASH-related HCC have underlying cirrhosis (Bugianesi et al., 2002) (Dongiovanni, Romeo, & Valenti, 2014). It is of note that the sequence of events and the time course of progression to HCC appear to be less predictable in NAFLD than in chronic hepatitis C or alcoholic liver disease (Zoller & Tilg, 2016).

4. SPHINGOLIPIDS AND THE PATHOGENESIS OF NONALCOHOLIC FATTY LIVER DISEASE

Simple steatosis is the hepatic manifestation of metabolic syndrome and it begins as a disease of deregulated lipid metabolism. Although essentially caused by the excess dietary fat supply, NAFLD etiology involves a complex interplay between diet, obesity, insulin resistance, genetic predisposition, and possibly other factors. Traditionally NAFLD is considered a disease of neutral glycerolipids, mostly triacylglycerols (TAGs) and diacylglycerols (DAGs), but more recently evidence has implicated another lipid class, namely sphingolipids, in the onset of metabolic syndrome in general and NAFLD in particular. Recent studies have found significant changes in the sphingolipid metabolism in conditions associated with hepatic steatosis (Bikman & Summers, 2011a, 2011b; Coen & Goodpaster, 2012). Glycosphingolipids, ceramide, and sphingomyelin (SM), which are the major sphingolipids, have been shown to accumulate in parallel to neutral lipids in livers of obese people and animals (Chaurasia & Summers, 2015; Iqbal et al., 2017). The enzymatic activities and/or mRNA expression levels of several sphingolipid metabolizing enzymes have been reported to be altered in the course of the disease, whereas various transgenic mouse models targeting sphingolipid-metabolizing enzymes have shown attenuation or exacerbation of specific NAFLD symptoms. It has been postulated that among the many lipids that accrue ceramide and its metabolites are amongst the most deleterious because they disrupt insulin sensitivity and mitochondrial metabolism, leading to metabolic derangement and cell death (Bharath et al., 2015; Park et al., 2008; Shimabukuro et al., 1998;

Summers, 2006). Bioactive sphingolipid metabolites, such as ceramide and sphingosine-1-phosphate (S1P), have been implicated in key steps of the pathophysiology of NAFLD, like the initiation of proinflammatory events leading to fibrosis and necroinflammation, as well as proliferative senescence (Fig. 1).

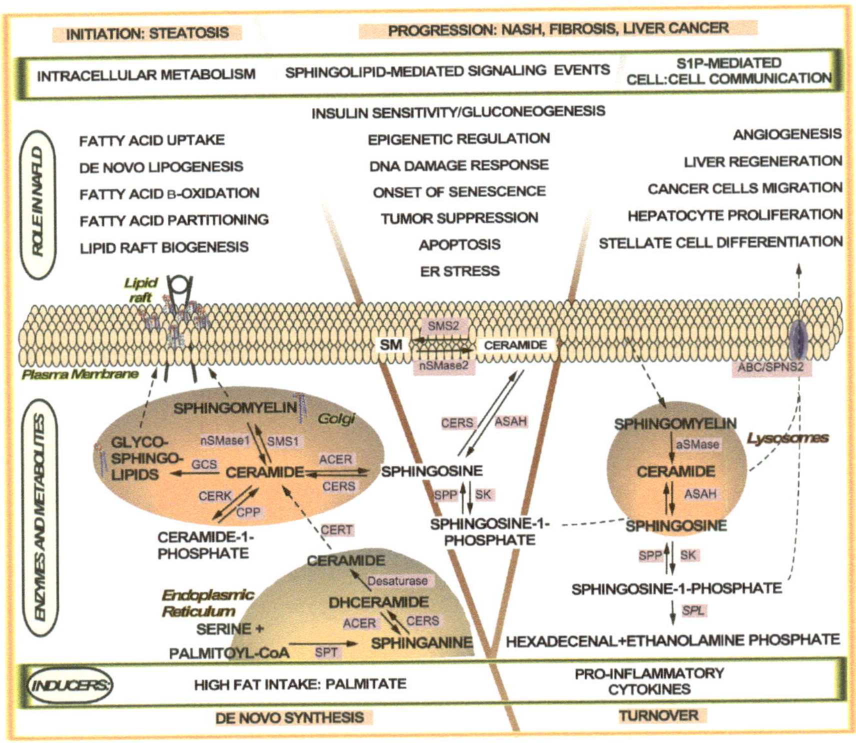

Figure 1 ***Sphingolipid metabolism and nonalcoholic fatty liver disease progression.*** The scheme illustrates key aspects of the known role of sphingolipid metabolic enzymes and their products in the development of steatosis, nonalcoholic steatohepatitis, fibrosis, and liver cancer (see text for detailed explanation). The main enzymatic reactions of sphingolipid synthesis and degradation are shown: Serine palmitoyltransferase (SPT), ceramide synthase (CerS), *sphingomyelin synthase* (*SMS*), glucosylceramide synthase (*GCS*), acid sphingomyelinase (aSMase), neutral sphingomyelinase (nSMase), alkaline ceramidase (ACER), neutral or acidic ceramidase (ASAH), sphingosine kinase (SK), sphingosine-1-phosphate phosphatase (SPP), sphingosine-1-phosphate lyase (SPL). In addition, the S1P exporter (ABC/spns2), which facilitates the release of S1P to the extracellular space and the ceramide transporter (CERT), which transfers ceramide from ER to the Golgi are included. The participation of SM and glycosphingolipids in the cholesterol-rich lipid rafts at the plasma membrane is also depicted. Sphingomyelin, SM; sphingosine-1-phosphate (S1P).

5. OVERVIEW OF SPHINGOLIPID METABOLISM AND FUNCTIONS

Sphingolipids are a class of lipid molecules that all contain a long-chain base called sphingoid backbone. Sphingosine is the most abundant long-chain base in mammalian cells. The synthesis of sphingolipids takes place at the endoplasmic reticulum (ER) and is catalyzed by serine palmitoyltransferase (SPT) (Merrill, Nixon, & Williams, 1985). The product of this reaction is further reduced to sphinganine by the action of 3-ketosphinganine reductase and then acylated to form dihydroceramide by the ceramide synthases (CerS). CerS differ in their specificity for fatty acids of particular acyl chain length and saturation and produce ceramide species that have different fatty acid composition and different bioactive properties (Fei et al., 2011). Ceramide is transferred from the ER to the Golgi apparatus and converted to more complex sphingolipids through the addition of either a phosphocholine group (for SM), glucose (for glucosylceramide) or a phosphate group (for ceramide-1-phosphate) (van Meer & Sprong, 2004) (Hanada et al., 2003). Sphingomyelin synthase (SMS) produces SM by transferring the phosphocholine group of phosphatidylcholine (PC) to ceramide and in the process DAG is generated. The SMS enzyme exists in two different isoforms, the Golgi-localized SMS1 and the plasma membrane–associated SMS2 (Holthuis & Luberto, 2010). Alternatively, ceramide can be glycosylated by glucosylceramide synthase (*GCS*) in the Golgi to form glucosylceramide. Further glycosylation of glucosylceramide can lead to the formation of over 300 different, more complex glycosphingolipids (Messner & Cabot, 2010). Complex glycosphingolipids and SM are main components of the cholesterol-rich lipid platforms on the plasma membrane that serve as signaling clusters and regulate signaling of many receptors.

The sphingomyelinases (SMases) are a family of enzymes all of which can remove the phosphocholine group of SM to produce ceramide. The SMases are classified based on their pH optimum and subcellular localization. The acid sphingomyelinase (aSMase) is localized in the lysosomes and hydrolyzes lysosomal SM delivered by the lipoproteins or through the endocytic pathways (Zeidan & Hannun, 2010). The neutral sphingomyelinase-1 (nSMase1) is localized in the ER/Golgi, whereas neutral sphingomyelinase-2 (nSMase2) facilitates SM hydrolysis at the cytosolic face of the plasma membrane, in the multilamellar bodies in the cytosol and at the nuclear envelope (Airola & Hannun, 2013). NSMase2 has been identified as a bona fide signaling enzyme that mediates cellular responses to stress and inflammation (Nikolova-Karakashian, Karakashian, & Rutkute, 2008; Shamseddine, Airola, & Hannun, 2015). Glycosphingolipids can also be hydrolyzed to ceramide but in contrast to SM, this happens only in the lysosomes by the conjoint activity of specific acid hydrolases. Ceramide is further broken down to sphingosine by a family of ceramidases (ACER1, 2, and 3, ASAH 1 and 2) that are expressed in various subcellular compartments (Coant et al., 2017). Sphingosine can then be phosphorylated to form S1P, a reaction catalyzed by sphingosine kinase 1 and 2 (SK1 and SK2). Consequently, S1P is either dephosphorylated by the S1P phosphatase back to sphingosine or degraded by the S1P lyase to ethanolamine phosphate and trans-2-hexadecenal, two products that cannot reenter the sphingolipid metabolic pathway. Many cells have the capacity to excrete S1P, which bound to HDL or albumin is transported in the serum or is released locally serving as an agonist of a family of G-protein coupled receptors named S1P receptors (S1PRs) (Olivera & Spiegel, 2001; Takabe et al., 2008).

Bioactive sphingolipids have emerged as important players in the regulation of a wide range of cell functions. S1P, sphingosine, and ceramide are the archetypical bioactive sphingolipids, which participate in cell-to-cell communications and in the intracellular signaling pathways (Bartke & Hannun, 2009; Zheng et al., 2006). The complex sphingolipids such as SM and glucosylceramides on the other hand are important for the formation of the signaling microdomains of the plasma membrane known as lipid rafts (Chiantia & London, 2013; Simons & Ikonen, 1997; Zhang et al., 2009). According to the current understanding of the functions of ceramide, various stress inducers, including proinflammatory cytokines, apoptotic stimuli, oxidative stress, and starvation, can cause transient ceramide increases through the activation of its de novo synthesis or via SM turnover (reviewed in (Airola & Hannun, 2013; Nikolova-Karakashian & Rozenova, 2010)). In turn, many stress responses depend on the regulated production of ceramide. Ceramide plays a key role in the onset of apoptosis, growth arrest, and senescence. It has also been implicated in the regulation of numerous cellular processes/functions such as autophagy, mitochondrial functions, DNA damage response, ER stress, and exosome formation. Sphingosine and S1P, which are ceramide metabolites, also have bioactive properties (Newton et al., 2015). S1P in particular has now been established as an important regulatory molecule in cancer on account of its ability to promote cell proliferation, migration, invasion, angiogenesis, and lymphangiogenesis, as well as resistance to chemotherapy, radiotherapy, and immunotherapy (Nagahashi et al., 2016).

6. SPHINGOLIPIDS AND SIMPLE STEATOSIS

6.1 Changes in Sphingolipid Metabolism During Simple Steatosis

Simple steatosis is defined as the presence of lipid droplets within the cytoplasm in more than 5% of the hepatocytes, in the absence of excessive alcohol consumption (less than 20 g/day for women and 30 g/day for men) (Brunt et al., 2015). Many studies in several animal models have established that a common feature in the onset of simple steatosis is activation in the liver of the first step in the de novo synthesis of sphingolipids, the step catalyzed by SPT. Mice fed proatherogenic diet (42% saturated fats and 1.5% cholesterol), develop steatosis associated with substantial (more than fivefold) increase in the expression of SPT1 (the main subunit of the SPT protein complex), as well as a corresponding increase in SPT activity, and the accumulation of all intermediates of the de novo synthesis of sphingolipids (dihydrosphingosine, dihydroceramide, ceramide, and SM) (Deevska et al., 2009). Other dietary studies did not find substantial net increase in hepatic ceramide levels but reported a significant shift in ceramide fatty acid makeup, away from very long chain species and towards the proinflammatory and proapoptotic C16:0 and C18:0 ceramides (Cinar et al., 2014; Kasumov et al., 2015). Feeding mice with methionine and choline deficient diet (MCD), which induces severe hepatic steatosis in the absence of systemic insulin resistance, was also shown to result in the hepatic accumulation of ceramide and SM, as well as the activation of acid sphingomyelinase (Mas et al., 2009). Genetic models of steatosis have also exhibited ceramide accumulation. Livers of methionine adenosyltransferase 1A knockout mice were characterized by high levels of tri- and diacylglycerols, as well as ceramides (Alonso et al., 2017). In the ob/ob mice, the hepatic levels of ceramide increase in correlation with the degree of steatosis (Yetukuri et al., 2007). At present it is unclear what, if anything, changes in the sphingolipid metabolism in humans with simple steatosis. This is due to the scarcity of biopsy samples of this early stage of NAFLD.

Solid evidence supports the notion that stimulation of sphingolipid de novo synthesis contributes to the onset of steatosis. Pharmacological inhibitors that target ceramide production and genetic ablation of particular sphingolipid enzymes in the mouse have an impact on the development of simple steatosis, albeit the underlying mechanisms are still elusive. Myriocin, a specific inhibitor of SPT, which decreases the levels of ceramide and other more complex sphingolipids, has been shown to alleviate steatosis in rats (Kurek et al., 2014), in genetically obese (ob/ob) mice, and in mice on a high-fat diet (Yang et al., 2009). In some cases, the suppressive effects seen on hepatic steatosis were associated with the overall change in body weight. Therefore, whether or not these effects of myriocine are intrinsic for hepatocytes is unclear.

6.2 Sphingolipid Regulation of Hepatic Lipid Accumulation

Increased influx of fatty acids from the adipose tissue, deregulation of hepatic de novo lipogenesis, defects in the β-oxidation of free fatty acids in the liver, and lipid dystrophy caused by defects in VLDL secretion are the main factors leading to the onset of steatosis (Arab, Arrese, & Trauner, 2018). Attempts to determine the origin of hepatic TAGs in NAFLD, using labeled stable isotopes, have indicated that serum free fatty acids (derived from lipolysis of TAG in the adipose tissue) account for up to 60% of the hepatic TAG in NAFLD patients. The contribution of fatty acids produced via the de novo lipogenesis (which has been shown to be up to 3 times higher in NAFLD patients as compared to controls) and fatty acids derived from the diet has been estimated to be 25% and 15%, respectively (Arab et al., 2018; Donnelly et al., 2005; Lambert et al., 2014).

6.2.1 Sphingolipid Role in Regulation of Fatty Acid Uptake by Hepatocytes

The onset of insulin resistance in adipose tissue of obese patients with diabetes is responsible for a surplus of available free fatty acids, which drives the initiation of steatosis. In NAFLD patients, insulin does not suppress the hormone sensitive lipase mRNA expression (and respectively, the adipocyte TAG lipolysis) to the same extent that it does in healthy individuals (Lewis et al., 2002). Sphingolipids, particularly ceramide, have been implicated in the onset of adipocyte insulin resistance as summarized in several excellent reviews on that topic (Bikman & Summers, 2011a, 2011b; Holland & Summers, 2008). For conciseness however, these results will not be discussed here. Notwithstanding, some complex sphingolipids, such as SM and glucosylceramide, have been shown to also influence the uptake of the adipocytes-derived fatty acids in hepatocytes. The latter express specialized transporters, such as FATP2, FATP5, caveolin-1, and CD36, which facilitate free fatty acid uptake (Berlanga et al., 2014). In morbidly obese patients with NAFLD, CD36 mRNA levels have been correlated with liver fat content and apoptosis (Bechmann et al., 2010;

Greco et al., 2008). Plasma membrane localized sphingolipids have been implicated in functional regulation of CD36 and caveolin 1. Mice deficient in the plasma membrane form of the SMS, SMS2, are protected from steatosis due to impaired CD36/FAT-mediated uptake of free fatty acids. It was postulated that the plasma membrane SM regulated the activity of CD36 in a manner involving lipid rafts formation (Mitsutake et al., 2011). Similarly, a shift in the sphingolipid fatty acid makeup away from long very long chains (C22:0 and longer) and toward long chain (C16:0) also reduced steatosis in the mouse due to selective downregulation of FATP5 and FABP1 and protein mislocalization (Park et al., 2014). Thus, sphingolipids apparently influence the fatty acid transport across the hepatocyte plasma membrane.

6.2.2 Sphingolipid Role in Regulation of De Novo Lipogenesis in Liver

Stimulation of de novo lipogenesis is another main source of fatty acids for hepatic TAG accumulation during steatosis (Postic & Girard, 2008). The selective insulin resistance of hepatocytes, combined with the hyperinsulinaemia that often accompanies type II diabetes, is the main culprits behind the increased de novo lipogenesis in NAFLD patients. In healthy hepatocytes, de novo lipogenesis is under the control of insulin, which stimulates acetyl-CoA carboxylase (ACC) and fatty acid synthase (FAS) mRNA expression in a SREBP1c-dependant manner. Simultaneously to these lipogenic effects, insulin also suppresses the expression of the rate-limiting step in gluconeogenesis, namely, PEPCK1 mRNA that is under the control of FoxO1 transcription factor. As a consequence of this mode of regulation, in nondiabetic animals and humans the release of insulin in fed state results is a decrease in hepatic glucose output, which helps to keep blood glucose low, and in an increase in TAG synthesis for secretion and storage in the adipose tissue. The concept of "selective insulin resistance" proposed by Goldstein and Brown (Brown & Goldstein, 2008) states that in the liver only the FoxO1 pathway becomes insulin resistant during type II diabetes; as consequence, the mRNAs for PEPCK and G6Pase remain high, and gluconeogenesis continues. The elevated blood glucose induces pancreatic secretion of more insulin, thus leading to hyperinsulinemia. In contrast, the hepatic insulin sensitivity is maintained in the SREBP-1c pathway (Shimomura et al., 2000), fatty acid synthesis is accelerated, and TAG accumulates in the liver. The paralleled hyperinsulinemia exacerbates the flux through the SREBP1c pathway. The presence of fructose in the diet fuels lipogenesis further via the activation of carbohydrate-responsive element-binding protein, which together with SREBPc induces ACC and FAS.

The causes of the selective insulin resistance of hepatocytes are not well understood. However, some of the changes in sphingolipid metabolism seen in steatosis might be involved. In a healthy state, the insulin-activated Akt phosphorylates FoxO-1 in the nucleus, resulting in Foxo-1 nuclear export and degradation. NSMase-2, which is sensitive to activation by a variety of cytokines present in obesity, has suppressive effect on the canonical IRS/PI3K/AKT/FOXO1 pathway. The product of nSMase-2, ceramide, has been shown to activate protein phosphatase 2A (PP2A) and inhibit Akt activity with subsequent sequestration from the cytoplasm to the plasma membrane (Holland & Summers, 2008) (Bikman & Summers, 2011a, 2011b). Exposure of cultured hepatocytes to short-chain ceramides blunts insulin-mediated activation of Akt supporting a role of ceramide in Akt deactivation during insulin resistance (Ribaux & Iynedjian, 2003). Furthermore, nSMase-2-derived ceramide induces directly nuclear retention and activation of FoxO-1 in hepatocytes via an Akt-independent mechanism (Dobierzewska et al., 2012). The elevated synthesis of complex glycosphingolipids during hepatic steatosis has also been implicated in the induction of insulin resistance by impacting the insulin receptor localization in the signaling compartments of the plasma membrane, the lipid rafts. The iminosugar N-(5′-adamantane-1′-yl-methoxy)-pentyl-1-deoxynojirimycin, an inhibitor of the enzyme GCS, is a potent enhancer of insulin signaling in rodent models for insulin resistance and type 2 diabetes (Bijl et al., 2009).

The NAFLD-associated increases in de novo lipogenesis are caused by increased SREBP1 mRNA expression and activity. SREBPs are present as membrane-bound inactive precursors at the ER whose maturation and activation are controlled by insulin and sterol depletion. The NH_2-terminal domain of SREBP is the active transcription factor region that is released proteolytically at the Golgi apparatus. SREBP cleavage activating protein (SCAP) and insulin-induced gene (INSIG) protein are the two proteins that control the transport of the SREBP1 precursor from ER to Golgi and the consequential release of the active transcription factor. Dissociation of INSIGs from the SCAP–SREBP complex is required for the SREBP–SCAP complex to move to the Golgi for the release of the active transcription factor by proteolysis. The mechanisms controlling SREBP1 maturation in response to insulin in a healthy state involve the PI3K/AMPK/mTOR pathway (Xu et al., 2013). Several studies have provided evidence that active sphingolipid synthesis in the ER is a prerequisite for the maturation of SREBP. Cells deficient in SPT fail to increase SREBP in response to sterol depletion, while inhibition of SPT or ceramide synthase decreases SREBP and SRE-mediated gene transcription causing an accumulation of precursor SREBP in the ER (Overturf et al., 1997) (Worgall, 2011). Vice versa, increasing sphingolipid synthesis by providing precursors or intermediaries of sphingolipid synthesis, a treatment that resemble closely the diet-induced stimulation of ceramide de novo synthesis increases SREBP and SRE-mediated gene transcription. Notably, the observed regulation of SREBP by sphingolipids occurs independently of cellular cholesterol concentration. These results suggest that the process of ongoing ceramide synthesis ostensibly may contribute to active processing of SREBP, perhaps affecting ER-to-Golgi trafficking (Worgall et al., 2004).

Another sphingolipid, S1P, appears to influence the regulation of SREBP1 levels via epigenetic mechanisms (Nagahashi et al., 2015). S1P is an endogenous inhibitor of histone deacetylases1 and 2 (Hait et al., 2009). A recent study found that SK2 is upregulated in hepatocytes of mice following high-fat diet leading to increased S1P production in the nuclei. Mice deficient in SK2 show substantially suppression of transcription of several SRE-dependent genes, suggesting that S1P may play a role in the onset or the progression of NAFLD.

6.2.3 Sphingolipid Role in Regulation of Mitochondrial β-Oxidation

Hepatic fatty acid oxidation occurs in the mitochondria, peroxisomes, or microsomes. During fasting, fatty acids enter into the citric cycle after being broken down into acetyl-CoA during β-oxidation. In the process, the long-chain fatty acids are activated to acyl-CoA in the cytosol, transported, and catalyzed in the outer mitochondrial membrane. The role of mitochondrial β-oxidation in NAFLD is not clear and contradictory reports show both reduced and increased rates of fatty oxidation in NAFLD, with some studies demonstrating an impairment with progression to NASH (Arab et al., 2018). Increased production of C16:0 ceramide has been associated with impaired β-oxidation resulting from inactivation of electron transport chain components and sufficient to increase hepatic propensity for steatohepatitis and insulin resistance (Raichur et al., 2014). In turn, the adiponectin receptor, which is known to increase free fatty acid oxidation and mitochondrial biogenesis on adiponectin activation, was recently reported to have intrinsic ceramidase activity, effectively decreasing cellular ceramide content (Arab et al., 2018; Holland et al., 2017).

6.2.4 Partitioning of Excess Fat Into Different Metabolic Pathways and the Onset of Steatosis

The partitioning of the excess fatty acids into different metabolic pathway lipid also has an impact on the development of steatosis in the liver. One of the earliest animal models of hepatic steatosis and NASH involved deprivation of mice or rats from choline and methionine (i.e., the MCD diet). As PC is essential for the secretion of TAG-rich VLDL particles, the decline in PC levels during MCD diet leads to defects in VLDL secretion and, respectively, steatosis. Lipoprotein secretion aside, more recent studies have shown that partitioning of excess fat in the neutral lipid, phospholipid, and sphingolipid metabolic pathways happens in a regulated fashion. Conceivably, the metabolic fate of the fatty acid along these different lipid classes might have an impact on the pathogenic potential of steatosis in view of the differences in properties of bioactive lipids generated along these pathways.

DAG is a precursor for the synthesis of all glycerolipids. TAG is synthesized following the acylation of DAG by acyl-CoA: diacylglycerol acyltransferase (DGAT) (Coleman & Lee, 2004; Declercq et al., 1984; Kennedy, 1957). The main phospholipids, PC and PE are produced via head group transfer catalyzed by choline/ethanolamine phosphotransferase 1 (CEPT1) (Vance, 2015). Several studies have shown that the DAG is being utilized for TAG and phospholipid synthesis pathways in a competitive manner (Fagone & Jackowski, 2013). For example, the enforced expression of CTP:phosphocholine cytidylyltransferase 1 (CCT1), which is the rate-limiting step in de novo PC synthesis and provides the CDP-choline substrate for CEPT1, has been shown to stimulate PC biosynthesis while reducing that of TAG (Jackowski, Wang, & Baburina, 2000). In contrast, the direct inhibition of CEPT1 or the genetic deletion of phosphoethanolamine cytidylyltransferase (which catalyzes the formation of CDP-ethanolamine for PE synthesis) has been shown to stimulate the synthesis of TAG by as much as 10-fold (Jackowski et al., 2000; Leonardi et al., 2009). Similar correlations are seen in vivo, as liver-specific deletion of CCT1 has been associated with the development of mild hepatic steatosis (Jacobs et al., 2004). The rate of sphingolipid synthesis influences the partitioning of fatty acids into TAG pathway (Aguilera-Romero, Gehin, & Riezman, 2014). Overexpression of the Golgi-localized form of SMS, SMS1, suppresses hepatocytes' ability to form and store TAG even when challenged with excess palmitic acid was also severely reduced (Deevska et al., 2017). Our earlier studies have shown that genetic ablation of aSMase in the

LDLR-deficient mice significantly reduces diet-induced hepatic steatosis due to disturbances of fatty acid metabolic flux (Deevska et al., 2009).

7. ROLE OF SPHINGOLIPIDS IN NONALCOHOLIC STEATOHEPATITIS

The progression from simple steatosis to the more aggressive form, NASH, is considered to be a sequential process, although the possibility that these two conditions are different entities has also been considered. A characteristic lipotoxic injury to the hepatocytes ensues early in the disease and leads to cell death, local inflammation, and fibrogenesis. Different factors, including fat quantity and type, genetic predisposition, and the presence of systemic inflammation, sustain and propagate the pathological process. At cellular level, lipoapoptosis, oxidative and/or ER stress, and deregulated autophagy have all been implicated in NASH (Engin, 2017). The detailed molecular mechanisms governing the transition from steatosis to steatohepatitis are not yet fully understood (Ilan, 2016).

7.1 Changes in Lipid Metabolism During Nonalcoholic Steatohepatitis

Lipids play key role in the progression of the disease phenotype in NASH. Excessive lipid accumulation drives hepatic injury; however, our knowledge of the specific lipids that initiate inflammation and fibrosis is still abstract. TAG is the dominant lipid that accumulates during hepatic steatosis. Later studies demonstrated significant alterations in long-chain fatty acyl coenzyme A, DAG, free cholesterol, ceramide, and other sphingolipids in NASH (Puri et al., 2007). A growing body of evidence suggests that the lipid compartmentalization in hepatocytes and the "quality" as opposed to the "quantity" of the accumulating lipids may play a central role in the risk for progressive disease (McClain, Barve, & Deaciuc, 2007). Moreover, considerable data have now indicated that TAG accumulation per se is not harmful to hepatocytes and may represent a protective mechanism against lipotoxicity. Saturated free fatty acids, whose level in the liver increases by almost threefold in biopsy samples of NASH patients, have been associated with the upregulation of death receptors such as CD95/Fas and tumor necrosis factor-related apoptosis-inducing ligand (TRAIL)-death receptor 5 (DR5) (Feldstein et al., 2003). Indeed, enhanced expression of these receptors is one potential mechanism by which free fatty acids may sensitize liver cells to apoptotic cell death. Both Fas and DR5 expression have been shown to increase in livers of patients with NASH (Kahraman et al., 2010). Free fatty acids have also been implicated in deregulation of mitochondrial–lysosomal pathway, the induction of ER stress and in activation of the Toll-like receptors (TLR), specifically TLR4, which may provide a basis of obesity-induced hepatic inflammation.

7.2 Sphingolipids as Mediators of Saturated Fatty Acid-Dependent Nonalcoholic Steatohepatitis

Sphingolipids, ceramide in particular, are the most recent addition to the plethora of bioactive lipids that have an impact on the progression of NASH. The dependence of ceramide synthesis on saturated fats, such as palmitate, makes this sphingolipid an appealing candidate metabolite for linking the excess saturated free fatty acids to the development of NASH (Alkhouri, Dixon, & Feldstein, 2009; Summers, 2006). Ceramide is a key mediator of cell stress and death ligand—induced cell death (Mari & Fernandez-Checa, 2007). Stimulation of de novo synthesis of ceramide is required for induction of apoptosis in response to TNFα and CD95/Fas (Alkhouri et al., 2009; Summers, 2006). Ceramide has also been implicated in regulation of autophagy, ER stress, and mitochondrial function (Mullen, Hannun, & Obeid, 2012). While the excess fat availability fuels ceramide generation by increasing the flux of precursors though the de novo synthetic pathway, sustained inflammation and GSH depletion, which is another feature of the steatotic liver, could increase ceramide generation further via activation of sphingomyelinase (Hardwick et al., 2010; Nikolova-Karakashian et al., 2008; Rutkute, Asmis, & Nikolova-Karakashian, 2007). Thus, ceramide is poised to play a key role in the "vicious cycle" of uncontrolled inflammation and injury that delineates the progression of NASH.

The case of ceramide involvement in NASH is well illustrated in studies using pharmacological or genetic targeting of ceramide-producing pathways in the liver. Chronic consumption of palmitic acid (which stimulates de novo ceramide synthesis) plus myristic acid-enriched diet for 6 months was found to induce NASH characterized with lipodystrophy, fibrosis, ER stress, and local inflammation in the mouse (Martinez et al., 2015). Treatment with myriocin, an SPT inhibitor, effectively prevented ceramide generation resulting in lower liver damage, attenuation of fibrosis, and a decreased expression of the inflammation markers IL-1β, IL-6, and TNFα. Myriocin was also effective in suppressing Western diet induced steatosis, inflammation, fibrosis, and apoptosis in $LDLR^{-/-}$ mice (Kasumov et al., 2015). ASMase has been similarly, associated with the progression of simple steatosis to NASH. ASMase mRNA levels increased threefold in patients with NASH compared with healthy controls (Moles et al., 2010), whereas ASMase-deficient mice are protected against dietary induced NASH via changes in autophagy and ER stress (Fucho et al., 2014). The pharmacological premise of targeting sphingolipid metabolism in human with NASH is underlined further by a study of patients undergoing a 1-year course of implementing lifestyle changes to delay NASH progression. Analyses of liver biopsies provided evidence that several genes involved in de novo ceramide biosynthesis and degradation were differentially expressed in hepatic tissue of subjects with NASH compared to controls. Weight reduction improved steatosis and hepatocellular injury and substantially changed the expression of ceramide-related genes in the liver. Changes in calorie intake and fat consumption, in particular cholesterol and saturated fat, were also significantly correlated with changes in ceramide-related gene expression (Promrat et al., 2011).

More recent studies suggest that the length and saturation of the fatty acid attached to the primary amine of the sphingoid base in ceramide is an important determinant of the "pathogenic" potential of ceramide. Haploinsufficiency for ceramide synthase 2 (CerS2), the dominant isoform in the liver that preferentially makes very-long-chain (C22/C24/C24:1) ceramides was shown to cause a compensatory increases in C16-ceramides and to confer susceptibility to diet-induced steatohepatitis (Raichur et al., 2014). C18- and C16-ceramides but not C24:1 have been associated with induction of apoptosis in hepatocytes (Osawa et al., 2005).

The ceramide derivative S1P seemingly mediates at least some of the profibrotic events during NASH pathogenesis. Hepatocyte-released S1P has been shown to induce macrophage chemotaxis in vitro (Kakazu et al., 2016), whereas FTY720, an S1PR antagonist, ameliorates NASH in C57BL/6 male mice feed a diet rich in fructose, saturated fat, and cholesterol for 22 weeks (Mauer et al., 2017). Daily intraperitoneal injections of FTY720 for the last 2 weeks of the diet were sufficient to improve liver histology with a significant reduction in hepatocyte ballooning and inflammatory foci and to reverse hepatomegaly, collagen accumulation, and macrophage infiltration (Mauer et al., 2017).

8. ADVANCED FIBROSIS AND CIRRHOSIS

Fibrosis is the wound-healing response of liver to chronic injury and involves myofibroblast—like cells derived from a number of cellular sources, including activated hepatic stellate cells (HSCs) that normally reside in the space of Disse, or bone marrow–derived mesenchymal stem cells (BMDSC) that migrate to the injured liver tissue (Friedman, 2008). S1P has emerged as key mediator of stellate cells activation by injured hepatocyte.

8.1 Changes in Sphingolipid Metabolism During Advanced Fibrosis

Advanced fibrosis in the liver is associated with significant alteration in S1P metabolism, according to animal and human studies. In fibrotic liver tissue from mice and human patients, hepatic S1P levels are elevated due to increased hepatic SK1 expression (King et al., 2017)(Li et al., 2011). The levels of two of the S1P receptors, S1P1 and S1P3, are also elevated in cholestasis-induced liver fibrosis and in human fibrotic samples, whereas the S1P2 levels are decreased (Xiu et al., 2015)(Li et al., 2011). The mRNA levels of the S1P transporter, Spns2, but not those of S1P lyase or S1P phosphatase, were reportedly enhanced in human liver with advanced fibrosis, indicating an increased export of S1P (Sato et al., 2016). Several studies have examined the association of circulating S1P to NAFLD and found conflicting results that may be attributed to different disease stage. In one study, serum S1P levels in patients with HCC were elevated as compared to those with cirrhosis (Ressom et al., 2012). Interestingly, analysis of 95 patients with end-stage liver disease showed that low plasma S1P concentrations are highly significantly associated with poor prognosis for end-stage liver disease (Becker et al., 2017). These apparent discrepancies suggest that S1P plays a complex role in pathogenesis of advanced hepatic fibrosis and cirrhosis.

According to the current concept, palmitate oversupply of hepatocytes increases intra- and extracellular S1P concentrations by activating SK1 (Al Fadel et al., 2016). In addition to hepatocytes, HSC, BMDSL, and the endothelial cells of hepatic vasculature can all generate excess S1P during advanced fibrosis (González-Fernández, 2017). This is most likely attributed to the presence of chronic inflammation, as proinflammatory cytokines, chemokines, and intercellular adhesion molecules have been known to activate SK.

8.2 Role of Sphingosine-1-Phosphate in Fibrosis, Angiogenesis, and Liver Regeneration

The functions of S1P during fibrosis have been associated mainly with the transdifferentiation of BMDSC into myofibroblast-like cells, the proliferation and activation of the HSCs, and the activation of synthesis and release of fibrous proteins, including smooth muscle α-actin, and collagen I and III. In the CCl_4-induced mouse model of liver injury, for example, TGF-β induces activation of the BMDSC via SK1-upregulation and modulation of S1P receptors expression. Pharmacological or siRNA-mediated inhibition of SK1 abrogates the prodifferentiating effect of TGF-β (Yang et al., 2012). The production of collagen α1 is increased by SK1 in human BMDSCs and in human hepatogenic profibrotic cells following TGF-β treatment (Xiu et al., 2015). In vitro experiments show that S1P directly participates in HSCs activation and differentiation to myofibroblasts and induces HSCs proliferation and migration (Ikeda et al., 2000). S1P exerts a powerful migratory action on human myofibroblasts (Li et al., 2011).

S1P is critically involved in liver regeneration and repair, which are beneficial for the outcome of NASH. S1P was found to facilitate hepatocytes regeneration after liver injury via the S1P2 (Ikeda et al., 2009). Hepatocytes-derived S1P-enriched exosomes have been shown to promote hepatocyte proliferation and liver regeneration in vivo (Nojima et al., 2016). S1P also promotes survival and proliferation of liver sinusoidal endothelial cells by activating ERK and Akt and by affecting the expression levels of Bcl-2, Bax, and cleaved caspase-3, as well as by promoting vascular endothelial growth factor production (Nowatari et al., 2015). S1P also plays a role in inducing hepatic angiogenesis during advanced fibrosis (Yang et al., 2013). In the injured liver, regeneration of the damaged sinusoidal vasculature is a requirement for the regrowth of a nonfibrotic liver mass and for hepatic restoration. In several animal models of fibrosis, including bile duct ligation and CCl_4 administration, as well as in human fibrotic samples, S1P levels in the liver change in correlation with mRNA expression of angiogenic markers, supporting a beneficial role of S1P in the repair of hepatic injury during advanced NAFLD (Yang et al., 2013).

Most of the effects of S1P are mediated via the S1P receptors, although S1P receptor-independent effects have also been observed (Ikeda et al., 2000). CCl_4-challenged S1P2-deficient mice manifest with less fibrosis, reduced expression of smooth-muscle alpha-actin-positive cells and decreased accumulation of HSCs than wild-type controls (Ikeda et al., 2009). The prodifferentiating effects of TGF-β1 and some angiogenic effects of S1P are seemingly mediated by S1PR1 and S1PR3 (Yang et al., 2012). The S1P migratory effects are effectively mimicked by SEW2871 (an S1P1 agonist) and blocked by suramin (an S1P3 antagonist). In contrast, JTE-013 (an S1P2 antagonist) enhances S1P-induced migration (Li et al.,

2011). S1PR1/3 antagonist significantly suppresses the upregulation of angiogenic markers in the injured liver and attenuated the extent of liver fibrosis, whereas S1PR2 antagonist had no effect on angiogenesis (Yang et al., 2013).

Earlier studies have shown that acid-sphingomyelinase-derived ceramide also plays a role in the advancement of fibrosis during NAFLD. Activation of acid sphingomyelinase has been observed in the process of HSC activation and shown to be required for the fibrotic response (Moles et al., 2010). Ostensibly, some of these effects could be due to changes in S1P production, which is downstream of aSMase.

9. SPHINGOLIPIDS IN HEPATOCELLULAR CARCINOMA

HCC is an increasingly recognized complication of NAFLD. The progression of chronic liver disease to cirrhosis and/or cancer is highly variable between individuals (Hardy & Mann, 2016). Genome-wide association studies have identified genetic variants that show association with fibrosis and HCC (Nowatari et al., 2015). Deregulated signaling cascades such as insulin growth factor, TGFβ, and wnt- and cadherin pathways have also been well documented. Of note, genetic changes in signaling pathways that link inflammation, cell cycle, and metabolic control via PI3K are present in just over 10% of HCC (Zoller & Tilg, 2016) Finally, aberrant DNA methylation has been shown to have a significant impact on disease progression. In human NAFLD, hepatic DNA hypomethylation has been found to be associated with more advanced forms of the disease (Murphy et al., 2013).

9.1 Changes in Ceramide Metabolism During Hepatocellular Carcinoma

Several recent studies have implicated ceramide in the onset and progression of HCC. Mass spectrometry—based analyses of serum concentrations of sphingolipid metabolites in 122 HCC patients and 127 age- and sex-matched patients with cirrhosis showed highly significant upregulation of long and very-long-chain ceramides (C16—C24) in the serum of patients with HCC as compared with patients with cirrhosis. In addition, dihydroceramides, sphingosine, S1P, and sphinganine-1-phosphate were all upregulated in patients with HCC (Grammatikos et al, 2015, 2016; Ressom et al., 2012). Increased de novo synthesis and secretion by the liver is one possible cause for these "across the board" changes seen in serum sphingolipids.

Importantly, C16-ceramide and S1P exhibited better diagnostic potential than α-fetoprotein suggesting that C16-ceramide and S1P may serve as novel diagnostic markers for the identification of HCC in patients with liver diseases.

Analyses of sphingolipid mass in liver tissues from HCC patients show the opposite trend to the changes seen in serum. Ceramide was markedly reduced in HCC tissues, whereas sphingomyelin content was elevated, suggesting an impaired sphingomyelinase activity in HCC (Krautbauer et al., 2016). Several studies have examined the role of neutral sphingomyelinase-2 (nSMase-2), a bona fide signaling enzyme, which in normal tissues is regulated by TNFα and IL-1β. An unbiased approach to study the role of aberrant DNA methylation of tumor suppressor genes in HCC (Hardy & Mann, 2016) lead to the discovery of 13 candidate genes (Revill et al., 2013), one of which was *smpd3, the gene encoding nSMase-2.* Knocking down smpd3 with small hairpin RNA was shown to promote cell invasion and migration in vitro and to increase tumor formation after subcutaneous injection or transplantation into mice (Revill et al., 2013). Reduced levels of *smpd3* were also associated with early recurrence of liver tumors following curative resection (Revill et al., 2013). Therefore, nSMase-2 was proposed to serve as a potent tumor suppressor gene, which is downregulated in HCC.

9.2 Sphingolipid-Targeted Therapeutic Approaches in Hepatocellular Carcinoma

Studies on ceramide-modifying enzymes provide further evidence of potentially critical role of ceramide in tumor cell survival. GCS, an enzyme that converts ceramide to more complex sphingolipids, essentially decreasing its cellular content, is upregulated by a chemotherapeutic agent, sorafenib, in sorafenib-resistant hepatoma cell lines. GCS silencing or pharmacological GCS inhibition sensitized these hepatoma cells to sorafenib exposure. Similar results are seen in a xenograft mouse model, suggesting that cellular ceramide homeostasis is important for HCC survival (Stefanovic et al., 2016). Exogenous delivery of nanoliposomal C6-ceramide to mice engrafted with SK-HEP-1 tumors reduced tumor vascularization and proliferation, induced tumor cell apoptosis, decreased phosphorylation of AKT and ultimately blocked tumor growth (Tagaram et al., 2011). In agreement with the emerging role of ceramide as tumor suppressor in HCC, CerS2-deficient mice were found to develop hepatocarcinomas within 7 months of age (Promrat et al., 2011).

9.3 Role of SK in HCC Progression

HCC is a highly aggressive neoplasm and sphingosine kinase and its product, S1P, seemingly contribute to this aggressiveness (Lu et al., 2017). Assessment of SK1/2 expression in 77 patients with HCC revealed higher mRNA levels of both sphingosine kinases, which were associated with poorer differentiation and microvascular invasion. Higher level of SK2 mRNA was identified as a risk factor for intra- and extrahepatic recurrence (Uranbileg et al., 2016). Similar changes were found in a separate study where the expression of SK1 protein was examined in human HCC tissues using immunohistochemistry, and its clinical implications and prognostic significance analyzed (Cai et al., 2017). The expression of SK1 in HCC tissue was significantly higher than in normal tissue and significantly associated with tumor size, tumor stage, and histological differentiation. The patients with low SK1 expression had higher overall survival and recurrence-free survival rates compared with patients with high SK1 expression. Interestingly, the natural prenylflavonoid icaritin, which has cytotoxic and proapoptotic effects on HCC, appears to do so by SK1 inhibition as primary mechanism (Lu et al., 2017).

Studies with FTY720 show that this antagonist of S1P signaling suppresses the invasiveness and proliferation of HCC in HCC-bearing rats by suppressing cancer cell migration and proliferation. FTY720 also leads to a significant prolongation of survival in vivo (Ushitora et al., 2009). Intracellular generation of S1P, via the nuclei-localized SK2 has been implicated in DNA damage response in HCC, whereas SphK1 and S1P were shown to transactivate EGFR promoting DNA damage repair (Nagahashi et al., 2016). Thus, in the context of HCC, SK, and S1P seemingly influence the carcinogenic process by induction of cell proliferation and DNA damage response.

10. THE ROLE OF SENESCENCE IN NONALCOHOLIC FATTY LIVER DISEASE

Cellular senescence plays a key role in controlling the balance between regenerative proliferation and the tumorigenic process. When liver damage persists, pathways that help hepatocyte proliferation are beneficial for achieving a sustained liver repair. At the same time, increased proliferative rates and the induction of DDR can favor a tumorigenic process. The onset of senescence in the regenerating hepatocytes during liver insult has been suggested to play important role in determining whether a hepatic injury will be contained or will progress to end-stage liver disease. Noteworthy, sphingolipids had emerged as potential novel mediators of senescence during NAFLD.

10.1 Mechanisms and Physiological Significance of Cellular Senescence

Initially used by Hayflick and colleagues to describe cells that ceased to divide in culture, the term cellular senescence now denotes a complex paradigm of cellular response to stress and damage from exogenous and endogenous sources. Senescent cells undergo a permanent cell cycle arrest and are identified by the presence of senescence markers such as senescence-associated β-galactosidase, $p16^{INK4A}$, senescence-associated DNA-damage and heterochromatin foci (Campisi & d'Adda di Fagagna, 2007). Telomere-dependent, oncogene-induced, and ROS-induced senescence are the three best-studied senescence pathways. They all involve activation of the DDR and lead to elevation of key cell cycle regulatory molecules including P53, p21, and P15INK4a, followed by growth arrest. In the case of telomere-dependent senescence, the DDR is a consequence of the telomeres shortening with each round of replication. Conversely, direct oxidative damage of the DNA is assumed to be the cause of ROS-induced senescence. DNA hyper-replication and consequently the appearance of partially replicated DNAs and DNA double-strand breaks have been implicated in oncogene-dependent senescence.

Activation of DDR and growth arrest is central for proliferative senescence. The prolong activation of DDR also activates molecules, such as ATM, NRS1, and CHK2, leading to the activation of inflammasome and, respectively, NF-KB and C/EBP transcription factor, which induce mRNA transcription of a plethora of proinflammatory cytokines (IL-1β, IL-1α, IL-6), chemokines (IL-8 and CVCR2), growth factors (IGF and IGFBPs), and matrix—remodeling proteins (PAI-1 and MM3), commonly referred to as SASP (Coppe et al., 2008). Replicative senescence provides a protective antitumorigenic mechanism for cellular defense. Studies in telomere and oncogene-dependent senescence have shown that escape from senescence is a basic feature of carcinogenesis. The SASP, in contrast, is detrimental for the surrounding cells by propagating inflammation, tumorigenicity, and metastasis. In SASP a low number of senescent cells in a tissue (typically <20%) may be sufficient to exert systemic effects (Palmer et al., 2015). The proteins comprising SASP can also exacerbate the senescent phenotype of the cells of their origin in an autocrine manner.

The mechanisms of cellular senescence have garnered significant interest over the last decade following the demonstration that senescence is not just only a cell culture phenomenon but also occurs in vivo. Senescent cells have been found in various organs of aging organisms, in precancerous tissues, and in other pathologies including liver disease and diabetes, suggesting that they might be contributing to disease progression. Of note, targeting senescent cells pharmacologically has been shown to be a promising therapeutic strategy (Marongiu et al., 2014). Despite significant advancement over the last decade, the mechanisms of cellular senescence remain not well understood.

10.2 Hepatocellular Senescence and Nonalcoholic Fatty Liver Disease Progression

Hepatocytes in the adult liver are quiescent cells; they are renewed slowly, approximately once a year (Aikata et al., 2000). However, the liver has powerful regenerative capacity as demonstrated in liver transplantation and liver injury. Importantly, as with other somatic cells, the capacity of hepatocytes to replicate is limited and when reached, it leads to the onset of cellular senescence. Decreased capacity for hepatocyte proliferation has been shown during normal aging (Gagliano, Grizzi, & Annoni, 2007) and in the cirrhosis stage of chronic liver disease (Delhaye et al., 1996; Michalopoulos, 2007). Convincing, in vivo evidence for the presence of senescent hepatocytes in aging and during liver disease comes from studies showing replicative senescence (as tested by β-galactosidase assay) increasing from 10% in normal liver to 84% in cirrhotic liver and to 60% in HCC (Paradis et al., 2001; Trak-Smayra et al., 2004; Wiemann et al., 2002).

It seems self-evident that entering premature senescence might limit the capacity of hepatocyte to regenerate and to repair the damage thus fueling the progression of the disease. Indeed, telomere shortening has been found to be a risk factor for cirrhosis (Carulli & Anzivino, 2014) and to correlate with the accumulation of senescent hepatocytes. Hepatocyte-specific telomere shortening associated with positive β-galactosidase staining has been shown in conditions associated with inflammation and during cirrhosis (Lechel et al., 2005). Telomere length in tissue from cirrhotic liver was shorter than in liver with chronic hepatitis and both were shorter than telomere length in normal liver tissue (Wiemann et al., 2002). The shortening of the telomeres was observed for hepatocytes but not stellate cells in areas of fibrosis, which had significantly longer telomere reserve (Rudolph et al., 2000). In turn, telomerase-deficient mice develop accelerated cirrhosis in response to chronic liver injury. Restoration of telomerase activity into the liver results in reduction of cirrhosis and improved liver function. Telomerase mutations in humans have been associated with short telomeres and shown to accelerate liver disease progression to cirrhosis (Calado et al., 2011; Hartmann et al., 2011). The possibility that the onset of senescence may predict worsening to NAFLD to cirrhosis and HCC was also supported by analyses of senescent markers in liver sections from 70 patients with NAFLD and 60 controls. The study found that hepatocyte senescence correlated closely with fibrosis stage and with the clinical outcome and hepatocyte p21 expression emerged as a valuable prognostic marker (Aravinthan et al., 2013b).

The telomere hypothesis of cancer holds that telomere shortening results in chromosomal instability, which then drives cancer initiation. Indeed, telomeres are shorter in HCC hepatocyte as compared with those in regenerative nodules and normal liver tissue within group of HCC and correlate with the ploidy grade, taken as a measure of chromosomal instability (Farazi et al., 2003). On the other hand, the progression of cirrhosis to HCC needs an efficient telomere signaling. Reactivation of telomerase activity and restored chromosomal stability to a level compatible with tumor cell viability has been shown for HCC. Telomerase activity was detected in nearly 90% of HCC as compared with only 21% of nontumor tissue and was paralleled by an increase of TERT mRNA levels(Lee et al., 2004) (Shimojima et al., 2004). The oncogenic potential of the shortened telomeres in HCC was associated with the state of another senescence marker, p53, which is required for the onset of the senescent program (Farazi et al., 2003). Noteworthy, a mutation/loss of heterozygosity of p53 is present in 90% of HCC patients (Thorgeirsson & Grisham, 2002). Together, these results have supported the concept that senescence-associated telomere shortening might be critical for the progression of NAFLD to HCC by limiting the capacity of hepatocytes to regenerate the injured tissue, and by instigating a protumorigenic transformation process.

Telomere-independent senescence, accompanied by inflammation, ROS, and DNA damage, has also been found in severe chronic liver disease (Farinati et al., 2007; Reuben, 2008; Shimoda et al., 1994). It appears to be especially significant for the progression of NAFLD when cellular proliferation and turnover is not prominent and therefore, telomeres of hepatocytes are still intact. Indeed, senescent hepatocytes have also been observed in fatty liver in dietary or genetic mouse models (Yang et al., 2004) and NASH (Aravinthan et al., 2013b) when liver regeneration had not been yet initiated. In these cases the senescence is manifested by activation of inflammasome, which brings about the release of SASP proteins. These senescence hepatocytes have been implicated in the development of obesity-associated HCC in mice (Yoshimoto et al., 2013).

The possibility that prodiabetic diets that induce fatty liver development might also induce senescence was first indicated by observations that palmitic acid can stimulate inflammasome formation in vitro (Csak et al., 2011). A recent in vivo study provided further evidence that the induction of senescence program in hepatocytes and steatosis is intimately linked. Hepatocyte-specific induction of senescence by a local DNA repair defect in the mouse resulted in liver steatosis. Vice versa, clearance of senescent cells by suicide gene-meditated ablation of $p16^{Ink4a}$-expressing senescent cells resulted in reduced overall hepatic steatosis in aged, obese, and diabetic mice (Ogrodnik et al., 2017).

SASP in nonhepatic cells also has been suggested to play a role in NAFLD progression. Senescent stellate cells have been found to contribute to HCC development in obese mice (Yoshimoto et al., 2013). IL-6, one of the SASP factors is a proinflammatory signaling protein that encourages tumor growth and exerts its oncogenic activity by triggering downstream STAT-3 and ERK pathways (Park et al., 2010).

10.3 Ceramide and the Induction of Cellular Senescence

Direct evidence of a role of ceramide and/or S1P in hepatocellular senescence is lacking mainly due to the limitation in the available in vitro models. Most hepatocellular cell lines are immortal, whereas primary hepatocytes do not proliferate ex vivo. Nevertheless, significant evidence from other cell types suggests that the accumulation of ceramide in NAFLD hepatocytes might be a driving force in the onset of premature senescence and, in the context of NAFLD, could mediate saturated fat induced senescence program (Ford, 2010; Palmer et al., 2015). Indeed, ceramide has been known to induce cellular senescence in vitro (Venable & Yin, 2009; Venable et al, 1995, 2006). In 1995, ceramide was first implicated in the onset of senescence in human diploid fibroblasts (Venable et al., 1995), an effect that was later extended to other cell types (Jadhav, Dungan, & Williamson, 2013). Consequent studies have defined a conserved signaling role for ceramide in growth arrest and DNA repair (Kolesnick & Fuks, 2003; Pettus, Chalfant, & Hannun, 2002; Summers, 2010; Zheng et al., 2006). Ceramide has been implicated in the regulation of PP2A, JNK, p38, telomerase activity (Wooten-Blanks et al., 2007), p21 (Lee, Bielawska, & Obeid, 2000), and other key signaling molecules that contribute to the senescence onset (Bartke & Hannun, 2009). Sphingolipid metabolizing enzymes regulating ceramide production have been shown to function both upstream and downstream of p53, another key molecule of hepatic cellular senescence (Heffernan-Stroud & Obeid, 2011). The telomeres are also influenced by ceramide metabolism. Ceramide has been found to inactivate c-Myc transcription factor, which otherwise activates the hTERT promoter of the telomerase (Heffernan-Stroud & Obeid, 2011). Ceramide also can cause telomerase-independent rapid shortening of telomere length (Heffernan-Stroud & Obeid, 2011). Ceramide has been associated also with oncogene-induced cellular senescence, which appears to involve the activation of PP1C, pRb dephosphorylation, and p53-induced growth arrest (Castro et al., 2008). Increases in ceramide concentrations in myoblast were associated with senescent phenotype, likely mediated via changes in p53 and p21protein expression. Ceramide also reduced myoblast proliferation that was associated with aberrant cell cycle regulation and a senescent phenotype. In the context of NAFLD, hepatic ceramide has been shown to initiate Nlrp3 inflammasome activation. It was proposed that inflammasome senses lipotoxicity-associated increases in intracellular ceramide to induce caspase-1 cleavage and thereby initiate SASP (Jadhav et al., 2013).

11. CONCLUSIONS

Growing body of evidence from animal models and human patients support a key role of altered sphingolipid metabolism in onset and progression NAFLD, which cover the entire disease spectrum, from simple steatosis to cirrhosis and HCC. The metabolic pathways governing ceramide and S1P production and clearance seem to be most impacted by the disease. Lipotoxocity, insulin response, and the onset of senescence in hepatocytes, together with the fibrotic responses of stellate cells are most intimately linked with these two key bioactive sphingolipids. Managing the production of ceramide and S1P during NAFLD emerges as promising new therapeutic approach to influence the disease progression at different stages.

ABBREVIATIONS

ACC Acetyl-CoA carboxylase
aSMase Acid sphingomyelinase
BMDSC Bone marrow–derived mesenchymal stem cells

CerS	Ceramide synthase
ChREBP	Carbohydrate-responsive element-binding protein
CEPT	CTP:phosphocholine cytidylyltransferase
DAG	Diacylglycerols
DDR	DNA damage response
DGAT	Acyl-CoA: diacylglycerol acyltransferases
ER	Endoplasmic reticulum
FAS	Fatty acid synthase
GCS	Glucosylceramide synthase
HCC	Hepatocellular carcinoma
HSC	Hepatic stellate cells
INSING	Insulin-induced genes
MCD	Methionine choline deficient diet
LDLR	LDL receptor
NAFLD	Nonalcoholic fatty liver disease
NASH	Nonalcoholic steatohepatitis
nSMase	Neutral sphingomyelinase
PC	Phosphatidylcholine
PE	Phosphatidylethanolamine
PS	Phosphatidylserine
S1P	Sphingosine 1-phosphate
S1P	Sphingosine 1-phosphate receptor
SK	Sphingosine kinase
SASP	Senescence-associated secretory proteins
SPT	Serine palmitoyltransferase
SM	Sphingomyelin
SMase	Sphingomyelinase
SMS	Sphingomyelin synthase
SREBP	Sterol regulatory element binding protein
SCAP	SREBP cleavage activating protein
TAG	Triacylglycerol

ACKNOWLEDGMENTS

The author wishes to acknowledge funding from the NIA AG019223 and CDMRP Discovery Award PR152294 and to thank Mr. A. Karakashian for help with document editing and critical reading.

REFERENCES

Aguilera-Romero, A., Gehin, C., & Riezman, H. (2014). Sphingolipid homeostasis in the web of metabolic routes. *Biochimica et Biophysica Acta, 1841*(5), 647–656.

Aikata, H., et al. (2000). Telomere reduction in human liver tissues with age and chronic inflammation. *Experimental Cell Research, 256*(2), 578–582.

Airola, M. V., & Hannun, Y. A. (2013). Sphingolipid metabolism and neutral sphingomyelinases. *Handbook of Experimental Pharmacology*, (215), 57–76.

Al Fadel, F., et al. (2016). Involvement of sphingosine 1-phosphate in palmitate-induced non-alcoholic fatty liver disease. *Cellular Physiology and Biochemistry: International Journal of Experimental Cellular Physiology, Biochemistry, and Pharmacology, 40*(6), 1637–1645.

Alison, M. R. (2005). Liver stem cells: Implications for hepatocarcinogenesis. *Stem Cell Reviews, 1*(3), 253–260.

Alkhouri, N., Dixon, L. J., & Feldstein, A. E. (2009). Lipotoxicity in nonalcoholic fatty liver disease: Not all lipids are created equal. *Expert Review of Gastroenterology & Hepatology, 3*(4), 445–451.

Alonso, C., et al. (2017). Metabolomic identification of subtypes of nonalcoholic steatohepatitis. *Gastroenterology, 152*(6), 1449–1461.e7.

Altekruse, S. F., et al. (2014). Changing hepatocellular carcinoma incidence and liver cancer mortality rates in the United States. *The American Journal of Gastroenterology, 109*(4), 542–553.

Arab, J. P., Arrese, M., & Trauner, M. (2018). Recent insights into the pathogenesis of nonalcoholic fatty liver disease. *Annual Review of Pathology, 13*(1), 321–350.

Aravinthan, A., et al. (2013a). Hepatocyte expression of the senescence marker p21 is linked to fibrosis and an adverse liver-related outcome in alcohol-related liver disease. *PLoS One, 8*(9), e72904.

Aravinthan, A., et al. (2013b). Hepatocyte senescence predicts progression in non-alcohol-related fatty liver disease. *Journal of Hepatology, 58*(3), 549–556.

Bartke, N., & Hannun, Y. A. (2009). Bioactive sphingolipids: metabolism and function. *The Journal of Lipid Research, 50*(Suppl.), S91–S96.

Bechmann, L. P., et al. (2010). Apoptosis is associated with CD36/fatty acid translocase upregulation in non-alcoholic steatohepatitis. *Liver International: Official Journal of the International Association for the Study of the Liver, 30*(6), 850–859.

Becker, S., et al. (2017). Low sphingosine-1-phosphate plasma levels are predictive for increased mortality in patients with liver cirrhosis. *PLoS One, 12*(3), e0174424.

Berlanga, A., et al. (2014). Molecular pathways in non-alcoholic fatty liver disease. *Clinical and Experimental Gastroenterology, 7*, 221–239.

Bharath, L. P., et al. (2015). Ceramide-initiated protein phosphatase 2A activation contributes to arterial dysfunction in vivo. *Diabetes, 64*(11), 3914–3926.

Bijl, N., et al. (2009). Modulation of glycosphingolipid metabolism significantly improves hepatic insulin sensitivity and reverses hepatic steatosis in mice. *Hepatology, 50*(5), 1431–1441.

Bikman, B. T., & Summers, S. A. (2011a). Ceramides as modulators of cellular and whole-body metabolism. *Journal of Clinical Investigation, 121*(11), 4222–4230.

Bikman, B. T., & Summers, S. A. (2011b). Sphingolipids and hepatic steatosis. *Advances in Experimental Medicine and Biology, 721*, 87–97.

Brown, M. S., & Goldstein, J. L. (2008). Selective versus total insulin resistance: A pathogenic paradox. *Cell Metabolism, 7*(2), 95–96.

Brunt, E. M., et al. (2015). Nonalcoholic fatty liver disease. *Nature Reviews Disease Primers, 1*, 15080.

Bugianesi, E. (2007). Non-alcoholic steatohepatitis and cancer. *Clinics in Liver Disease, 11*(1), 191–207.

Bugianesi, E., et al. (2002). Expanding the natural history of nonalcoholic steatohepatitis: From cryptogenic cirrhosis to hepatocellular carcinoma. *Gastroenterology, 123*(1), 134–140.

Cai, H., et al. (2017). Sphingosine kinase 1: A novel independent prognosis biomarker in hepatocellular carcinoma. *Oncology Letters, 13*(4), 2316–2322.

Calado, R. T., et al. (2011). Constitutional telomerase mutations are genetic risk factors for cirrhosis. *Hepatology, 53*(5), 1600–1607.

Calzadilla Bertot, L., & Adams, L. A. (2016). The natural course of non-alcoholic fatty liver disease. *International Journal of Molecular Sciences, 17*(5), 774.

Campisi, J., & d'Adda di Fagagna, F. (2007). Cellular senescence: When bad things happen to good cells. *Nature Reviews. Molecular Cell Biology, 8*(9), 729–740.

Carulli, L., & Anzivino, C. (2014). Telomere and telomerase in chronic liver disease and hepatocarcinoma. *World Journal of Gastroenterology: WJG, 20*(20), 6287–6292.

Castro, M. E., et al. (2008). PPP1CA contributes to the senescence program induced by oncogenic Ras. *Carcinogenesis, 29*(3), 491–499.
Chaurasia, B., & Summers, S. A. (2015). Ceramides - lipotoxic inducers of metabolic disorders. *Trends in Endocrinology and Metabolism, 26*(10), 538–550.
Chiantia, S., & London, E. (2013). Sphingolipids and membrane domains: Recent advances. *Handbook of Experimental Pharmacology*, (215), 33–55.
Cinar, R., et al. (2014). Hepatic CB(1) receptors mediate diet-induced insulin resistance by increasing de novo synthesis of long chain ceramides. *Hepatology, 59*(1).
Coant, N., et al. (2017). Ceramidases, roles in sphingolipid metabolism and in health and disease. *Advances in Biological Regulation, 63*, 122–131.
Coen, P. M., & Goodpaster, B. H. (2012). Role of intramyocelluar lipids in human health. *Trends in Endocrinology and Metabolism, 23*(8), 391–398.
Coleman, R. A., & Lee, D. P. (2004). Enzymes of triacylglycerol synthesis and their regulation. *Progress in Lipid Research, 43*(2), 134–176.
Coppe, J. P., et al. (2008). Senescence-associated secretory phenotypes reveal cell-nonautonomous functions of oncogenic RAS and the p53 tumor suppressor. *PLoS Biology, 6*(12), 2853–2868.
Csak, T., et al. (2011). Fatty acid and endotoxin activate inflammasomes in mouse hepatocytes that release danger signals to stimulate immune cells. *Hepatology, 54*(1), 133–144.
Declercq, P. E., et al. (1984). Rat liver dihydroxyacetone-phosphate acyltransferases and their contribution to glycerolipid synthesis. *Journal of Biological Chemistry, 259*(14), 9064–9075.
Deevska, G. M., et al. (2009). Acid sphingomyelinase deficiency prevents diet-induced hepatic triacylglycerol accumulation and hyperglycemia in mice. *Journal of Biological Chemistry, 284*(13), 8359–8368.
Deevska, G. M., et al. (2017). Novel interconnections in lipid metabolism revealed by overexpression of sphingomyelin synthase-1. *Journal of Biological Chemistry, 292*(12), 5110–5122.
Delhaye, M., et al. (1996). Relationship between hepatocyte proliferative activity and liver functional reserve in human cirrhosis. *Hepatology, 23*(5), 1003–1011.
Dobierzewska, A., et al. (2012). Interleukin 1beta regulation of FoxO1 protein content and localization: Evidence for a novel ceramide-dependent mechanism. *Journal of Biological Chemistry, 287*(53), 44749–44760.
Dongiovanni, P., Romeo, S., & Valenti, L. (2014). Hepatocellular carcinoma in nonalcoholic fatty liver: Role of environmental and genetic factors. *World Journal of Gastroenterology: WJG, 20*(36), 12945–12955.
Donnelly, K. L., et al. (2005). Sources of fatty acids stored in liver and secreted via lipoproteins in patients with nonalcoholic fatty liver disease. *Journal of Clinical Investigation, 115*(5), 1343–1351.
Duan, X. Y., et al. (2014). NAFLD leads to liver cancer: Do we have sufficient evidence? *Cancer Letters, 345*(2), 230–234.
Engin, A. (2017). Non-alcoholic fatty liver disease. *Advances in Experimental Medicine and Biology, 960*, 443–467.
Fagone, P., & Jackowski, S. (2013). Phosphatidylcholine and the CDP-choline cycle. *Biochimica et Biophysica Acta, 1831*(3), 523–532.
Farazi, P. A., et al. (2003). Differential impact of telomere dysfunction on initiation and progression of hepatocellular carcinoma. *Cancer Research, 63*(16), 5021–5027.
Farinati, F., et al. (2007). Hepatitis C virus: From oxygen free radicals to hepatocellular carcinoma. *Journal of Viral Hepatitis, 14*(12), 821–829.
Fei, W., et al. (2011). A role for phosphatidic acid in the formation of "Supersized" lipid droplets. *PLoS Genetics, 7*(7), e1002201.
Feldstein, A. E., et al. (2003). Diet associated hepatic steatosis sensitizes to Fas mediated liver injury in mice. *Journal of Hepatology, 39*(6), 978–983.

Ford, J. H. (2010). Saturated fatty acid metabolism is key link between cell division, cancer, and senescence in cellular and whole organism aging. *Age (Dordr), 32*(2), 231–237.
Friedman, S. L. (2008). Mechanisms of hepatic fibrogenesis. *Gastroenterology, 134*(6), 1655–1669.
Fucho, R., et al. (2014). ASMase regulates autophagy and lysosomal membrane permeabilization and its inhibition prevents early stage non-alcoholic steatohepatitis. *Journal of Hepatology, 61*(5), 1126–1134.
Gagliano, N., Grizzi, F., & Annoni, G. (2007). Mechanisms of aging and liver functions. *Digestive Diseases, 25*(2), 118–123.
Global Burden of Disease Liver Cancer Collaboration. (2017). The burden of primary liver cancer and underlying etiologies from 1990 to 2015 at the global, regional, and national level: Results from the global burden of disease study 2015. *JAMA Oncology, 3*(12), 1683–1691.
González-Fernández, B. (2017). Sphingosine 1-phosphate signaling as a target in hepatic fibrosis therapy. *Frontiers in Pharmacology, 8*.
Grammatikos, G., et al. (2015). Serum sphingolipid variations associate with hepatic decompensation and survival in patients with cirrhosis. *PLoS One, 10*(9), e0138130.
Grammatikos, G., et al. (2016). Serum sphingolipidomic analyses reveal an upregulation of C16-ceramide and sphingosine-1-phosphate in hepatocellular carcinoma. *Oncotarget*, 7(14), 18095–18105.
Greco, D., et al. (2008). Gene expression in human NAFLD. *American Journal of Physiology. Gastrointestinal and Liver Physiology, 294*(5), G1281–G1287.
Hait, N. C., et al. (2009). Regulation of histone acetylation in the nucleus by sphingosine-1-phosphate. *Science, 325*(5945), 1254–1257.
Hanada, K., et al. (2003). Molecular machinery for non-vesicular trafficking of ceramide. *Nature, 426*(6968), 803–809.
Hardwick, R. N., et al. (2010). Diversity in antioxidant response enzymes in progressive stages of human nonalcoholic fatty liver disease. *Drug Metabolism and Disposition: the Biological Fate of Chemicals, 38*(12), 2293–2301.
Hardy, T., & Mann, D. A. (2016). Epigenetics in liver disease: From biology to therapeutics. *Gut, 65*(11), 1895–1905.
Hartmann, D., et al. (2011). Telomerase gene mutations are associated with cirrhosis formation. *Hepatology, 53*(5), 1608–1617.
Heffernan-Stroud, L. A., & Obeid, L. M. (2011). p53 and regulation of bioactive sphingolipids. *Advances in Enzyme Regulation, 51*(1), 219–228.
Holland, W. L., et al. (2017). Inducible overexpression of adiponectin receptors highlight the roles of adiponectin-induced ceramidase signaling in lipid and glucose homeostasis. *Molecular Metabolism, 6*(3), 267–275.
Holland, W. L., & Summers, S. A. (2008). Sphingolipids, insulin resistance, and metabolic disease: New insights from in vivo manipulation of sphingolipid metabolism. *Endocrine Reviews, 29*(4), 381–402.
Holthuis, J. C., & Luberto, C. (2010). Tales and mysteries of the enigmatic sphingomyelin synthase family. *Advances in Experimental Medicine and Biology, 688*, 72–85.
Ikeda, H., et al. (2000). Biological activities of novel lipid mediator sphingosine 1-phosphate in rat hepatic stellate cells. *American Journal of Physiology. Gastrointestinal and Liver Physiology, 279*(2), G304–G310.
Ikeda, H., et al. (2009). Sphingosine 1-phosphate regulates regeneration and fibrosis after liver injury via sphingosine 1-phosphate receptor 2. *The Journal of Lipid Research, 50*(3), 556–564.
Ilan, Y. (2016). Compounds of the sphingomyelin-ceramide-glycosphingolipid pathways as secondary messenger molecules: New targets for novel therapies for fatty liver disease and insulin resistance. *American Journal of Physiology. Gastrointestinal and Liver Physiology, 310*(11), G1102–G1117.

Iqbal, J., et al. (2017). Sphingolipids and lipoproteins in health and metabolic disorders. *Trends in Endocrinology and Metabolism, 28*(7), 506–518.
Jackowski, S., Wang, J., & Baburina, I. (2000). Activity of the phosphatidylcholine biosynthetic pathway modulates the distribution of fatty acids into glycerolipids in proliferating cells. *Biochimica et Biophysica Acta, 1483*(3), 301–315.
Jacobs, R. L., et al. (2004). Targeted deletion of hepatic CTP: phosphocholine cytidylyltransferase alpha in mice decreases plasma high density and very low density lipoproteins. *Journal of Biological Chemistry, 279*(45), 47402–47410.
Jadhav, K. S., Dungan, C. M., & Williamson, D. L. (2013). Metformin limits ceramide-induced senescence in C2C12 myoblasts. *Mechanisms of Ageing and Development, 134*(11–12), 548–559.
Kahraman, A., et al. (2010). Major histocompatibility complex class I-related chains A and B (MIC A/B): A novel role in nonalcoholic steatohepatitis. *Hepatology, 51*(1), 92–102.
Kakazu, E., et al. (2016). Hepatocytes release ceramide-enriched pro-inflammatory extracellular vesicles in an IRE1alpha-dependent manner. *The Journal of Lipid Research, 57*(2), 233–245.
Kasumov, T., et al. (2015). Ceramide as a mediator of non-alcoholic fatty liver disease and associated atherosclerosis. *PLoS One, 10*(5), e0126910.
Kennedy, E. P. (1957). Metabolism of lipides. *Annual Review of Biochemistry, 26*, 119–148.
King, A., et al. (2017). Sphingosine-1-phosphate prevents egress of hematopoietic stem cells from liver to reduce fibrosis. *Gastroenterology, 153*(1), 233–248.e16.
Kolesnick, R., & Fuks, Z. (2003). Radiation and ceramide-induced apoptosis. *Oncogene, 22*(37), 5897–5906.
Krautbauer, S., et al. (2016). Ceramide and polyunsaturated phospholipids are strongly reduced in human hepatocellular carcinoma. *Biochimica et Biophysica Acta, 1861*(11), 1767–1774.
Kurek, K., et al. (2014). Inhibition of ceramide de novo synthesis reduces liver lipid accumulation in rats with nonalcoholic fatty liver disease. *Liver International: Official Journal of the International Association for the Study of the Liver, 34*(7), 1074–1083.
Lambert, J. E., et al. (2014). Increased de novo lipogenesis is a distinct characteristic of individuals with nonalcoholic fatty liver disease. *Gastroenterology, 146*(3), 726–735.
Lechel, A., et al. (2005). The cellular level of telomere dysfunction determines induction of senescence or apoptosis in vivo. *EMBO Reports, 6*(3), 275–281.
Lee, C. M., et al. (2004). Telomerase activity and telomerase catalytic subunit in hepatocellular carcinoma. *Hepato-Gastroenterology, 51*(57), 796–800.
Lee, J. Y., Bielawska, A. E., & Obeid, L. M. (2000). Regulation of cyclin-dependent kinase 2 activity by ceramide. *Experimental Cell Research, 261*(2), 303–311.
Leonardi, R., et al. (2009). Elimination of the CDP-ethanolamine pathway disrupts hepatic lipid homeostasis. *Journal of Biological Chemistry, 284*(40), 27077–27089.
Lewis, G. F., et al. (2002). Disordered fat storage and mobilization in the pathogenesis of insulin resistance and type 2 diabetes. *Endocrine Reviews, 23*(2), 201–229.
Li, C., et al. (2011). Sphingosine 1-phosphate (S1P)/S1P receptors are involved in human liver fibrosis by action on hepatic myofibroblasts motility. *Journal of Hepatology, 54*(6), 1205–1213.
Lu, P. H., et al. (2017). Identification of sphingosine kinase 1 (SphK1) as a primary target of icaritin in hepatocellular carcinoma cells. *Oncotarget, 8*(14), 22800–22810.
Mari, M., & Fernandez-Checa, J. C. (2007). Sphingolipid signalling and liver diseases. *Liver International: Official Journal of the International Association for the Study of the Liver, 27*(4), 440–450.
Marongiu, F., et al. (2014). Clearance of senescent hepatocytes in a neoplastic-prone microenvironment delays the emergence of hepatocellular carcinoma. *Aging, 6*(1), 26–34.

Martinez, L., et al. (2015). Myristic acid potentiates palmitic acid-induced lipotoxicity and steatohepatitis associated with lipodystrophy by sustaning de novo ceramide synthesis. *Oncotarget, 6*(39), 41479–41496.

Mas, E., et al. (2009). IL-6 deficiency attenuates murine diet-induced non-alcoholic steatohepatitis. *PLoS One, 4*(11), e7929.

Matteoni, C. A., et al. (1999). Nonalcoholic fatty liver disease: A spectrum of clinical and pathological severity. *Gastroenterology, 116*(6), 1413–1419.

Mauer, A. S., et al. (2017). Inhibition of sphingosine 1-phosphate signaling ameliorates murine nonalcoholic steatohepatitis. *American Journal of Physiology. Gastrointestinal and Liver Physiology, 312*(3), G300–G313.

McClain, C. J., Barve, S., & Deaciuc, I. (2007). Good fat/bad fat. *Hepatology, 45*(6), 1343–1346.

McPherson, S., et al. (2015). Evidence of NAFLD progression from steatosis to fibrosing-steatohepatitis using paired biopsies: Implications for prognosis and clinical management. *Journal of Hepatology, 62*(5), 1148–1155.

van Meer, G., & Sprong, H. (2004). Membrane lipids and vesicular traffic. *Current Opinion in Cell Biology, 16*(4), 373–378.

Merrill, A. H., Nixon, D. W., & Williams, R. D. (1985). Activities of serine palmitoyltransferase (3-ketosphinganine synthase) in microsomes from different rat tissues. *Journal of Lipid Research, 26*(5), 617–622.

Messner, M. C., & Cabot, M. C. (2010). Glucosylceramide in humans. *Advances in Experimental Medicine and Biology, 688*, 156–164.

Michalopoulos, G. K. (2007). Liver regeneration. *Journal of Cellular Physiology, 213*(2), 286–300.

Mitsutake, S., et al. (2011). Dynamic modification of sphingomyelin in lipid microdomains controls development of obesity, fatty liver, and type 2 diabetes. *Journal of Biological Chemistry, 286*(32), 28544–28555.

Moles, A., et al. (2010). Acidic sphingomyelinase controls hepatic stellate cell activation and in vivo liver fibrogenesis. *The American Journal of Pathology, 177*(3), 1214–1224.

Mullen, T. D., Hannun, Y. A., & Obeid, L. M. (2012). Ceramide synthases at the centre of sphingolipid metabolism and biology. *The Biochemical Journal, 441*(3), 789–802.

Murphy, S. K., et al. (2013). Relationship between methylome and transcriptome in patients with nonalcoholic fatty liver disease. *Gastroenterology, 145*(5), 1076–1087.

Nagahashi, M., et al. (2015). Conjugated bile acid-activated S1P receptor 2 is a key regulator of sphingosine kinase 2 and hepatic gene expression. *Hepatology, 61*(4), 1216–1226.

Nagahashi, M., et al. (2016). DNA damage response and sphingolipid signaling in liver diseases. *Surgery Today, 46*(9), 995–1005.

Newton, J., et al. (2015). Revisiting the sphingolipid rheostat: Evolving concepts in cancer therapy. *Experimental Cell Research, 333*(2), 195–200.

Nikolova-Karakashian, M., Karakashian, A., & Rutkute, K. (2008). Role of neutral sphingomyelinases in aging and inflammation. *Subcellular Biochemistry, 49*, 469–486.

Nikolova-Karakashian, M. N., & Rozenova, K. A. (2010). Ceramide in stress response. *Advances in Experimental Medicine and Biology, 688*, 86–108.

Nojima, H., et al. (2016). Hepatocyte exosomes mediate liver repair and regeneration via sphingosine-1-phosphate. *Journal of Hepatology, 64*(1), 60–68.

Nowatari, T., et al. (2015). Sphingosine 1-phosphate has anti-apoptotic effect on liver sinusoidal endothelial cells and proliferative effect on hepatocytes in a paracrine manner in human. *Hepatology Research, 45*(11), 1136–1145.

Ogrodnik, M., et al. (2017). Cellular senescence drives age-dependent hepatic steatosis. *Nature Communications, 8*, 15691.

Olivera, A., & Spiegel, S. (2001). Sphingosine kinase: A mediator of vital cellular functions. *Prostaglandins & Other Lipid Mediators, 64*(1–4), 123–134.

Osawa, Y., et al. (2005). Roles for C16-ceramide and sphingosine 1-phosphate in regulating hepatocyte apoptosis in response to tumor necrosis factor-alpha. *Journal of Biological Chemistry, 280*(30), 27879–27887.

Overturf, K., et al. (1997). Serial transplantation reveals the stem-cell-like regenerative potential of adult mouse hepatocytes. *The American Journal of Pathology, 151*(5), 1273–1280.

Palmer, A. K., et al. (2015). Cellular senescence in type 2 diabetes: A therapeutic opportunity. *Diabetes, 64*(7), 2289–2298.

Paradis, V., et al. (2001). Replicative senescence in normal liver, chronic hepatitis C, and hepatocellular carcinomas. *Human Pathology, 32*(3), 327–332.

Park, E. J., et al. (2010). Dietary and genetic obesity promote liver inflammation and tumorigenesis by enhancing IL-6 and TNF expression. *Cell, 140*(2), 197–208.

Park, T. S., et al. (2008). Ceramide is a cardiotoxin in lipotoxic cardiomyopathy. *The Journal of Lipid Research, 49*(10), 2101–2112.

Park, W. J., et al. (2014). Hepatic fatty acid uptake is regulated by the sphingolipid acyl chain length. *Biochimica et Biophysica Acta, 1841*(12), 1754–1766.

Pettus, B. J., Chalfant, C. E., & Hannun, Y. A. (2002). Ceramide in apoptosis: An overview and current perspectives. *Biochimica et Biophysica Acta, 1585*(2–3), 114–125.

Postic, C., & Girard, J. (2008). Contribution of de novo fatty acid synthesis to hepatic steatosis and insulin resistance: Lessons from genetically engineered mice. *The Journal of Clinical Investigation, 118*(3), 829–838.

Promrat, K., et al. (2011). Weight loss amelioration of non-alcoholic steatohepatitis linked to shifts in hepatic ceramide expression and serum ceramide levels. *Hepatology Research, 41*(8), 754–762.

Puri, P., et al. (2007). A lipidomic analysis of nonalcoholic fatty liver disease. *Hepatology, 46*(4), 1081–1090.

Raichur, S., et al. (2014). CerS2 haploinsufficiency inhibits beta-oxidation and confers susceptibility to diet-induced steatohepatitis and insulin resistance. *Cell Metabolism, 20*(4), 687–695.

Ressom, H. W., et al. (2012). Utilization of metabolomics to identify serum biomarkers for hepatocellular carcinoma in patients with liver cirrhosis. *Analytica Chimica Acta, 743c*, 90–100.

Reuben, A. (2008). Alcohol and the liver. *Current Opinion in Gastroenterology, 24*(3), 328–338.

Revill, K., et al. (2013). Genome-wide methylation analysis and epigenetic unmasking identify tumor suppressor genes in hepatocellular carcinoma. *Gastroenterology, 145*(6), 1424–1435.e1-25.

Ribaux, P. G., & Iynedjian, P. B. (2003). Analysis of the role of protein kinase B (cAKT) in insulin-dependent induction of glucokinase and sterol regulatory element-binding protein 1 (SREBP1) mRNAs in hepatocytes. *The Biochemical Journal, 376*(Pt 3), 697–705.

Ringelhan, M., et al. (2018). The immunology of hepatocellular carcinoma. *Nature Immunology*.

Rudolph, K. L., et al. (2000). Inhibition of experimental liver cirrhosis in mice by telomerase gene delivery. *Science, 287*(5456), 1253–1258.

Rutkute, K., Asmis, R. H., & Nikolova-Karakashian, M. N. (2007). Regulation of neutral sphingomyelinase-2 by GSH: A new insight to the role of oxidative stress in aging-associated inflammation. *The Journal of Lipid Research, 48*(11), 2443–2452.

Sato, M., et al. (2016). Sphingosine kinase-1, S1P transporter spinster homolog 2 and S1P2 mRNA expressions are increased in liver with advanced fibrosis in human. *Scientific Reports, 6*, 32119.

Shamseddine, A. A., Airola, M. V., & Hannun, Y. A. (2015). Roles and regulation of neutral sphingomyelinase-2 in cellular and pathological processes. *Advances in Biological Regulation, 57*, 24–41.

Shimabukuro, M., et al. (1998). Fatty acid-induced beta cell apoptosis: A link between obesity and diabetes. *Proceedings of the National Academy of Sciences of the United States of America, 95*(5), 2498–2502.
Shimoda, R., et al. (1994). Increased formation of oxidative DNA damage, 8-hydroxydeoxyguanosine, in human livers with chronic hepatitis. *Cancer Research, 54*(12), 3171–3172.
Shimojima, M., et al. (2004). Detection of telomerase activity, telomerase RNA component, and telomerase reverse transcriptase in human hepatocellular carcinoma. *Hepatology Research, 29*(1), 31–38.
Shimomura, I., et al. (2000). Decreased IRS-2 and increased SREBP-1c lead to mixed insulin resistance and sensitivity in livers of lipodystrophic and ob/ob mice. *Molecular Cell, 6*(1), 77–86.
Simons, K., & Ikonen, E. (1997). Functional rafts in cell membranes. *Nature, 387*(6633), 569–572.
Stefanovic, M., et al. (2016). Targeting glucosylceramide synthase upregulation reverts sorafenib resistance in experimental hepatocellular carcinoma. *Oncotarget,* 7(7), 8253–8267.
Summers, S. A. (2006). Ceramides in insulin resistance and lipotoxicity. *Progress in Lipid Research, 45*(1), 42–72.
Summers, S. A. (2010). Sphingolipids and insulin resistance: The five Ws. *Current Opinion in Lipidology, 21*(2), 128–135.
Tagaram, H. R., et al. (2011). Nanoliposomal ceramide prevents in vivo growth of hepatocellular carcinoma. *Gut, 60*(5), 695–701.
Takabe, K., et al. (2008). "Inside-Out" signaling of sphingosine-1-phosphate: Therapeutic targets. *Pharmacological Reviews, 60*(2), 181–195.
Thorgeirsson, S. S., & Grisham, J. W. (2002). Molecular pathogenesis of human hepatocellular carcinoma. *Nature Genetics, 31*(4), 339–346.
Torre, L. A., et al. (2015). Global cancer statistics, 2012. *CA Cancer Journal for Clinicians, 65*(2), 87–108.
Trak-Smayra, V., et al. (2004). Role of replicative senescence in the progression of fibrosis in hepatitis C virus (HCV) recurrence after liver transplantation. *Transplantation,* 77(11), 1755–1760.
Uranbileg, B., et al. (2016). Increased mRNA levels of sphingosine kinases and S1P lyase and reduced levels of S1P were observed in hepatocellular carcinoma in association with poorer differentiation and earlier recurrence. *PLoS One, 11*(2), e0149462.
Ushitora, Y., et al. (2009). Suppression of hepatocellular carcinoma recurrence after rat liver transplantation by FTY720, a sphingosine-1-phosphate analog. *Transplantation, 88*(8), 980–986.
Vance, J. E. (2015). Phospholipid synthesis and transport in mammalian cells. *Traffic, 16*(1), 1–18.
Venable, M. E., & Yin, X. (2009). Ceramide induces endothelial cell senescence. *Cell Biochemistry and Function, 27*(8), 547–551.
Venable, M. E., et al. (1995). Role of ceramide in cellular senescence. *Journal of Biological Chemistry, 270*(51), 30701–30708.
Venable, M. E., et al. (2006). Shift in sphingolipid metabolism leads to an accumulation of ceramide in senescence. *Mechanisms of Ageing and Development, 127*(5), 473–480.
Wiemann, S. U., et al. (2002). Hepatocyte telomere shortening and senescence are general markers of human liver cirrhosis. *Federation of American Societies for Experimental Biology Journal, 16*(9), 935–942.
Wooten-Blanks, L. G., et al. (2007). Mechanisms of ceramide-mediated repression of the human telomerase reverse transcriptase promoter via deacetylation of Sp3 by histone deacetylase 1. *FASEB Journal: Official Publication of the Federation of American Societies for Experimental Biology, 21*(12), 3386–3397.

Worgall, T. S. (2011). Sphingolipid synthetic pathways are major regulators of lipid homeostasis. *Advances in Experimental Medicine and Biology, 721*, 139–148.
Worgall, T. S., et al. (2004). Ceramide synthesis correlates with the posttranscriptional regulation of the sterol-regulatory element-binding protein. *Arteriosclerosis, Thrombosis, and Vascular Biology, 24*(5), 943–948.
Xiu, L., et al. (2015). Intracellular sphingosine 1-phosphate contributes to collagen expression of hepatic myofibroblasts in human liver fibrosis independent of its receptors. *The American Journal of Pathology, 185*(2), 387–398.
Xu, X., et al. (2013). Transcriptional control of hepatic lipid metabolism by SREBP and ChREBP. *Seminars in Liver Disease, 33*(4), 301–311.
Yang, G., et al. (2009). Central role of ceramide biosynthesis in body weight regulation, energy metabolism, and the metabolic syndrome. *American Journal of Physiology. Endocrinology and Metabolism, 297*(1), E211–E224.
Yang, L., et al. (2012). Bone marrow-derived mesenchymal stem cells differentiate to hepatic myofibroblasts by transforming growth factor-beta1 via sphingosine kinase/sphingosine 1-phosphate (S1P)/S1P receptor axis. *The American Journal of Pathology, 181*(1), 85–97.
Yang, L., et al. (2013). Sphingosine kinase/sphingosine 1-phosphate (S1P)/S1P receptor axis is involved in liver fibrosis-associated angiogenesis. *Journal of Hepatology, 59*(1), 114–123.
Yang, S., et al. (2004). Oval cells compensate for damage and replicative senescence of mature hepatocytes in mice with fatty liver disease. *Hepatology, 39*(2), 403–411.
Yetukuri, L., et al. (2007). Bioinformatics strategies for lipidomics analysis: Characterization of obesity related hepatic steatosis. *BMC Systems Biology, 1*, 12.
Yoshimoto, S., et al. (2013). Obesity-induced gut microbial metabolite promotes liver cancer through senescence secretome. *Nature, 499*(7456), 97–101.
Zeidan, Y. H., & Hannun, Y. A. (2010). The acid sphingomyelinase/ceramide pathway: Biomedical significance and mechanisms of regulation. *Current Molecular Medicine, 10*(5), 454–466.
Zhang, Y., et al. (2009). Ceramide-enriched membrane domains—structure and function. *Biochimica et Biophysica Acta, 1788*(1), 178–183.
Zheng, W., et al. (2006). Ceramides and other bioactive sphingolipid backbones in health and disease: Lipidomic analysis, metabolism and roles in membrane structure, dynamics, signaling and autophagy. *Biochimica et Biophysica Acta, 1758*(12), 1864–1884.
Zoller, H., & Tilg, H. (2016). Nonalcoholic fatty liver disease and hepatocellular carcinoma. *Metabolism, 65*(8), 1151–1160.

CHAPTER SEVEN

Ceramide Signaling and p53 Pathways

Kristen A. Jeffries* **and Natalia I. Krupenko***,§,1,a
*Nutrition Research Institute, UNC Chapel Hill, Kannapolis, NC, United States
§Department of Nutrition, UNC Chapel Hill, Chapel Hill, NC, United States
[1]Corresponding author: E-mail: natalia_krupenko@unc.edu

Contents

Abstract

Ceramides, important players in signal transduction, interact with multiple cellular pathways, including p53 pathways. However, the relationship between ceramide and p53 is very complex, and mechanisms underlying their coregulation are diverse and not fully characterized. The role of p53, an important cellular regulator and a transcription factor, is linked to its tumor suppressor function. Ceramides are involved in the regulation of fundamental processes in cancer cells including cell death, proliferation, autophagy, and drug resistance. This regulation, however, can be pro-death or pro-survival depending on cancer type, the balance between ceramide species, the rate of their synthesis and utilization, and the availability of a specific array of downstream targets. This chapter highlights the central role of ceramide in sphingolipid metabolism, its role in cancer, specific effectors in ceramide pathways controlled by p53, and coregulation

[a] Senior Author.

Advances in Cancer Research, Volume 140
ISSN 0065-230X
https://doi.org/10.1016/bs.acr.2018.04.011

of ceramide and p53 signaling. We discuss the recent studies, which underscore the function of p53 in the regulation of ceramide pathways and the reciprocal regulation of p53 by ceramide. This complex relationship is based on several molecular mechanisms including the p53-dependent transcriptional regulation of enzymes in sphingolipid pathways, the activation of mutant p53 through ceramide-mediated alternative splicing, as well as modulation of the p53 function through direct and indirect effects on p53 coregulators and downstream targets. Further insight into the connections between ceramide and p53 will allow simultaneous targeting of the two pathways with a potential to yield more efficient anticancer therapeutics.

1. INTRODUCTION

Ceramides, members of the second largest class of membrane lipids, represent a family of more than 200 structurally related sphingolipids, compounds formed by a sphingoid base and a fatty acid linked via the amide bond (Hannun & Obeid, 2011, 2017). Over the past decades numerous studies have demonstrated that depending on their molecular structure, ceramides confer unique biophysical properties to the biological membranes (Castro, Prieto, & Silva, 2014). While the mechanisms of ceramide function in biological membranes and the biological consequences of ceramide alterations for cellular processes are still poorly understood, overwhelming data point to the role of these sphingolipids as important cellular regulators. Over the last 3 decades ceramides were established as second messengers regulating key cellular processes including cytoskeleton dynamics, endocytosis, protein transport and subcellular localization, cell cycle, autophagy, and apoptosis (Castro et al., 2014; Coant, Sakamoto, Mao, & Hannun, 2017; Dany & Ogretmen, 2015; Mullen, Hannun, & Obeid, 2012). Therefore, ceramides are indispensable for cellular homeostasis and control of fundamental functions, such as proliferation, migration, differentiation, adaptation to stress, survival, and senescence (Hannun & Obeid, 2008, 2017).

2. CERAMIDE SIGNALING—BIOLOGY

Ceramide, a central molecule in sphingolipid metabolism (Hannun & Obeid, 2008; Levy & Futerman, 2010), can be generated by the de novo pathway, which starts from the serine and palmitoyl-CoA condensation. It can be also produced by the salvage pathway, whereby sphingolipids are hydrolyzed in the lysosomes to sphingosine, which can be reacylated back to ceramide by one of the ceramide synthases (CerSs) (Coant et al., 2017). Finally, it can originate from the direct hydrolysis of complex sphingolipids such as sphingomyelin and glycosphingolipids (Hannun & Obeid, 2017). Overall, ceramide biosynthesis and the downstream function are interconnected with other lipid pathways, highly compartmentalized within the cell and also engaged in communication with organelle-specific metabolic pathways (Coant et al., 2017; Hannun & Obeid, 2011; Hoeferlin, Fekry, Ogretmen, Krupenko, & Krupenko, 2013). Collectively, these aspects of ceramide metabolism call for a "Many Ceramides" approach (Hannun & Obeid, 2011) for investigating ceramide effects and their significance for biological processes. Such an approach requires characterization of the temporal and spatial dynamics of different ceramide species, identification of pathways regulating such dynamics, and the close look at specific cellular compartments where the changes take place. Understanding of the complex aspects of ceramide biology will provide a comprehensive picture of the ceramide function, allow more precise predictions of the end point effects of ceramide alterations, and enable the development of therapeutic means to modulate ceramide-dependent processes (Hannun & Obeid, 2011). Ultimately, not only ceramides but also sphingosine 1-phosphate (S1P), as well as related complex sphingolipids, determine the overall cellular response to different stimuli, a concept represented by a "sphingolipid rheostat" mechanism, originally suggested in 1998 (Mandala et al., 1998). Importantly, increased ceramide metabolism by certain enzymes such as glucosylceramide synthase (GCS), sphingomyelin synthase, ceramide kinase, ceramidases, and sphingosine kinases (SPHKs) has been shown for cancer cells, underscoring the role of sphingolipid metabolism in cancer. Toward this end, the increase in sphingolipids with prosurvival function emerged as a novel target for anticancer therapy (Ogretmen, 2018).

3. CERAMIDE AND CANCER

Cancers are frequently characterized by altered levels of sphingolipids, including ceramide, but mechanisms by which these alterations result in tumor formation and progression are largely unknown. Ceramide-producing enzymes CerS2 (ceramide synthase 2) and CerS6 (ceramide synthase 6) were shown to be elevated in cancerous breast tissues as compared with normal breast tissues (Erez-Roman, Pienik, & Futerman, 2010). Endogenous C_{16}-, C_{24}-, and $C_{24:1}$-ceramide levels were increased in human head and neck squamous cell carcinomas, compared to normal tissues levels (Karahatay et al., 2007). Importantly, C_{16}- and C_{18}-ceramides were demonstrated to play two opposing roles in human head and neck squamous cell carcinomas, prosurvival and proapoptotic, respectively (Senkal, Ponnusamy, Bielawski, Hannun, & Ogretmen, 2010). Over the years numerous studies presented evidence of the ceramide function in the signal transduction of cancer cells and in their response to nutrient and oxidative stress, chemotherapy, and radiation (Hannun & Obeid, 2017; Ogretmen & Hannun, 2004; Saddoughi et al., 2013). While initially ceramides were regarded as death-promoting signaling molecules, it is now believed that effect of the ceramide elevation depends on the specific ceramide structure and the cellular context (Mesicek et al., 2010; Schiffmann et al., 2009).

Ceramides also emerged as important biomarkers in cancer therapy. For example, increase of total plasma ceramide in the patients undergoing radiation therapy in combination with irinotecan chemotherapy was associated with favorable tumor response (Dubois et al., 2016). Analogously, CerS6 was identified as a member of the 19-gene-signature predicting survival of patients with colorectal cancer (Abdul Aziz et al., 2016). Mechanistic basis for the distinct functions of endogenous ceramides is not known; however, interaction of ceramides with specific proteins was suggested to be responsible for this, at least in part (Saddoughi et al., 2013).

Another important aspect of altered ceramide levels in cancer, specific elevation of ceramide-metabolizing enzymes, is connected to cancer resistance, a major obstacle in cancer therapy. Thus, elevation of SPHK1 and GCS in prostate cancer cells was found to be responsible for resistance to paclitaxel (Aoyama et al., 2017). Elevated GCS expression has been demonstrated to correlate with progression of breast, urinary, ovarian cancers, and leukemia (Liu, Hill, & Li, 2013), whereas upregulation of acid ceramidase on irradiation conferred prostate cancer resistance (Cheng et al., 2013). This opens an opportunity to target these enzymes for treatment of resistant tumors (Ogretmen, 2018).

4. P53: THE BEGINNING

The p53 transcription factor, originally discovered as an endogenous cellular protein bound to a simian virus 40 large T antigen (Lane & Crawford, 1979; Linzer & Levine, 1979), has been initially described as an oncogene. Later studies demonstrated that a normal p53 gene is not required for embryonic development, but its absence predisposes mice to neoplastic disease, and that the oncogenic mutant forms of p53 are not required for development of many types of cancer (Donehower et al., 1992). Thus, the p53 protein was established as tumor suppressor. It is also commonly referred to as the guardian of the genome, due to its strong ability to respond to DNA damage or to stress stimuli capable of the induction of such damage (Lane, 1992). *TP53*, the gene encoding for the p53 protein, is the most often mutated gene in human cancers with the mutation rate estimated to be above 50% across all cancer types (Shirole et al., 2016). Due to the fact that p53 protein does not easily respond to pharmacological intervention, it has been considered "undruggable," and therapies capable of overcoming the oncogenic effects of mutant p53 have not been developed yet, almost 40 years after discovery of the protein function in cancer (Sabapathy & Lane, 2018). Nevertheless, the line of research evaluating p53 as a chemotherapeutic target tremendously expanded our understanding of the protein structure and function.

5. P53 IS A TRANSCRIPTION FACTOR

Regardless of all the nicknames of the p53 protein, first and foremost, it is a transcription factor controlling expression of hundreds of targets (Fischer, 2017). At the transcriptional level, p53 acts exclusively as a direct transcriptional activator, whereas inhibitory effects on some targets are usually indirect and mediated by cyclin-dependent kinase inhibitor p21 (CDKN1A or Waf1) interfering with phosphorylation of the retinoblastoma-like proteins RBL1 (p107) and RBL2 (p130) (Quaas, Muller, & Engeland, 2012; Sullivan, Galbraith, Andrysik, & Espinosa, 2018). In turn, hypophosphorylated RBLs stabilize a multicomponent repressor DREAM (including dimerization proteins, RBLs, E2F4, and multivulval class B complex [MUVB]), which inhibits transcription by engaging E2F or cell cycle homology region sites and suppresses transcription (Fischer, 2017). In addition, p53 may indirectly suppress transcription by a variety of mechanisms, such as binding to and disabling specific transcription factors, outcompeting specific activators from promoters due to overlapping binding sites, recruiting chromatin-modifying complexes that block gene expression, and via regulation of both miRNAs and lincRNAs (Barsotti & Prives, 2010l; Huarte et al., 2010).

The p53 protein is a highly labile and tightly controlled transcription factor present in nonstressed cells at low level due to its binding by E3 ubiquitin ligase MDM2, which enables p53 ubiquitination and proteasomal degradation (Hock & Vousden, 2014). On a variety of cellular cues, p53 escapes the control by MDM2, accumulates in the nuclei, and activates multiple genes to regulate diverse cellular processes including progression through the cell cycle, DNA damage repair, apoptosis, metabolic alterations, epigenomic modifications, shortening of telomeres, ferroptosis, etc (Haupt & Haupt, 2017). While numerous data indicate that p53 can regulate a wide array of biological processes, it should be noted that cell type, context, and specific stimuli define not just the different levels of gene expression but rather affect the transcription of very different sets of genes (Kastenhuber & Lowe, 2017). Thus, it is not surprising that depending on the cell type, p53 can generate opposite responses to the same stressor. Still, in many cases, mechanisms for such differences are not known. It has been suggested that context-dependent posttranslational modifications of p53 may dictate the interaction with a specific gene target (Kumari, Kohli, & Das, 2014). In addition, the kinetics of the p53 activation and the architecture of core promoters play a role in defining cellular response in a specific case: short p53 activation with rapid assembly of preinitiation complex and few rounds of RNA polymerase II reinitiation taking place at the p21 core promoter, and sustained activation with slow assembly of preinitiation complex that supports multiple rounds of transcription in case of Fas/APO1 and PUMA (Gomes & Espinosa, 2010; Morachis, Murawsky, & Emerson, 2010). The presence of epigenetic marks and other transcription factors at the target promoters may further attenuate the gene response to p53 activation (Itahana et al., 2016; Kastenhuber & Lowe, 2017). Thus, p53 functions in a complex network of regulators that support flexibility of cellular response to changing environment, depending on the cell type and state.

6. P53 AS A METABOLIC REGULATOR

Despite the canonical p53 definition as a tumor suppressor, one should keep in mind that ancestral members of the p53 gene family, found in unicellular protists and short-lived multicellular organisms (*Drosophila*) that do not develop cancers, function only in germ stem cells as protectors from DNA damage and uncontrolled proliferation (in the absence of DNA damage) (Wylie, Lu, D'Brot, Buszczak, & Abrams, 2014). These findings point to the cancer prevention by p53 as a later development of an ancient ability to prevent DNA damage and regulate growth of stem cells. It is also believed that the p53 regulatory network acquired diverse physiological roles before it was engaged in tumor suppression (Kastenhuber & Lowe, 2017). A few examples demonstrating the breadth of the p53 regulatory network include the response to various types of cellular stress, such as DNA damage (Williams & Schumacher, 2016), nutrient deprivation (Hoeferlin et al., 2013), hypoxia (Humpton & Vousden, 2016), and ribosomal stress (Deisenroth, Franklin, & Zhang, 2016). In addition, p53 regulates metabolic activities of the cell (Floter, Kaymak, & Schulze, 2017), the epigenome (Haupt & Haupt, 2017), and miRNAs (Liu, Zhang, Zhao, & Feng, 2017). Importantly, ceramide elevation is also observed in response to DNA damage, nutrient stress, senescence, autophagy, and apoptosis (Hannun & Obeid, 2017), pointing to a possible connection between the two pathways.

As a transcription factor, p53 regulates a variety of metabolic pathways, including glycolysis, pentose phosphate pathway, serine metabolism, mitochondrial metabolism, redox status, and lipid metabolism (Floter et al., 2017). By limiting NADPH production, p53 slows down many anabolic processes, including fatty acid biosynthesis. At the same time, wild-type (WT) p53 upregulates fatty acid oxidation by elevating expression of lipin 1 and guanidinoacetate methyltransferase (Liu, Zhang, Hu, & Feng, 2015). Moreover, via direct transcriptional control of such targets as sterol-regulatory-binding protein-1 (SREBP-1), acetyl-CoA dehydrogenase family member 11 (ACAD11), LIPIN1, malonyl-CoA decarboxylase (MCD), dehydrogenase/reductase 3 (DHRS3), carnitine palmitoyltransferase 1C (CPT1C) and caveolin 1, p53 controls multiple nodes in lipid metabolism (Parrales & Iwakuma, 2016).

Additional role of p53, different from the transcription activation, is known (Green & Kroemer, 2009; Vaseva & Moll, 2009). It is based on the ability of p53 to interact with myriads of cellular proteins thus affecting their functions. Importantly, this nontransactivation function of p53 can be fulfilled by WT protein and by its mutant variants. For example, while WT p53 suppresses the SREBP-1 levels via yet unidentified mechanisms, the R273H and R280K p53 mutants can bind to SREBP increasing its activity. Because p53 is frequently mutated in cancers, this further expands the processes controlled by p53. Another example includes the binding of mutant p53 to AMPK that inhibits this kinase activity (Zhou et al., 2014). The consequences of this effect can be profound taking into consideration that AMPK has recently emerged as a key regulator of cellular metabolism. Finally, it should be mentioned that p53 has yet another function associated with its translocation to mitochondria. This pathway involves the p53 interaction with members of the Bcl2 family and regulates apoptosis (Marchenko & Moll, 2014; Vaseva & Moll, 2009).

7. P53 IN TUMOR SUPPRESSION

Since the discovery of a tumor suppressor function of p53, it was believed to exert its cancer suppressor role by inducing senescence and apoptosis. Later it was demonstrated that p53 proficient mice, contrary to mice deficient in this protein, were completely suppressing development of thymic lymphomas even in the absence of p21, PUMA, and NOXA, critical players in p53-dependent senescence and programmed cell death (Valente et al., 2013). Moreover, mice expressing the $p53^{3KR}$ mutant, deficient in apoptotic and growth arrest functions but proficient in noncanonical and metabolic-regulatory functions, were still protected from tumor formation (Wang et al., 2016). These studies underscore the significance of the p53-mediated transcription for tumor suppression. However, identification of the p53 target genes and responses contributing to tumor suppression is very challenging, and only a few of already identified mechanisms have been tested in vivo (Kaiser & Attardi, 2018).

8. MUTANT P53 FUNCTION

The variety and frequency of p53 mutations in cancer deserve special consideration. Simplistically, the 393-amino-acid protein is organized in several functional domains, the amino terminal transactivation domain (residues 1–92), the DNA-binding domain (DBD, residues 101–306), the oligomerization domain (residues 307–355), and the carboxy-terminal domain (residues 356–393). Of note, tetramerization of p53 is crucial for its activity as a transcription factor (Lomax, Barnes, Hupp, Picksley, & Camplejohn, 1998). While the high rate of the *TP53* mutations affecting more than 300 amino acid residues creates an impediment for development of the p53-targeting therapeutics, a novel paradigm of "a rainbow of mutants" may lead to the development of efficient, mutant type–specific therapeutics (Sabapathy & Lane, 2018). The idea of the p53 mutants possessing functional differences suggests differential approach for targeting tumors with specific types of the protein mutations. Thus, mutations in the amino terminal domain often result in the utilization of an alternative start codon and expression of a shorter version of the protein, p47, defective in induction of the cell cycle arrest, but retaining selective transactivation function and ability to induce apoptotic genes (Phang et al., 2015). These mutations predispose organisms to cancer but predict better survival on therapeutic intervention. Distinct from the amino terminal domain, mutations in oligomerization domain result in the loss of function by p53 proteins, inability to form homotetramers, limited or absent transactivation capability, and absent tumor suppressor activity (Giacomazzi et al., 2014). Importantly, approximately 85% of all *TP53* mutations are found in the central 205-amino-acid DBD (Bouaoun et al., 2016). Most of the p53 DBD mutations lead to the loss of the ability to activate genes targeted by nonmutated p53 (Lee et al., 2012). Indeed, the rate of tumorigenesis is similar in *TP53*$^{-/-}$ mice and mice homozygous for mutant *TP53*, confirming the loss of tumor suppression function by the mutant. Interestingly, the dominant-negative effect of the p53 mutations makes WT allele dysfunctional when the cells are stressed by chemotherapeutics or irradiation (Lee et al., 2012; Willis, Jung, Wakefield, & Chen, 2004). So, an approach to downregulate only mutant p53 keeps promise for improvement of the tumor sensitivity to cancer drugs.

In the past decades numerous studies confirmed the "gain of function" hypothesis, which proposed that some DBD mutants may not just have lost the WT p53 activity but rather acquired an oncogenic activity (gained function), independent of their effects on the WT protein (Freed-Pastor & Prives, 2012). The addiction of tumors to mutant p53 was demonstrated in the mutant *TP53* knock-in mice, as the tamoxifen-induced mutant p53 ablation resulted in reduction of tumor growth, induction of apoptosis, and tumor regression, leading to 37% increase in animal survival (Alexandrova et al., 2015). It is believed that "gain of function" characteristics of the mutant p53 could be achieved via one of the two major mechanisms: a mutation conferring ability to transactivate the completely new set of genes (possibly via interaction with other transcription factors and chromatin remodeling proteins) and/or the interaction with other cellular proteins/signaling pathways (Freed-Pastor & Prives, 2012; Kim & Lozano, 2018). Precise details of both of these mechanisms will be dictated by the cell type and context and may be rather dynamic. Not all but only some of the p53 mutants display "gain of function" properties, and these properties seem to manifest in cancer cells only (Sabapathy & Lane, 2018).

Thus, accumulating evidence points to a potential benefit of therapeutic approaches targeting p53, as long as they take into account the cancer type, p53 status, and cellular context.

9. P53 AND CERAMIDE PATHWAYS

Cancer research in the past decades consistently implicated both the p53 and ceramide pathways in the regulation of cell growth, cell cycle arrest, senescence, and apoptosis. Therefore, it is not surprising that the connection between the two has been regularly reviewed (Carroll, Donaldson, & Obeid, 2015; Hage-Sleiman, Esmerian, Kobeissy, & Dbaibo, 2013; Heffernan-Stroud & Obeid, 2011).

Even the first study, which discovered a dose- and time-dependent elevation of the endogenous ceramide following the p53 activation by actinomycin D or gamma irradiation, observed a complex relationship between p53 and ceramide. The authors concluded that both p53-dependent and p53-independent stress response pathways could be activated by DNA damage, and ceramide accumulation could be a common feature in both of them (Dbaibo et al., 1998). Since then, numerous reports have demonstrated ceramide function both, downstream of p53 activation, upstream of p53, or independent of p53, in different cell types and under different stimuli (Heffernan-Stroud & Obeid, 2011). Interestingly, in *Caenorhabditis elegans*, radiation-induced apoptosis of germ cells was induced by the parallel pathways of CEP-1 (worm homologue of p53)-mediated elevation of the BH3-domain protein EGL-1, and CerS-mediated ceramide accumulation that converged at the mitochondrial membrane (Deng et al., 2008), thus representing yet another, synergistic, relationship between the p53 and ceramide.

Several sphingolipid-metabolizing enzymes, which produce or degrade ceramide, are regulated by p53 (Fig. 1). Thus, activation of the DNA damage pathway by doxorubicin resulted in the increase of neutral sphingomyelinase 2 (nSMase 2) activity and elevation of ceramide levels, as well as in induction of cell growth arrest. Activation of the nSMase was a result of direct transcriptional activation by p53 at a novel start site upstream of exon 3 (Shamseddine et al., 2015). In contrast, there is practically no evidence on acid sphingomyelinase (aSMase) activation by p53. In one of the recent studies, a derivative of PRIMA-1 (p53 reactivation and induction of massive apoptosis, compound 8b) was found to activate aSMase in H1299 cells (null for p53). PRIMA-1 is a synthetic compound that forms adducts with the p53 thiol groups and facilitates proper protein folding, thus correcting effects of p53-disabling mutations (Soans, Evans, Cipolla, & Fernandes, 2014). However, when the activated targets in H1299 p53$^{+/+}$ and H1299

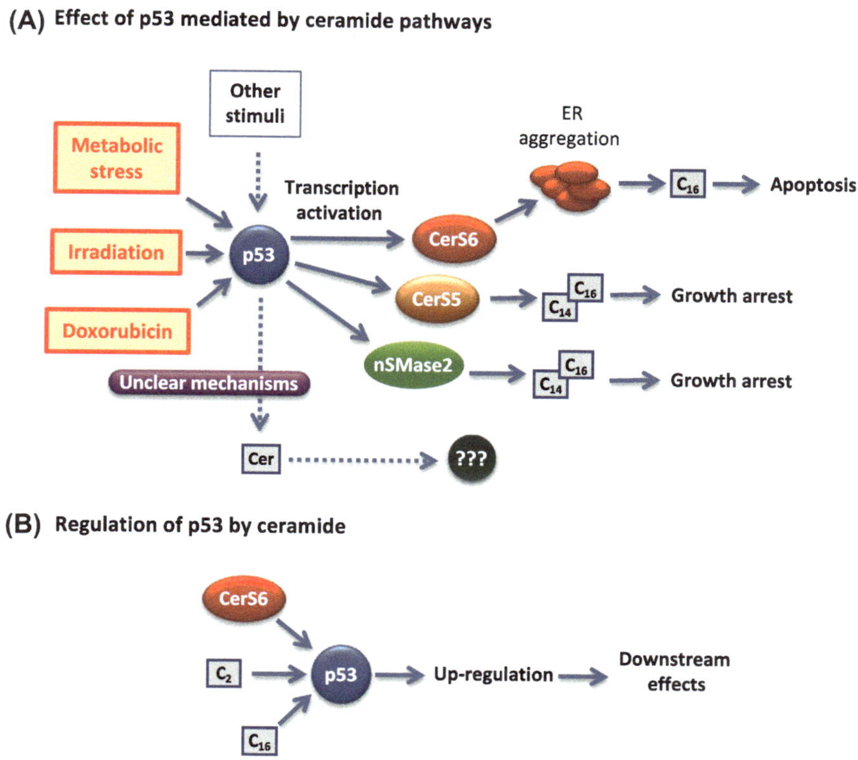

Figure 1 ***Circuits connecting the p53 function and ceramide signaling.*** (A) The induction of p53 has been associated with transcriptional activation of several enzymes of ceramide metabolism including CerS6 (in response to folate-mediated metabolic stress); CerS5 (in response to γ-irradiation); and nSMase (in response to doxorubicin). In all cases, C_{14}- and/or C_{16}-ceramides accumulated that caused antiproliferative effects (growth arrest or apoptosis). Other stress stimuli, which induce p53, can cause accumulation of certain ceramides, thus implicating p53 in this outcome, though precise underlying mechanisms remain unclear. (B). As part of signal transduction, ceramides themselves could activate p53 as a downstream target. Such effect has been demonstrated for C_2- and C_{16}-ceramides as well as for CerS6 (perhaps through the C_{16} generation as an intermediate step).

cells were compared, this led to a conclusion that the compound demonstrated an off-target effect and activated the TNF receptor superfamily pathway in these cells, arguing against p53 involvement. In a different study, p53 was shown to activate the nSMase but not aSMase (Sawada et al., 2002). Studies in different cell lines, leukemic and colon carcinoma, revealed that, on p53 upregulation, the de novo pathway is a major contributor to ceramide accumulation. Interestingly, the increase in ceramide was associated with the transcriptional upregulation of CerS5 (ceramide synthase 5) and not with change in serine palmitoyltransferase (Panjarian et al., 2008).

Another ceramide synthase, CerS6, was recently found to be directly transcriptionally activated by p53 via a noncanonical binding site, 11 nucleotides upstream of transcription start site (Fekry, Jeffries, et al., 2016). In contrast to the upregulation of ceramide synthases, SPHK1, a sphingolipid enzyme balancing the ceramide/S1P rheostat (Baran et al., 2007), was shown to be downregulated by p53 (Heffernan-Stroud et al., 2012; Taha et al., 2004). This regulation involves both canonical effector caspases of the mitochondrial apoptotic pathway and other, noncaspase/lysosomal proteases, such as cathepsins. More recent data revealed that p53-mediated activation of caspase-2 was required for SPHK1 proteolysis. In the triple-negative breast cancer cells with mutated p53, caspase-2 was not activated by doxorubicin treatment, and SPHK1 activity was preserved. However, suppression of the CHK1 was able to activate both caspase-2 and SPHK1 degradation (Carroll, Bonica, Shamseddine, Hannun, & Obeid, 2018).

Overall, the information regarding the regulatory effects of p53 on sphingolipid enzymes and the mechanisms involved in this regulation is rather scarce.

10. ACTIVATION OF P53 BY CERAMIDE

The relationship between ceramide and p53 is often interrogated by treatment of cells with exogenous ceramides. Because natural ceramides are highly hydrophobic and are not soluble in aqueous media, either short-chain analogs (C_2- or C_6-ceramides) or soluble derivatives of the ceramide carrying pyridinium group have been used in these studies (Hou et al., 2011). For example, C_2-ceramide was shown to induce SKN-SH cell death via elevation of p53, subsequent increase in Bax/Bcl-2 ratio and caspase activation (Kim et al., 2002). Similarly, in murine myeloid NSF.H7 cells, C_2-ceramide activated the protein phosphatase 2A (PP2A) with a consequent dephosphorylation of the Bcl-2, that allowed for direct binding of Bcl-2 and p53, thus preventing Bcl-2 from disabling the proapoptotic Bad and Bax proteins and leading to apoptosis (Deng, Gao, & May, 2009). In this case, p53 function was not dependent on ceramide but contributed to its downstream mechanisms. On the other hand, treatment of HCT116 adenocarcinoma cells with exogenous C_{16}-ceramide upregulated expression of p53 and, additionally, modulated expression of 50 cellular proteins (Renert et al., 2009). Experiments in which mouse fibroblasts were treated by alkylating agent MNNG (1-methyl-3-nitro-1-nitrosoguanidine) demonstrated that DNA damage response triggered in these cells was mediated by ceramide independently of p53 (Yang & Duerksen-Hughes, 2001). Nevertheless, the authors concluded that although the pathways were separable, they were interacting with each other, as the effects of ceramide on p53 levels and stability were noted. Another study concluded, based on the differences in morphology and DNA staining, as well as on the timing of the apoptotic death, that radiation-induced apoptosis was p53 dependent and that ceramide-induced apoptosis was not (Shi, Wuergler, Blattmann, & Crompton, 2001). Furthermore, a complex relationship between p53 and ceramide was also found on investigation of the cellular response to folate stress (Hoeferlin et al., 2013). Stressing A549 cells by either depletion of folate from culture media or by ectopic expression of a major folate-regulatory enzyme caused p53-dependent activation of de novo ceramide biosynthesis and CerS6-dependent C_{16}-ceramide elevation followed by apoptosis. Furthermore, overexpression of CerS6 by transient transfection, without any additional stress, resulted in strong p53 elevation and apoptosis, suggesting a feedback mechanism between C_{16}-ceramide and the tumor suppressor (Fig. 1).

Connection between p53 and ceramide pathways is often exploited in search for the combinational therapeutics. Coadministration of C_6-ceramide with chemotherapeutic drug vincristine caused necrosis and apoptosis in multiple cell lines. Significant activation of AMPK followed by p53 activation, inactivation of mTORC1, and downregulation of Bcl-1/Hif-1α were observed in these cells. Activated p53 translocated to mitochondria and bound to cyclophilin D, which led to mitochondrial permeability pore opening and cell necrosis. The authors suggested that p53 activation by these synergistic drugs activates both necroptosis and apoptosis (Chen et al., 2015). Although the number of studies demonstrating ceramide effects on p53 is steadily increasing, our knowledge of the mechanisms of ceramide effects on p53 levels and activity is still rudimentary.

11. ACTIVATION OF OTHER GENES BY CERAMIDE

Ceramide has been reported to regulate expression of several genes. Early investigations of the ceramide function identified several genes regulated by ceramide, such as cyclooxygenase (Ballou, Chao, Holness, Barker, & Raghow, 1992) and IL-6 (Laulederkind, Bielawska, Raghow, Hannun, & Ballou, 1995). MMP-1 expression was shown to be induced by endogenous or short-chain exogenous ceramides via activation of three distinct MAPK pathways in fibroblasts (Reunanen et al., 1998). Ceramide treatment of human hepatocarcinoma cells induced p53-independent elevation of the p21 mRNA and protein, as well as G1 cell cycle arrest (Kim, Kang, Kim, & Choi, 2000). Furthermore, both exogenous and endogenous ceramides were shown to reduce the hTERT (human telomerase reverse transcriptase) promoter activity (Wooten & Ogretmen, 2005). The mechanism of TERT inhibition involved reduction of acetylation of the Sp3 transcription factor, increasing the ratio of Sp3/Sp1 occupying the promoter and reducing the recruitment of RNA polymerase II to the promoter. Interestingly, a sphingolipid enzyme GCS implicated in the development of resistance to cancer chemotherapy was found to be transcriptionally upregulated by the ceramide, with the Sp1 transcription factor being essential for this regulation (Liu et al., 2008). In a different study, cannabinoid treatment—induced de novo ceramide generation activated transcription and increased protein levels of the stress-regulated protein 8 (p8) and its downstream targets Atf-4 and TRB3, leading to apoptosis of the pancreatic cancer cells (Carracedo et al., 2006). As in the case of p53, the mechanisms by which ceramide affects expression of its numerous targets are not fully understood.

12. CERAMIDE SIGNALING: MECHANISMS

Over the years, extensive research presented compelling evidence that changes in cellular ceramide levels are followed by activation of downstream effectors, which results in cell cycle arrest and adaptation, senescence, or programmed cell death.

Because ceramide is a hydrophobic molecule synthesized in and localized to the membranes, one of the plausible mechanisms of its effects on the downstream signaling molecules could be the changes in the biophysical properties of the membranes (lipid rafts formation, fluidity, etc.) where the changes in ceramide occur. These changes affect both the arrangement and properties of the membrane proteins, such as receptors, transporters, and signaling complexes that will transmit the signal further down the line (Edmond et al., 2015; Lockshon et al., 2012).

In addition, both short- and long-chain ceramides were able to form stable ceramide channels in planar membranes in vitro, though the experimental evidence for this process in vivo is still missing (Hernandez-Corbacho, Salama, Canals, Senkal, & Obeid, 2017; Siskind & Colombini, 2000). Alternatively, ceramide elevation resulting from treatment of A549 cells with methotrexate caused ER (endoplasmic reticulum) membrane aggregation leading to apoptosis (Fekry, Esmaeilniakooshkghazi, Krupenko, & Krupenko, 2016).

Another mechanism could be the direct binding of ceramide to the components of a signaling pathway that changes their activity. Direct binding of ceramide has been shown for protein kinases PKCα, PKCδ, PKCz, KSR, and c-Raf. PKCα was shown to be activated on ceramide binding, and its translocation from cytosol to the membrane was delayed, whereas PKCδ showed no change in activity on binding but rather decreased autophosphorylation (Huwiler, Fabbro, & Pfeilschifter, 1998). PKCz interaction with ceramide via the cysteine-rich domain was found to elevate its activity and induce the interaction with stress-activated protein kinase (SAPK) complex, resulting in growth suppression (Bourbon, Yun, & Kester, 2000). Similarly, both endogenous and exogenous ceramides were able to bind to the KSR cysteine-rich domain C1 and activate it via autophosphorylation that promoted binding to and transactivation of Raf-1 (Zhang et al., 1997) (Yin et al., 2009). Curiously, using the photoactivatable iodine-labeled ceramide, c-Raf itself was identified as a ceramide-binding protein, which was activated on ligand binding and further activated MAPK1 and MAPK3 signaling cascade (Huwiler et al., 1996). Importantly, the regulation by direct physical binding to its target is not limited to ceramide. Recently, S1P was demonstrated to function as a cofactor of TRAF2 (TNF receptor—associated factor 2), which, on binding, stimulated the ubiquitin-ligase activity of the protein toward RIP1 (Alvarez et al., 2010).

Recent searches for ceramide-binding proteins using bifunctional lipids with a photoactivatable diazirine group as a bait revealed 20 candidates among cellular cytosolic proteins that were cross-linked to the labeled ceramide (Bockelmann et al., 2018). Identified targets are involved in a variety of cellular functions, including ceramide transport, DNA damage response (DNA damage binding protein 1, and DnaJ homologue DnaJc7), protein ubiquitination (COP9 signalosome subunit 3), protein sorting (sorting nexin-1 and 2), trafficking (unc-119 A and B), and metabolism (GLO1, ALDH7A1, ADO, PTGR2, HMGCS2, AKR1C13, CTPS1, PAOX). This underscores the broad spectrum of processes regulated by ceramide

(Bockelmann et al., 2018). However, none of the ceramide-metabolizing enzymes have been identified in this work; this is likely due to exclusion of all cellular membranes from the experiments.

Another plausible mechanism for regulation of gene expression by ceramide could be the activation of protein phosphatase 1 (PP1) and PP2A by ceramide. While PP1 was shown to bind ceramide on its own both in vitro and in vivo (Sumanasekera et al., 2012), the activation of PP2A was mediated by the direct binding of ceramide to the inhibitor of PP2A (IPP2A/SET (SET nuclear protoco-oncogene)), as demonstrated by pull-down of SET on biotinylated C_6-ceramide and avidin column (Mukhopadhyay et al., 2009). Moreover, specificity toward C_{18}-ceramide, and not to C_{16}-ceramide, was described for the PP2A inhibitor. Ceramide binding by PP1 inhibited the enzyme, whereas binding of ceramide by SET resulted in activation of PP2A accompanied by proteasomal degradation of c-Myc and tumor suppression (Mukhopadhyay et al., 2009; Oaks & Ogretmen, 2014). Change in the activity of either major cellular phosphatase may also affect phosphorylation status of multiple targets, such as transcription factors, signaling proteins, etc., and contribute to regulation of gene activity as well.

In line with these findings, it has been shown that in HEK293 cells, natural C_{16}- and C_{24}-ceramides bind and inhibit PP1, thus preserving phosphorylation of a group of splicing-regulatory proteins containing the evolutionarily conserved motif RVxF recognized by PP1 (Sumanasekera et al., 2012). On the other hand, in multidrug resistant NCI/ADR-RES ovarian cancer cells expressing mutant p53 (five-amino-acid deletion in exon 5), the synthetic pyridinium C_6-ceramide activated PP1 thus causing dephosphorylation of serine/arginine-rich splicing-factor 1 and inducing its translocation to the nucleus (Patwardhan et al., 2014). This enabled the alternative splicing of the pre-mRNA encoding the p53 deletion mutant and restoration of the mRNA splice variant translated into WT p53. In a similar manner, recovery of the WT p53 mRNA expression on GCS knockdown and ceramide accumulation in OVCAR-8 and NCI/ADR-RES cells (both heterozygous for the p53 exon 5 deletion) led to the elevation of the p53 transcriptional targets p21, Bax, and PUMA, as well as induction of apoptosis (Liu et al., 2011). The same group has also reported that the inhibition of GCS, accompanied by the elevation of intracellular ceramide, restored the WT p53 activity in the cancer cells heterozygous for a missense R273H mutant. This further sensitized cells toward doxorubicin and increased apoptosis (Hosain et al., 2016). Ceramide-mediated alternative splicing by the function of the splice factor SRp30a was also shown for Bcl-x and caspase 9 (Chalfant et al., 2002; Massiello & Chalfant, 2006).

Thus the number of ceramide-binding protein targets is increasing as we learn more about the signaling functions of ceramide, yet we still do not understand how ceramide is delivered to specific targets (especially to cytosolic proteins), how specificity of the binding is achieved, and what mechanisms modulate the targets activity. The downstream effectors of ceramide-binding proteins activated/inhibited by ceramide are also often unknown.

13. CONCLUSIONS

Ceramides emerged as important novel signaling entities in the early 90s (Hannun, Obeid, & Wolff, 1993), and the past 25 years of research have revealed significant structural variability of ceramides, their involvement in vast array of biological processes, as well as very complex and intertwined relationships with their targets. An important role for ceramide in cancer development and response to therapeutic interventions has been also established, although our understanding of precise mechanisms guiding the cell type–specific and context-depending responses generated by ceramide alterations is still limited.

Tumor suppressor protein p53, discovered about 15 years earlier, is now, perhaps, the most studied protein, with the number of publications considering it approaching 90,000. These studies revealed many details about the protein structure, regulation, biological function, and its critical role for cancer suppression and therapy. Interestingly, similar to ceramide, the mechanisms of the p53 function are also cell type–specific and stimulus-specific, and highly context-dependent. Despite the years of intensive research, our understanding of these mechanisms is still incomplete.

In recent years, numerous studies indicated connections between the ceramide and p53 pathways (Fig. 1). Indeed, both of the pathways have been implicated in the critical cell fate decision processes, such as cell cycle arrest, senescence, and apoptosis. In these processes, ceramide and p53 can function upstream or downstream of each other, making the overall relationship quite complex. In this regard, due to cell type–dependent and context- and stimulus-dependent nature of both pathways, their intersection increases the variability of the response exponentially, making the repertoire of adaptive changes significantly larger. Accordingly, the task of examining these adaptations is extremely complex. Further investigation of the sphingolipid enzyme structures and regulation, identification of ceramide-specific protein targets, and changes that ceramide binding confers, are necessary to significantly improve our ability to untangle the mixed and sometimes contradictory data regarding the connection between p53 and ceramide. As our knowledge of both the ceramide and p53 function expands, it will reveal the regulatory nodes that can be efficiently targeted for therapeutic intervention in cancers.

REFERENCES

Abdul Aziz, N. A., Mokhtar, N. M., Harun, R., Mollah, M. M., Mohamed Rose, I., Sagap, I., et al. (2016). A 19-Gene expression signature as a predictor of survival in colorectal cancer. *BMC Medical Genomics, 9*(1), 58. https://doi.org/10.1186/s12920-016-0218-1.

Alexandrova, E. M., Yallowitz, A. R., Li, D., Xu, S., Schulz, R., Proia, D. A., et al. (2015). Improving survival by exploiting tumour dependence on stabilized mutant p53 for treatment. *Nature, 523*(7560), 352–356. https://doi.org/10.1038/nature14430.

Alvarez, S. E., Harikumar, K. B., Hait, N. C., Allegood, J., Strub, G. M., Kim, E. Y., et al. (2010). Sphingosine-1-phosphate is a missing cofactor for the E3 ubiquitin ligase TRAF2. *Nature, 465*(7301), 1084–1088. https://doi.org/10.1038/nature09128.

Aoyama, Y., Sobue, S., Mizutani, N., Inoue, C., Kawamoto, Y., Nishizawa, Y., et al. (2017). Modulation of the sphingolipid rheostat is involved in paclitaxel resistance of the human prostate cancer cell line PC3-PR. *Biochemical and Biophysical Research Communications, 486*(2), 551–557. https://doi.org/10.1016/j.bbrc.2017.03.084.

Ballou, L. R., Chao, C. P., Holness, M. A., Barker, S. C., & Raghow, R. (1992). Interleukin-1-mediated PGE2 production and sphingomyelin metabolism. Evidence for the regulation of cyclooxygenase gene expression by sphingosine and ceramide. *Journal of Biological Chemistry, 267*(28), 20044–20050.

Baran, Y., Salas, A., Senkal, C. E., Gunduz, U., Bielawski, J., Obeid, L. M., et al. (2007). Alterations of ceramide/sphingosine 1-phosphate rheostat involved in the regulation of resistance to imatinib-induced apoptosis in K562 human chronic myeloid leukemia cells. *Journal of Biological Chemistry, 282*(15), 10922–10934. https://doi.org/10.1074/jbc.M610157200.

Barsotti, A. M., & Prives, C. (2010). Noncoding RNAs: The missing "linc" in p53-mediated repression. *Cell, 142*(3), 358–360. https://doi.org/10.1016/j.cell.2010.07.029.

Bockelmann, S., Mina, J. G. M., Korneev, S., Hassan, D. G., Mueller, D., Hilderink, A., et al. (2018). A search for ceramide binding proteins using bifunctional lipid analogs yields CERT-related protein StarD7. *The Journal of Lipid Research*. https://doi.org/10.1194/jlr.M082354.

Bouaoun, L., Sonkin, D., Ardin, M., Hollstein, M., Byrnes, G., Zavadil, J., et al. (2016). TP53 variations in human cancers: New lessons from the IARC TP53 database and genomics data. *Human Mutation, 37*(9), 865–876. https://doi.org/10.1002/humu.23035.

Bourbon, N. A., Yun, J., & Kester, M. (2000). Ceramide directly activates protein kinase C zeta to regulate a stress-activated protein kinase signaling complex. *Journal of Biological Chemistry, 275*(45), 35617–35623. https://doi.org/10.1074/jbc.M007346200.

Carracedo, A., Gironella, M., Lorente, M., Garcia, S., Guzman, M., Velasco, G., et al. (2006). Cannabinoids induce apoptosis of pancreatic tumor cells via endoplasmic reticulum stress-related genes. *Cancer Research, 66*(13), 6748–6755. https://doi.org/10.1158/0008-5472.CAN-06-0169.

Carroll, B., Donaldson, J. C., & Obeid, L. (2015). Sphingolipids in the DNA damage response. *Advances in Biological Regulation, 58*, 38–52. https://doi.org/10.1016/j.jbior.2014.11.001.

Carroll, B. L., Bonica, J., Shamseddine, A. A., Hannun, Y. A., & Obeid, L. M. (2018). A role for caspase-2 in sphingosine kinase 1 proteolysis in response to doxorubicin in breast cancer cells - implications for the CHK1-suppressed pathway. *FEBS Open Biology, 8*(1), 27–40. https://doi.org/10.1002/2211-5463.12344.

Castro, B. M., Prieto, M., & Silva, L. C. (2014). Ceramide: A simple sphingolipid with unique biophysical properties. *Progress in Lipid Research, 54*, 53–67. https://doi.org/10.1016/j.plipres.2014.01.004.

Chalfant, C. E., Rathman, K., Pinkerman, R. L., Wood, R. E., Obeid, L. M., Ogretmen, B., et al. (2002). De novo ceramide regulates the alternative splicing of caspase 9 and Bcl-x in A549 lung adenocarcinoma cells. Dependence on protein phosphatase-1. *Journal of Biological Chemistry, 277*(15), 12587–12595. https://doi.org/10.1074/jbc.M112010200.

Chen, M. B., Jiang, Q., Liu, Y. Y., Zhang, Y., He, B. S., Wei, M. X., et al. (2015). C6 ceramide dramatically increases vincristine sensitivity both in vivo and in vitro, involving AMP-activated protein kinase-p53 signaling. *Carcinogenesis, 36*(9), 1061–1070. https://doi.org/10.1093/carcin/bgv094.

Cheng, J. C., Bai, A., Beckham, T. H., Marrison, S. T., Yount, C. L., Young, K., et al. (2013). Radiation-induced acid ceramidase confers prostate cancer resistance and tumor relapse. *Journal of Clinical Investigation, 123*(10), 4344–4358. https://doi.org/10.1172/JCI64791.

Coant, N., Sakamoto, W., Mao, C., & Hannun, Y. A. (2017). Ceramidases, roles in sphingolipid metabolism and in health and disease. *Advances in Biological Regulation, 63*, 122–131. https://doi.org/10.1016/j.jbior.2016.10.002.

Dany, M., & Ogretmen, B. (2015). Ceramide induced mitophagy and tumor suppression. *Biochimica et Biophysica Acta, 1853*(10 Pt B), 2834–2845. https://doi.org/10.1016/j.bbamcr.2014.12.039.

Dbaibo, G. S., Pushkareva, M. Y., Rachid, R. A., Alter, N., Smyth, M. J., Obeid, L. M., et al. (1998). p53-dependent ceramide response to genotoxic stress. *Journal of Clinical Investigation, 102*(2), 329–339. https://doi.org/10.1172/JCI1180.

Deisenroth, C., Franklin, D. A., & Zhang, Y. (2016). The evolution of the ribosomal protein-MDM2-p53 pathway. *Cold Spring Harbor Perspectives in Medicine, 6*(12). https://doi.org/10.1101/cshperspect.a026138.

Deng, X., Gao, F., & May, W. S. (2009). Protein phosphatase 2A inactivates Bcl2's antiapoptotic function by dephosphorylation and up-regulation of Bcl2-p53 binding. *Blood, 113*(2), 422–428. https://doi.org/10.1182/blood-2008-06-165134.

Deng, X., Yin, X., Allan, R., Lu, D. D., Maurer, C. W., Haimovitz-Friedman, A., et al. (2008). Ceramide biogenesis is required for radiation-induced apoptosis in the germ line of *C. elegans*. *Science, 322*(5898), 110–115. https://doi.org/10.1126/science.1158111.

Donehower, L. A., Harvey, M., Slagle, B. L., McArthur, M. J., Montgomery, C. A., Jr., Butel, J. S., et al. (1992). Mice deficient for p53 are developmentally normal but susceptible to spontaneous tumours. *Nature, 356*(6366), 215–221. https://doi.org/10.1038/356215a0.

Dubois, N., Rio, E., Ripoche, N., Ferchaud-Roucher, V., Gaugler, M. H., Campion, L., et al. (2016). Plasma ceramide, a real-time predictive marker of pulmonary and hepatic metastases response to stereotactic body radiation therapy combined with irinotecan. *Radiotherapy and Oncology: Journal of the European Society for Therapeutic Radiology and Oncology, 119*(2), 229–235. https://doi.org/10.1016/j.radonc.2016.03.014.

Edmond, V., Dufour, F., Poiroux, G., Shoji, K., Malleter, M., Fouque, A., et al. (2015). Downregulation of ceramide synthase-6 during epithelial-to-mesenchymal transition reduces plasma membrane fluidity and cancer cell motility. *Oncogene, 34*(8), 996–1005. https://doi.org/10.1038/onc.2014.55.

Erez-Roman, R., Pienik, R., & Futerman, A. H. (2010). Increased ceramide synthase 2 and 6 mRNA levels in breast cancer tissues and correlation with sphingosine kinase expression. *Biochemical and Biophysical Research Communications, 391*(1), 219–223. https://doi.org/10.1016/j.bbrc.2009.11.035.

Fekry, B., Esmaeilniakooshkghazi, A., Krupenko, S. A., & Krupenko, N. I. (2016). Ceramide synthase 6 is a novel target of methotrexate mediating its antiproliferative effect in a p53-dependent manner. *PLoS One, 11*(1), e0146618. https://doi.org/10.1371/journal.pone.0146618.

Fekry, B., Jeffries, K. A., Esmaeilniakooshkghazi, A., Ogretmen, B., Krupenko, S. A., & Krupenko, N. I. (2016). CerS6 is a novel transcriptional target of p53 protein activated by non-genotoxic stress. *Journal of Biological Chemistry, 291*(32), 16586–16596. https://doi.org/10.1074/jbc.M116.716902.

Fischer, M. (2017). Census and evaluation of p53 target genes. *Oncogene, 36*(28), 3943–3956. https://doi.org/10.1038/onc.2016.502.

Floter, J., Kaymak, I., & Schulze, A. (2017). Regulation of metabolic activity by p53. *Metabolites,* 7(2). https://doi.org/10.3390/metabo7020021.

Freed-Pastor, W. A., & Prives, C. (2012). Mutant p53: One name, many proteins. *Genes & Development, 26*(12), 1268–1286. https://doi.org/10.1101/gad.190678.112.

Giacomazzi, J., Graudenz, M. S., Osorio, C. A., Koehler-Santos, P., Palmero, E. I., Zagonel-Oliveira, M., et al. (2014). Prevalence of the TP53 p.R337H mutation in breast cancer patients in Brazil. *PLoS One, 9*(6), e99893. https://doi.org/10.1371/journal.pone.0099893.

Gomes, N. P., & Espinosa, J. M. (2010). Disparate chromatin landscapes and kinetics of inactivation impact differential regulation of p53 target genes. *Cell Cycle, 9*(17), 3428–3437. https://doi.org/10.4161/cc.9.17.12998.

Green, D. R., & Kroemer, G. (2009). Cytoplasmic functions of the tumour suppressor p53. *Nature, 458*(7242), 1127–1130. https://doi.org/10.1038/nature07986.

Hage-Sleiman, R., Esmerian, M. O., Kobeissy, H., & Dbaibo, G. (2013). p53 and ceramide as collaborators in the stress response. *International Journal of Molecular Sciences, 14*(3), 4982–5012. https://doi.org/10.3390/ijms14034982.

Hannun, Y. A., & Obeid, L. M. (2008). Principles of bioactive lipid signalling: Lessons from sphingolipids. *Nature Reviews. Molecular Cell Biology, 9*(2), 139–150. https://doi.org/10.1038/nrm2329.

Hannun, Y. A., & Obeid, L. M. (2011). Many ceramides. *Journal of Biological Chemistry, 286*(32), 27855–27862. https://doi.org/10.1074/jbc.R111.254359.

Hannun, Y. A., & Obeid, L. M. (2017). Sphingolipids and their metabolism in physiology and disease. *Nature Reviews. Molecular Cell Biology.* https://doi.org/10.1038/nrm.2017.107.

Hannun, Y. A., Obeid, L. M., & Wolff, R. A. (1993). The novel second messenger ceramide: Identification, mechanism of action, and cellular activity. *Advances in Lipid Research, 25,* 43–64.

Haupt, S., & Haupt, Y. (2017). P53 at the start of the 21st century: Lessons from elephants. *F1000Research, 6,* 2041. https://doi.org/10.12688/f1000research.12682.1.

Heffernan-Stroud, L. A., Helke, K. L., Jenkins, R. W., De Costa, A. M., Hannun, Y. A., & Obeid, L. M. (2012). Defining a role for sphingosine kinase 1 in p53-dependent tumors. *Oncogene, 31*(9), 1166–1175. https://doi.org/10.1038/onc.2011.302.

Heffernan-Stroud, L. A., & Obeid, L. M. (2011). p53 and regulation of bioactive sphingolipids. *Advances in Enzyme Regulation, 51*(1), 219–228. https://doi.org/10.1016/j.advenzreg.2010.10.003.

Hernandez-Corbacho, M. J., Salama, M. F., Canals, D., Senkal, C. E., & Obeid, L. M. (2017). Sphingolipids in mitochondria. *Biochimica et Biophysica Acta, 1862*(1), 56–68. https://doi.org/10.1016/j.bbalip.2016.09.019.

Hock, A. K., & Vousden, K. H. (2014). The role of ubiquitin modification in the regulation of p53. *Biochimica et Biophysica Acta, 1843*(1), 137–149. https://doi.org/10.1016/j.bbamcr.2013.05.022.

Hoeferlin, L. A., Fekry, B., Ogretmen, B., Krupenko, S. A., & Krupenko, N. I. (2013). Folate stress induces apoptosis via p53-dependent de novo ceramide synthesis and up-regulation of ceramide synthase 6. *Journal of Biological Chemistry, 288*(18), 12880–12890. https://doi.org/10.1074/jbc.M113.461798.

Hosain, S. B., Khiste, S. K., Uddin, M. B., Vorubindi, V., Ingram, C., Zhang, S., et al. (2016). Inhibition of glucosylceramide synthase eliminates the oncogenic function of p53 R273H mutant in the epithelial-mesenchymal transition and induced pluripotency of colon cancer cells. *Oncotarget,* 7(37), 60575–60592. https://doi.org/10.18632/oncotarget.11169.

Hou, Q., Jin, J., Zhou, H., Novgorodov, S. A., Bielawska, A., Szulc, Z. M., et al. (2011). Mitochondrially targeted ceramides preferentially promote autophagy, retard cell growth, and induce apoptosis. *The Journal of Lipid Research, 52*(2), 278–288. https://doi.org/10.1194/jlr.M012161.

Huarte, M., Guttman, M., Feldser, D., Garber, M., Koziol, M. J., Kenzelmann-Broz, D., et al. (2010). A large intergenic noncoding RNA induced by p53 mediates global gene repression in the p53 response. *Cell, 142*(3), 409–419. https://doi.org/10.1016/j.cell.2010.06.040.

Humpton, T. J., & Vousden, K. H. (2016). Regulation of cellular metabolism and hypoxia by p53. *Cold Spring Harbor Perspectives in Medicine, 6*(7). https://doi.org/10.1101/cshperspect.a026146.

Huwiler, A., Brunner, J., Hummel, R., Vervoordeldonk, M., Stabel, S., van den Bosch, H., et al. (1996). Ceramide-binding and activation defines protein kinase c-Raf as a ceramide-activated protein kinase. *Proceedings of the National Academy of Sciences of the United States of America, 93*(14), 6959–6963.

Huwiler, A., Fabbro, D., & Pfeilschifter, J. (1998). Selective ceramide binding to protein kinase C-alpha and -delta isoenzymes in renal mesangial cells. *Biochemistry, 37*(41), 14556–14562. https://doi.org/10.1021/bi981401i.

Itahana, Y., Zhang, J., Goke, J., Vardy, L. A., Han, R., Iwamoto, K., et al. (2016). Histone modifications and p53 binding poise the p21 promoter for activation in human embryonic stem cells. *Scientific Reports, 6,* 28112. https://doi.org/10.1038/srep28112.

Kaiser, A. M., & Attardi, L. D. (2018). Deconstructing networks of p53-mediated tumor suppression in vivo. *Cell Death and Differentiation, 25*(1), 93–103. https://doi.org/10.1038/cdd.2017.171.

Karahatay, S., Thomas, K., Koybasi, S., Senkal, C. E., Elojeimy, S., Liu, X., et al. (2007). Clinical relevance of ceramide metabolism in the pathogenesis of human head and neck squamous cell carcinoma (HNSCC): Attenuation of C(18)-ceramide in HNSCC tumors correlates with lymphovascular invasion and nodal metastasis. *Cancer Letters, 256*(1), 101–111. https://doi.org/10.1016/j.canlet.2007.06.003.

Kastenhuber, E. R., & Lowe, S. W. (2017). Putting p53 in context. *Cell, 170*(6), 1062–1078. https://doi.org/10.1016/j.cell.2017.08.028.

Kim, M. P., & Lozano, G. (2018). Mutant p53 partners in crime. *Cell Death and Differentiation, 25*(1), 161–168. https://doi.org/10.1038/cdd.2017.185.

Kim, S. S., Chae, H. S., Bach, J. H., Lee, M. W., Kim, K. Y., Lee, W. B., et al. (2002). P53 mediates ceramide-induced apoptosis in SKN-SH cells. *Oncogene, 21*(13), 2020–2028. https://doi.org/10.1038/sj.onc.1205037.

Kim, W. H., Kang, K. H., Kim, M. Y., & Choi, K. H. (2000). Induction of p53-independent p21 during ceramide-induced G1 arrest in human hepatocarcinoma cells. *Biochemistry and Cell Biology, 78*(2), 127–135.

Kumari, R., Kohli, S., & Das, S. (2014). p53 regulation upon genotoxic stress: intricacies and complexities. *Molecular & Cellular Oncology, 1*(3), e969653. https://doi.org/10.4161/23723548.2014.969653.

Lane, D. P. (1992). Cancer. p53, guardian of the genome. *Nature, 358*(6381), 15–16. https://doi.org/10.1038/358015a0.

Lane, D. P., & Crawford, L. V. (1979). T antigen is bound to a host protein in SV40-transformed cells. *Nature, 278*(5701), 261–263.

Laulederkind, S. J., Bielawska, A., Raghow, R., Hannun, Y. A., & Ballou, L. R. (1995). Ceramide induces interleukin 6 gene expression in human fibroblasts. *The Journal of Experimental Medicine, 182*(2), 599–604.

Lee, M. K., Teoh, W. W., Phang, B. H., Tong, W. M., Wang, Z. Q., & Sabapathy, K. (2012). Cell-type, dose, and mutation-type specificity dictate mutant p53 functions in vivo. *Cancer Cell, 22*(6), 751–764. https://doi.org/10.1016/j.ccr.2012.10.022.

Levy, M., & Futerman, A. H. (2010). Mammalian ceramide synthases. *IUBMB Life, 62*(5), 347–356. https://doi.org/10.1002/iub.319.

Linzer, D. I., & Levine, A. J. (1979). Characterization of a 54K dalton cellular SV40 tumor antigen present in SV40-transformed cells and uninfected embryonal carcinoma cells. *Cell, 17*(1), 43–52.

Liu, J., Zhang, C., Hu, W., & Feng, Z. (2015). Tumor suppressor p53 and its mutants in cancer metabolism. *Cancer Letters, 356*(2 Pt A), 197–203. https://doi.org/10.1016/j.canlet.2013.12.025.

Liu, J., Zhang, C., Zhao, Y., & Feng, Z. (2017). MicroRNA control of p53. *Journal of Cellular Biochemistry, 118*(1), 7–14. https://doi.org/10.1002/jcb.25609.

Liu, Y. Y., Hill, R. A., & Li, Y. T. (2013). Ceramide glycosylation catalyzed by glucosylceramide synthase and cancer drug resistance. *Advances in Cancer Research, 117*, 59–89. https://doi.org/10.1016/B978-0-12-394274-6.00003-0.

Liu, Y. Y., Patwardhan, G. A., Bhinge, K., Gupta, V., Gu, X., & Jazwinski, S. M. (2011). Suppression of glucosylceramide synthase restores p53-dependent apoptosis in mutant p53 cancer cells. *Cancer Research, 71*(6), 2276–2285. https://doi.org/10.1158/0008-5472.CAN-10-3107.

Liu, Y. Y., Yu, J. Y., Yin, D., Patwardhan, G. A., Gupta, V., Hirabayashi, Y., et al. (2008). A role for ceramide in driving cancer cell resistance to doxorubicin. *FASEB Journal, 22*(7), 2541–2551. https://doi.org/10.1096/fj.07-092981.

Lockshon, D., Olsen, C. P., Brett, C. L., Chertov, A., Merz, A. J., Lorenz, D. A., et al. (2012). Rho signaling participates in membrane fluidity homeostasis. *PLoS One, 7*(10), e45049. https://doi.org/10.1371/journal.pone.0045049.

Lomax, M. E., Barnes, D. M., Hupp, T. R., Picksley, S. M., & Camplejohn, R. S. (1998). Characterization of p53 oligomerization domain mutations isolated from Li-Fraumeni and Li-Fraumeni like family members. *Oncogene, 17*(5), 643–649. https://doi.org/10.1038/sj.onc.1201974.

Mandala, S. M., Thornton, R., Tu, Z., Kurtz, M. B., Nickels, J., Broach, J., et al. (1998). Sphingoid base 1-phosphate phosphatase: A key regulator of sphingolipid metabolism and stress response. *Proceedings of the National Academy of Sciences of the United States of America, 95*(1), 150–155.

Marchenko, N. D., & Moll, U. M. (2014). Mitochondrial death functions of p53. *Molecular & Cellular Oncology, 1*(2), e955995. https://doi.org/10.1080/23723548.2014.955995.

Massiello, A., & Chalfant, C. E. (2006). SRp30a (ASF/SF2) regulates the alternative splicing of caspase-9 pre-mRNA and is required for ceramide-responsiveness. *The Journal of Lipid Research, 47*(5), 892–897. https://doi.org/10.1194/jlr.C600003-JLR200.

Mesicek, J., Lee, H., Feldman, T., Jiang, X., Skobeleva, A., Berdyshev, E. V., et al. (2010). Ceramide synthases 2, 5, and 6 confer distinct roles in radiation-induced apoptosis in HeLa cells. *Cellular Signalling, 22*(9), 1300–1307. https://doi.org/10.1016/j.cellsig.2010.04.006.

Morachis, J. M., Murawsky, C. M., & Emerson, B. M. (2010). Regulation of the p53 transcriptional response by structurally diverse core promoters. *Genes & Development, 24*(2), 135–147. https://doi.org/10.1101/gad.1856710.

Mukhopadhyay, A., Saddoughi, S. A., Song, P., Sultan, I., Ponnusamy, S., Senkal, C. E., et al. (2009). Direct interaction between the inhibitor 2 and ceramide via sphingolipid-protein binding is involved in the regulation of protein phosphatase 2A activity and signaling. *FASEB Journal, 23*(3), 751–763. https://doi.org/10.1096/fj.08-120550.

Mullen, T. D., Hannun, Y. A., & Obeid, L. M. (2012). Ceramide synthases at the centre of sphingolipid metabolism and biology. *The Biochemical Journal, 441*(3), 789–802. https://doi.org/10.1042/BJ20111626.

Oaks, J., & Ogretmen, B. (2014). Regulation of PP2A by sphingolipid metabolism and signaling. *Frontiers in Oncology, 4*, 388. https://doi.org/10.3389/fonc.2014.00388.

Ogretmen, B. (2018). Sphingolipid metabolism in cancer signalling and therapy. *Nature Reviews Cancer, 18*(1), 33–50. https://doi.org/10.1038/nrc.2017.96.

Ogretmen, B., & Hannun, Y. A. (2004). Biologically active sphingolipids in cancer pathogenesis and treatment. *Nature Reviews Cancer, 4*(8), 604–616.

Panjarian, S., Kozhaya, L., Arayssi, S., Yehia, M., Bielawski, J., Bielawska, A., et al. (2008). De novo N-palmitoylsphingosine synthesis is the major biochemical mechanism of ceramide accumulation following p53 up-regulation. *Prostaglandins & Other Lipid Mediators, 86*(1–4), 41–48. https://doi.org/10.1016/j.prostaglandins.2008.02.004.

Parrales, A., & Iwakuma, T. (2016). p53 as a regulator of lipid metabolism in cancer. *International Journal of Molecular Sciences, 17*(12). https://doi.org/10.3390/ijms17122074.

Patwardhan, G. A., Hosain, S. B., Liu, D. X., Khiste, S. K., Zhao, Y., Bielawski, J., et al. (2014). Ceramide modulates pre-mRNA splicing to restore the expression of wild-type tumor suppressor p53 in deletion-mutant cancer cells. *Biochimica et Biophysica Acta, 1841*(11), 1571–1580. https://doi.org/10.1016/j.bbalip.2014.08.017.

Phang, B. H., Othman, R., Bougeard, G., Chia, R. H., Frebourg, T., Tang, C. L., et al. (2015). Amino-terminal p53 mutations lead to expression of apoptosis proficient p47 and prognosticate better survival, but predispose to tumorigenesis. *Proceedings of the National Academy of Sciences of the United States of America, 112*(46), E6349–E6358. https://doi.org/10.1073/pnas.1510043112.

Quaas, M., Muller, G. A., & Engeland, K. (2012). p53 can repress transcription of cell cycle genes through a p21(WAF1/CIP1)-dependent switch from MMB to DREAM protein complex binding at CHR promoter elements. *Cell Cycle, 11*(24), 4661–4672. https://doi.org/10.4161/cc.22917.

Renert, A. F., Leprince, P., Dieu, M., Renaut, J., Raes, M., Bours, V., et al. (2009). The proapoptotic C16-ceramide-dependent pathway requires the death-promoting factor Btf in colon adenocarcinoma cells. *Journal of Proteome Research, 8*(10), 4810–4822. https://doi.org/10.1021/pr9005316.

Reunanen, N., Westermarck, J., Hakkinen, L., Holmstrom, T. H., Elo, I., Eriksson, J. E., et al. (1998). Enhancement of fibroblast collagenase (matrix metalloproteinase-1) gene expression by ceramide is mediated by extracellular signal-regulated and stress-activated protein kinase pathways. *Journal of Biological Chemistry, 273*(9), 5137–5145.

Sabapathy, K., & Lane, D. P. (2018). Therapeutic targeting of p53: All mutants are equal, but some mutants are more equal than others. *Nature Reviews. Clinical Oncology, 15*(1), 13–30. https://doi.org/10.1038/nrclinonc.2017.151.

Saddoughi, S. A., Gencer, S., Peterson, Y. K., Ward, K. E., Mukhopadhyay, A., Oaks, J., et al. (2013). Sphingosine analogue drug FTY720 targets I2PP2A/SET and mediates lung tumour suppression via activation of PP2A-RIPK1-dependent necroptosis. *EMBO Molecular Medicine, 5*(1), 105–121. https://doi.org/10.1002/emmm.201201283.

Sawada, M., Nakashima, S., Kiyono, T., Yamada, J., Hara, S., Nakagawa, M., et al. (2002). Acid sphingomyelinase activation requires caspase-8 but not p53 nor reactive oxygen species during Fas-induced apoptosis in human glioma cells. *Experimental Cell Research, 273*(2), 157–168. https://doi.org/10.1006/excr.2001.5437.

Schiffmann, S., Sandner, J., Birod, K., Wobst, I., Angioni, C., Ruckhaberle, E., et al. (2009). Ceramide synthases and ceramide levels are increased in breast cancer tissue. *Carcinogenesis, 30*(5), 745–752. https://doi.org/10.1093/carcin/bgp061.

Senkal, C. E., Ponnusamy, S., Bielawski, J., Hannun, Y. A., & Ogretmen, B. (2010). Antiapoptotic roles of ceramide-synthase-6-generated C16-ceramide via selective regulation of the ATF6/CHOP arm of ER-stress-response pathways. *FASEB Journal, 24*(1), 296–308. https://doi.org/10.1096/fj.09-135087.

Shamseddine, A. A., Clarke, C. J., Carroll, B., Airola, M. V., Mohammed, S., Rella, A., et al. (2015). P53-dependent upregulation of neutral sphingomyelinase-2: Role in doxorubicin-induced growth arrest. *Cell Death & Disease, 6*, e1947. https://doi.org/10.1038/cddis.2015.268.

Shirole, N. H., Pal, D., Kastenhuber, E. R., Senturk, S., Boroda, J., Pisterzi, P., et al. (2016). TP53 exon-6 truncating mutations produce separation of function isoforms with pro-tumorigenic functions. *Elife, 5*. https://doi.org/10.7554/eLife.17929.

Shi, Y. Q., Wuergler, F. E., Blattmann, H., & Crompton, N. E. (2001). Distinct apoptotic phenotypes induced by radiation and ceramide in both p53-wild-type and p53-mutated lymphoblastoid cells. *Radiation and Environmental Biophysics, 40*(4), 301–308.

Siskind, L. J., & Colombini, M. (2000). The lipids C2- and C16-ceramide form large stable channels. Implications for apoptosis. *Journal of Biological Chemistry, 275*(49), 38640–38644. https://doi.org/10.1074/jbc.C000587200.

Soans, E., Evans, S. C., Cipolla, C., & Fernandes, E. (2014). Characterizing the sphingomyelinase pathway triggered by PRIMA-1 derivatives in lung cancer cells with differing p53 status. *Anticancer Research, 34*(7), 3271–3283.

Sullivan, K. D., Galbraith, M. D., Andrysik, Z., & Espinosa, J. M. (2018). Mechanisms of transcriptional regulation by p53. *Cell Death and Differentiation, 25*(1), 133–143. https://doi.org/10.1038/cdd.2017.174.

Sumanasekera, C., Kelemen, O., Beullens, M., Aubol, B. E., Adams, J. A., Sunkara, M., et al. (2012). C6 pyridinium ceramide influences alternative pre-mRNA splicing by inhibiting protein phosphatase-1. *Nucleic Acids Research, 40*(9), 4025–4039. https://doi.org/10.1093/nar/gkr1289.

Taha, T. A., Osta, W., Kozhaya, L., Bielawski, J., Johnson, K. R., Gillanders, W. E., et al. (2004). Down-regulation of sphingosine kinase-1 by DNA damage: Dependence on proteases and p53. *Journal of Biological Chemistry, 279*(19), 20546–20554. https://doi.org/10.1074/jbc.M401259200.

Valente, L. J., Gray, D. H., Michalak, E. M., Pinon-Hofbauer, J., Egle, A., Scott, C. L., et al. (2013). p53 efficiently suppresses tumor development in the complete absence of its cell-cycle inhibitory and proapoptotic effectors p21, Puma, and Noxa. *Cell Reports, 3*(5), 1339–1345. https://doi.org/10.1016/j.celrep.2013.04.012.

Vaseva, A. V., & Moll, U. M. (2009). The mitochondrial p53 pathway. *Biochimica et Biophysica Acta, 1787*(5), 414–420. https://doi.org/10.1016/j.bbabio.2008.10.005.

Wang, S. J., Li, D., Ou, Y., Jiang, L., Chen, Y., Zhao, Y., et al. (2016). Acetylation is crucial for p53-mediated ferroptosis and tumor suppression. *Cell Reports, 17*(2), 366–373. https://doi.org/10.1016/j.celrep.2016.09.022.

Williams, A. B., & Schumacher, B. (2016). p53 in the DNA-Damage-Repair Process. *Cold Spring Harbor Perspectives in Medicine, 6*(5). https://doi.org/10.1101/cshperspect.a026070.

Willis, A., Jung, E. J., Wakefield, T., & Chen, X. (2004). Mutant p53 exerts a dominant negative effect by preventing wild-type p53 from binding to the promoter of its target genes. *Oncogene, 23*(13), 2330–2338. https://doi.org/10.1038/sj.onc.1207396.

Wooten, L. G., & Ogretmen, B. (2005). Sp1/Sp3-dependent regulation of human telomerase reverse transcriptase promoter activity by the bioactive sphingolipid ceramide. *Journal of Biological Chemistry, 280*(32), 28867–28876. https://doi.org/10.1074/jbc.M413444200.

Wylie, A., Lu, W. J., D'Brot, A., Buszczak, M., & Abrams, J. M. (2014). p53 activity is selectively licensed in the Drosophila stem cell compartment. *Elife, 3*, e01530. https://doi.org/10.7554/eLife.01530.

Yang, J., & Duerksen-Hughes, P. J. (2001). Activation of a p53-independent, sphingolipid-mediated cytolytic pathway in p53-negative mouse fibroblast cells treated with N-methyl-N-nitro-N-nitrosoguanidine. *Journal of Biological Chemistry, 276*(29), 27129–27135. https://doi.org/10.1074/jbc.M100729200.

Yin, X., Zafrullah, M., Lee, H., Haimovitz-Friedman, A., Fuks, Z., & Kolesnick, R. (2009). A ceramide-binding C1 domain mediates kinase suppressor of ras membrane translocation. *Cellular Physiology and Biochemistry: International Journal of Experimental Cellular Physiology, Biochemistry, and Pharmacology, 24*(3–4), 219–230. https://doi.org/10.1159/000233248.

Zhang, Y., Yao, B., Delikat, S., Bayoumy, S., Lin, X. H., Basu, S., et al. (1997). Kinase suppressor of Ras is ceramide-activated protein kinase. *Cell, 89*(1), 63–72.

Zhou, G., Wang, J., Zhao, M., Xie, T. X., Tanaka, N., Sano, D., et al. (2014). Gain-of-function mutant p53 promotes cell growth and cancer cell metabolism via inhibition of AMPK activation. *Molecular Cell, 54*(6), 960–974. https://doi.org/10.1016/j.molcel.2014.04.024.

CHAPTER EIGHT

The Role of Ceramide 1-Phosphate in Tumor Cell Survival and Dissemination

Antonio Gomez-Muñoz[1]
Department of Biochemistry and Molecular Biology, Faculty of Science and Technology, University of the Basque Country (UPV/EHU), Bilbao, Spain
[1]Corresponding author: E-mail: antonio.gomez@ehu.es

Contents

Abstract

Ceramide 1-phosphate (C1P) is a pleiotropic bioactive sphingolipid metabolite capable of regulating key physiologic cell functions and promoting pathologic processes. Concerning pathology, C1P or ceramide kinase (CerK), the enzyme responsible for its biosynthesis in mammalian cells, has been implicated in cancer cell growth, survival, and dissemination and is involved in inflammatory responses associated with different types of cancer cells. The mechanisms or signaling pathways mediating these C1P actions have only been partially described. This chapter reviews recent progress in identifying signal transduction pathways involved in the promotion of cancer cell growth, survival, and dissemination by CerK and C1P.

1. INTRODUCTION

The bioactivity of ceramide 1-phosphate (C1P) has been known for over two decades. Although the existence of ceramide kinase (CerK), the only enzyme so far known to generate C1P in mammalian cells, had already

Advances in Cancer Research, Volume 140
ISSN 0065-230X
https://doi.org/10.1016/bs.acr.2018.04.012

been reported in brain tissue (Bajjalieh, Martin, & Floor, 1989), the presence of C1P in cells was first observed in human leukemia HL60 monocytes (Dressler & Kolesnick, 1990). C1P is synthesized in the Golgi apparatus through phosphorylation of ceramides by CerK. The ceramide enzyme substrate, which is produced de novo in the endoplasmic reticulum (ER), is transported by a specific ceramide transfer protein that has the ability to move from the ER to the Golgi where CerK resides (reviewed in Presa et al., 2016) (Fig. 1).

CerK is also present in the cytosol, nucleus, and the plasma membrane of cells where it may phosphorylate ceramides produced by pathways other than the de novo pathway. In fact, two other major pathways by which ceramides can be synthesized outside the ER include the sphingomyelinase (SMase) pathway, which can generate ceramides at the plasma membrane and some cytosolic compartments, including lysosomes or mitochondrial-associated membranes, and the salvage pathway, which includes the

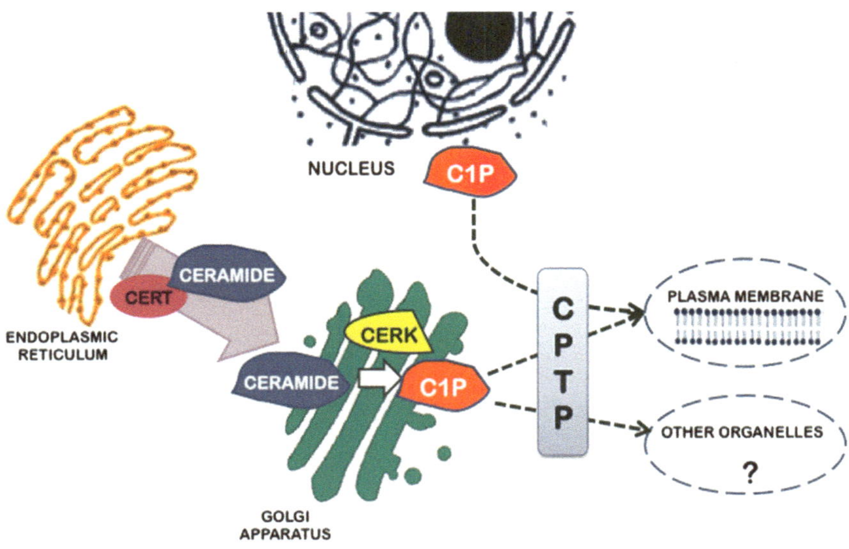

Figure 1 Biosynthesis and transport of ceramide 1-phosphate (C1P). Ceramides are generated by de novo synthesis in the endoplasmic reticulum and are transported to the Golgi apparatus by a ceramide transfer protein (CERT). Ceramides can be phosphorylated by ceramide kinase (CerK) in the Golgi to generate C1P. A C1P transfer protein (CPTP) will transport C1P to the plasma membrane and probably other organelles. *Taken from Presa, N., Gomez-Larrauri, A., Rivera, I. G., Ordonez, M., Trueba, M. & Gomez-Munoz, A. (2016). Regulation of cell migration and inflammation by ceramide 1-phosphate. Biochimica et Biophysica Acta, 1861, 402–409, with permission.*

recycling of sphingosine that is formed by metabolism of complex sphingolipids (Fig. 2).

Once synthesized, C1P can be transported to the plasma membrane or to other organelles by a recently identified specific C1P transfer protein that is located in the cytosol and has the ability to move to the Golgi network to collect C1P (Simanshu et al., 2013) (Fig. 1). Nonetheless, a role of C1P in regulating cell biology went unnoticed until the discovery that it had mitogenic properties. First, it was shown that synthetic

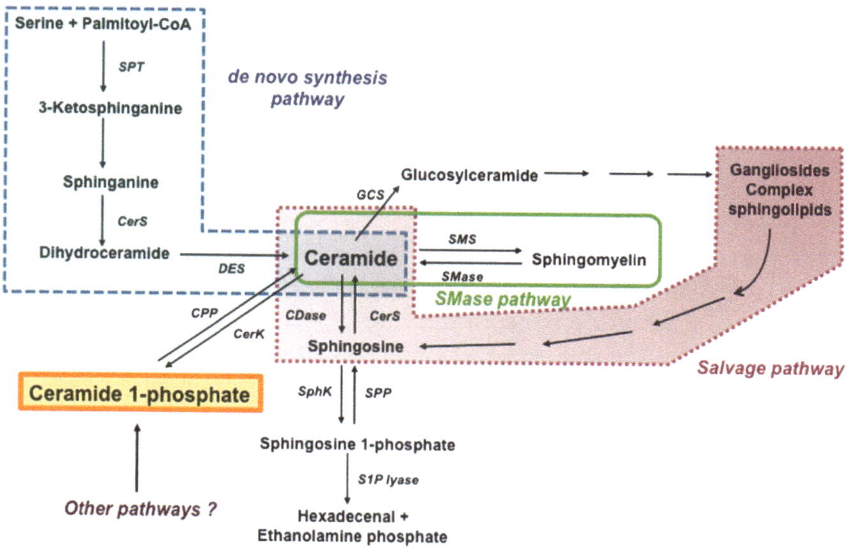

Figure 2 Ceramide/sphingolipid biosynthetic pathways. Ceramides are the central core of sphingolipid metabolism. They can be synthesized mainly by three major pathways: (1) the de novo synthesis pathway involves the concerted actions of serine palmitoyl-CoA transferase (SPT), ceramide synthase (CerS), and dihydroceramide desaturase (DES) activities (blue box); (2) the sphingomyelinase (SMase) pathway generates ceramide directly through degradation of sphingomyelin (SM) by SMases. The reversed reaction is catalyzed by SM synthases to generate SM (green box); (3) the salvage pathway recycles sphingosine derived from the metabolism of complex sphingolipids to form ceramide. Once generated, ceramide can be phosphorylated by the action of ceramide kinase (CerK) to generate ceramide 1-phosphate or it can be used for synthesis of complex sphingolipids (purple box). Additionally, ceramidases can degrade ceramides to form sphingosine, which can in turn be converted into S1P by sphingosine kinases (SphK). The reverse reaction is catalyzed by sphingosine phosphate phosphatases (SPP). S1P can then be degraded by S1P lyase to produce 2-trans hexadecenal and ethanolamine phosphate. *GCS*, glycosylceramide synthase. *Taken from A. Gomez-Muñoz et al. Prog. Lipid Res. 2016, with permission.*

short-chain (acetyl, C2- or octanoyl, C8-) C1P stimulated DNA synthesis as determined by the incorporation of [^{3}H]thymidine into DNA and that it increased cell division in rat fibroblasts (Gomez-Munoz, Waggoner, O'Brien, & Brindley, 1995). Subsequent studies demonstrated that long-chain (natural) C1P was also able to stimulate cell growth (Gómez-Muñoz, Frago, Alvarez, & Varela-Nieto, 1997). Interestingly, the mitogenic effects of C1P were counteracted by nonphosphorylated short-chain ceramides (Gómez-Muñoz et al., 1995; Gómez-Muñoz et al., 1997). At least part of the mechanism by which ceramides inhibited C1P-stimulated cell proliferation involved enhancement of C1P dephosphorylation. This action was largely caused by activation of the magnesium-independent form of phosphatidate phosphohydrolase, an enzyme that was later shown to dephosphorylate a variety of lipid phosphate metabolites including phosphatidic acid, lysophosphatidic acid (LPA), and sphingosine 1-phosphate, besides C1P (Waggoner, Gomez-Muñoz, Dewald, & Brindley, 1996), and is now known as lipid phosphate phosphatase (Tang, Benesch, & Brindley, 2015). Dephosphorylation of C1P would on one hand decrease the levels of mitogenic C1P and on the other hand would increase the concentration of ceramides, which are potent inhibitors of cell proliferation and inducers of apoptosis (Hannun & Obeid, 1995, 2002, 2008; Kolesnick & Golde, 1994; Kolesnick, Goñi, & Alonso, 2002; Kondo et al., 2002; Merrill, 1991, 2002; Merrill et al., 1997; Merrill, Sullards, Wang, Voss, & Riley, 2001; Okazaki, Bell, & Hannun, 1989, 1990), thereby leading to potent arrest of cell growth and potentiation of cell death processes. In addition to counteracting C1P actions, we observed that short-chain ceramides induced the production of endogenous (long-chain) ceramides (Gómez-Muñoz et al., 1995), which would further enhance the inhibitory effects of the short-chain ceramides on C1P-stimulated cell growth. The mechanism by which short-chain ceramides led to the synthesis of endogenous ceramides was later described by Ogretmen and coworkers (Ogretmen et al., 2002) who demonstrated that generation of long-chain ceramides was due to recycling of the sphingosine backbone of the short-chain ceramides via deacylation/reacylation reactions and not by elongation of their fatty acid moiety.

In addition to stimulating cell proliferation, C1P has also been shown to inhibit cell death and to promote migration and invasion of different cancer cell types. In addition, many studies have implicated C1P in inflammatory responses, which are closely associated with the establishment or progression of a variety of pathologies including cancer. All of these actions place

C1P or CerK, the enzyme responsible for its biosynthesis, as novel targets for counteracting tumorigenesis and tumor spreading. These crucial aspects of C1P biology are discussed further.

2. REGULATION OF CELL PROLIFERATION BY CERAMIDE 1-PHOSPHATE

Initial studies demonstrated the ability of C1P to stimulate cell proliferation in rat fibroblasts (Gómez-Muñoz et al., 1995; Gómez-Muñoz et al., 1997). This observation was later extended to other cell types, including primary (normal) and transformed macrophages (Gangoiti, Granado, Arana, Ouro, & Gomez-Munoz, 2010; Gangoiti, Wang, et al., 2008; Ouro et al., 2017), C2C12 myoblasts (Bernacchioni et al., 2018; Gangoiti et al., 2012), primary rat aortic vascular smooth muscle cells (Kim et al., 2011), primary photoreceptor progenitors (Miranda, Abrahan, Agnolazza, Politi, & Rotstein, 2011), and different types of cancer cells including A549 human lung adenocarcinoma cells (Mitra et al., 2007), NCI-H358 human epithelial lung/bronchiole cancer cells, MCF-7 breast cancer cells (Pastukhov, Romer et al., 2014; Pastukhov, Zangemeister-Wittke, et al., 2014), and mouse leukemia RAW264.7 cells (Ouro et al., 2017). The mechanisms by which C1P stimulated cell proliferation involved activation of different kinases and transcription factors of different signaling pathways. In particular, C1P-stimulated proliferation of bone marrow-derived macrophage (BMDM) involves stimulation of mitogen-activated protein kinase kinase (MEK) and its downstream targets extracellularly regulated kinases 1 and 2 (ERK1-2), phosphatidylinositol 3-kinase (PI3K), and its downstream target Akt (also known as protein kinase B), the Akt target glycogen synthase kinase 3-beta (GSK-3β), the GSK-3β downstream targets cyclin D-1 and c-Myc, c-Jun N-terminal kinase, and the transcription factor NF-κB (nuclear factor-kappa B) (Gangoiti, Wang, et al., 2008). Subsequent studies demonstrated that C1P-stimulated BMDM proliferation was also mediated by activation of protein kinase C alpha, which activity was stimulated by diacylglycerol derived from sphingomyelin synthase activation in the absence of phosphatidylinositol-dependent phospholipase C (PLC) or phosphatidylcholine-dependent PLC activation (Gangoiti et al., 2010). Moreover, stimulation of BMDM proliferation involved phosphorylation and activation of the mammalian target of rapamycin complex 1 (mTORC1) and its downstream target p70S6K/RhoA/ROCK pathway (Gangoiti et al., 2011), as well as production of relatively low concentrations of reactive oxygen species (ROS) (Arana et al., 2012). Stimulation of myoblast and aortic vascular smooth muscle cell proliferation by C1P implicated hyperphosphorylation of the product of retinoblastoma and prevented cyclin D1 from being degraded in these cells (Gangoiti et al., 2012; Kim et al., 2011). Interestingly, stimulation of C2C12 mouse myoblasts by C1P was blocked by inhibitors of the LPA1 and LPA3 receptors following prior activation of type IV cytosolic phospholipase A_2 ($cPLA_2$), thereby suggesting participation of LPA in the mitogenic effect of C1P in these cells (Bernacchioni et al., 2018). In addition, C1P-stimulated proliferation of mouse leukemia cells was indirect involving the release of vascular endothelial growth factor (VEGF) and the subsequent interaction with its VEGFR2 receptor in an autocrine manner (Ouro et al., 2017). The physiological relevance of the mitogenic effect of C1P was underscored by the demonstration that the macrophage growth factor M-CSF (macrophage-colony stimulating factor) increased the intracellular concentration of C1P in primary macrophages (Gangoiti, Wang, et al., 2008), and by the observation that platelet-derived growth factor—induced rat aortic smooth muscle cell proliferation involved upregulation of CerK (Kim et al., 2011). Moreover, inhibition of CerK activity resulted in substantial reduction of DNA synthesis in C2C12 myoblasts further emphasizing the physiological relevance of C1P in the regulation of cell proliferation (Gangoiti et al., 2012).

3. REGULATION OF CELL DEATH AND SURVIVAL BY CERAMIDE 1-PHOSPHATE

Although deregulated cell proliferation is a key feature of cancer cells, suppression of cell death processes is equally as important. In fact, impaired apoptosis is a fundamental feature in the pathogenesis of cancer. Key regulators of apoptosis include SMase-derived ceramides, which can be produced by activation of different SMases. Specifically, activation of acid SMase (ASMase) is a major pathway implicated in the promotion of apoptosis by a large number of proapoptotic signals in different cell types (Hannun, 1996; Hannun & Obeid, 2008; Hundal et al., 2003). However, in studies using ASMase −/− mice we showed that BMDM incubated in the absence of M-CSF, a condition known to trigger apoptosis in these cells, still produced significant amounts of ceramides and died by apoptosis (Wang et al., 2015). We demonstrated that the alternative pathway for synthesis of ceramide in ASMase −/− BMDM was the de novo synthesis pathway, as the activity of serine palmitoyltransferase (SPT), the rate limiting enzyme for this pathway, was potently increased under these conditions (Wang et al., 2015). Another relevant finding concerning C1P biology was the promotion of cell survival, which occurred through inhibition of apoptosis. This action involved potent reduction of proapoptotic ceramide levels through mechanisms involving inhibition of ASMase (Gómez-Muñoz, Kong, Salh, & Steinbrecher, 2004) or blockade of SPT activation (Granado, Gangoiti, Ouro, Arana, & Gomez-Munoz, 2009; Wang et al., 2015). In addition, C1P inhibited stimulation of Bax, which is a proapoptotic Bcl2 protein family member that causes permeabilization of the outer mitochondrial membrane resulting in leakage of cytochrome c and blocked cytochrome c release into the cytosol (Granado, Gomez-Munoz, et al., 2009). Moreover, C1P prevented the activation of initiator caspase 9 and executioner caspase 3 in cells incubated under apoptotic conditions thereby leading to blockade of DNA fragmentation and the promotion of cell survival (Gómez-Muñoz et al., 2004). The physiological relevance of the antiapoptotic effect of C1P was emphasized by the observation that the intracellular levels of C1P were significantly reduced in apoptotic cells. This might result in the release of ASMase from inhibition and a subsequent increase of ceramide levels thereby triggering apoptotic cell death (Gómez-Muñoz et al., 2004). As mentioned above, activation of the PI3K/Akt pathway is a major mechanism by which C1P stimulates cell proliferation. However, this pathway is also implicated in the prosurvival actions of many growth factors and oncogenes (Marte & Downward, 1997). We found that inhibition of PI3K or Akt completely abolished the antiapoptotic effects of C1P, suggesting that the PI3K/Akt pathway is essential in this process (Gómez-Muñoz et al., 2005). In addition, the prosurvival effect of C1P required activation of the NF-κB, which is a transcription factor downstream of PI3K/Akt, as well as upregulation of the antiapoptotic Bcl-2 family member Bcl-X_L (Gómez-Muñoz et al., 2005).

Another relevant metabolite with signaling properties is nitric oxide (NO). This is a gas compound that plays a dual role in cell biology as it causes apoptosis when produced at high concentrations while promoting cell survival at low concentrations and is also implicated in angiogenesis and tumor metastasis (Ghimire, Altmann, Straub, & Isenberg, 2017; Mandal, 2017). In this connection, C1P was shown to upregulate the inducible form of nitric oxide synthase (iNOS) and subsequent production of relatively low concentrations of NO leading to cell survival. Interestingly,

C1P-stimulated iNOS expression and NO production were blocked by inhibitors of PI3K or NF-κB further emphasizing the importance of the PI3K/Akt/NF-κB pathway in the stimulation of cell survival by C1P (Gangoiti, Granado, Arana, Ouro, & Gomez-Munoz, 2008). Concerning the implication of CerK in cell survival, in experiments using specific siRNA to silence the gene encoding this kinase, Mitra et al. (2007) found that CerK is crucial for survival of human lung adenocarcinoma A549 cells. Knockdown of CerK in the lung cancer cells resulted in activation of caspase 3 (Mitra et al., 2007) and the subsequent promotion of apoptosis. CerK upregulation is also profoundly important in the development of human breast cancer. Noteworthy, analysis of gene expression profiles from a cohort of over 2200 patients revealed that increased CerK expression was associated with an elevated risk of tumor recurrence in women with breast cancer. In addition, CerK was spontaneously upregulated during tumor recurrence of multiple mouse models of breast cancer, thereby identifying a functional role of CerK in this type of malignancy (Payne, Pant, Pan, & Chodosh, 2014). CerK upregulation also protected tumor cells from apoptosis on HER2/neu downregulation in vitro (Payne et al., 2014). Further work showed that CerK downregulation and subsequent reduction of C1P levels was a major mechanism by which vitamin D3 exerted its antiproliferative actions in human neuroblastoma cells (Bini et al., 2012). The latter findings suggest that clinical agents targeted against the CerK/C1P axis may be useful as anticancer therapy against human cancer.

Another important aspect of C1P biology is that in addition to being generated intracellularly, C1P can be also found in the extracellular milieu. Interestingly, plasma C1P concentrations were shown to vary and be dependent on the nutritional state of the organism. Values found in vivo have been reported to be in the range 0.5–20 μM (Hammad et al., 2010; Kim et al., 2013; Mietla et al., 2013). In studies using photosensitive caged C1P analogues (4-bromo-5-hydroxy-2-nitrobenzhydryl or 7-(*N*,*N*-diethylamino) coumarin), we demonstrated that both stimulation of cell proliferation and inhibition of cell death required intracellular accumulation of C1P (Gomez-Munoz, Gangoiti, et al., 2016). The caged C1P analogues are cell permeable, do not disturb cell integrity in any manner, and bypass plasma membrane receptors (Lankalapalli, Ouro, Arana, Gomez-Munoz, & Bittman, 2009). We found that on light irradiation of specific wavelengths, C1P can be released in the cytosol and activate mitogenic or antiapoptotic signaling pathways leading to cell growth and survival (Arana et al., 2012; Gomez-Munoz, Gangoiti, et al., 2016; Lankalapalli et al., 2009).

The observation that C1P existed in extracellular pools provided a new dimension to C1P biology. Although C1P might be incorporated by cells through mechanisms involving flippases or scramblases to increase its intracellular levels, the existence of an extracellular C1P pool would allow its interaction with cell membranes and the possible unfolding of additional molecular (signaling) events. Indeed, a major biological function that was specifically attributed to extracellular C1P was the regulation of cell migration, as discussed further.

4. REGULATION OF CELL MIGRATION AND INVASION BY CERAMIDE 1-PHOSPHATE

Cell migration is a complex physiological process that is absolutely necessary for organogenesis, embryogenesis, tissue regeneration, wound healing, or immune responses. However, migration of cells to inappropriate sites in the organism can be detrimental for cell and tissue homeostasis and often results in the development of chronic inflammatory diseases including asthma, inflammatory bowel disease (IBD) (namely Crohn's disease and ulcerative colitis), rheumatoid arthritis, multiple sclerosis, or development of malignant tumors (Beltman, Maree, & de Boer, 2009; Biro, Munoz, & Weninger, 2014). Concerning cancer, cell migration contributes to tumor cell spreading or metastasis, for which cells develop invasive properties characterized by elevated expression or activity of proteolytic enzymes including different types of matrix metalloproteinases (MMPs). Upregulation of these enzymes causes disruption of the extracellular matrix, which is the physical barrier for cell migration, leading to cell detachment and invasion (Kessenbrock, Plaks, & Werb, 2010). Of interest, at least some MMPs have been shown to also play important roles in cancer cell growth and survival (Kessenbrock et al., 2010; Masson et al., 1998), actions that are known be regulated by C1P, as discussed earlier.

We previously showed that treatment of cells with exogenous C1P potently stimulated cell migration/invasion in a variety of cell types, including cancer cells (Gomez-Munoz, Presa, et al., 2016; Presa et al., 2016). In particular, exogenous C1P was first found to stimulate migration of mouse leukemia RAW264.7 monocytes (Granado, Gonzalez, et al., 2009), J774.A1 mouse reticulum cell sarcoma, and acute monocytic human leukemia THP-1 cells (Arana et al., 2013). In mouse leukemia cells, C1P triggered migration fairly rapidly. Optimal cell migration occurred after 4 h of treatment with exogenous C1P. This observation led to identification of a specific putative C1P receptor, which on ligation with C1P resulted in activation of the MEK/ERK1-2 and PI3K/Akt signaling pathways and upregulation of the transcription factor NF-κB (Arana et al., 2013; Granado, Gonzalez, et al., 2009). Interestingly, pertussis toxin (PTX), which inhibits Gi/o proteins and is often used for studying the biology of G protein-coupled receptors (GPCR), completely inhibited C1P-stimulated migration of the leukemia cells suggesting that the putative C1P receptor belongs to the GPCR family of receptors (Arana et al., 2013; Granado, Gonzalez, et al., 2009). The stimulation of cell migration by C1P was confirmed by Kim et al. using hematopoietic stem progenitor cells (Kim et al., 2013, 2012; Ratajczak, Suszynska, Borkowska, Ratajczak, & Schneider, 2014) and by Kester et al. using both coronary artery macrovascular endothelial cells and retinal microvascular endothelial cells (Hankins et al., 2013). The latter authors also showed that C1P elicited cell migration/invasion through interaction with the annexin a2-p11 heterotetrameric protein complex, which serves as receptor platform for multiple proteins implicated in vascular invasion through the extracellular matrix and wound healing. In addition, Chalfant et al. (Wijesinghe et al., 2014) found that the levels of C1P increase during the early stages of wound healing in human patients, and that CerK-derived C1P is required for migration of fibroblasts into wound sites. More recently, we found that extracellular C1P is a potent chemoattractant for human pancreatic ductal adenocarcinoma cells (Rivera et al., 2016). Pancreatic cancer is a very aggressive disease characterized by invasiveness, rapid progression, and profound resistance to treatment. In addition, the poor prognosis of this type of cancer is due to the difficulty of diagnosis at early stages of the disease. We found that inhibition or knockdown of ERK1-2, PI3K, Akt1, or mTORC1 resulted in potent blockade of C1P-stimulated pancreatic cancer cell migration and invasion in human pancreatic cancer PANC-1 and MIA PaCa-2 cells (Rivera et al., 2016). In addition, like for mouse or human leukemia cells (Arana et al., 2013; Gomez-Munoz, Presa, et al., 2016), C1P-stimulated pancreatic cancer cell migration/invasion was abolished by PTX, suggesting the participation of a GPCR in this process (Rivera et al., 2016). PTX also inhibited the stimulation of glucose uptake and metabolism that was elicited by C1P in the macrophages, suggesting that this was also a receptor-mediated effect of C1P (Ouro et al., 2013). However, although C1P-stimulated macrophage migration involved prior release of macrophage chemoattractant protein-1, this was not the case for stimulation of pancreatic cancer cells by C1P (I. G. Rivera, and A. Gomez-Muñoz, unpublished work). It should also be pointed out that pancreatic cancer PANC-1 and MIA PaCa-2 cells can migrate spontaneously, in the absence of exogenous chemoattractants (Rivera et al., 2016). In this connection, another relevant observation was that inhibition of CerK activity by the selective inhibitor NVP-231 or knockdown of CerK with specific siRNA to silence the gene encoding this kinase substantially reduced spontaneous migration of the pancreatic cancer cells. By contrast, overexpression of CerK increased the levels of C1P (Rivera et al., 2016) and potently enhanced migration of the pancreatic cancer cells suggesting that intracellular formation of C1P plays a critical role in this process and pointing to a relevant role of the CerK/C1P axis in pancreatic cancer metastasis. Exogenous C1P also enhanced the adhesion and motility of other types of cancer cells such as human rhabdomyosarcoma cells (Schneider et al., 2013) and human lung cancer cells, including nonsmall cell lung cancer NSCLC, A549, HTB177, HTB183, and CRL5803, as well as small cell lung cancer SCLC, CRL2062, and CRL5853 cells (Schneider et al., 2017). Further investigation into the mechanisms by which C1P regulates cell migration/invasion revealed that MMP 2 and 9 are key factors in this process, and that C1P potently increased actin polymerization and paxillin phosphorylation, which are essential regulators of cell migration (Ordonez, Rivera, Presa, & Gomez-Munoz, 2016). Interestingly, upregulation of MMP 2 and 9 expression or activity was blocked by pharmacological inhibitors of MEK (PD98059) or PI3K (LY294002) or by specific siRNA to silence the genes encoding these kinases, in agreement with previous work (Gomez-Munoz, Presa, et al., 2016). It was also observed that PTX inhibited Akt, ERK1-2, and paxillin phosphorylation leading to inhibition of cell migration, actions that are in agreement with the notion that C1P-stimulated macrophage migration is a receptor-mediated effect. Upregulation of MMP2 and MMP9 by C1P was also observed in multipotent stroma cells incubated with relatively low concentrations (0.1–1 μM) of exogenous C1P (Kim et al., 2013). Although these two MMPs are most likely implicated in the stimulation of multipotent stroma cell motility, this was not addressed in the latter report. C1P also stimulated cell migration in noncancer cells including endothelial progenitor cells, which participate in new vessel formation (Ratajczak et al., 2014), bone marrow-derived stroma cells, and human umbilical vein endothelial cells (Kim et al., 2013). It is therefore clear that both exogenous and endogenous C1Ps are key regulators of cell migration/invasion processes, although this is dependent on cell type.

5. IMPLICATION OF CERAMIDE 1-PHOSPHATE IN INFLAMMATORY RESPONSES

The implication of C1P in inflammatory responses is well documented. Initial work by Chalfant et al. showed that C1P was able to stimulate $cPLA_2$ and subsequent increases in the levels of arachidonic acid and proinflammatory eicosanoids in a variety of cell types, including cancer cells such as A549 human lung adenocarcinoma cells (Chalfant & Spiegel, 2005; Pettus et al., 2003, 2004). Many cancer risk factors are associated with chronic inflammation, although not all inflammatory diseases may cause cancer. In particular, it is unclear why rheumatoid arthritis or psoriasis, which affects the joints and skin, respectively, does not promote tumorigenesis, in comparison with IBD and chronic hepatitis, which are clearly protumorigenic diseases (Grivennikov, Greten, & Karin, 2010). Also of importance is the fact that the tumor microenvironment contains innate immune cells, including macrophages, besides cancer cells and surrounding stromal tissue. Macrophages secrete a variety of cytokines and chemokines, which can act in an autocrine or paracrine manner to control the growth of tumors and/or their dissemination. In fact, the role played by tumor-associated macrophages in malignant cell invasion and tumor metastasis is well recognized (Grivennikov et al., 2010). In addition, activated inflammatory cells produce ROS and nitrogen intermediates including NO, which can result in DNA damage and the promotion of tumorigenesis. In this regard, two of the mechanisms by which C1P promotes cell growth and survival include the production on relatively low levels of ROS and iNOS-derived NO (Arana et al., 2012; Gangoiti, Arana, et al., 2008), thereby reinforcing the notion that C1P may be relevant for controlling both tumor mass of primary tumors and formation of secondary tumors through cell migration/invasion events. From a clinical point of view, metastasis is probably the most critical feature of tumor formation. Many metastatic mechanisms involved activation of the NF-κB transcription factor, which is also potently induced by C1P (Gangoiti et al., 2011; Gangoiti et al., 2012; Gangoiti, Wang, et al., 2008; Gómez-Muñoz et al., 2005).

In addition to the above, it is known that at least 20% of cancers are linked to obesity, which is a low grade inflammatory disease, in which infiltrating monocytes in the adipose tissue differentiate into proinflammatory M1 macrophages. Using mouse 3T3-L1 preadipocytes, we have recently shown that CerK is upregulated during adipogenesis (Ordonez, Presa, Trueba, & Gomez-Munoz, 2017), suggesting that CerK/C1P may also be a relevant factor in the onset or development of obesity. Nonetheless, although the proinflammatory actions of intracellular C1P or CerK are fairly well documented, it is not clear whether exogenous C1P may cause different and at times opposing effects in cells, regarding inflammation. Specifically, administration of C1P exogenously attenuated lipopolysaccharide (LPS)-induced acute lung inflammation and potently reduced emphysema in mouse models of chronic obstructive pulmonary disease or mice exposed to cigarette smoke (Baudiss et al., 2015, 2016). A putative antiinflammatory action of C1P was previously reported by Kester and coworkers who found that C1P potently inhibited LPS-activated transcription of NF-κB and abrogated the production of the proinflammatory cytokines interleukin (IL)-6, IL-8, and IL-1β in human peripheral blood mononuclear cells (Hankins, Fox, Barth, Unrath, & Kester, 2011).

6. CONCLUDING REMARKS

C1P is well recognized as key regulator of cell growth, survival, and migration of different types of nonmalignant and malignant cells. Appropriate balances between cell growth and death, the production of pro- and antiinflammatory cytokines, and regulation of cell motility are fundamental aspects that strongly influence cell fate. Still many aspects of C1P as regulator of cell growth, survival, inflammation, and cell migration/invasion processes remain to be discovered, but it seems fairly clear that CerK and C1P are potential relevant targets for attenuating tumor formation and spreading. It may also be acknowledged that further characterization and cloning of the putative C1P receptor(s) may be relevant for developing new inhibitors and novel therapeutic strategies for treatment of cancer and metastasis.

ACKNOWLEDGMENTS

Work supported in part by "Departamento de Educación, Universidades e Investigación del Gobierno Vasco," Basque Country, Spain (Grant IT1106-16), and "Ministerio de Economía y Competitividad," Madrid, Spain (Grant SAF2016-79695-R).

APPENDIX

Permissions for reusing both figures have been granted by Biochimica et Biophysica Acta (Fig. 1) and Progress in Lipid Research (Fig. 2). Please see corresponding URL for each particular permission:

Biochimica et Biophysica Acta: https://s100.copyright.com/CustomerAdmin/PLF.jsp?ref=8086daaa-ba89-4726-a966-9601d5f1cf88.

Prog. Lipd Res: https://s100.copyright.com/CustomerAdmin/PLF.jsp?ref=ec768530-cce0-4311-a185-8d94cbfba965.

REFERENCES

Arana, L., Gangoiti, P., Ouro, A., Rivera, I. G., Ordonez, M., Trueba, M., et al. (2012). Generation of reactive oxygen species (ROS) is a key factor for stimulation of macrophage proliferation by ceramide 1-phosphate. *Experimental Cell Research, 318*, 350–360.

Arana, L., Ordonez, M., Ouro, A., Rivera, I. G., Gangoiti, P., Trueba, M., et al. (2013). Ceramide 1-phosphate induces macrophage chemoattractant protein-1 release: Involvement in ceramide 1-phosphate-stimulated cell migration. *American Journal of Physiology. Endocrinology and Metabolism, 304*, E1213–E1226.

Bajjalieh, S. M., Martin, T. F. J., & Floor, E. (1989). Synaptic vesicle ceramide kinase. A calcium-stimulated lipid kinase that copurifies with brain synaptic vesicles. *Journal of Biological Chemistry, 264*, 14354–14360.

Baudiss, K., Ayata, C. K., Lazar, Z., Cicko, S., Beckert, J., Meyer, A., et al. (2015). Ceramide-1-phosphate inhibits cigarette smoke-induced airway inflammation. *The European Respiratory Journal, 45*, 1669–1680.

Baudiss, K., de Paula Vieira, R., Cicko, S., Ayata, K., Hossfeld, M., Ehrat, N., et al. (2016). C1P attenuates lipopolysaccharide-induced acute lung injury by preventing NF-kappaB activation in neutrophils. *The Journal of Immunology: Official Journal of the American Association of Immunologists*.

Beltman, J. B., Maree, A. F., & de Boer, R. J. (2009). Analysing immune cell migration. *Nature Reviews. Immunology, 9*, 789–798.

Bernacchioni, C., Cencetti, F., Ouro, A., Bruno, M., Gomez-Munoz, A., Donati, C., et al. (2018). Lysophosphatidic acid signaling axis mediates ceramide 1-phosphate-induced proliferation of C2C12 myoblasts. *International Journal of Molecular Sciences, 19*.

Bini, F., Frati, A., Garcia-Gil, M., Battistini, C., Granado, M., Martinesi, M., et al. (2012). New signalling pathway involved in the anti-proliferative action of vitamin D(3) and its analogues in human neuroblastoma cells. A role for ceramide kinase. *Neuropharmacology, 63*, 524–537.

Biro, M., Munoz, M. A., & Weninger, W. (2014). Targeting Rho-GTPases in immune cell migration and inflammation. *British Journal of Pharmacology, 171*, 5491–5506.

Chalfant, C., & Spiegel, S. (2005). Sphingosine-1-phosphate and ceramide-1-phosphate: Expanding roles in cell signaling. *Journal of Cell Science, 118*, 4605–4612.

Dressler, K. A., & Kolesnick, R. N. (1990). Ceramide-1-phosphate, a novel phospholipid in human leukemia (HL-60) cells. *Journal of Biological Chemistry, 265*, 14917–14921.

Gangoiti, P., Arana, L., Ouro, A., Granado, M. H., Trueba, M., & Gomez-Munoz, A. (2011). Activation of mTOR and RhoA is a major mechanism by which Ceramide 1-phosphate stimulates macrophage proliferation. *Cellular Signalling, 23*, 27–34.

Gangoiti, P., Bernacchioni, C., Donati, C., Cencetti, F., Ouro, A., Gomez-Munoz, A., et al. (2012). Ceramide 1-phosphate stimulates proliferation of C2C12 myoblasts. *Biochimie.*

Gangoiti, P., Granado, M. H., Arana, L., Ouro, A., & Gomez-Munoz, A. (2008). Involvement of nitric oxide in the promotion of cell survival by ceramide 1-phosphate. *FEBS Letters, 582*, 2263–2269.

Gangoiti, P., Granado, M. H., Arana, L., Ouro, A., & Gomez-Munoz, A. (2010). Activation of protein kinase C-alpha is essential for stimulation of cell proliferation by ceramide 1-phosphate. *FEBS Letters, 584*, 517–524.

Gangoiti, P., Granado, M. H., Wang, S. W., Kong, J. Y., Steinbrecher, U. P., & Gomez-Munoz, A. (2008). Ceramide 1-phosphate stimulates macrophage proliferation through activation of the PI3-kinase/PKB, JNK and ERK1/2 pathways. *Cellular Signalling, 20*, 726–736.

Ghimire, K., Altmann, H. M., Straub, A. C., & Isenberg, J. S. (2017). Nitric oxide: what's new to NO? *American Journal of Physiology. Cell Physiology, 312*, C254–C262.

Gómez-Muñoz, A., Duffy, P. A., Martin, A., O'Brien, L., Byun, H. S., Bittman, R., et al. (1995). Short-chain ceramide 1-phosphates are novel stimulators of DNA synthesis and cell division: Antagonism by cell-permeable ceramides. *Molecular Pharmacology, 47*, 883–889.

Gómez-Muñoz, A., Frago, L., Alvarez, L., & Varela-Nieto, I. (1997). Stimulation of DNA synthesis by natural ceramide 1-phosphate. *The Biochemical Journal, 325*, 435–440.

Gomez-Munoz, A., Gangoiti, P., Rivera, I. G., Presa, N., Gomez-Larrauri, A., & Ordonez, M. (2016). Caged ceramide 1-phosphate (C1P) analogs: Novel tools for studying C1P biology. *Chemistry and Physics of Lipids, 194*, 79–84.

Gómez-Muñoz, A., Kong, J. Y., Parhar, K., Wang, S. W., Gangoiti, P., González, M., et al. (2005). Ceramide-1-phosphate promotes cell survival through activation of the phosphatidylinositol 3-kinase/protein kinase B pathway. *FEBS Letters, 579*, 3744–3750.

Gómez-Muñoz, A., Kong, J. Y., Salh, B., & Steinbrecher, U. P. (2004). Ceramide-1-phosphate blocks apoptosis through inhibition of acid sphingomyelinase in macrophages. *The Journal of Lipid Research, 45*, 99–105.

Gomez-Munoz, A., Presa, N., Gomez-Larrauri, A., Rivera, I. G., Trueba, M., & Ordonez, M. (2016). Control of inflammatory responses by ceramide, sphingosine 1-phosphate and ceramide 1-phosphate. *Progress in Lipid Research, 61*, 51–62.

Gomez-Munoz, A., Waggoner, D. W., O'Brien, L., & Brindley, D. N. (1995). Interaction of ceramides, sphingosine, and sphingosine 1-phosphate in regulating DNA synthesis and phospholipase D activity. *Journal of Biological Chemistry, 270*, 26318–26325.

Granado, M. H., Gangoiti, P., Ouro, A., Arana, L., & Gomez-Munoz, A. (2009). Ceramide 1-phosphate inhibits serine palmitoyltransferase and blocks apoptosis in alveolar macrophages. *Biochimica et Biophysica Acta, 1791*, 263–272.

Granado, M. H., Gangoiti, P., Ouro, A., Arana, L., Gonzalez, M., Trueba, M., et al. (2009). Ceramide 1-phosphate (C1P) promotes cell migration Involvement of a specific C1P receptor. *Cellular Signalling, 21*, 405–412.

Grivennikov, S. I., Greten, F. R., & Karin, M. (2010). Immunity, inflammation, and cancer. *Cell, 140*, 883–899.

Hammad, S. M., Pierce, J. S., Soodavar, F., Smith, K. J., Al Gadban, M. M., Rembiesa, B., et al. (2010). Blood sphingolipidomics in healthy humans: Impact of sample collection methodology. *The Journal of Lipid Research, 51*, 3074–3087.

Hankins, J. L., Fox, T. E., Barth, B. M., Unrath, K. A., & Kester, M. (2011). Exogenous ceramide-1-phosphate reduces lipopolysaccharide (LPS)-mediated cytokine expression. *Journal of Biological Chemistry, 286*, 44357–44366.

Hankins, J. L., Ward, K. E., Linton, S. S., Barth, B. M., Stahelin, R. V., Fox, T. E., et al. (2013). Ceramide 1-phosphate mediates endothelial cell invasion via the annexin a2-p11 heterotetrameric protein complex. *Journal of Biological Chemistry, 288*, 19726–19738.

Hannun, Y. A. (1996). Functions of ceramide in coordinating cellular responses to stress. *Science, 274*, 1855–1859.
Hannun, Y. A., & Obeid, L. M. (1995). Ceramide: An intracellular signal for apoptosis. *Trends in Biochemical Sciences, 20*, 73–79.
Hannun, Y. A., & Obeid, L. M. (2002). The ceramide-centric universe of lipid-mediated cell regulation: Stress encounters of the lipid kind. *Journal of Biological Chemistry, 277*, 25847–25850.
Hannun, Y. A., & Obeid, L. M. (2008). Principles of bioactive lipid signalling: Lessons from sphingolipids. *Nature Reviews. Molecular Cell Biology, 9*, 139–150.
Hundal, R. S., Gómez-Muñoz, A. J., Kong, Y., Salh, B., Marotta, A., Duronio, V., et al. (2003). Oxidized low density lipoprotein inhibits macrophage apoptosis by blocking ceramide generation thereby maintaining PKB activation and Bcl-XL levels. *Journal of Biological Chemistry, 278*, 24399–24408.
Kessenbrock, K., Plaks, V., & Werb, Z. (2010). Matrix metalloproteinases: Regulators of the tumor microenvironment. *Cell, 141*, 52–67.
Kim, C., Schneider, G., Abdel-Latif, A., Mierzejewska, K., Sunkara, M., Borkowska, S., et al. (2013). Ceramide-1-phosphate regulates migration of multipotent stromal cells and endothelial progenitor cells–implications for tissue regeneration. *Stem Cells, 31*, 500–510.
Kim, C. H., Wu, W., Wysoczynski, M., Abdel-Latif, A., Sunkara, M., Morris, A., et al. (2012). Conditioning for hematopoietic transplantation activates the complement cascade and induces a proteolytic environment in bone marrow: A novel role for bioactive lipids and soluble C5b-C9 as homing factors. *Leukemia, 26*, 106–116.
Kim, T. J., Kang, Y. J., Lim, Y., Lee, H. W., Bae, K., Lee, Y. S., et al. (2011). Ceramide 1-phosphate induces neointimal formation via cell proliferation and cell cycle progression upstream of ERK1/2 in vascular smooth muscle cells. *Experimental Cell Research, 317*, 2041–2051.
Kolesnick, R., & Golde, D. W. (1994). The sphingomyelin pathway in tumor necrosis factor and interleukin-1 signaling. *Cell, 77*, 325–328.
Kolesnick, R. N., Goñi, F. M., & Alonso, A. (2002). Compartmentalization of ceramide signalling. Physical foundations and biological effects. *Journal of Cellular Physiology, 184*, 285–300.
Kondo, T., Kitano, T., Iwai, K., Watanabe, M., Taguchi, Y., Yabu, T., et al. (2002). Control of ceramide-induced apoptosis by IGF-1: Involvement of PI3-kinase, caspase-3 and catalase. *Cell Death and Differentiation, 9*, 682–692.
Lankalapalli, R. S., Ouro, A., Arana, L., Gomez-Munoz, A., & Bittman, R. (2009). Caged ceramide 1-phosphate analogues: Synthesis and properties. *The Journal of Organic Chemistry, 74*, 8844–8847.
Mandal, P. (2017). Molecular signature of nitric oxide on major cancer hallmarks of colorectal carcinoma. *Inflammopharmacology*.
Marte, B. M., & Downward, J. (1997). PKB/Akt: Connecting phosphoinositide 3-kinase to cell survival and beyond. *Trends in Biochemical Sciences, 22*, 355–358.
Masson, R., Lefebvre, O., Noel, A., Fahime, M. E., Chenard, M. P., Wendling, C., et al. (1998). In vivo evidence that the stromelysin-3 metalloproteinase contributes in a paracrine manner to epithelial cell malignancy. *Journal of Cell Biology, 140*, 1535–1541.
Merrill, A. H. (1991). Cell regulation by sphingosine and more complex sphingolipids. *Journal of Bioenergetics and Biomembranes, 23*, 83–104.
Merrill, A. H. (2002). De novo sphingolipid biosynthesis: A necessary, but dangerous, pathway. *Journal of Biological Chemistry, 277*, 25843–25846.
Merrill, A. H., Schmeltz, E. M., Dillehay, D. L., Spiegel, S., Shayman, J. A., Schroeder, J. J., et al. (1997). Sphingolipids–the enigmatic lipid class: Biochemistry, physiology, and pathophysiology. *Toxicology and Applied Pharmacology, 142*, 208–225.

Merrill, A. H., Sullards, M. C., Wang, E., Voss, K. A., & Riley, R. T. (2001). Sphingolipid metabolism: Roles in signal transduction and disruption by fumonisins. *Environmental Health Perspectives, 109*, 283–289.

Mietla, J. A., Wijesinghe, D. S., Hoeferlin, L. A., Shultz, M. D., Natarajan, R., Fowler, A. A., 3rd, et al. (2013). Characterization of eicosanoid synthesis in a genetic ablation model of ceramide kinase. *The Journal of Lipid Research, 54*, 1834–1847.

Miranda, G. E., Abrahan, C. E., Agnolazza, D. L., Politi, L. E., & Rotstein, N. P. (2011). Ceramide-1-phosphate, a new mediator of development and survival in retina photoreceptors. *Investigative Ophthalmology & Visual Science, 52*, 6580–6588.

Mitra, P., Maceyka, M., Payne, S. G., Lamour, N., Milstien, S., Chalfant, C. E., et al. (2007). Ceramide kinase regulates growth and survival of A549 human lung adenocarcinoma cells. *FEBS Letters, 581*, 735–740.

Ogretmen, B., Pettus, B. J., Rossi, M. J., Wood, R., Usta, J., Szulc, Z., et al. (2002). Biochemical mechanisms of the generation of endogenous long chain ceramide in response to exogenous short chain ceramide in the A549 human lung adenocarcinoma cell line. Role for endogenous ceramide in mediating the action of exogenous ceramide. *Journal of Biological Chemistry, 277*, 12960–12969.

Okazaki, T., Bell, R. M., & Hannun, Y. A. (1989). Sphingomyelin turnover induced by vitamin D3 in HL-60 cells. Role in cell differentiation. *Journal of Biological Chemistry, 264*, 19076–19080.

Okazaki, T., Bielawska, A., Bell, R. M., & Hannun, Y. A. (1990). Role of ceramide as a lipid mediator of 1 alpha,25-dihydroxyvitamin D3-induced HL-60 cell differentiation. *Journal of Biological Chemistry, 265*, 15823–15831.

Ordonez, M., Presa, N., Trueba, M., & Gomez-Munoz, A. (2017). Implication of ceramide kinase in adipogenesis. *Mediators of Inflammation, 2017*, 9374563.

Ordonez, M., Rivera, I. G., Presa, N., & Gomez-Munoz, A. (2016). Implication of matrix metalloproteinases 2 and 9 in ceramide 1-phosphate-stimulated macrophage migration. *Cellular Signalling, 28*, 1066–1074.

Ouro, A., Arana, L., Gangoiti, P., Rivera, I. G., Ordonez, M., Trueba, M., et al. (2013). Ceramide 1-phosphate stimulates glucose uptake in macrophages. *Cellular Signalling, 25*, 786–795.

Ouro, A., Arana, L., Riazy, M., Zhang, P., Gomez-Larrauri, A., Steinbrecher, U., et al. (2017). Vascular endothelial growth factor mediates ceramide 1-phosphate-stimulated macrophage proliferation. *Experimental Cell Research, 361*, 277–283.

Pastukhov, O., Schwalm, S., Romer, I., Zangemeister-Wittke, U., Pfeilschifter, J., & Huwiler, A. (2014). Ceramide kinase contributes to proliferation but not to prostaglandin E2 formation in renal mesangial cells and fibroblasts. *Cellular Physiology and Biochemistry: International Journal of Experimental Cellular Physiology, Biochemistry, and Pharmacology, 34*, 119–133.

Pastukhov, O., Schwalm, S., Zangemeister-Wittke, U., Fabbro, D., Bornancin, F., Japtok, L., et al. (2014). The ceramide kinase inhibitor NVP-231 inhibits breast and lung cancer cell proliferation by inducing M phase arrest and subsequent cell death. *British Journal of Pharmacology, 171*, 5829–5844.

Payne, A. W., Pant, D. K., Pan, T. C., & Chodosh, L. A. (2014). Ceramide kinase promotes tumor cell survival and mammary tumor recurrence. *Cancer Research, 74*, 6352–6363.

Pettus, B. J., Bielawska, A., Spiegel, S., Roddy, P., Hannun, Y. A., & Chalfant, C. E. (2003). Ceramide kinase mediates cytokine- and calcium ionophore-induced arachidonic acid release. *Journal of Biological Chemistry, 278*, 38206–38213.

Pettus, B. J., Bielawska, A., Subramanian, P., Wijesinghe, D. S., Mayceka, M., Leslie, C. C., et al. (2004). Ceramide-1-phosphate is a direct activator of cytosolic phospholipase A_2. *Journal of Biological Chemistry, 279*, 11320–11326.

Presa, N., Gomez-Larrauri, A., Rivera, I. G., Ordonez, M., Trueba, M., & Gomez-Munoz, A. (2016). Regulation of cell migration and inflammation by ceramide 1-phosphate. *Biochimica et Biophysica Acta, 1861*, 402–409.

Ratajczak, M. Z., Suszynska, M., Borkowska, S., Ratajczak, J., & Schneider, G. (2014). The role of sphingosine-1 phosphate and ceramide-1 phosphate in trafficking of normal stem cells and cancer cells. *Expert Opinion on Therapeutic Targets, 18*, 95–107.

Rivera, I. G., Ordonez, M., Presa, N., Gangoiti, P., Gomez-Larrauri, A., Trueba, M., et al. (2016). Ceramide 1-phosphate regulates cell migration and invasion of human pancreatic cancer cells. *Biochemical Pharmacology, 102*, 107–119.

Schneider, G., Bryndza, E., Abdel-Latif, A., Ratajczak, J., Maj, M., Tarnowski, M., et al. (2013). Bioactive lipids S1P and C1P are prometastatic factors in human rhabdomyosarcoma, and their tissue levels increase in response to radio/chemotherapy. *Molecular Cancer Research, 11*, 793–807.

Schneider, G., Sellers, Z. P., Bujko, K., Kakar, S. S., Kucia, M., & Ratajczak, M. Z. (2017). Novel pleiotropic effects of bioactive phospholipids in human lung cancer metastasis. *Oncotarget, 8*, 58247–58263.

Simanshu, D. K., Kamlekar, R. K., Wijesinghe, D. S., Zou, X., Zhai, X., Mishra, S. K., et al. (2013). Non-vesicular trafficking by a ceramide-1-phosphate transfer protein regulates eicosanoids. *Nature, 500*, 463–467.

Tang, X., Benesch, M. G., & Brindley, D. N. (2015). Lipid phosphate phosphatases and their roles in mammalian physiology and pathology. *The Journal of Lipid Research, 56*, 2048–2060.

Waggoner, D. W., Gomez-Muñoz, A., Dewald, J., & Brindley, D. N. (1996). Phosphatidate phosphohydrolase catalyzes the hydrolysis of ceramide-1-phosphate, lysophosphatidate, and sphingosine-1-phosphate. *Journal of Biological Chemistry, 271*, 16506–16509.

Wang, S. W., Hojabrpour, P., Zhang, P., Kolesnick, R. N., Steinbrecher, U. P., Gomez-Munoz, A., et al. (2015). Regulation of ceramide generation during macrophage apoptosis by ASMase and de novo synthesis. *Biochimica et Biophysica Acta, 1851*, 1482–1489.

Wijesinghe, D. S., Brentnall, M., Mietla, J. A., Hoeferlin, L. A., Diegelmann, R. F., Boise, L. H., et al. (2014). Ceramide kinase is required for a normal eicosanoid response and the subsequent orderly migration of fibroblasts. *The Journal of Lipid Research, 55*, 1298–1309.

CHAPTER NINE

The Onus of Sphingolipid Enzymes in Cancer Drug Resistance

Samy A.F. Morad* and Myles C. Cabot§,[1]
*Department of Pharmacology, Faculty of Veterinary Medicine, South Valley University, Qena, Egypt
§Department of Biochemistry and Molecular Biology, East Carolina University, Brody School of Medicine, and the East Carolina Diabetes and Obesity Institute, Greenville, SC, United States
[1]Corresponding author: E-mail: cabotm@ecu.edu

Contents

Abstract

Chemotherapy resistance, inherent or acquired, represents a serious barrier to the successful treatment of cancer. Although drug efflux, conducted by plasma membrane-resident proteins, detoxification enzymes, cell death inhibition, and DNA damage repair are ensemble players in this unwanted biology, a full understanding of the many in

Advances in Cancer Research, Volume 140
ISSN 0065-230X
https://doi.org/10.1016/bs.acr.2018.04.013

concert molecular mechanisms driving drug resistance is lacking. Recent discoveries in sphingolipid (SL) metabolism have provided significant insight into the role of these lipids in cancer growth; however, considerably less is known with respect to SLs and the drug-resistant phenotype. One exception here is enhanced ceramide glycosylation, a hallmark of multidrug resistance that is believed responsible, in part, for diminishing ceramides tumor-suppressor potential. This chapter will review various aspects of SL biology that relate to chemotherapy resistance and extend this topic to acknowledge the role of chemotherapy selection pressure in promoting dysregulated SL metabolism, a characteristic in cancer and an exploitable target for therapy.

1. ETUDE ON CHEMOTHERAPY RESISTANCE

The genesis of drug resistance arising in biological systems is ever-present and widespread. Owing to evolutionary processes and/or sudden selection pressures, nature has evolved defenses, some beneficial, some problematic. These defenses protect from toxins of all kinds, including anticancer drugs. Increasingly, drug-resistant bacteria and viruses, the current H3N2 influenza A virus scourge (Dugan et al., 2017; Hussain, Galvin, Haw, Nutsford, & Husain, 2017), pesticide-resistant insects, and herbicide-resistant weeds pose growing threats to human health and the ecosystem. Drugs used to treat cancer are no exception; drug resistance is the major cause of cancer treatment failure.

Chemotherapy resistance can be intrinsic, a form of resistance characteristic in ovarian, pancreatic, and melanoma cancers, or acquired, as with prostate and breast cancers (Gottesman, Lavi, Hall, & Gillet, 2016; Holohan, Van Schaeybroeck, Longley, & Johnston, 2013; Housman et al., 2014; Keating, Cambrosio, Nelson, Mogoutov, & Cointet, 2013; Lippert, Ruoff, & Volm, 2008; Niero et al., 2014). With intrinsic resistance, factors preexist in cancer cells that render treatment ineffective. Acquired chemotherapy resistance develops during treatment; and in these cases, patients initially respond and often experience periods of apparent complete remission. A classic example of acquired resistance was illustrated in the high-dose, poly-chemotherapy trials for treatment of pediatric leukemia (Frei et al., 1965), which were met initially with outstanding success. Most patients, however, eventually relapsed and presented with more virulent, drug-resistant disease. The mutations and adaptive responses arising during treatment and tumor heterogeneity, an increasingly recognized characteristic of cancer, can serve to generate chemotherapy-induced selection of resistant clones. This type of selection pressure and its impact on sphingolipid (SL) metabolism will be discussed in this review.

Many factors contribute to chemotherapy resistance (Fig. 1). The most studied factors include drug efflux driven by membrane transport proteins, drug activation/inactivation, DNA damage repair, and cell death inhibition, also known as "inhibition of apoptosis." This review will encroach on established territory by introducing a new player in the drug resistance repertoire, dysfunctional SL metabolism, hopefully not as a deceptive cadence in this somewhat harmonious area of study.

Overexpression of the membrane efflux transporter P-glycoprotein (P-gp), an ATP-binding cassette (ABC) transporter protein, is one of the most consistent biological alterations in drug resistance (Bradley & Ling, 1994; Gottesman, 2002; Gottesman et al., 2016; Housman et al., 2014; Sun, Patel, Kumar, & Chen, 2012). Some noted members of this family include P-gp (ABCB1), MRP1 (multidrug resistance protein, ABCC1), and BCRP (breast cancer resistance protein, ABCG2). These proteins reduce the intracellular concentration of antitumor agents via ATP-dependent drug extrusion, thus limiting effective dosing at the intended sites. P-gp is expressed in many types of untreated malignancies, including breast, colorectal, renal, and ovarian cancers, neuroblastoma, acute myeloid leukemia (AML), non-Hodgkin's lymphoma, and multiple myeloma (Sandor, Fojo, & Bates, 1998). Exposure to chemotherapy can also upregulate P-gp expression, as occurs in acquired drug resistance. In

Figure 1 Factors contributing to chemotherapy resistance in cancer. The most prominent factors promoting resistance to chemotherapy are shown. The inclusion of altered sphingolipid metabolism marks the first time that this biology has been included in a list of factors contributing to chemotherapy resistance. As noted in the text, modifications in sphingolipid metabolism work in concert with the other contributors, most notably, the ABC transporters that direct chemotherapy efflux.

breast cancer, for example, a study by Rudas et al. (2003) revealed that expression of P-gp was 55% before chemotherapy and 100% after chemotherapy; and interestingly, the expression of MRP1 and lung resistance protein was also heightened in postchemotherapy specimens. Despite production of a wide array of efflux transporter antagonists and compelling in vitro results, results from clinical trials have been disappointing, providing little progress in treating drug resistance (Callaghan, Luk, & Bebawy, 2014).

Drug activation and inactivation also play major roles in the development of chemotherapy resistance. Cytarabine (Ara-c), for example, used in treatment of AML, must be phosphorylated (activated) to be effective. Resistance to Ara-c can occur when cancer cells downregulate deoxycytidine kinase, the rate-limiting kinase involved in Ara-c activation, a step that leads to buildup of the inactive prodrug (Abraham et al., 2015; Candelaria et al., 2016; Galmarini et al., 2003). Counterpoint to drug activation is drug inactivation, which is also contributory to drug resistance, and in this scenario GSH conjugation factors into the picture. GSH conjugation to platinum drugs generates products that are substrates for the drug efflux pumps (Meijer et al., 1992), discussed above; this is stealth drug resistance biology considering that platinum drugs are not substrates of the ABC transporters. Other GSH-related factors are as well involved in resistance to cisplatin (Chen & Kuo, 2010).

Certain chemotherapy drugs can either directly or indirectly damage DNA, their intended purpose in rapidly dividing cancer cells (Woods & Turchi, 2013). DNA damage repair, as in the other mechanisms of drug resistance, protects cancer cells from what would otherwise be a debilitating *sforzando*. A noteworthy example is resistance to platinum-based anticancer agents, which produce deleterious DNA cross-links that are remedied by nucleotide excision repair and homologous recombination (Bonanno, Favaretto, & Rosell, 2014). Many players are involved in the DNA damage response, as reviewed by Woods and Turchi (2013).

Lastly, in this brief overview, comes the topic of resistance to apoptosis. Most anticancer compounds, including the tumor-toxic SL ceramide (Morad & Cabot, 2013), trigger apoptosis; however, an unwanted characteristic in cancer is inherent activation of antiapoptotic programs that translate to proliferation, resistance, and disease recurrence. A cacophonous array of performers such as Bcl-2 and Mcl-1 proteins, necrosis, and necroptosis processes involved in autophagy, heat shock protein signaling, and the proteasome pathway contribute to apoptosis resistance in cancer, virtuosos working much like a busy Mahler orchestration as opposed to a Bach two-part Invention. Summoning an attack against such powerful biology may seem insurmountable; however, a broad range of strategies to overcome apoptosis resistance are on the horizon (Mohammad et al., 2015). These strategies present a myriad of opportunities for exploration and exploitation. For in-depth reviews on chemotherapy resistance in cancer, the reader is directed to several excellent works (Gottesman et al., 2016; Holohan et al., 2013; Housman et al., 2014; Kartal-Yandim, Adan-Gokbulut, & Baran, 2016; Keating et al., 2013; Mohammad et al., 2015; Niero et al., 2014; Zahreddine & Borden, 2013).

2. A PRELUDE TO SPHINGOLIPID METABOLISM

As this volume is devoted to SLs, and the contributions of other authors will offer a detailed picture, the topic of SL metabolism will be reviewed *presto*. Ceramide is the diradyl building block of most SLs (Zheng et al., 2006); this lipid serves both structural roles in biomembranes and has wide impact on signal transduction and regulation of cell function (Morad & Cabot, 2013). Major among these actions is the potentiation of signaling cascades that lead to cell death, hence the epithet, tumor-suppressor. Synthesis of ceramide is initiated by serine palmitoyltransferase (Fig. 2), which generates a sphingoid base, sphinganine, whose acylation is conducted by ceramide synthases, a sextet of enzymes performing *vivace* to produce an astonishing array of chemically diverse dihydroceramides, via ceramide synthase preferences for specific acyl-CoAs. This process generates molecular species of ceramides (a desaturase step catalyzed by dihydroceramide desaturase competes ceramide production through addition of a 4,5-*trans* double bond) with carbon chain lengths ranging from C14 to C32, with varying degrees of saturation. And, as nature would have it, many of the actions of ceramide are molecular species specific in biology ranging from cancer to diabetes to coronary artery disease (Dany et al., 2016; Grosch, Schiffmann, & Geisslinger, 2012; Helke et al., 2018; Koch et al., 2017; Mullen, Hannun, & Obeid, 2012; Saleem et al., 2017; Tidhar et al., 2012; Wang et al., 2017). Apropos here is work by Siddique et al. (2012), illustrating fine molecular species nuances, in this case the 4,5-*trans* double bond that is required for cancer cell sensitivity to etoposide, a semi-synthetic derivative of podophyllotoxin from the rhizome of the wild

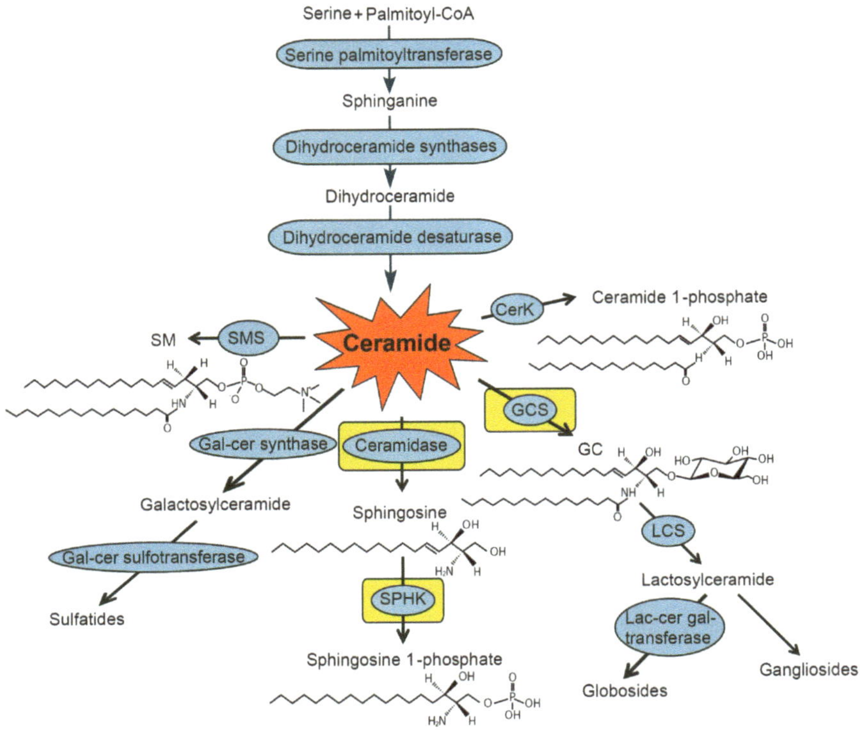

Figure 2 Formation of ceramide de novo and subsequent metabolism. Ceramide formation is initiated de novo by condensation of serine and palmitoyl-CoA, catalyzed by serine palmitoyltransferase, to form sphinganine. Once formed, ceramide can be converted into sphingomyelin by sphingomyelin synthase (SMS), ceramide 1-phosphate, by ceramide kinase (CerK), glucosylceramide (GC), by glucosylceramide synthase (GCS), or hydrolyzed by ceramidases (there are five ceramidases) to produce sphingosine. Sphingosine can then be phosphorylated via the action of sphingosine kinase (SPHK) (there are two isoforms). The enzymes of relevance in chemotherapy resistance are highlighted in yellow [GCS, ceramidase (acid ceramidase, AC), SPHK). *gal-cer synthase*, galactosylceramide synthase; *Lac-cer gal-transferase*, lactosylceramide galactosyltransferase; *LCS*, lactosylceramide synthase.

mandrake that is used in the treatment of Kaposi's sarcoma and Ewing's sarcoma, lung and testicular cancer, and glioblastoma.

Unfortunately, the SL-metabolizing machinery of cancer cells can transpose, up or down a pitch, the tumor suppressor benefits of ceramide (Fig. 2). For example, the conversion of ceramide into glucosylceramide (GC) by glucosylceramide synthase (GCS) is a primary anabolic passage that is used by cancer cells to subdue downstream death signals orchestrated by ceramide (Bleicher & Cabot, 2002; Liu, Han, Giuliano, Ichikawa, et al., 1999). The impact of chemotherapy drugs on ceramide glycosylation will be discussed in detail in a later section. The cancer-suppressor properties of ceramide can also be dampened via the action of sphingomyelin synthases, ceramidases, and to a lesser extent by ceramide kinase (Bornancin, 2011; Gomez-Munoz, 2006; Kolesnick & Hemer, 1990; Meng et al., 2004; Mitra et al., 2007) (Fig. 2). Even though hydrolysis by ceramidase is an effective means to destroy ceramide, this step has repercussions because it generates sphingosine, a cytotoxic element of merit on its own (Beljanski, Knaak, Zhuang, & Smith, 2011; Sun et al., 2009; Wang et al., 2008) and the substrate of sphingosine kinase (SPHK) that catalyzes production of mitogenic sphingosine-1-phosphate (S1P) (Pyne & Pyne, 2010). Of the quintet of ceramidases, acid ceramidase (AC) discovered by Gatt (1963) is of importance in cancer and a target in exciting approaches to therapy (Flowers et al., 2012; Gouaze-Andersson et al., 2011; Liu et al., 2009, 2008; Morad et al., 2012; Pyne & Pyne, 2010; Tan et al., 2016). In summary, a trio consisting of GCS, AC, and SPHK (Fig. 2) tightly regulate ceramide metabolism in cancer cells. Too much ceramide can be deleterious, a good thing, whereas too little can portend unwelcome proliferative responses. Controlling the levels of intracellular ceramide and S1P is a complex balancing act, but this regulation is key in cancer growth control.

3. IN THE BEGINNING

Historically, some of the early evidence of cancer drug interaction with SL metabolism came from the landmark works of Bose et al. (1995), in New York City, and Jaffrezou et al. (1996), in Toulouse. These investigations showed that the chemotherapy drug, daunorubicin, stimulated ceramide formation in leukemia cells and elicited apoptosis via two distinct pathways (Fig. 3). In the work by Bose et al. (1995), ceramide elevation was shown to result from activation of de novo synthesis via the action of ceramide synthase (see Fig. 2). Fumonisin B1, a pan ceramide synthase inhibitor, diminished daunorubicin-governed ceramide generation and downstream apoptosis. Work by the Toulouse group (Jaffrezou et al., 1996) showed that daunorubicin-induced increases in ceramide levels, which preceded apoptosis, were mediated by neutral sphingomyelinase

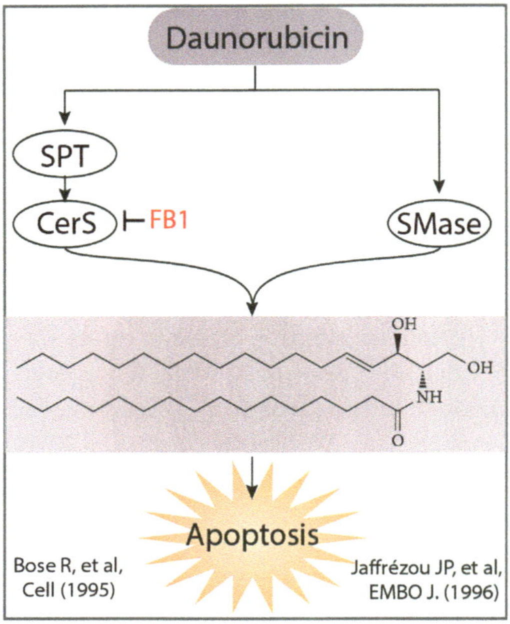

Figure 3 Schematic depicting the impact of daunorubicin on ceramide generation in cancer cells. This depiction was constructed from the work of Bose et al. and Jaffrezou et al., who discovered that the anthracycline, daunorubicin, activates ceramide formation in leukemia cell models via activation of the de novo pathway, left, and via hydrolysis of sphingomyelin by sphingomyelinase (SMase), right. Fumonisin B1 (FB1), a pan caspase inhibitor, halted ceramide generation and suppressed downstream apoptosis. *CerS*, ceramide synthase; *SPT*, serine palmitoyltransferase.

and unperturbed by Fumonisin B1. These studies emphasize the importance of ceramide in chemotherapy-induced apoptosis, in this instance the anthracycline daunorubicin, a chemical cousin of doxorubicin. With ceramide playing "first violin" in the induction of apoptosis in a myriad of cancer types, it is easy to envision that cancer cell defense mechanisms might employ anabolic and catabolic avenues to regulate ceramide levels in a prosurvival direction. With regard to anabolic directions, upregulated ceramide glycosylation has been shown to be a hallmark of drug resistance in numerous cancer types (Gouaze et al., 2004; Hajj, Becker-Flegler, & Haimovitz-Friedman, 2015; Lavie, Cao, Bursten, Giuliano, & Cabot, 1996; Liu, Han, Giuliano, & Cabot, 1999, 2001; Liu et al., 2011; Morad, Davis, Kester, Loughran, & Cabot, 2015; Xie et al., 2008).

4. GLUCOSYLCERAMIDE SYNTHASE—CULPABILITY (OR NOT) IN CANCER CELL RESISTANCE TO CHEMOTHERAPY

4.1 Glucosylceramide in Multidrug Resistance

Until the mid-1990s, relatively few works had been directed at studies of glycolipids in drug resistance. Early studies showed that multidrug-resistant cancer cells accumulate GC (Lavie et al., 1996; Lucci et al., 1998). This characteristic was demonstrated in multidrug-resistant human ovarian cancer cells, NCI-ADR/RES (formally known as MCF-7-AdrR cells), KB-V-1 adenocarcinoma (HeLa subclones), NIH: OVCAR-3 cells, established from a patient resistant to doxorubicin, melphalan, and cisplatin, and in drug-resistant ovarian cancer cells, A2780AD, compared with the drug-sensitive wild-type counterparts (Lucci et al., 1998; Morjani et al., 2001; Veldman et al., 2002). Moreover, lipid analysis of tumors from cancer patients who failed to respond to chemotherapy demonstrated elevated GC levels (Lucci et al., 1998). Although these studies point to the utility of GC as a marker of multidrug resistance, the more intriguing question is why does GC accumulate in multidrug-resistant cancer and what are the ramifications. Knowing what we know of ceramide's tumor-suppressor properties, it is likely that this mechanism was devised to hold ceramide at bay. Because many anticancer agents are ceramide generators (Senchenkov, Litvak, & Cabot, 2001), ceramide glycosylation becomes a stealth maneuver in the genesis of drug resistance. Lastly, early work by Itoh et al. (2003) poses the idea of using ceramide as an indicator of chemoresistance.

4.2 Consequences of Manipulating GCS Expression—GCS Crescendo and Decrescendo and Response to Chemotherapy

Augmenting or diminishing the activity of GCS, either chemically or genetically, has been instructive in illustrating the influence of ceramide glycosylation on response to chemotherapy. In breast and ovarian cancer cells, for example, sensitivity to anthracyclines, *Vinca* alkaloids, and taxanes could be increased or decreased by manipulating GCS expression and/or activity (Liu, Cabot, et al., 1999; Liu, Han, Giuliano, & Cabot, 2001; Liu, Han, Giuliano, Hansen, & Cabot, 2000). Transfection of drug-sensitive MCF-7 cells with GCS cDNA using a Tet-on expression system

(Liu & Cabot, 2004) conferred resistance to doxorubicin (Liu, Cabot, et al., 1999); the same manipulation also potentiated resistance to TNF-α (Liu, Ichikawa, et al., 1999), which regulates apoptosis via ceramide generation from sphingomyelin (Pena, Fuks, & Kolesnick, 1997). Conversely, GCS decrescendo in a multidrug-resistant ovarian cancer model increased cell sensitivity to doxorubicin, vinblastine, and paclitaxel (Liu et al., 2001) that was consistent with the resumption of ceramide-caspase-apoptotic signaling. Limiting GCS activity has also been shown to decrease the growth rate of neuroepithelioma cells and reduce tumorigenicity in melanoma (Deng, Li, Guerrera, Liu, & Ladisch, 2002; Di Sano et al., 2002). In melanoma, Deng et al. (2002) successfully employed GCS antisense to inhibit melanoma formation in an animal model, with a focus on ganglioside GM3 synthesis, a downstream product formed from lactosylceramide. Overall, these studies illustrate the importance of GCS in cancer cell response to chemotherapy and suggest there is room for engagement in this area of research.

4.3 Glucosylceramide Synthase and P-Glycoprotein—"It Takes Two to Tango"

Overexpression of GCS and P-gp has been shown to coincide. Work in multidrug-resistant breast cancer, leukemia, melanoma, colon cancer, and adenocarcinoma (Gouaze et al., 2004) supports a view of an enzyme duet, as do studies in an array of multidrug-resistant adenocarcinoma cells, KB-V.01, KB-V.1, and KB-V1, listed in order of increasing resistance to vinblastine. In this model, GC levels, GCS mRNA, GCS protein, and P-gp expression increased with increased vinblastine selection pressure. Thus, selection pressure for natural product drug resistance not only upregulates MDR1/P-gp (Chaudhary & Roninson, 1993; Gottesman, 2002; Kohno, Sato, Takano, Matsuo, & Kuwano, 1989) but also enhances ceramide metabolism through GCS. Klappe, Hinrichs, Kroesen, Sietsma, and Kok (2004) and Kok et al. (2000) reported similar findings in HT-29 colon cancer cells selected for colchicine resistance. In that study, increases in GC levels occurred concomitantly with upregulation of MRP1, perhaps revealing the promiscuous nature of the ABC transporters for ceramides with sugars. On the other side of the coin, GCS inhibition via GCS antisense, siRNA, or the employ of small molecule inhibitors of GCS downregulated MDR1/P-gp expression in drug-resistant cancer cells

(Gouaze et al., 2005). What is the attraction underlying this symbiotic duet? Ceramide is a substrate for GCS, and the product, GC, is a substrate for P-gp, as discussed below. Therefore, GCS produces substrate for P-gp, and the "it takes two to tango" relationship thrives. Interestingly, however, P-gp has been known to have relationships with other SL enzymes, a possible suggestion of infidelity.

Work by Tan et al. (2014) provides data suggesting that an accord exists between AC and P-gp. For example, enforced overexpression of AC in HL-60 cells also upregulated P-gp expression, conferring resistance to mitoxantrone and daunorubicin, whereas AC knockdown in HL-60/VCR (vincristine-resistant) decreased P-gp expression. These data support a role for AC in regulating P-gp expression in AML and present a new target for suppressing drug resistance. Another ABC transporter, BCRP, has also been shown to have a curious rapport with SLs (Kong et al., 2015). In this work, C6-ceramide induced secretion of BCRP-associated exosomes, conferring retention of and sensitivity to doxorubicin in triple negative MDA-MB-231 human breast cancer cells, indeed a novel mechanism by which ceramide might enhance the efficacy of chemotherapy. On a similar note, Storch, Ehehalt, Haefeli, and Weiss (2007) found that BCRP was localized in lipid rafts/caveolae and that its activity was modulated not by SLs but by cholesterol. Immunoprecipitation experiments indicated a physical interaction between BCRP and caveolin-1. MRP1 is also enriched in lipid rafts during multidrug resistance acquisition in colon cancer cells (Klappe et al., 2004). Looking back, perhaps the partnering that exists between SLs and P-gp is not exclusive because both BCRP and MRP1 have similar liaisons. Excellent insight on the function of SLs, ABCB1 upregulation, and the conferring of multidrug resistance is available in a recent work by Lee and Kolesnick (2017).

In summary, "it takes two to tango" might also be phrased, "show me a substrate, and I will show you an enzyme." This theme is exemplified in work by Gouaze-Andersson et al. (2007), who demonstrated that exposure of cultured breast cancer cells to either ceramide or GC upregulated P-gp expression. As ceramide is converted into GC and GC is a substrate for the catalytic flippase activity inherent in Golgi-resident P-gp (van Helvoort et al., 1996), it makes sense that presentation of excess substrate (GC, in this instance) would signal for elevated P-gp expression and activity.

4.4 Does GCS Inhibition Take Center Stage in Ceramide Sensitization or Is P-gp an Essential Performer in This Opus?

Work implicating GC in the biochemistry of multidrug resistance (Lavie et al., 1996; Lucci et al., 1999), studies revealing that elevated levels of GC in cancer cells correlate with overexpression of P-gp (Gouaze et al., 2004), the finding that commonly employed P-gp antagonists suppresses GC synthesis in drug-resistant cancer cells (Lavie et al., 1997), and landmark research from The Netherlands showing that Golgi-resident P-gp can function as a sphingomyelin and GC transmembrane flippase (van Helvoort, Giudici, Thielemans, & van Meer, 1997; van Helvoort et al., 1996) promoted the novel idea that P-gp played a role in ceramide metabolism and may also be involved in the *piu mosso* tempo of ceramide-orchestrated cell death. The experiments of De Rosa, Sillence, Ackerley, and Lingwood (2004) also demonstrated that drug resistance proteins assist in glycosphingolipid biosynthesis. To this end, enlightening works by Smyth, Krasovskis, Sutton, and Johnstone (1998) showed that P-gp protected drug-resistant leukemia cells from caspase-dependent apoptosis mediated by cytotoxic drugs and by ligation of Fas (a ceramide-linked death cascade), implying that P-gp exerts a role in regulating caspase-dependent apoptotic pathways. In a theme and variations on the subject, Shabbits and Mayer (2002) demonstrated that P-gp controlled ceramide-mediated sensitivity to paclitaxel, a ceramide generating taxane (Charles et al., 2001), suggesting that ceramide metabolism and apoptosis are regulated not only by GCS but also by P-gp. Moreover, studies in leukemia showed that resistance to C6-ceramide-induced apoptosis in P-gp-positive cell lines could be averted by exposure to P-gp antagonists (Turzanski, Grundy, Shang, Russell, & Pallis, 2005). Lastly, the reader must consider the work of Sietsma, Dijkhuis, Kamps, and Kok (2002) on the role of SLs in drug resistance. This insightful review, focusing on neuroblastoma, discusses refined ensemble between SLs and the ABC transporters that may serve to stanchion the drug-resistant phonotype.

The idea that P-gp and perhaps other ABC transporters protect cancer cells from ceramide-orchestrated cytotoxicity has been supported by an informative set of studies in HeLa cells with conditional expression of P-gp (Chapman, Gouaze-Andersson, & Cabot, 2010), a clean, uncomplicated system with which to dissect out the role of this renown drug pump in ceramide metabolism and cytotoxicity. That study showed that

P-gp-expressing cells converted ceramide into GC at 4-times the rate of the P-gp devoid cell counterpart, a finding that clearly implicates P-gp in the production of hexosylceramides. Interestingly, P-gp-expressing cells were resistant to ceramide (C6-ceramide), whereas P-gp-devoid cells were ceramide sensitive. The interesting denouement to this story is that ceramide resistance, inherent in P-gp-expressing cells, was reversed by inclusion of P-gp antagonists, a response that was accompanied by a block in GC production. Of significance, P-gp antagonists had no effect on GC synthesis in cell-free assays, indicating that the machinery of the intact cell is required. These works suggest that P-gp can function as a guardian of ceramide cytotoxicity and thus advocate new and novel indications for P-gp antagonists (Morad et al., 2017).

The role of P-gp as ceramide sentry was further explored in work by Chapman et al. (2010) who demonstrated that multidrug-resistant ovarian cancer cells have an enhanced capacity, over wild-type, for conversion of ceramide into GC; albeit, the levels of GCS enzyme activity in cell-free assays were similar. Furthermore, whereas the addition of low-dose GCS inhibitors potently blocked the conversion of C6-ceramide into C6-GC, this recipe did not enhance C6-ceramide cytotoxicity, suggesting that inhibition of GCS is not what matters. However, when the concentration of GCS inhibitor was increased from 0.2 to 5.0 μM, C6-ceramide cytotoxicity ensued, a hint that at higher doses, GCS inhibitors interact with P-gp in much the same fashion as chemical antagonists of P-gp. This idea was strengthened by the work of Sietsma et al. (2000), who demonstrated that GCS inhibitors decreased paclitaxel and vincristine efflux in neuroblastoma cells, thus acting as P-gp antagonists, a keen, informative finding. In similar studies, Chai et al. (2011) showed that the chemosensitizing properties of GCS inhibitors was mediated mainly through modulation of P-gp function, which brings us, as in sonata *allegro* form, to the recapitulation of the original theme: P-gp potentiates ceramide glycosylation, but if antagonized, augments ceramide sensitivity, features previously ascribed to GCS. This notion was bolstered by work showing that inhibition of GCS was not sufficient to reverse drug resistance in cancer cells (Norris-Cervetto, Callaghan, Platt, Dwek, & Butters, 2004). However, that study employed the GCS inhibitor NB-DNJ (N-butyl-deoxynojirimycin, miglustat), a water-soluble imido sugar with little to no affection for P-gp.

Although it may be an oversimplification, the upshot from these studies is that P-gp confers resistance to ceramide-regulated cell death by ramping up ceramide conversion to GC. GC, synthesized on the cytosolic surface of the Golgi, can be flipped into the Golgi by P-gp, where it is converted into higher glycosphingolipids. In our opinion, the efficiency of this "handoff" contributes to ceramide resistance by providing a driving force to perpetuate ceramide metabolism via glycosylation. Resistance in this instance can be to ceramide, if used directly as a therapeutic, or to the many drugs that generate and utilize ceramide signaling (Senchenkov et al., 2001). In summation, selection pressure for resistance to natural product anticancer drugs not only activates the *MDR1* gene (Chaudhary & Roninson, 1993; Gottesman, 2002; Kohno et al., 1989) but also enhances ceramide metabolism via glycosylation.

5. THE ROLE OF GCS IN CHEMOTHERAPY RESISTANCE—*BASSO PROFUNDO OR DIMINUENDO?*

Upregulated expression of GCS has been linked to tumor progression and to an overall reduction in intracellular ceramide levels. By converting proapoptotic ceramide into GC, GCS can confer "ceramide resistance" in cancer cells and contribute to the multidrug-resistant phenotype (Bleicher & Cabot, 2002; Liu et al., 2001), an impressive performance. As previously shown, daunorubicin, used in treatment of AML, activates ceramide formation in leukemia cells (Bose et al., 1995; Jaffrezou et al., 1996). Thus, an interesting parallel can be drawn here between ceramide-generating chemotherapy drugs and the enhancement of GCS expression and activity, a cause and effect relationship. Interestingly, similar processes have been illustrated in vivo. For example, in breast cancer patients, C16: 0 ceramide levels were shown to be lower post chemotherapy (Che, Huang, Xu, & Zhang, 2017), suggesting that ceramide depletion is a response to chemotherapy. Mechanistically, it was shown that ceramide levels were suppressed via upregulated expression of GCS, thus upregulated GCS is involved in breast cancer drug resistance, in vivo. This supports earlier work showing that GCS, a factor in modulation of drug resistance, is overexpressed in metastatic breast carcinoma (Liu et al., 2011). In invasive ductal carcinoma, J. Liu et al. (2016) demonstrated that changes in GCS methylation of the GCS promoter correlated with acquired drug resistance. As little work has been done on the role of methylation in transcriptional regulation of GCS, this direction merits further exploration.

In hepatocellular carcinoma, GCS is upregulated in response to sorafenib therapy, whereas targeting GCS by silencing or use of inhibitors reverted resistance to this tyrosine kinase inhibitor (Stefanovic et al., 2016). This suggests that targeting GCS may be a strategy for increasing sorafenib efficacy. A GCS-targeted approach has also been researched in imatinib resistance in chronic myeloid leukemia, where it was shown that chemical inhibition of GCS enhanced ceramide accumulation and increased cell death in response to imatinib (Baran, Bielawski, Gunduz, & Ogretmen, 2011). In head and neck cancer, a rather novel indication for GCS was shown in sensitization to cisplatin (Roh, Kim, Park, & Kim, 2015). These scientists found that GCS and P-gp expression were associated with cisplatin resistance and that GCS inhibition enhanced cisplatin-induced cell death by P-gp downregulation and proapoptotic protein activation, factors evident in vitro and in vivo. Even though cisplatin has in a few instances been shown to increase ceramides (Dupre et al., 2017; Rebillard et al., 2007), the findings of Roh et al. (2015) are rather unique because platinum drugs are poor substrates for the ABC transporters. Furthermore, overexpression of P-gp in cisplatin resistance is unusual, although this could be related to a generalized stress response (Stordal et al., 2012). In HCT-8 cells, a colon cancer cell model, Song, Zang, Zhang, Cao, and Yang (2012) showed that GCS and P-gp levels decreased on stable transfection of HCT-8/VCR cells (vincristine-resistant) with UCGC (GCS) shRNA, resulting in suppressed cell proliferation and restoration of drug sensitivity. These results support the scenario that GCS and P-gp enjoy close communication, ideas that are also mirrored in work showing that resistance to C6-ceramide is linked to increases in GCS, P-gp, and MRP-1 (Gutierrez-Iglesias, Hurtado, Palma-Lara, & Lopez-Marure, 2014).

As noted above, overexpression of GCS contributes to multidrug resistance in many cancer types. From the standpoint of chemotherapy resistance biology (see Fig. 1), this represents a unique addition to the field and an opportunity for development of new therapeutic approaches. However, GCS is only one player in this opus, lest we forget AC and SPHK, important virtuosos as well. The literature is rich with works showing ceramide generation in response to cytotoxic agents is central to cancer cell death. Just as an overabundance of ceramide can lead to enhanced GCS (P-gp) expression for conversion of ceramide into GC, high intracellular ceramide levels might also promote AC expression and activity. From here, a chain reaction could continue with the production of sphingosine, which is likely to call SPHK into action for production of S1P, an unwanted mitogenic entity in cancer. The following sections will touch on these enzymes in the context of drug resistance in cancer; however, for more details on AC, S1P, and SPHK in cancer, the reader is obliged to consult chapters herein, by experts in these fields.

6. ACID CERAMIDASE AND CHEMOTHERAPY RESISTANCE

Few works have explored the effects of chemotherapy drugs on AC expression and activity in cancer. The prize here goes to another type of therapy, ionizing radiation. Interest in the role of AC in cancer had its genesis in findings demonstrating characteristically elevated levels of this protein in prostate cancer and in leukemia (Liu et al., 2009; Mahdy et al., 2009; Tan et al., 2016). Elevated levels of AC in prostate cancer confer resistance to irradiation, suggesting that inhibition of AC might sensitize cells to radiation therapy (Liu et al., 2009; Mahdy et al., 2009). It was subsequently shown that AC inhibitors, LCL204 and LCL385, were cytotoxic in prostate cancer (Holman et al., 2008; Mahdy et al., 2009). In an in-depth study on AC in prostate cancer, which is upregulated in 60% of primary cases, Saad et al. (2007) showed that enforced AC overexpression in DU145 cells enhanced cell proliferation and migration, yielded larger tumor volumes in mice, compared with wild-type cells, and conferred resistance to doxorubicin, cisplatin, etoposide, gemcitabine, and C6-ceramide, whereas knockdown of AC restored chemotherapy sensitivity. A clinical investigation by Cheng et al. (2013) showed that radiation-induced AC expression conferred prostate cancer resistance and tumor relapse, suggesting a key role for AC in radiotherapy failure and disease recurrence. In summary, AC appears a central player in the regulation of prostate cancer growth and a formidable target in treatment. AC also plays a role in chemotherapy resistance in other types of cancer.

Constitutive expression of AC was shown to be a component in T-cell large granular lymphocyte leukemia, wherein inhibition of the enzyme led to apoptotic cell demise (Shah et al., 2008). AC also exerts a role in chronic myelogenous leukemia (CML), where the enzyme was shown to be regulated by IRF regulatory factor 8 (IRF8) (Hu et al., 2011), a key transcription factor for myeloid cell differentiation. Regulation of AC expression by IRF8 mediates apoptosis and suppresses CML. In AML, Tan et al. (2016) demonstrated that AC is upregulated and key in AML cell survival via altered SL metabolism, where it was shown that AC increased the expression of antiapoptotic Mcl-1. AC expression in AML also impacts the expression of P-gp and drug sensitivity (Tan et al., 2014). A recent review supports the concept that AC represents a novel and realistic therapeutic target in AML (Tan, Pearson, Feith, & Loughran, 2017), and this mirrors the ideas set forth in work by Liu et al. (2008) who proposed that inhibition of AC might provide a promising strategy in cancer therapy.

Using CrispR-Cas9-mediated gene editing to delete the gene encoding for AC (ASAH1), Lai et al. (2017) demonstrated that AC ablation prevented cancer-initiating cell formation in melanoma, more compelling data supporting a role for AC in the regulation of neoplasia. Thus, AC, like GCS, appears a critical regulator of cancer progression and chemotherapy sensitivity, a unique common denominator for two different enzymes in the SL pathway, perhaps exemplifying the tight ensemble practiced by these players.

7. SPHK1, SPHK2, AND SPHINGOSINE 1-PHOSPHATE IN CHEMOTHERAPY RESISTANCE

There are two isoforms of SPHK, SPHK1 and SPHK2. These enzymes have identical kinase domains but different subcellular localizations and distinct kinetic properties. SPHK1 and SPHK2 and their product, S1P, have become central figures in studies on cancer growth, metastasis, and response to chemotherapy (Giussani, Tringali, Riboni, Viani, & Venerando, 2014; Ponnusamy et al., 2010; Selvam & Ogretmen, 2013). Overexpression of SPHK can promote chemotherapy resistance, and as discussed below, selection of cells for growth in chemotherapy-containing culture media can upregulate SPHK expression. Thus, to increase survival, cancer cells counterbalance ceramide accumulation in response to a myriad of ceramide-generating agents (Senchenkov et al., 2001), via GCS and AC metabolic routes, and increase production of S1P and GC, markers of survival.

Changes in expression and activity of enzymes of SL metabolism that may occur during chemotherapy selection pressure is of interest because this biology can imitate the in vivo type of acquired resistance that results from chemotherapy exposure in chemotherapy-naïve patients. Models for acquired resistance can be replicated in the laboratory by growing cancer cells with chemotherapy drugs and cloning the survivors. For example, using this method, Aoyama et al. (2017) showed that paclitaxel resistance in PC-3 prostate cancer cells was accompanied by upregulation of both SPHK1 and GCS. A look at the mechanics involved in resistance demonstrated that paclitaxel treatment suppressed cell proliferation and increased ceramide levels in wild-type cells but not in the paclitaxel-resistant counterpart. The introduction of siRNA or pharmacological inhibitors of SPHK1 and GCS restored sensitivity to this taxane in drug-resistant cells, demonstrating the role of the GCS-SPHK1 enzyme partnership in the resistance process. It is notable that the induction of paclitaxel resistance was accompanied by upregulation of two important junctures in SL metabolism: the conversion of ceramide into GC, by GCS, and the phosphorylation of sphingosine to generate S1P, by the action of SPHK1. Similar responses have been observed in gastroesophageal cancer, employing cell lines selected for resistance to oxaliplatin, cisplatin, and docetaxel. Notably, most of the genes differentially expressed in the chemoresistant cell lines were involved in SL and glycosphingolipid metabolism, and drug resistance was shown to correlate with increased SPHK1 expression, S1P levels, and decreased S1P lyase expression (Matula et al., 2015). In human ovarian cancer cells cultured to tolerate *N*-(4-hydroxyphenyl)retinamide (4-HPR, fenretinide), S1P production, and SPHK1 expression were significantly higher compared with 4-HPR-naïve A2780 cells (Illuzzi et al., 2010), representing yet another example of the interrelatedness of chemotherapy sensitivity and SPHK expression in selection pressure.

Overexpression of SPHK, regardless of the driving force behind this biology, has been shown to contribute to chemotherapy resistance in many cancer types. In colorectal cancer, overexpression of SPHK1 promoted cetuximab resistance, whereas the inclusion of FTY720, an S1P receptor antagonist, resensitized to cetuximab in vitro and in vivo, resulting in inhibition of tumor growth (Rosa et al., 2013). In colorectal cancer and in nonsmall cell lung cancer, overexpression of SPHK2 contributed to all-trans retinoic acid resistance and gefitinib resistance, respectively (W. Liu et al., 2016; Shi et al., 2017). In a mechanism-based study in gastric cancer cells, Xiong et al. (2014) showed that SPHK1 bestowed antiapoptotic influence via the Akt/FoxO3a/Bim pathway. Exciting research abounds in the fascinating, clinical-translational area of SPHKs and cancer; for further insight into the realm of SPHK as a therapeutic target, the reader may consult several excellent reviews (Gao, Gao, Chen, Tian, & Zhao, 2015; Giussani et al., 2014; Ogretmen, 2018; Ponnusamy et al., 2010; Pyne, El Buri, Adams, & Pyne, 2017; Pyne, Adams, & Pyne, 2016; Pyne & Pyne, 2013; Selvam & Ogretmen, 2013).

8. CYTARABINE AND DAUNORUBICIN SELECTION PRESSURE IN AML CELLS UPREGULATES A TRIO OF SL ENZYMES, AND P-GP, OF COURSE

A chemotherapy regimen consisting of daunorubicin and cytarabine (Ara-c) is the cornerstone of induction therapy in AML, a hematological cancer marked by the accumulation of large numbers of immature myoblasts in the bone marrow. Prognosis in AML is poor. In older patients, unfavorable karyotypes are common with antecedent myelodysplastic syndromes that portend even worse prognosis. In patients receiving therapy with curative intent, less than one-half achieve long-term survival. Chemotherapy resistance, which can be primary or acquired in AML, is the major cause of treatment failure.

Apoptosis appears to be the principle mechanism underlying the anticancer impact of Ara-c and daunorubicin. The most prominent mechanisms of resistance to these agents in AML include diminished daunorubicin intracellular retention, promoted by the ABC efflux transporters, and with Ara-c, resistance is often linked to reduced capacity of the activating enzyme deoxycytidine kinase and overexpression of the inactivating 5′-nucleotidase, cytoplasmic 5′-nucleotidase II (Galmarini et al., 2002; McLornan, McMullin, Johnston, & Longley, 2007; Negoro et al., 2011; Urasaki et al., 1996). Whereas we have made great strides unraveling the fine points of chemotherapy-induced apoptosis and the downstream performers, the initial signaling events of chemotherapy action remain unclear. Nonetheless, a curious commonality between the actions of Ara-c and daunorubicin is that both increase intracellular ceramide levels in leukemia (Bezombes et al., 2001; Bradshaw, Ella, Thomas, Qi, & Meier, 1996; Jaffrezou et al., 1996; Terrisse, Bezombes, Lerouge, Laurent, & Jaffrezou, 2002). This commonality in response to structurally diverse chemotherapy drugs positions SLs at the gateway to the apoptosis initiative process. From what we have read herein, we might forecast that if ceramide is generated, processes within cancer cells will kick in and counteract. This scenario is clearly illustrated in HL-60 cells selected for growth in Arc-c- and daunorubicin-containing culture media, wherein members of an enzyme trio consisting of GCS, AC, and SPHK1 were globally upregulated (Fig. 4) (unpublished data, personal communication, the authors' laboratory). As expected, daunorubicin selection pressure resulted in upregulated expression of P-gp; however, that this was accompanied by GCS, AC, and SPHK1 upregulation should again draw our

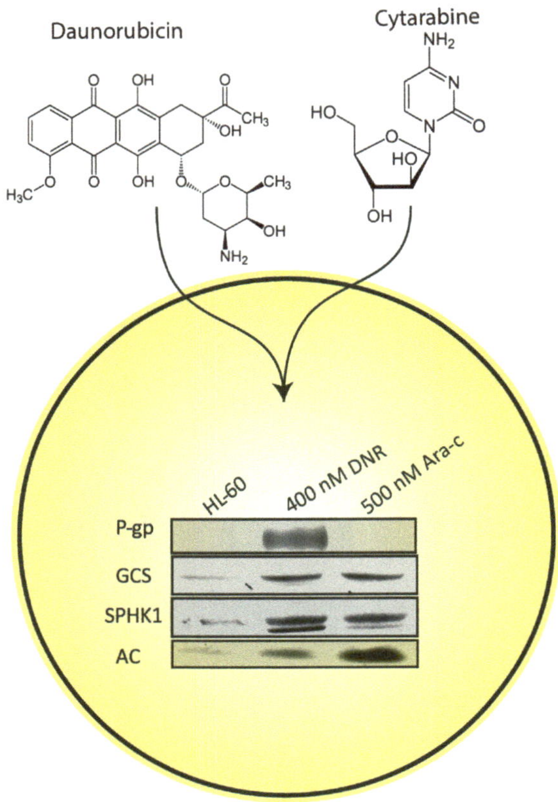

Figure 4 The effect of chemotherapy selection pressure on the enzymes of sphingolipid metabolism. HL-60 cells (human acute myeloid leukemia cell line) were grown with increasing concentrations of Cytarabine (Ara-c) and daunorubicin (DNR) in the culture medium, over a period of 4 months. Cells in the experiment shown here were selected for growth in medium containing 400 nM DNR and in medium containing 500 nM Ara-c. Protein extracts from HL-60 wild-type and DNR- and Ara-c-grown cells were analyzed by Western blot. *AC*, acid ceramidase. Chemical structure of the chemotherapy drugs is shown at the top; *GCS*, glucosylceramide synthase; *P-gp*, P-glycoprotein; *SPHK1*, sphingosine kinase 1. *Unpublished data from the authors' laboratory (East Carolina University, Greenville, NC, USA).*

attention to possible clandestine relationships between ABC transporters and SLs. Regrettably, the purpose of this global upregulation, we might surmise, is to more effectively eliminate ceramide, via glycosylation, hydrolysis, and conversion of sphingosine into S1P (see Fig. 2).

9. CODA

The onus of SL enzymes in cancer drug resistance: Of the many performers involved in initiation and development of multidrug resistance in cancer, we believe that an ensemble of SL enzymes, most notably, GCS, AC, and SPHK1/2, deserves a place "on stage." This group does not have the renown and status of the ABC transporters, the DNA damage repairers, or the virtuosos involved in orchestrating resistance to apoptosis; however, the curious links between multidrug resistance and SLs, too many to be coincidental, suggest that consideration be given to the role of SL metabolism in chemotherapy resistance and to the enzymes that present exploitable targets for novel approaches to treatment. Thus, from a strategic point, controlling the metabolism of ceramide offers new opportunities for regulating cancer growth, regardless of the drug-resistant phenotype (Barth, Cabot, & Kester, 2011; Truman, Garcia-Barros, Obeid, & Hannun, 2014).

REFERENCES

Abraham, A., Devasia, A. J., Varatharajan, S., Karathedath, S., Balasubramanian, P., & Mathews, V. (2015). Effect of cytosine arabinoside metabolizing enzyme expression on drug toxicity in acute myeloid leukemia. *Annals of Hematology, 94*(5), 883–885.

Aoyama, Y., Sobue, S., Mizutani, N., Inoue, C., Kawamoto, Y., Nishizawa, Y., et al. (2017). Modulation of the sphingolipid rheostat is involved in paclitaxel resistance of the human prostate cancer cell line PC3-PR. *Biochemical and Biophysical Research Communications, 486*(2), 551–557.

Baran, Y., Bielawski, J., Gunduz, U., & Ogretmen, B. (2011). Targeting glucosylceramide synthase sensitizes imatinib-resistant chronic myeloid leukemia cells via endogenous ceramide accumulation. *Journal of Cancer Research and Clinical Oncology, 137*(10), 1535–1544.

Barth, B. M., Cabot, M. C., & Kester, M. (2011). Ceramide-based therapeutics for the treatment of cancer. *Anti-Cancer Agents in Medicinal Chemistry, 11*(9), 911–919.

Beljanski, V., Knaak, C., Zhuang, Y., & Smith, C. D. (2011). Combined anticancer effects of sphingosine kinase inhibitors and sorafenib. *Investigational New Drugs, 29*(6), 1132–1142.

Bezombes, C., Plo, I., Mansat-De Mas, V., Quillet-Mary, A., Negre-Salvayre, A., Laurent, G., et al. (2001). Oxidative stress-induced activation of Lyn recruits sphingomyelinase and is requisite for its stimulation by Ara-C. *The FASEB Journal, 15*(9), 1583–1585.

Bleicher, R. J., & Cabot, M. C. (2002). Glucosylceramide synthase and apoptosis. *Biochimica et Biophysica Acta, 1585*(2–3), 172–178.

Bonanno, L., Favaretto, A., & Rosell, R. (2014). Platinum drugs and DNA repair mechanisms in lung cancer. *Anticancer Research, 34*(1), 493–501.
Bornancin, F. (2011). Ceramide kinase: The first decade. *Cellular Signalling, 23*(6), 999–1008.
Bose, R., Verheij, M., Haimovitz-Friedman, A., Scotto, K., Fuks, Z., & Kolesnick, R. (1995). Ceramide synthase mediates daunorubicin-induced apoptosis: An alternative mechanism for generating death signals. *Cell, 82*(3), 405–414.
Bradley, G., & Ling, V. (1994). P-glycoprotein, multidrug resistance and tumor progression. *Cancer and Metastasis Reviews, 13*(2), 223–233.
Bradshaw, C. D., Ella, K. M., Thomas, A. L., Qi, C., & Meier, K. E. (1996). Effects of Ara-C on neutral sphingomyelinase and mitogen- and stress-activated protein kinases in T-lymphocyte cell lines. *Biochemistry & Molecular Biology International, 40*(4), 709–719.
Callaghan, R., Luk, F., & Bebawy, M. (2014). Inhibition of the multidrug resistance P-glycoprotein: Time for a change of strategy? *Drug Metabolism and Disposition: the Biological Fate of Chemicals, 42*(4), 623–631.
Candelaria, M., Corrales-Alfaro, C., Gutierrez-Hernandez, O., Diaz-Chavez, J., Labardini-Mendez, J., Vidal-Millan, S., et al. (2016). Expression levels of human equilibrative nucleoside transporter 1 and deoxycytidine kinase enzyme as prognostic factors in patients with acute myeloid leukemia treated with cytarabine. *Chemotherapy, 61*(6), 313–318.
Chai, L., McLaren, R. P., Byrne, A., Chuang, W. L., Huang, Y., Dufault, M. R., et al. (2011). The chemosensitizing activity of inhibitors of glucosylceramide synthase is mediated primarily through modulation of P-gp function. *International Journal of Oncology, 38*(3), 701–711.
Chapman, J. V., Gouaze-Andersson, V., & Cabot, M. C. (2010). Expression of P-glycoprotein in HeLa cells confers resistance to ceramide cytotoxicity. *International Journal of Oncology, 37*(6), 1591–1597.
Charles, A. G., Han, T. Y., Liu, Y. Y., Hansen, N., Giuliano, A. E., & Cabot, M. C. (2001). Taxol-induced ceramide generation and apoptosis in human breast cancer cells. *Cancer Chemotherapy and Pharmacology, 47*(5), 444–450.
Chaudhary, P. M., & Roninson, I. B. (1993). Induction of multidrug resistance in human cells by transient exposure to different chemotherapeutic drugs. *Journal of the National Cancer Institute, 85*(8), 632–639.
Che, J., Huang, Y., Xu, C., & Zhang, P. (2017). Increased ceramide production sensitizes breast cancer cell response to chemotherapy. *Cancer Chemotherapy and Pharmacology, 79*(5), 933–941.
Chen, H. H., & Kuo, M. T. (2010). Role of glutathione in the regulation of Cisplatin resistance in cancer chemotherapy. *Metal-based Drugs, 2010*.
Cheng, J. C., Bai, A., Beckham, T. H., Marrison, S. T., Yount, C. L., Young, K., et al. (2013). Radiation-induced acid ceramidase confers prostate cancer resistance and tumor relapse. *Journal of Clinical Investigation, 123*(10), 4344–4358.
Dany, M., Gencer, S., Nganga, R., Thomas, R. J., Oleinik, N., Baron, K. D., et al. (2016). Targeting FLT3-ITD signaling mediates ceramide-dependent mitophagy and attenuates drug resistance in AML. *Blood, 128*(15), 1944–1958.
De Rosa, M. F., Sillence, D., Ackerley, C., & Lingwood, C. (2004). Role of multiple drug resistance protein 1 in neutral but not acidic glycosphingolipid biosynthesis. *Journal of Biological Chemistry, 279*(9), 7867–7876.
Deng, W., Li, R., Guerrera, M., Liu, Y., & Ladisch, S. (2002). Transfection of glucosylceramide synthase antisense inhibits mouse melanoma formation. *Glycobiology, 12*(3), 145–152.
Di Sano, F., Di Bartolomeo, S., Fazi, B., Fiorentini, C., Matarrese, P., Spinedi, A., et al. (2002). Antisense to glucosylceramide synthase in human neuroepithelioma affects cell growth but not apoptosis. *Cell Death and Differentiation, 9*(6), 693–695.

Dugan, V. G., Blanton, L., Elal, A. I. A., Alabi, N., Barnes, J., Brammer, L., et al. (2017). Update: Influenza activity - United States, October 1—November 25, 2017. *MMWR. Morbidity and Mortality Weekly Report, 66*(48), 1318—1326.

Dupre, T. V., Doll, M. A., Shah, P. P., Sharp, C. N., Siow, D., Megyesi, J., et al. (2017). Inhibiting glucosylceramide synthase exacerbates cisplatin-induced acute kidney injury. *The Journal of Lipid Research, 58*(7), 1439—1452.

Flowers, M., Fabrias, G., Delgado, A., Casas, J., Abad, J. L., & Cabot, M. C. (2012). C6-ceramide and targeted inhibition of acid ceramidase induce synergistic decreases in breast cancer cell growth. *Breast Cancer Research and Treatment, 133*(2), 447—458.

Frei, E., 3rd, Karon, M., Levin, R. H., Freireich, E. J., Taylor, R. J., Hananian, J., et al. (1965). The effectiveness of combinations of antileukemic agents in inducing and maintaining remission in children with acute leukemia. *Blood, 26*(5), 642—656.

Galmarini, C. M., Thomas, X., Calvo, F., Rousselot, P., Rabilloud, M., El Jaffari, A., et al. (2002). In vivo mechanisms of resistance to cytarabine in acute myeloid leukaemia. *British Journal of Haematology, 117*(4), 860—868.

Galmarini, C. M., Thomas, X., Graham, K., El Jafaari, A., Cros, E., Jordheim, L., et al. (2003). Deoxycytidine kinase and cN-II nucleotidase expression in blast cells predict survival in acute myeloid leukaemia patients treated with cytarabine. *British Journal of Haematology, 122*(1), 53—60.

Gao, Y., Gao, F., Chen, K., Tian, M. L., & Zhao, D. L. (2015). Sphingosine kinase 1 as an anticancer therapeutic target. *Drug Design, Development and Therapy, 9*, 3239—3245.

Gatt, S. (1963). Enzymic hydrolysis and synthesis of ceramides. *Journal of Biological Chemistry, 238*, 3131—3133.

Giussani, P., Tringali, C., Riboni, L., Viani, P., & Venerando, B. (2014). Sphingolipids: Key regulators of apoptosis and pivotal players in cancer drug resistance. *International Journal of Molecular Sciences, 15*(3), 4356—4392.

Gomez-Munoz, A. (2006). Ceramide 1-phosphate/ceramide, a switch between life and death. *Biochimica et Biophysica Acta, 1758*(12), 2049—2056.

Gottesman, M. M. (2002). Mechanisms of cancer drug resistance. *Annual Review of Medicine, 53*, 615—627.

Gottesman, M. M., Lavi, O., Hall, M. D., & Gillet, J. P. (2016). Toward a better understanding of the complexity of cancer drug resistance. *Annual Review of Pharmacology and Toxicology, 56*, 85—102.

Gouaze, V., Liu, Y. Y., Prickett, C. S., Yu, J. Y., Giuliano, A. E., & Cabot, M. C. (2005). Glucosylceramide synthase blockade down-regulates P-glycoprotein and resensitizes multidrug-resistant breast cancer cells to anticancer drugs. *Cancer Research, 65*(9), 3861—3867.

Gouaze, V., Yu, J. Y., Bleicher, R. J., Han, T. Y., Liu, Y. Y., Wang, H., et al. (2004). Overexpression of glucosylceramide synthase and P-glycoprotein in cancer cells selected for resistance to natural product chemotherapy. *Molecular Cancer Therapeutics, 3*(5), 633—639.

Gouaze-Andersson, V., Flowers, M., Karimi, R., Fabrias, G., Delgado, A., Casas, J., et al. (2011). Inhibition of acid ceramidase by a 2-substituted aminoethanol amide synergistically sensitizes prostate cancer cells to N-(4-hydroxyphenyl) retinamide. *The Prostate, 71*(10), 1064—1073.

Gouaze-Andersson, V., Yu, J. Y., Kreitenberg, A. J., Bielawska, A., Giuliano, A. E., & Cabot, M. C. (2007). Ceramide and glucosylceramide upregulate expression of the multidrug resistance gene MDR1 in cancer cells. *Biochimica et Biophysica Acta, 1771*(12), 1407—1417.

Grosch, S., Schiffmann, S., & Geisslinger, G. (2012). Chain length-specific properties of ceramides. *Progress in Lipid Research, 51*(1), 50—62.

Gutierrez-Iglesias, G., Hurtado, Y., Palma-Lara, I., & Lopez-Marure, R. (2014). Resistance to the antiproliferative effect induced by a short-chain ceramide is associated with an increase of glucosylceramide synthase, P-glycoprotein, and multidrug-resistance gene-1 in cervical cancer cells. *Cancer Chemotherapy and Pharmacology, 74*(4), 809–817.

Hajj, C., Becker-Flegler, K. A., & Haimovitz-Friedman, A. (2015). Novel mechanisms of action of classical chemotherapeutic agents on sphingolipid pathways. *Biological Chemistry, 396*(6–7), 669–679.

Helke, K., Angel, P., Lu, P., Garrett-Mayer, E., Ogretmen, B., Drake, R., et al. (2018). Ceramide synthase 6 deficiency enhances inflammation in the DSS model of colitis. *Scientific Reports, 8*(1), 1627.

van Helvoort, A., Giudici, M. L., Thielemans, M., & van Meer, G. (1997). Transport of sphingomyelin to the cell surface is inhibited by brefeldin A and in mitosis, where C6-NBD-sphingomyelin is translocated across the plasma membrane by a multidrug transporter activity. *Journal of Cell Science, 110*(Pt 1), 75–83.

van Helvoort, A., Smith, A. J., Sprong, H., Fritzsche, I., Schinkel, A. H., Borst, P., et al. (1996). MDR1 P-glycoprotein is a lipid translocase of broad specificity, while MDR3 P-glycoprotein specifically translocates phosphatidylcholine. *Cell, 87*(3), 507–517.

Holman, D. H., Turner, L. S., El-Zawahry, A., Elojeimy, S., Liu, X., Bielawski, J., et al. (2008). Lysosomotropic acid ceramidase inhibitor induces apoptosis in prostate cancer cells. *Cancer Chemotherapy and Pharmacology, 61*(2), 231–242.

Holohan, C., Van Schaeybroeck, S., Longley, D. B., & Johnston, P. G. (2013). Cancer drug resistance: An evolving paradigm. *Nature Reviews Cancer, 13*(10), 714–726.

Housman, G., Byler, S., Heerboth, S., Lapinska, K., Longacre, M., Snyder, N., et al. (2014). Drug resistance in cancer: An overview. *Cancers, 6*(3), 1769–1792.

Hu, X., Yang, D., Zimmerman, M., Liu, F., Yang, J., Kannan, S., et al. (2011). IRF8 regulates acid ceramidase expression to mediate apoptosis and suppresses myelogeneous leukemia. *Cancer Research, 71*(8), 2882–2891.

Hussain, M., Galvin, H. D., Haw, T. Y., Nutsford, A. N., & Husain, M. (2017). Drug resistance in influenza A virus: The epidemiology and management. *Infection and Drug Resistance, 10*, 121–134.

Illuzzi, G., Bernacchioni, C., Aureli, M., Prioni, S., Frera, G., Donati, C., et al. (2010). Sphingosine kinase mediates resistance to the synthetic retinoid N-(4-hydroxyphenyl) retinamide in human ovarian cancer cells. *Journal of Biological Chemistry, 285*(24), 18594–18602.

Itoh, M., Kitano, T., Watanabe, M., Kondo, T., Yabu, T., Taguchi, Y., et al. (2003). Possible role of ceramide as an indicator of chemoresistance: Decrease of the ceramide content via activation of glucosylceramide synthase and sphingomyelin synthase in chemoresistant leukemia. *Clinical Cancer Research, 9*(1), 415–423.

Jaffrezou, J. P., Levade, T., Bettaieb, A., Andrieu, N., Bezombes, C., Maestre, N., et al. (1996). Daunorubicin-induced apoptosis: Triggering of ceramide generation through sphingomyelin hydrolysis. *The EMBO Journal, 15*(10), 2417–2424.

Kartal-Yandim, M., Adan-Gokbulut, A., & Baran, Y. (2016). Molecular mechanisms of drug resistance and its reversal in cancer. *Critical Reviews in Biotechnology, 36*(4), 716–726.

Keating, P., Cambrosio, A., Nelson, N. C., Mogoutov, A., & Cointet, J. P. (2013). Therapy's shadow: A short history of the study of resistance to cancer chemotherapy. *Frontiers in Pharmacology, 4*, 58.

Klappe, K., Hinrichs, J. W., Kroesen, B. J., Sietsma, H., & Kok, J. W. (2004). MRP1 and glucosylceramide are coordinately over expressed and enriched in rafts during multidrug resistance acquisition in colon cancer cells. *International Journal of Cancer, 110*(4), 511–522.

Koch, A., Grammatikos, G., Trautmann, S., Schreiber, Y., Thomas, D., Bruns, F., et al. (2017). Vitamin D supplementation enhances C18(dihydro)ceramide levels in type 2 diabetes patients. *International Journal of Molecular Sciences, 18*(7).

Kohno, K., Sato, S., Takano, H., Matsuo, K., & Kuwano, M. (1989). The direct activation of human multidrug resistance gene (MDR1) by anticancer agents. *Biochemical and Biophysical Research Communications, 165*(3), 1415–1421.

Kok, J. W., Veldman, R. J., Klappe, K., Koning, H., Filipeanu, C. M., & Muller, M. (2000). Differential expression of sphingolipids in MRP1 overexpressing HT29 cells. *International Journal of Cancer, 87*(2), 172–178.

Kolesnick, R. N., & Hemer, M. R. (1990). Characterization of a ceramide kinase activity from human leukemia (HL-60) cells. Separation from diacylglycerol kinase activity. *Journal of Biological Chemistry, 265*(31), 18803–18808.

Kong, J. N., He, Q., Wang, G., Dasgupta, S., Dinkins, M. B., Zhu, G., et al. (2015). Guggulsterone and bexarotene induce secretion of exosome-associated breast cancer resistance protein and reduce doxorubicin resistance in MDA-MB-231 cells. *International Journal of Cancer, 137*(7), 1610–1620.

Lai, M., Realini, N., La Ferla, M., Passalacqua, I., Matteoli, G., Ganesan, A., et al. (2017). Complete Acid Ceramidase ablation prevents cancer-initiating cell formation in melanoma cells. *Scientific Reports, 7*(1), 7411.

Lavie, Y., Cao, H., Bursten, S. L., Giuliano, A. E., & Cabot, M. C. (1996). Accumulation of glucosylceramides in multidrug-resistant cancer cells. *Journal of Biological Chemistry, 271*(32), 19530–19536.

Lavie, Y., Cao, H., Volner, A., Lucci, A., Han, T. Y., Geffen, V., et al. (1997). Agents that reverse multidrug resistance, tamoxifen, verapamil, and cyclosporin A, block glycosphingolipid metabolism by inhibiting ceramide glycosylation in human cancer cells. *Journal of Biological Chemistry, 272*(3), 1682–1687.

Lee, W. K., & Kolesnick, R. N. (2017). Sphingolipid abnormalities in cancer multidrug resistance: Chicken or egg? *Cellular Signalling, 38*, 134–145.

Lippert, T. H., Ruoff, H. J., & Volm, M. (2008). Intrinsic and acquired drug resistance in malignant tumors. The main reason for therapeutic failure. *Arzneimittelforschung, 58*(6), 261–264.

Liu, X., Cheng, J. C., Turner, L. S., Elojeimy, S., Beckham, T. H., Bielawska, A., et al. (2009). Acid ceramidase upregulation in prostate cancer: Role in tumor development and implications for therapy. *Expert Opinion on Therapeutic Targets, 13*(12), 1449–1458.

Liu, X., Elojeimy, S., Turner, L. S., Mahdy, A. E., Zeidan, Y. H., Bielawska, A., et al. (2008). Acid ceramidase inhibition: A novel target for cancer therapy. *Frontiers in Bioscience, 13*, 2293–2298.

Liu, Y. Y., & Cabot, M. C. (2004). Development of a mammalian tet-on expression cell line: Glucosylceramide synthase regulates TNF-alpha-induced apoptosis. *Methods in Molecular Biology, 249*, 177–192.

Liu, Y. Y., Han, T. Y., Giuliano, A. E., & Cabot, M. C. (1999). Expression of glucosylceramide synthase, converting ceramide to glucosylceramide, confers adriamycin resistance in human breast cancer cells. *Journal of Biological Chemistry, 274*(2), 1140–1146.

Liu, Y. Y., Han, T. Y., Giuliano, A. E., & Cabot, M. C. (2001). Ceramide glycosylation potentiates cellular multidrug resistance. *The FASEB Journal, 15*(3), 719–730.

Liu, Y. Y., Han, T. Y., Giuliano, A. E., Hansen, N., & Cabot, M. C. (2000). Uncoupling ceramide glycosylation by transfection of glucosylceramide synthase antisense reverses adriamycin resistance. *Journal of Biological Chemistry, 275*(10), 7138–7143.

Liu, Y. Y., Han, T. Y., Giuliano, A. E., Ichikawa, S., Hirabayashi, Y., & Cabot, M. C. (1999). Glycosylation of ceramide potentiates cellular resistance to tumor necrosis factor-alpha-induced apoptosis. *Experimental Cell Research, 252*(2), 464–470.

Liu, W., Ning, J., Li, C., Hu, J., Meng, Q., Lu, H., et al. (2016). Overexpression of Sphk2 is associated with gefitinib resistance in non-small cell lung cancer. *Tumour Biology, 37*(5), 6331–6336.

Liu, Y. Y., Patwardhan, G. A., Xie, P., Gu, X., Giuliano, A. E., & Cabot, M. C. (2011). Glucosylceramide synthase, a factor in modulating drug resistance, is overexpressed in metastatic breast carcinoma. *International Journal of Oncology, 39*(2), 425–431.

Liu, J., Zhang, X., Liu, A., Zhang, D., Su, Y., Liu, Y., et al. (2016). Altered methylation of glucosylceramide synthase promoter regulates its expression and associates with acquired multidrug resistance in invasive ductal breast cancer. *Oncotarget,* 7(24), 36755–36766.

Lucci, A., Cho, W. I., Han, T. Y., Giuliano, A. E., Morton, D. L., & Cabot, M. C. (1998). Glucosylceramide: A marker for multiple-drug resistant cancers. *Anticancer Research, 18*(1B), 475–480.

Lucci, A., Giuliano, A. E., Han, T. Y., Dinur, T., Liu, Y. Y., Senchenkov, A., et al. (1999). Ceramide toxicity and metabolism differ in wild-type and multidrug-resistant cancer cells. *International Journal of Oncology, 15*(3), 535–540.

Mahdy, A. E., Cheng, J. C., Li, J., Elojeimy, S., Meacham, W. D., Turner, L. S., et al. (2009). Acid ceramidase upregulation in prostate cancer cells confers resistance to radiation: AC inhibition, a potential radiosensitizer. *Molecular Therapy: the Journal of the American Society of Gene Therapy, 17*(3), 430–438.

Matula, K., Collie-Duguid, E., Murray, G., Parikh, K., Grabsch, H., Tan, P., et al. (2015). Regulation of cellular sphingosine-1-phosphate by sphingosine kinase 1 and sphingosine-1-phopshate lyase determines chemotherapy resistance in gastroesophageal cancer. *BMC Cancer, 15*, 762.

McLornan, D. P., McMullin, M. F., Johnston, P., & Longley, D. B. (2007). Molecular mechanisms of drug resistance in acute myeloid leukaemia. *Expert Opinion on Drug Metabolism & Toxicology, 3*(3), 363–377.

Meijer, C., Mulder, N. H., Timmer-Bosscha, H., Sluiter, W. J., Meersma, G. J., & de Vries, E. G. (1992). Relationship of cellular glutathione to the cytotoxicity and resistance of seven platinum compounds. *Cancer Research, 52*(24), 6885–6889.

Meng, A., Luberto, C., Meier, P., Bai, A., Yang, X., Hannun, Y. A., et al. (2004). Sphingomyelin synthase as a potential target for D609-induced apoptosis in U937 human monocytic leukemia cells. *Experimental Cell Research, 292*(2), 385–392.

Mitra, P., Maceyka, M., Payne, S. G., Lamour, N., Milstien, S., Chalfant, C. E., et al. (2007). Ceramide kinase regulates growth and survival of A549 human lung adenocarcinoma cells. *FEBS Letters, 581*(4), 735–740.

Mohammad, R. M., Muqbil, I., Lowe, L., Yedjou, C., Hsu, H. Y., Lin, L. T., et al. (2015). Broad targeting of resistance to apoptosis in cancer. *Seminars in Cancer Biology, 35*(Suppl.), S78–S103.

Morad, S. A., & Cabot, M. C. (2013). Ceramide-orchestrated signalling in cancer cells. *Nature Reviews Cancer, 13*(1), 51–65.

Morad, S. A., Davis, T. S., Kester, M., Loughran, T. P., Jr., & Cabot, M. C. (2015). Dynamics of ceramide generation and metabolism in response to fenretinide—Diversity within and among leukemia. *Leukemia Research, 39*(10), 1071–1078.

Morad, S. A. F., Davis, T. S., MacDougall, M. R., Tan, S. F., Feith, D. J., Desai, D. H., et al. (2017). Role of P-glycoprotein inhibitors in ceramide-based therapeutics for treatment of cancer. *Biochemical Pharmacology, 130*, 21–33.

Morad, S. A., Levin, J. C., Shanmugavelandy, S. S., Kester, M., Fabrias, G., Bedia, C., et al. (2012). Ceramide—antiestrogen nanoliposomal combinations—novel impact of hormonal therapy in hormone-insensitive breast cancer. *Molecular Cancer Therapeutics, 11*(11), 2352–2361.

Morjani, H., Aouali, N., Belhoussine, R., Veldman, R. J., Levade, T., & Manfait, M. (2001). Elevation of glucosylceramide in multidrug-resistant cancer cells and accumulation in cytoplasmic droplets. *International Journal of Cancer, 94*(2), 157–165.
Mullen, T. D., Hannun, Y. A., & Obeid, L. M. (2012). Ceramide synthases at the centre of sphingolipid metabolism and biology. *The Biochemical Journal, 441*(3), 789–802.
Negoro, E., Yamauchi, T., Urasaki, Y., Nishi, R., Hori, H., & Ueda, T. (2011). Characterization of cytarabine-resistant leukemic cell lines established from five different blood cell lineages using gene expression and proteomic analyses. *International Journal of Oncology, 38*(4), 911–919.
Niero, E. L., Rocha-Sales, B., Lauand, C., Cortez, B. A., de Souza, M. M., Rezende-Teixeira, P., et al. (2014). The multiple facets of drug resistance: One history, different approaches. *Journal of Experimental & Clinical Cancer Research, 33*, 37.
Norris-Cervetto, E., Callaghan, R., Platt, F. M., Dwek, R. A., & Butters, T. D. (2004). Inhibition of glucosylceramide synthase does not reverse drug resistance in cancer cells. *Journal of Biological Chemistry, 279*(39), 40412–40418.
Ogretmen, B. (2018). Sphingolipid metabolism in cancer signalling and therapy. *Nature Reviews Cancer, 18*(1), 33–50.
Pena, L. A., Fuks, Z., & Kolesnick, R. (1997). Stress-induced apoptosis and the sphingomyelin pathway. *Biochemical Pharmacology, 53*(5), 615–621.
Ponnusamy, S., Meyers-Needham, M., Senkal, C. E., Saddoughi, S. A., Sentelle, D., Selvam, S. P., et al. (2010). Sphingolipids and cancer: Ceramide and sphingosine-1-phosphate in the regulation of cell death and drug resistance. *Future Oncology, 6*(10), 1603–1624.
Pyne, N. J., El Buri, A., Adams, D. R., & Pyne, S. (2017). Sphingosine 1-phosphate and cancer. *Advances in Biological Regulation*.
Pyne, N. J., & Pyne, S. (2010). Sphingosine 1-phosphate and cancer. *Nature Reviews Cancer, 10*(7), 489–503.
Pyne, S., Adams, D. R., & Pyne, N. J. (2016). Sphingosine 1-phosphate and sphingosine kinases in health and disease: Recent advances. *Progress in Lipid Research, 62*, 93–106.
Pyne, S., & Pyne, N. J. (2013). New perspectives on the role of sphingosine 1-phosphate in cancer. *Handbook of Experimental Pharmacology, 216*, 55–71.
Rebillard, A., Tekpli, X., Meurette, O., Sergent, O., LeMoigne-Muller, G., Vernhet, L., et al. (2007). Cisplatin-induced apoptosis involves membrane fluidification via inhibition of NHE1 in human colon cancer cells. *Cancer Research, 67*(16), 7865–7874.
Roh, J. L., Kim, E. H., Park, J. Y., & Kim, J. W. (2015). Inhibition of glucosylceramide synthase sensitizes head and neck cancer to cisplatin. *Molecular Cancer Therapeutics, 14*(8), 1907–1915.
Rosa, R., Marciano, R., Malapelle, U., Formisano, L., Nappi, L., D'Amato, C., et al. (2013). Sphingosine kinase 1 overexpression contributes to cetuximab resistance in human colorectal cancer models. *Clin Cancer Research, 19*(1), 138–147.
Rudas, M., Filipits, M., Taucher, S., Stranzl, T., Steger, G. G., Jakesz, R., et al. (2003). Expression of MRP1, LRP and Pgp in breast carcinoma patients treated with preoperative chemotherapy. *Breast Cancer Research and Treatment, 81*(2), 149–157.
Saad, A. F., Meacham, W. D., Bai, A., Anelli, V., Elojeimy, S., Mahdy, A. E., et al. (2007). The functional effects of acid ceramidase overexpression in prostate cancer progression and resistance to chemotherapy. *Cancer Biology & Therapy, 6*(9), 1455–1460.
Saleem, M., Herrmann, N., Dinoff, A., Mielke, M. M., Oh, P. I., Shammi, P., et al. (2017). A lipidomics approach to assess the association between plasma sphingolipids and verbal memory performance in coronary artery disease patients undertaking cardiac rehabilitation: A C18:0 signature for cognitive response to exercise. *Journal of Alzheimer's Disease, 60*(3), 829–841.

Sandor, V., Fojo, T., & Bates, S. E. (1998). Future perspectives for the development of P-glycoprotein modulators. *Drug Resistance Updates, 1*(3), 190–200.
Selvam, S. P., & Ogretmen, B. (2013). Sphingosine kinase/sphingosine 1-phosphate signaling in cancer therapeutics and drug resistance. *Handbook of Experimental Pharmacology, 216*, 3–27.
Senchenkov, A., Litvak, D. A., & Cabot, M. C. (2001). Targeting ceramide metabolism—a strategy for overcoming drug resistance. *Journal of the National Cancer Institute, 93*(5), 347–357.
Shabbits, J. A., & Mayer, L. D. (2002). P-glycoprotein modulates ceramide-mediated sensitivity of human breast cancer cells to tubulin-binding anticancer drugs. *Molecular Cancer Therapeutics, 1*(3), 205–213.
Shah, M. V., Zhang, R., Irby, R., Kothapalli, R., Liu, X., Arrington, T., et al. (2008). Molecular profiling of LGL leukemia reveals role of sphingolipid signaling in survival of cytotoxic lymphocytes. *Blood, 112*(3), 770–781.
Shi, W. N., Cui, S. X., Song, Z. Y., Wang, S. Q., Sun, S. Y., Yu, X. F., et al. (2017). Overexpression of SphK2 contributes to ATRA resistance in colon cancer through rapid degradation of cytoplasmic RXRalpha by K48/K63-linked polyubiquitination. *Oncotarget, 8*(24), 39605–39617.
Siddique, M. M., Bikman, B. T., Wang, L., Ying, L., Reinhardt, E., Shui, G., et al. (2012). Ablation of dihydroceramide desaturase confers resistance to etoposide-induced apoptosis in vitro. *PLoS One*, 7(9), e44042.
Sietsma, H., Dijkhuis, A. J., Kamps, W., & Kok, J. W. (2002). Sphingolipids in neuroblastoma: Their role in drug resistance mechanisms. *Neurochemical Research, 27*(7–8), 665–674.
Sietsma, H., Veldman, R. J., Kolk, D., Ausema, B., Nijhof, W., Kamps, W., et al. (2000). 1-phenyl-2-decanoylamino-3-morpholino-1-propanol chemosensitizes neuroblastoma cells for taxol and vincristine. *Clinical Cancer Research, 6*(3), 942–948.
Smyth, M. J., Krasovskis, E., Sutton, V. R., & Johnstone, R. W. (1998). The drug efflux protein, P-glycoprotein, additionally protects drug-resistant tumor cells from multiple forms of caspase-dependent apoptosis. *Proceedings of the National Academy of Sciences of the United States of America, 95*(12), 7024–7029.
Song, M., Zang, W., Zhang, B., Cao, J., & Yang, G. (2012). GCS overexpression is associated with multidrug resistance of human HCT-8 colon cancer cells. *Journal of Experimental & Clinical Cancer Research, 31*, 23.
Stefanovic, M., Tutusaus, A., Martinez-Nieto, G. A., Barcena, C., de Gregorio, E., Moutinho, C., et al. (2016). Targeting glucosylceramide synthase upregulation reverts sorafenib resistance in experimental hepatocellular carcinoma. *Oncotarget*, 7(7), 8253–8267.
Storch, C. H., Ehehalt, R., Haefeli, W. E., & Weiss, J. (2007). Localization of the human breast cancer resistance protein (BCRP/ABCG2) in lipid rafts/caveolae and modulation of its activity by cholesterol in vitro. *The Journal of Pharmacology and Experimental Therapeutics, 323*(1), 257–264.
Stordal, B., Hamon, M., McEneaney, V., Roche, S., Gillet, J. P., O'Leary, J. J., et al. (2012). Resistance to paclitaxel in a cisplatin-resistant ovarian cancer cell line is mediated by P-glycoprotein. *PLoS One*, 7(7), e40717.
Sun, W., Hu, W., Xu, R., Jin, J., Szulc, Z. M., Zhang, G., et al. (2009). Alkaline ceramidase 2 regulates beta1 integrin maturation and cell adhesion. *The FASEB Journal, 23*(2), 656–666.
Sun, Y. L., Patel, A., Kumar, P., & Chen, Z. S. (2012). Role of ABC transporters in cancer chemotherapy. *Chinese Journal of Cancer, 31*(2), 51–57.

Tan, S.-F., Liu, X., Doi, K., Wang, H.-G., Cabot, M., Feith, D., et al. (2014). Abstract 761: Acid ceramidase promotes drug resistance in acute myeloid leukemia through P-gp upregulation mediated by NF- kB activation. *Cancer Research, 74*(19 Supplement), 761.

Tan, S. F., Liu, X., Fox, T. E., Barth, B. M., Sharma, A., Turner, S. D., et al. (2016). Acid ceramidase is upregulated in AML and represents a novel therapeutic target. *Oncotarget,* 7(50), 83208–83222.

Tan, S. F., Pearson, J. M., Feith, D. J., & Loughran, T. P., Jr. (2017). The emergence of acid ceramidase as a therapeutic target for acute myeloid leukemia. *Expert Opinion on Therapeutic Targets, 21*(6), 583–590.

Terrisse, A. D., Bezombes, C., Lerouge, S., Laurent, G., & Jaffrezou, J. P. (2002). Daunorubicin- and Ara-C-induced interphasic apoptosis of human type II leukemia cells is caspase-8-independent. *Biochimica et Biophysica Acta, 1584*(2–3), 99–103.

Tidhar, R., Ben-Dor, S., Wang, E., Kelly, S., Merrill, A. H., Jr., & Futerman, A. H. (2012). Acyl chain specificity of ceramide synthases is determined within a region of 150 residues in the Tram-Lag-CLN8 (TLC) domain. *Journal of Biological Chemistry, 287*(5), 3197–3206.

Truman, J. P., Garcia-Barros, M., Obeid, L. M., & Hannun, Y. A. (2014). Evolving concepts in cancer therapy through targeting sphingolipid metabolism. *Biochimica et Biophysica Acta, 1841*(8), 1174–1188.

Turzanski, J., Grundy, M., Shang, S., Russell, N., & Pallis, M. (2005). P-glycoprotein is implicated in the inhibition of ceramide-induced apoptosis in TF-1 acute myeloid leukemia cells by modulation of the glucosylceramide synthase pathway. *Experimental Hematology, 33*(1), 62–72.

Urasaki, Y., Ueda, T., Yoshida, A., Fukushima, T., Takeuchi, N., Tsuruo, T., et al. (1996). Establishment of a daunorubicin-resistant cell line which shows multi-drug resistance by multifactorial mechanisms. *Anticancer Research, 16*(2), 709–714.

Veldman, R. J., Klappe, K., Hinrichs, J., Hummel, I., van der Schaaf, G., Sietsma, H., et al. (2002). Altered sphingolipid metabolism in multidrug-resistant ovarian cancer cells is due to uncoupling of glycolipid biosynthesis in the Golgi apparatus. *The FASEB Journal, 16*(9), 1111–1113.

Wang, H., Maurer, B. J., Liu, Y. Y., Wang, E., Allegood, J. C., Kelly, S., et al. (2008). N-(4-Hydroxyphenyl)retinamide increases dihydroceramide and synergizes with dimethylsphingosine to enhance cancer cell killing. *Molecular Cancer Therapeutics,* 7(9), 2967–2976.

Wang, Z., Wen, L., Zhu, F., Wang, Y., Xie, Q., Chen, Z., et al. (2017). Overexpression of ceramide synthase 1 increases C18-ceramide and leads to lethal autophagy in human glioma. *Oncotarget, 8*(61), 104022–104036.

Woods, D., & Turchi, J. J. (2013). Chemotherapy induced DNA damage response: Convergence of drugs and pathways. *Cancer Biology & Therapy, 14*(5), 379–389.

Xie, P., Shen, Y. F., Shi, Y. P., Ge, S. M., Gu, Z. H., Wang, J., et al. (2008). Overexpression of glucosylceramide synthase in associated with multidrug resistance of leukemia cells. *Leukemia Research, 32*(3), 475–480.

Xiong, H., Wang, J., Guan, H., Wu, J., Xu, R., Wang, M., et al. (2014). SphK1 confers resistance to apoptosis in gastric cancer cells by downregulating Bim via stimulating Akt/FoxO3a signaling. *Oncology Reports, 32*(4), 1369–1373.

Zahreddine, H., & Borden, K. L. (2013). Mechanisms and insights into drug resistance in cancer. *Frontiers in Pharmacology, 4,* 28.

Zheng, W., Kollmeyer, J., Symolon, H., Momin, A., Munter, E., Wang, E., et al. (2006). Ceramides and other bioactive sphingolipid backbones in health and disease: Lipidomic analysis, metabolism and roles in membrane structure, dynamics, signaling and autophagy. *Biochimica et Biophysica Acta, 1758*(12), 1864–1884.

CHAPTER TEN

Interdiction of Sphingolipid Metabolism Revisited: Focus on Prostate Cancer

Christina Voelkel-Johnson[1], James S. Norris and Shai White-Gilbertson
Department of Microbiology and Immunology, Medical University of South Carolina, Charleston, SC, United States
[1]Corresponding author: E-mail: johnsocv@musc.edu

Contents

Advances in Cancer Research, Volume 140
ISSN 0065-230X
https://doi.org/10.1016/bs.acr.2018.04.014

Abstract

Sphingolipid metabolism is known to play a role in cell death, survival, and therapy resistance in cancer. Sphingolipids, particularly dihydroceramide and ceramide, are associated with antiproliferative or cell death responses, respectively, and are central to effective cancer therapy. Within the last decade, strides have been made in elucidating many intricacies of sphingolipid metabolism. New information has emerged on the mechanisms by which sphingolipid metabolism is dysregulated during malignancy and how cancer cells survive and/or escape therapeutic interventions. This chapter focuses on three main themes: (1) sphingolipid enzymes that are dysregulated in cancer, particularly in prostate cancer; (2) inhibitors of sphingolipid metabolism that antagonize prosurvival responses; and (3) sphingolipid-driven escape mechanisms that allow cancer cells to evade therapies. We explore clinical and preclinical approaches to interdict sphingolipid metabolism and provide a rationale for combining strategies to drive the generation of antiproliferative ceramides with prevention of ceramide clearance.

1. INTRODUCTION

Prostate cancer is the most diagnosed cancer among American men and the second leading cause of cancer deaths in the United States. Over 160,000 new cases will be diagnosed in 2018 and nearly 30,000 deaths are anticipated (Siegel, Miller, & Jemal, 2018). Treatment for prostate cancer varies, but advanced disease typically involves surgical removal of the gland followed by androgen ablation therapy, radiation, or chemotherapy. Sphingolipid metabolism is known to play a role in cell death, survival, and therapy resistance in cancer, including prostate cancer (Beckham, Cheng, Marrison, Norris, & Liu, 2013; Ogretmen, 2018). Ceramide, a central molecule in sphingolipid metabolism, is generally associated with antiproliferative responses while sphingosine 1-phosphate (S1P) promotes survival, angiogenesis, inflammation, and resistance. Cancer therapy is therefore aimed at promoting ceramide generation and inhibiting S1P levels.

Within the last decade, strides have been made in elucidating the intricacies of sphingolipid metabolism, but even early studies identified correlations between therapy responses in prostate cancer and modulation of intracellular ceramides. For example, hormone withdrawal in androgen-sensitive LNCaP prostate cancer cells led to cell cycle arrest and apoptosis and was accompanied by an increase in C_{16}-ceramide that could be blocked by inhibitors of the de novo pathway of sphingolipid metabolism (Eto et al., 2003). Nearly 20 years ago, researchers suggested that the relative levels of sphingolipid metabolites may also play a role in susceptibility to radiation therapy and that the enhancement of ceramide and sphingosine generation could be of therapeutic value (Nava et al., 2000). The role of sphingolipids in radiation therapy has been comprehensively reviewed (Hajj & Haimovitz-Friedman, 2013). Radiation resistance or treatment failure remains a major obstacle in advanced prostate cancer. While radiation exposure induces ceramide-mediated antitumor effects, ceramide itself induces a protective response in a subset of cells resulting not only in resistance to radiation therapy but also in cross-resistance to chemotherapy (Cheng et al., 2013; Mahdy et al., 2009). This chapter discusses sphingolipid enzymes, inhibitors, and mechanisms of therapy resistance with a focus on prostate cancer.

2. SPHINGOLIPID ENZYMES

2.1 Ceramide Synthases

Ceramide is a central molecule in sphingolipid metabolism and consists of a sphingoid base that, via an amide bond, is attached to an acyl chain. Differences in acyl chain length and saturation give rise to a variety of ceramide species that endow cells or tissues with a "ceramide profile." The discovery of six different ceramide synthases (CerSs) with tissue-specific distribution and acyl chain preferences was key in beginning to understand the physiology and pathologies associated with alterations in ceramide profiles. Because several comprehensive reviews on CerSs have recently been published, we will only briefly introduce these enzymes (Mullen, Hannun, & Obeid, 2012; Park, Park, & Futerman, 2014; Wegner, Schiffmann, Parnham, Geisslinger, & Grosch, 2016).

CerSs are primarily localized to the endoplasmic reticulum, but some evidence suggests they can also be detected in mitochondria (and associated membranes) and the nucleus (and perinuclear structures). CerSs acetylate either dihydrosphingosine (dhSph) in the de novo pathway to generate dihydroceramide (dhCer) or sphingosine in the salvage pathway to generate ceramide (Fig. 1). Levy and Futerman (2010) analyzed mRNA distribution of CerSs across tissues. The mRNA for CerS1, which preferentially generates C_{18}-ceramide, is strongly expressed in the brain and skeletal

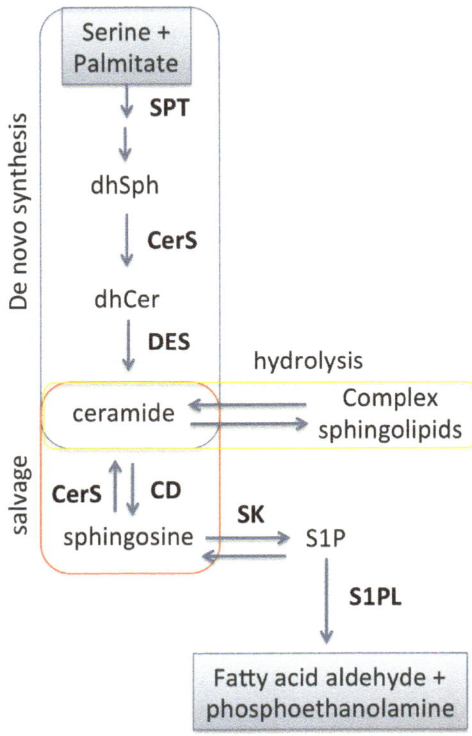

Figure 1 ***Sphingolipid metabolism and key enzymes.*** Ceramide is central to sphingolipid metabolism and can be generated de novo (blue), through hydrolysis of complex sphingolipids (yellow) or via the salvage pathway (red). In the de novo pathway, serine palmitoyltransferase (SPT) catalyzes the first reaction resulting in 3-ketosphinganine, which is reduced to dihydrosphingosine (dhSph), further metabolized to dihydroceramide (dhCer) by ceramide synthases (CerSs), and desaturated by dihydroceramide desaturase (DES) to generate ceramide. Ceramide is used as a building block for complex sphingolipids such as sphingomyelin, glycosphingolipids, and gangliosides. These complex sphingolipids can be hydrolyzed to regenerate ceramide. Ceramide is hydrolyzed by ceramidases (CD) to generate sphingosine, which serves as substrates for CerSs or SK1 to regenerate ceramide via the salvage pathway or to generate sphingosine 1-phosphate (S1P). Further hydrolysis of S1P by S1P lyase represents the exit from sphingolipid metabolism. This schematic is focused on the enzymes discussed in the text and does not include all enzymes involved in sphingolipid metabolism.

muscle, whereas mRNA for CerS3, which generates very-long-chain ceramides (C_{22}–C_{26}-ceramides), is primarily detected in the testis. The substrate preference of CerS2 is similar to CerS3, but in contrast to CerS3, CerS2 mRNA is widely expressed. CerS4 generates C_{18}- and C_{20}-ceramides and is strongly expressed in the skin, heart, and spleen. CerS5 and CerS6 both generate C_{16}-ceramides, but their mRNAs differ in tissue distribution. Regulation of CerS activity is incompletely understood but complex. Members of the CerS family are regulated individually at the genetic and epigenetic level as well as through posttranslational mechanisms and stability. Activity can also be influenced through homodimer or heterodimer formation of family members (Laviad, Kelly, Merrill, & Futerman, 2012).

2.1.1 Ceramide Synthase Expression in Prostate Cancer

Little is known about the role of CerSs in prostate cancer. However, some information on expression can be gleaned from the Protein Atlas (https://www.proteinatlas.org). Expression of CerS1, CerS3, and CerS6 was either not detected or not detected in the majority of prostate cancer samples examined. Although CerS6 immunohistochemistry yielded negative results, we have detected expression of this enzyme whole cell lysates of prostate cancer cells such as PPC1, 22Rv1, Du145, and LNCaP by Western blotting (Voelkel-Johnson, unpublished). However, in the prostate, mRNA for CerS5 is more prevalent than CerS6, suggesting a role for CerS5 rather than CerS6 in this organ (Levy & Futerman, 2010). Protein for CerS5 as well as CerS2 and CerS4 is detected at low to medium levels in all prostate cancers. Staining of these enzymes is primarily cytoplasmic/membranous. Nuclear staining was detected only for CerS5 in a fraction of samples (3/8). The relevance of CerS expression level and subcellular pattern remains to be determined.

2.2 Dihydroceramide Desaturase

Dihydroceramide (dhCer) generated by CerSs in the de novo pathway is further metabolized by dihydroceramide desaturase (DES), an enzyme that is responsible for inserting the 4,5-trans-double bond into the sphingolipid backbone of dihydroceramide to generate ceramide (Fig. 1). Two isoforms of DES have been identified. DES2 has dual functions as desaturase and hydroxylase and is preferentially expressed in the kidney, small intestine, and skin, where generation of phytoceramides is essential. In contrast, DES1 is ubiquitously expressed with high DES activity and very low hydroxylase activity (Siddique, Li, Chaurasia, Kaddai, & Summers, 2015). Expression of DES1 has been associated as an unfavorable marker in renal, liver, endometrial, and urothelial cancers (https://www.proteinatlas.org/ENSG00000143753-DEGS1/pathologys) and potentially plays an important role in prostate cancer.

2.2.1 Dihydroceramide Desaturase 1 in Prostate Cancer

DES1 has recently been identified as a target gene of the androgen receptor (McNair et al., 2017). Analysis by chromatin immunoprecipitation coupled to detection by quantitative real-time PCR confirmed that the androgen receptor binds to the DES1 regulatory region early in the G1 phase of the cell cycle and knockdown of DES1 significantly decreased migration of castration-resistant C4-2 prostate cancer cells. Cohort analysis of 780 patients with localized, high-risk prostate cancer revealed a correlation between high DES1 expression and poor prognosis (McNair et al., 2017). Although the recognition of DES1 as a potential therapeutic target in prostate cancer is novel, several compounds associated with antitumor activity in this disease, including fenretinide, celecoxib, resveratrol, and curcumin, were identified as inhibitors of DES1 (Siddique et al., 2015).

2.3 Ceramidases

Five ceramidases with different pH optimums hydrolyze ceramide to generate sphingosine (Fig. 1). There are three alkaline ceramidases (ACER1, ACER2, and ACER3), acid ceramidase (ASAH1), and neutral ceramidase (ASAH2) (Coant, Sakamoto, Mao, & Hannun, 2017). Among the ceramidases, ASAH1 and ASAH2 have been associated with cancer. ASAH2, which is localized to the plasma membrane and primarily expressed in the small intestine and colon, is involved in digestion and has been implicated in colon carcinogenesis (Camp, Patterson, Kester, & Voelkel-Johnson, 2017; Coant et al., 2017). ASAH1 is primarily found in the lysosomal compartment, and mutations in its gene result in rare genetic disorders and other diseases (Koch et al., 1996; Park & Schuchman, 2006). Knockout of ASAH1 leads to death during the two-cell stage of embryonic development, indicating that other ceramidases are unable to compensate for its enzymatic activity (Eliyahu, Park, Shtraizent, He, & Schuchman, 2007). The human ASAH1 gene is located in 8p22, a region that is frequently affected in cancer.

2.3.1 ASAH1 in Cancer

ASAH1 is dysregulated in many types of cancer with a majority of studies linking elevated ASAH1 expression to proliferation and cancer cell survival. Increased expression of ASAH1 in head and neck cancer reduces responsiveness to cisplatin (Mehta et al., 2000). Genetic or pharmacological inhibition of ASAH1 using shRNA or N-oleoyl-ethanolamine, respectively, increased susceptibility of head and neck cancer cells to cisplatin in vitro and in vivo (Roh, Park, Kim, & Jang, 2016). In the skin, melanocytes and melanoma cells express higher levels of ASAH1 than other cell types but with different subcellular localization between normal and malignant cells. In melanocytes, ASAH1 localizes to both the nucleus and the cytosol, whereas expression in melanoma cells was limited to the cytosol and associated with reduced ceramide levels (Realini et al., 2016). In MCF7 breast cancer cells, treatment with the ASAH1 inhibitor Ceranib-2 decreases viability in a dose- and time-dependent manner through modulation of the mitochondrial membrane potential (Vejselova, Kutlu, & Kus, 2016). Interestingly, in luminal subtypes of breast cancer and epithelial ovarian cancer, elevated expression of ASAH1 correlated with a better prognosis (Hanker et al., 2013; Sanger et al., 2014). The correlation between ASAH1 expression and prognosis may be context dependent and determined by the relative expression of CerSs and sphingosine kinases (SKs) that compete for ASAH1-generated sphingosine as a substrate (Camp et al., 2017).

2.3.2 ASAH1 in Prostate Cancer

A role for ASAH1 in prostate cancer was suggested as early as 2000 when Seelan et al. reported that this enzyme was frequently overexpressed but not mutated in prostate cancer. An increase in ASAH1 was found in almost half (15/36) of patient-matched prostate tumor tissues and was also detected in commonly used prostate cancer cell lines Du145, LNCaP, and PC3 (Seelan et al., 2000). Two other groups independently confirmed that ASAH1 expression is increased in prostate cancer (Camacho et al., 2013; Norris et al., 2006).

In prostate cancer ASAH1 can be induced transcriptionally or regulated at the posttranscriptional level by the androgen-stimulated deubiquitinase USP2, which stabilizes protein half-life (Cheng et al., 2013; Mizutani et al., 2015). Overexpression of ASAH1 in Du145 prostate cancer cells increases tumor burden in vivo and was associated with resistance to a wide range of chemotherapeutic agents (Saad et al., 2007). Turner et al. demonstrated that overexpression of ASAH1 increases lysosomal density and autophagy. The autophagy inhibitor 3-methyl-adenine restored susceptibility to exogenous ceramide, suggesting a connection between autophagy and resistance to ceramide (Turner et al., 2011). In human prostate tumors elevated expression of ASAH1 was associated with increased phosphorylation of the oncogenic kinase Akt. Mechanistic studies indicate that ASAH1 activates sphingosine kinase 1 (SK1) to generate S1P, which then via S1P receptor 2 stimulates PI3K (Beckham, Cheng, Lu, Shao, et al., 2013). In addition, S1P-mediated Akt signaling can promote nuclear export of the tumor suppressor PTEN (Beckham, Cheng, Lu, Marrison, et al., 2013). Increased expression of ASAH1 has also been shown to promote Ets1 nuclear expression and binding to the matrix-degrading protease cathepsin B. The increased abundance of cathepsin B in ASAH1 overexpressing prostate cancer cells enhanced their ability to invade through a collagen matrix, which was abrogated by cathepsin B inhibition (Beckham et al., 2012). The apparent link between ASAH1 and tumor aggression in prostate cancer prompted the pursuit of ASAH1 as a novel therapeutic target in this disease.

2.4 Sphingosine Kinases

As part of a superfamily of lipid signaling kinases, SKs phosphorylate sphingosine to generate the soluble lipid second messenger S1P (Wattenberg, Pitson, & Raben, 2006) (Fig. 1). Overexpression of SKs has been observed in the development and progression of many different cancer types (Haddadi, Lin, Simpson, Nassif, & McGowan, 2017). The two SK isoforms, SK1 and SK2, have multiple splice variants, although roles and functions of these variants are incompletely understood (Haddadi et al., 2017). SK1 and SK2 share high sequence similarity but have distinct subcellular distribution, regulation, and function. SK1 is a cytosolic enzyme that when activated translocates to the plasma membrane where its product S1P is released extracellularly and in an autocrine or a paracrine manner can bind to S1P receptors to promote tumorigenesis (Newton, Lima, Maceyka, & Spiegel, 2015). SK2 localizes to the nucleus and has been shown to bind and inhibit histone deacetylases (HDACs) (Fu et al., 2018; Neubauer & Pitson, 2013). The enzyme has also been detected in other subcellular compartments (Neubauer et al., 2016).

2.4.1 Sphingosine Kinase in Prostate Cancer

SK activity in prostate cancer is approximately twofold higher relative to patient-matched normal tissue and corresponded to higher levels of Prostate

Specific Antigen. Analysis of SK1 in 30 patients revealed that elevated expression levels were associated with tumor volume, positive margins, and surgical failure (Malavaud et al., 2010). To gain additional insight into SK expression and its potential association with tumor aggressiveness, we analyzed a data set from Oncomine of a study that included samples from 59 patients with localized prostate cancer and 35 patients with metastatic castration-resistant disease (Grasso et al., 2012). Using the Mann–Whitney test we determined that mRNA for both SK1 and SK2 was significantly elevated in metastatic prostate cancer compared to localized disease ($P = .0035$ for SK1 and $P = .00038$ for SK2). Among the 35 patients with advanced disease 19 had an increase in both isoforms of SKs, 10 had an increase in either SK1 or SK2 but not both, and only 6 showed no increase relative to the median value of patients with localized disease. Although further studies at the protein level are needed, these data suggest that an increase in either one or both SK isoforms may contribute to tumor aggressiveness in prostate cancers.

The expression of SK1 in nonmalignant tissue has also been shown to play a role in cancer. In an elegant study Ponnusamy et al. demonstrated that systemic SK1-generated S1P is critical for communication with malignant cells. In the TRAMP model of prostate cancer, survival of mice was increased from 10 to 12.5 months in SK1-deficient hosts (Ponnusamy et al., 2012). Additional experiments in the MB49 bladder cancer model demonstrated that the growth and metastasis of SK1 + MB49 cells were significantly reduced in SK1-deficient hosts, suggesting an important role for nontumor-derived S1P in cancer promotion and dissemination (Ponnusamy et al., 2012). Thus although circulating S1P may not necessarily be tumor derived, targeting the generation of S1P remains a key strategy in cancer therapy.

3. SPHINGOLIPID ENZYMES AS TARGETS IN PROSTATE CANCER

3.1 Inhibition of Dihydroceramide Desaturase

Dihydroceramides (dhCers) do not have the biological effects of ceramide, which led to the dogma that they are inert sphingolipids. In 2006 a group led by Al Merrill developed novel technologies to measure sphingolipids and observed that dhCers rather than ceramides accumulated in Du145 prostate cancer cells treated with fenretinide

(4-HPR) (Zheng et al., 2006). This seminal study was the first to associate dhCers with autophagy and led to increased interest in DES. Three comprehensive reviews on physiological and pathological functions of dhCers, DES biochemistry, and DES inhibitors have recently been published (Casasampere, Ordonez, Pou, & Casas, 2016; Rodriguez-Cuenca, Barbarroja, & Vidal-Puig, 2015; Siddique et al., 2015). The identification of DES1 as a target gene of the androgen receptor revealed a previously unknown link between these two proteins, and the observation that high levels of DES1 correspond to poor overall survival in prostate cancer suggests DES1 as a novel therapeutic target in this disease (McNair et al., 2017). Interestingly, compounds that inhibit DES activity exert antitumor effects in prostate cancer (Fig. 2).

3.1.1 Fenretinide

Fenretinide, a synthetic retinoid with chemopreventive effects on prostate cancer in rodent models, directly inhibits DES (Pienta, Nguyen, & Lehr, 1993; Pollard, Luckert, & Sporn, 1991; Rahmaniyan, Curley, Obeid, Hannun, & Kraveka, 2011; Slawin et al., 1993) Mechanistic studies in prostate cancer cell lines demonstrate that fenretinide interferes with FAK and Akt activation and promotes degradation of β-catenin

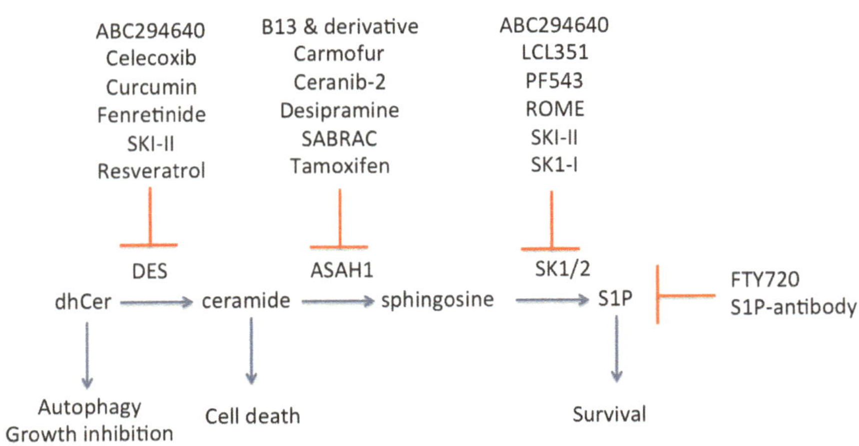

Figure 2 ***Inhibitors of sphingolipid metabolism.*** Inhibitors of DES promote autophagy and growth inhibition. Drugs that target sphingolipid metabolism downstream of ceramide have the potential to result in accumulation of ceramide to promote cell death and/or to prevent sphingosine 1-phosphate (S1P)—mediated survival signaling. Details of the inhibitors shown are discussed in the text. *ROME*, (R)-FTY720 methyl ether; *SK1*, sphingosine kinase 1.

(Benelli, Monteghirfo, Vene, Tosetti, & Ferrari, 2010). Fenretinide preferentially accumulates in fatty tissue and is effective against breast cancer (Formelli et al., 1993; Sabichi et al., 2003). However, accumulation of fenretinide in the human prostate was significantly lower than in mouse prostate or in human breast tissue, which may explain its lack of clinical success in prostate cancer patients (Cheung et al., 2009; Pienta, Esper, Zwas, Krzeminski, & Flaherty, 1997; Thaller et al., 2000). Subsequent efforts have focused on combining fenretinide with an inhibitor of acid ceramidase, which increased dhCer and reactive oxygen species resulting in cytotoxicity against PC3 and Du145 prostate cancer cells. However, the role of ceramide synthesis remains elusive because blocking de novo synthesis of ceramide did not block the cytotoxic response (Gouaze-Andersson et al., 2011).

3.1.2 Celecoxib

Celecoxib is a COX2 inhibitor that decreases the mean PSA velocity in prostate cancer patients (Pruthi et al., 2006; Smith et al., 2006). Analysis of gene expression in prostate cancer patients who had taken celecoxib for 4 weeks before prostatectomy revealed changes that were consistent with increased apoptosis and tumor suppressor function (Sooriakumaran, Macanas-Pirard, et al., 2009). A parallel study of tissue sections supported finding of the gene expression pattern and showed that celecoxib decreased tumor cell proliferation, microvessel density, and angiogenesis while enhancing apoptosis (Sooriakumaran, Coley, et al., 2009). In cell lines, celecoxib inhibits DES with a half maximal inhibitory concentration (IC_{50}) of $78.9 \pm 1.5\ \mu M$ and significantly increases dhCers (C_{16}, C_{24}, $C_{24:1}$) while decreasing C_{24}- and $C_{24:1}$-ceramides in a time- and dose-dependent manner (Schiffmann et al., 2009). Inhibitors of DES (8-CPPC) or CerS (FB1) prevented the increase in ceramides, and the serine palmitoyltransferase (SPT) inhibitor myriocin reduced the antiproliferative effect of celecoxib, suggesting that de novo synthesis of ceramide is required for the antiproliferative effect of celecoxib. Recently, Maeng et al. demonstrated that celecoxib induces SPT (mRNA, protein and activity) and CerS1/2/3/5/6 mRNA in HepG2 cells, whereas DES mRNA expression was unaltered. Changes in gene expression were accompanied by increased dhCers (C_{16}, C_{24}, $C_{24:1}$), increases in long-chain ceramides ($C_{14}-C_{20}$), and a specific decrease in C_{16}-sphingomyelin (SM) (Maeng et al., 2017). Celecoxib has also been shown to stimulate CerS activity nearly threefold and activates CerS6-mediated generation of C_{16}-ceramide via the salvage pathway (Schiffmann et al., 2010). Taken together, these studies suggest a model in which celecoxib stimulates the expression and activity of SPT as well as CerSs to drive de novo synthesis of sphingolipids. However, without concomitant induction of DES mRNA potentially in combination with DES enzymatic inhibition, dhCers accumulate. Hydrolysis of complex sphingolipids such as SM or activation of the salvage pathway could be responsible for the increase in ceramide observed in response to celecoxib.

3.1.3 Natural Compounds

Analyses of dietary habits and prostate cancer have identified resveratrol and curcumin as natural compounds associated with reduced risk of prostate cancer (Bemis, Katz, & Buttyan, 2006). Resveratrol has numerous effects on cancer cells and signaling pathways, including sphingolipid enzymes (Casasampere et al., 2016). Inhibition of DES and induction of autophagy have been shown in a gastric cancer cells, although the mechanism may be mediated indirectly through modulation of redox status (Gagliostro et al., 2012; Shin et al., 2012; Signorelli et al., 2009). Resveratrol supplementation in animal models has shown variable results, and to date, there is no clinical evidence to make recommendations for cancer prevention or therapy (Carter, D'Orazio, & Pearson, 2014; Jasinski, Jasinska, & Ogrodowczyk, 2013).

Similar to COX2 inhibitors, curcumin can reduce PSA levels when used in conjunction with isoflavones (Ide et al., 2010). In PC3 prostate cancer cells, curcumin activated MAPK, JNK, and caspases, inducing the release of cytochrome *c* and AIF from mitochondria and resulted in accumulation of ceramides. Because neither kinase nor caspase inhibitors prevented curcumin-induced cytotoxicity, the authors speculated that the curcumin-induced increase in ceramide and caspase-independent cell death is responsible for the effects (Hilchie et al., 2010). Curcumin has been shown to induce rapid stimulation of ceramide synthesis through inducing dimerization of CerSs (Laviad et al., 2012). Curcumin can also reduce DES1 activity by 28% at 10 μM in a human gastric adenocarcinoma cell line (Fabrias et al., 2012). Because curcumin induces autophagy primarily in cancer models, it is possible that increased CerS activity through dimerization in combination with partial, indirect inhibition of DES leads to an accumulation of dhCer that promotes autophagy. The in vivo use of curcumin is hampered by poor bioavailability, although nanoparticle encapsulation was shown to enhance its therapeutic effect in prostate cancer (Gupta, Patchva, & Aggarwal, 2013; Thangavel, Yoshitomi, Sakharkar, & Nagasaki, 2015).

3.1.4 Sphingosine Kinase Inhibitors

Recently, two SK inhibitors, SKI-II and ABC294640, have been shown to potently inhibit DES (Cingolani et al., 2014; Venant et al., 2015). SKI-II was identified by the Smith laboratory in a high-throughput screen as a dual SK inhibitor of SK1 and SK2 (French et al., 2003, 2006). SKI-II inhibits the growth of PC3 prostate cancer cells, and analysis of structure-activity relationship suggests that DES rather than SK is responsible for this effect (Aurelio et al., 2016).

The Smith laboratory also identified the SK2-selective inhibitor ABC294640 and spearheaded its clinical translation (Britten et al., 2017; French et al., 2010). ABC294640 significantly inhibits DES activity in TRAMP prostate cancer cells and increases dhCer (Venant et al., 2015). The increase in dhCer was observed even in SK2-deficient murine embryonic fibroblasts (MEFs), suggesting the possibility that ABC294640 directly inhibits DES. Treatment of mice with 50 mg/kg ABC294640 also increased levels of dhCer in TRAMP subcutaneous tumors, and in the recent clinical trial, ABC294640 significantly elevated plasma dhC_{16}-ceramide (the only dhCer measured) at 8 and 12 h (Britten et al., 2017; Venant et al., 2015). These data suggest that the drug may inhibit DES at clinically relevant concentrations.

3.2 Inhibitors of Acid Ceramidase

The observations that expression of ASAH1 is increased in several types of cancer and is associated with resistance have fueled an interest in targeting this enzyme therapeutically. The first-generation ceramidase inhibitors such as *N*-Oleoylethanolamide (NOE) were ceramide mimetics (Sugita, Willians, Dulaney, & Moser, 1975). However, low potency and poor selectivity precluded clinical development. The NOE analogue DM102 inhibits ASAH1 with an IC_{50} of ~15 μM in cells and can kill A549 lung cancer cells with a median lethal dose (LD_{50}) of ~40 μM (Bedia et al., 2008). DM102 has been evaluated in combination with fenretinide in PC3 and Du145 prostate cancer cell lines, which synergistically decreased cell viability. Fenretinide, which inhibits DES, increased dhCer sixfold, and DM102 increased reactive oxygen species 30-fold compared to the control (Gouaze-Andersson et al., 2011). However, the mechanism of synergy remains elusive. Additional ASAH1 inhibitors have been identified in library screens or through structure-activity relationship analyses, some of which are further discussed below (Saied & Arenz, 2016).

3.2.1 SABRAC

PC3/Mc is a highly tumorigenic derivative of the PC3 prostate cancer cell line that has increased expression of ASAH1, metastatic potential, and chemoresistance. Genetic targeting of ASAH1 increased ceramide and decreased clonogenicity. SABRAC (*N*-[(2*S*,3*R*)-1,3-dihydroxyoctadecan-2-yl]2-bromoacetamide) is a novel ceramide analogue with an in vitro IC_{50} of 52 nM that irreversibly inhibits ASAH1. Treatment of PC3/Mc prostate cancer cells with SABRAC increased ceramide in a dose-dependent fashion and similar to genetic targeting of ASAH1 reduced growth and clonogenic potential (Camacho et al., 2013).

3.2.2 Ceranib-2

Ceranib-2 is a more potent analogue of Ceranib-1, a ceramidase inhibitor that was identified during a screen of a small-molecule library (Draper et al., 2011). This small-molecule inhibitor was identified as a quinolinone compound and has an IC_{50} of 28 μM. Ceranib-2 delayed tumor growth in a syngeneic model of breast cancer without apparent toxic side effects and decreased viability in LNCaP and Du145 prostate cancer cells in a time- and dose-dependent manner (Draper et al., 2011; Kus, Kabadere, Uyar, & Kutlu, 2015). Depending on the cell line studied, ceranib-2 induces apoptotic or nonapoptotic cell death, activates stress and mitogen-induced kinases JNK and p38 kinases, and inhibits the oncogenic kinase Akt (Kus et al., 2015; Vethakanraj et al., 2018).

3.2.3 B13 and Its Derivatives

B13 is a potent and selective ASAH1 inhibitor with an IC_{50} of about 10 μM and induces apoptosis in a variety of cancer cell lines (Bielawska, Linardic, & Hannun, 1992; Samsel et al., 2004). In PC3 prostate cancer xenografts, B13 also sensitized cells to radiation therapy (Samsel et al., 2004). However, as a neutral lipophilic molecule, B13 does not reach the acid lysosomal compartment where ASAH1 resides (Bai et al., 2014).

LCL204 is a lysosomotropic analogue of B13 and consistent with ASAH1 inhibition, treatment of Du145 prostate cancer cells selectively increased C_{14}-, C_{16}-, and C_{18}-ceramides and decreased sphingosine. Exposure to LCL204 resulted in cathepsin-dependent degradation of ASAH1, rapid destabilization of lysosomes, and cytoplasmic release of lysosomal proteases, which led to depolarization of mitochondria, activation of caspases, and apoptotic cell death (Holman et al., 2008). LCL204 indirectly inhibited ASAH1 activity through lysosomal destabilization leading to proteolytic degradation of the enzyme, which prompted the development of a second series of lysosomotropic ω-*N*-aminoacyl analogues using the B13 scaffold (Bai et al., 2009). The ASAH1 inhibitor LCL385 sensitized PPC1 prostate cancer cells to radiation and significantly reduced xenograft growth in vivo (Mahdy et al., 2009).

A series of dimethylglycine (DMG)-B13 ester prodrugs (LCL521, LCL522, LCL596) have also been developed (Bai et al., 2014). The addition of the DMG ester significantly enhanced the cellular effectiveness of B13, and LCL521 was identified as the most promising ASAH1 inhibitor. Combination of radiation therapy with LCL521 completely inhibited radiation treatment failure in PPC1 prostate cancer xenografts (Cheng et al., 2013). Interestingly, the addition of LCL521 did not significantly improve the response to radiation treatment because tumors in all groups (radiation alone, radiation plus vehicle, or radiation plus LCL521) experienced a similar decrease in volume by day 40. However, mice treated with radiation or radiation plus vehicle relapsed with a survival rate of only 25% at week 35. In contrast, 100% of mice treated with radiation and LCL521 survived. LCL521 has also been evaluated in photodynamic therapy (PDT). Pretreatment with LCL521 greatly enhanced efficacy of PDT in vitro and an adjuvant in a vaccination approach. LCL521 was able to significantly reduce two populations of immune suppressor cells (T regulatory cells and myeloid-derived suppressor cells), which are known to interfere with the efficacy of PDT vaccines (Korbelik et al., 2016). In myeloid-derived suppressor cells, LCL521 increased autophagic vesicles and induced endoplasmic reticulum stress (Liu et al., 2016). Thus LCL521 may impact tumor cells directly or indirectly through modification of the tumor microenvironment (or both).

3.2.4 FDA-Approved Drugs With "Off-Target" ASAH1 Inhibitory Effects

Some drugs that were already used clinically were also identified to inhibit ASAH1. These include the tricyclic antidepressant desipramine, the prodrug carmofur, and the antiestrogen tamoxifen. Desipramine was shown to induce downregulation of ASAH1 in a time- and dose-dependent manner (Elojeimy et al., 2006). Carmofur is an orally administered prodrug that is taken up by the intestine to be intracellularly converted into 5-fluorouracil, which inhibits thymidylate synthase. This was believed to be the primary mechanism of action, and the use of carmofur as adjuvant chemotherapy for colorectal cancer patients in China, Japan, and Finland does extend survival (Sakamoto, Oba, Matsui, & Kobayashi, 2006). Interestingly, during screening of a commercial chemical library, carmofur was identified to inhibit recombinant rat ASAH1 with an IC_{50} of 29 nM (Realini et al., 2013). Carmofur increases intracellular levels of ceramide in cancer cells, including in LNCaP prostate cancer cells, and the uracil carboxamide scaffold was subsequently used as a platform for the development of novel ASAH1 inhibitors (Realini et al., 2013). A group led by Myles Cabot demonstrated that a low micromolar dose of tamoxifen inhibited ASAH1 in a spectrum of cancer cells, including prostate cancer. Further analysis of the mechanism by which tamoxifen inhibits ASAH1 pointed to an indirect effect because enzyme activity was not affected in cell-free assays. In intact cells, tamoxifen increased lysosomal permeability, which in a cathepsin B—dependent manner resulted in downregulation of ASAH1 protein levels. The structurally similar antiestrogen toremifene similarly inhibited ASAH1 (Morad et al., 2013).

3.3 Inhibition of Sphingosine Kinases

SKs generate the soluble, bioactive S1P that plays an important role in angiogenesis, tumor cell survival, metastasis, and therapy resistance. Thus targeting SK activity has been a major therapeutic effort. However, interpreting the effects of SK inhibitors in cells or in preclinical models is complicated by lack of specificity for SK isoforms and off-target effects. SKI-II was identified as a dual SK inhibitor (Gao, Peterson, Smith, & Smith, 2012). Recently, this inhibitor was shown to also target DES1, and structure-activity relationship analysis of SKI-II targets suggests that DES1 rather than SK inhibition is primarily responsible for the antiproliferative effects of SKI-II and its analogues (Aurelio et al., 2016; Cingolani et al., 2014). Additional caution is warranted because SKI-II is sometimes referred to as an SK1 inhibitor in the literature. Because SK1 and SK2 exhibit structural differences in their binding pocket, attempts have been made to generate isoform-specific inhibitors (Gao et al., 2012).

3.3.1 Sphingosine Kinase 1 Inhibitors

SK1-I is a sphingosine analogue and a competitive, specific inhibitor of SK1 that is not to be confused with the dual SK inhibitor SKI-II (Paugh et al., 2008). SK1-I decreases S1P, increases ceramide, and inhibits tumor cell growth in vitro and in vivo (Kapitonov et al., 2009; Nagahashi et al., 2012; Paugh et al., 2008; Song et al., 2011). SK1-I reduces phosphorylation of Akt in glioblastoma, sensitizes lungs cancer cells to chemotherapy, and decreases lymph node and lung metastasis in a breast cancer model through inhibition of hemangiogenesis and lymphangiogenesis.

PF543 is a potent inhibitor of SK1 with an IC_{50} of 2–3.6 nM (vs. 356 nM for SK2) and prevents the phosphorylation of sphingosine in cancer cells. At 10 μM, PF543 did not significantly inhibit the activity of other protein and lipid kinases or bind to S1P receptors (Schnute et al., 2012). Cingolani et al. compared the effects observed with SKI-II and PF543. Although both SKI-II and PF543 decreased S1P to nearly undetectable levels, only SKI-II but not PF543 reduced proliferation, G0/G1 inhibition, and autophagy (Cingolani et al., 2014). Rex et al. (2013) also failed to detect a cytotoxic effect of SK inhibitors on cancer cells. The lack of anticancer efficacy by specific SK inhibitors may have been an underlying factor for abandoning the pursuit of patents (Lynch, Thorpe, & Santos, 2016).

Nevertheless, as a soluble mediator that can be produced by noncancerous cells to influence malignant cell behavior, extracellular S1P generated by SK1 remains an important target. For example, a recent study using cell-specific SK1 knockout mice demonstrated that macrophage-derived S1P plays an important role in colon carcinogenesis (Furuya et al., 2017). In prostate and bladder cancer, nontumor-derived S1P plays an important role in metastasis (Ponnusamy et al., 2012). The novel SK1-specific inhibitor LCL351 decreases the migration rate of human Du145 prostate cells (Sharma, 2011). LCL351 also decreases the inflammatory response and reduces colitis-induced development of colorectal cancer (Furuya et al., 2017).

3.3.2 SK2 Inhibitors

The SK2-selective inhibitor, (R)-FTY720 methyl ether (ROME), has been evaluated in LNCaP prostate cancer cells. Treatment with ROME increases sphingosine and decreases S1P without affecting ceramide levels or inducing apoptosis (Watson et al., 2013). ROME was also compared with SKI-II for its ability to reduce expression of the androgen receptor but no SK2-specific effect was found (Tonelli et al., 2013). The effect of SKI-II on the androgen receptor was independent of inhibitor-induced proteasomal degradation of SK1 and could not be reversed with the pan-CerS inhibitor Fumonisin B1.

Another SK2-selective inhibitor is ABC294640, which, similar to SKI-II, inhibits the activity of DES but not SK1 or other kinases (Gao et al., 2012; Venant et al., 2015). Although ABC294640 has been reported to induce apoptosis in some cell models, in prostate cancer, it promotes lysosomal acidification, autophagy and growth inhibition (Beljanski, Knaak, & Smith, 2010; Qin et al., 2014; Venant et al., 2015). It also downregulates signaling pathways including STAT3, Akt, ERK, p21, p53, and FAK (Gao et al., 2012). Two studies have shown that ABC294640 decreases c-Myc and androgen receptor expression in prostate cancer (Schrecengost, Keller, Schiewer, Knudsen, & Smith, 2015; Venant et al., 2015). Genetic targeting of SK2 confirmed that the effect of ABC294640 on c-Myc requires SK2 expression. However, the effect of ABC294640 on the androgen receptor remains to be elucidated and could be mediated through inhibition of DES or indirectly via c-Myc (Schrecengost et al., 2015; Venant et al., 2015). ABC294640 exerts antitumor efficacy in numerous preclinical models (Antoon et al., 2010; Chumanevich et al., 2010; Dai, Smith, Foroozesh, Miele, & Qin, 2017; Lewis, Voelkel-Johnson, & Smith, 2016; Qin et al., 2014; Schrecengost et al., 2015; Venkata et al., 2014) and has recently been evaluated in a Phase I clinical trial. The study, which involved 21 patients with advanced solid tumors, demonstrated that a dose of 500 mg ABC294640 was well tolerated and resulted in plasma sphingolipid changes that may have utility as pharmacodynamics markers of drug activity (Britten et al., 2017). The Phase I trial did not include any prostate cancer patients. However, ABC294640 may be particularly relevant for the treatment of prostate cancer because both c-Myc and the androgen receptor are frequently dysregulated in this disease.

3.3.3 Targeting the Downstream Effects of Sphingosine Kinase

The primary goal of targeting SK is to prevent the protumor function of its product S1P. Two strategies that interfere with S1P and S1P signaling are S1P antibodies and SK substrate analogues. An antibody against S1P known as Sphingomab has been evaluated in prostate cancer models. Prostate cancer preferentially metastasizes to the bone, and S1P generated by osteoblasts has been hypothesized to influence prostate cancer cell behavior. Brizuela et al. (2014) investigated whether paracrine S1P signaling from osteoblastic MC3T3 cells played a role in the proliferation of PC3 and C4-2 prostate cancer cells and demonstrated that genetic knockdown of SK1 or neutralization of S1P using Sphingomab significantly inhibited cancer cell proliferation. A subsequent study revealed that Sphingomab was able to transiently normalize tumor vasculature, reducing tumor hypoxia, and increasing efficacy of chemotherapy as evidenced by inhibiting tumor growth and metastasis (Ader et al., 2015). However, despite proof of concept, promising results in preclinical models, and safety of a humanized version of the antibody, therapeutic efficacy of S1P antibodies fell short of expectations in clinical trials for the treatment of macular degeneration (NCT01334255) and advanced renal cell carcinoma (NCT01762033).

FTY720 (Fingolimod, Gilenya) is a sphingosine analogue that is phosphorylated by SK to FTY720-P, which unlike S1P cannot signal S1P receptors. As a S1P receptor modulator, FTY720 is currently approved as an immune-modulating drug for the treatment of multiple sclerosis. Depending on culture conditions, FTY720 can also modulate ceramide synthesis in cells. When dhSph concentrations are high (500 nM−5 μM), FTY720 can significantly inhibit ceramide synthesis, but this effect was not observed when dhSph was <200 nM (Lahiri et al., 2009). In contrast, treatment of cells with 25 μM FTY720 increased C_{18}−C_{22}-ceramides within 90 min, which points to a potentially complex interaction between FTY720 and CerSs (Lahiri et al., 2009). FTY720 may also exert antitumor effects in a S1P receptor−independent manner because silencing of S1P receptors had no effect on cell death in PC3 cells (Pchejetski et al., 2010).

Although the exact mode of action for FTY720 remains to be further elucidated, several studies point to an anticancer effect in models of prostate cancer. High doses of FTY720 induce apoptosis (40 μM) or G1-cell cycle arrest (20 μM) in Du145 prostate cancer cells (Permpongkosol et al., 2002; Wang et al., 1999). In CWR22R prostate tumors, administration of 10 mg/kg FTY720 suppressed tumor growth in vivo and corresponded to a reduction in serum PSA levels, suppression of angiogenesis, reversal of epithelial mesenchymal transition through restoration of E-cadherin expression, and an increase in apoptotic cells (Chua et al., 2005). A dose of 1.5 μM FTY720 was sufficient to decrease prostate cancer cell migration and invasion in vitro and caused a significant reduction in actin stress fibers and filopodia through inhibition of RhoA-GTPase activity (Zhou et al., 2006). A low dose of FTY720 also decreased phosphorylation of Akt, reduced Runx2 and N-cadherin, which plays an important role in prostate cancer metastasis (Chua et al., 2009). Similar to other modulators of sphingolipid metabolism, FTY720 can sensitize tumor cells to radiation therapy in vitro and in vivo (Pchejetski et al., 2010).

4. ESCAPE FROM THERAPEUTIC TARGETING

It is known that the malignant cell population is heterogeneous and genomically unstable. Transformed cells acquire the ability to evade immune surveillance and survive in the hostile tumor microenvironment eventually developing into detectable malignant lesions. Cellular stress induced by cancer therapy, such as radiation or chemotherapy, increases intracellular levels of ceramide to induce cell death. However, invariably, a small, resistant subset of malignant cells cannot be eradicated and the cancer stem cell (CSC) hypothesis posits that this small cell population can regenerate tumors with even more heterogeneity and resistance, which makes recurrent cancer extremely difficult to treat.

Therapy may select for a preexisting resistant subset of cells, but evidence also points to therapy-induced resistance as a consequence of ceramide stress. Chronic exposure to ceramide or glucosylceramide increases the expression of the multidrug resistance gene that codes for P-glycoprotein, which enhances resistance to doxorubicin and paclitaxel (Gouaze-Andersson et al., 2007). Daunorubicin has been shown to posttranscriptionally activate ASAH1 in liver cancer cells (Morales et al., 2007). In colon cancer cells overexpression of CerS6 increases intracellular C_{16}-ceramide and transcriptionally induces ASAH1 through a c-Jun N-terminal kinase—mediated mechanism (Tirodkar et al., 2015). Cheng et al. showed that radiation-induced ceramide stress also transcriptionally activates the expression of ASAH1 in a wide variety of other cancer cell types. Clinical data from prostate cancer patients for whom a preradiation and postradiation biopsy was available support the hypothesis that ASAH1 expression becomes elevated following treatment (Cheng et al., 2013).

4.1 Sphingolipids in Epithelial Mesenchymal Transition

A key to cancer progression in solid tumors is epithelial mesenchymal transition (EMT), a process during which epithelial cells lose polarization and adhesion, gain the ability to migrate and invade, and acquire pluripotency (Nieto, Huang, Jackson, & Thiery, 2016). The EMT program is driven by several families of transcription factors that directly or indirectly suppress the expression of E-cadherin resulting in a so-called "cadherin switch" during which E-cadherin is lost and expression of the mesenchymal filament protein vimentin is gained. Treatment of prostate cancer with androgen deprivation therapy on the one hand promotes accumulation of C_{16}-ceramide and apoptosis but on the other hand induces EMT through loss of androgen receptor—mediated suppression of the chemokine CCL2 (Eto et al., 2003; Tsai et al., 2018). Whether a direct cause and effect exists between androgen withdrawal—induced ceramide and EMT is not yet known, but a role of sphingolipids in EMT has recently been explored. A study using a 35-gene sphingolipid-EMT signature linked EMT to a significant reduction in ceramide and an increase in S1P (Meshcheryakova et al., 2016).

4.1.1 Reduced Ceramide Generation

CerS4 and CerS6 mRNA was found to be decreased with induction of EMT (Meshch-eryakova et al., 2016). CerS6 preferentially generates C_{16}-ceramide, a ceramide species that has shown to be preferentially increased during apoptosis (Tirodkar & Voelkel-Johnson, 2012). Reduced expression of CerS6 enhances therapy resistance in colon cancer cells and increases invasive potential in melanoma cells (Tang et al., 2016; White-Gilbertson et al., 2009). Both cell death resistance and elevated metastatic potential are characteristics of EMT. Edmond et al. investigated CerS6 expression in the context of EMT in the NCI-60 panel of cancer cells. Grouping cells into "epithelial" or "mesenchymal" phenotypes based on E-cadherin/vimentin ratios, they showed a significant decrease in CerS6 mRNA in the mesenchymal phenotype (Edmond et al., 2015). Analysis of CerS6 expression at the protein level in a molecularly defined model of EMT confirmed that reduced CerS6 expression coincides with the "cadherin switch" (Edmond et al., 2015). Thus a reduction in ceramide, perhaps specifically C_{16}-ceramide, may be associated with EMT.

4.1.2 Mechanisms of Ceramide Clearance

Ceramide levels can also be decreased through enhanced clearance. Enzymes that clear ceramide and have been associated with therapy resistance include glucosylceramide synthase (GCS) in breast cancer, ASAH1 in prostate cancer, and ASAH2 in colorectal cancer. GCS clears ceramide through adding glucose, which is the first step in the synthesis of glycosphingolipids. Glucosylceramide forms the backbone of hundreds of structurally different glycosphingolipids including gangliosides that are found on the outer leaflet of the plasma membrane where they interact with various other molecules and play a role in adhesion, proliferation, and cell recognition. Expression of GCS is elevated in metastatic breast cancer, and tamoxifen, which inhibits ceramide glycosylation, plays an important role as part of breast cancer therapy (Liu et al., 2011; Morad & Cabot, 2015).

The ceramidase product sphingosine can either be recycled to ceramide by CerS in the salvage pathway or serve as substrate for SKs. A shift in the EMT sphingolipid signature, including a strong increase in SK1, suggests that sphingosine would be preferentially metabolized to S1P (Meshcheryakova et al., 2016). Indeed cells with a mesenchymal phenotype have a twofold increase in S1P, and analysis of the sphingolipid-EMT signature could be used to differentiate the epithelial cancer stage from cells that had undergone EMT. Recent studies suggest a key role for S1P in the maintenance of CSCs (Lewis, Powell, & Pitson, 2017).

4.2 Sphingolipids and Stemness

The concept of CSCs is still debated within the scientific community, but it is generally accepted that a small population of malignant cells can lie dormant, even for many years. Unless this population is eradicated, the cancer has the potential to recur. There is now ample evidence that sphingolipid metabolism has multiple intersections with stem cell signaling and function in cancer (Lewis et al., 2017). The literature supports a model in which cancer therapy induces the generation of ceramide and cell death in a majority of cells but also induces a shift toward ceramide clearance and enhanced S1P signaling in a fraction of cells. S1P plays an important role in the tumor microenvironment that acts as a chemotactic factor and promoter of metastasis.

The rational combination of agents that interfere with ceramide clearance should in theory be able to improve therapy outcomes. Proof of principal for this hypothesis comes from combining tamoxifen, which has an off-target effect on GCS, with chemotherapy in breast cancer. In mice, targeting GCS can prevent the accumulation of breast CSCs during doxorubicin chemotherapy (Morad & Cabot, 2015). A similar approach may be of value for the treatment of prostate cancer in which ASAH1 is strongly upregulated. As discussed previously, genetic targeting or pharmacological inhibition of ASAH1 completely abrogated radiation treatment failure in a preclinical model of prostate cancer (Cheng et al., 2013). Prostate cancer is among the malignancies described to harbor CSCs, and the presence of CSCs in the PPC1 model used by Cheng et al. has been established (Liu et al., 2015). Although the effect on PPC1 CSCs was not directly tested, the lack of relapse suggests that all malignant cells were eradicated. To date, only two studies specifically link ASAH1 and CSCs. Lai et al. (2017) targeted ASAH1 using CRISPR-Cas9-mediated approach and demonstrated that the formation of cancer-initiating cells in melanoma cells could be prevented. In glioblastoma, S1P in the tumor microenvironment has been shown to fuel the survival of tumor stem cells (Marfia et al., 2014). Three different inhibitors of ASAH1 (carmofur, ARN14988, and NOE) were able to kill the tumor stem cell population in four different glioblastoma cell lines (Doan et al., 2017).

5. CONCLUSIONS

In prostate cancer, all standard therapeutic modalities such as androgen ablation therapy, radiation, and chemotherapy increase ceramide. ASAH1 is frequently upregulated in prostate cancer, and treatment-induced generation of ceramide has the potential to further increase ASAH1 expression, which could lead to cross-resistance. Effective therapy for advanced castration-resistant prostate cancer remains a major clinical challenge, and combination with agents that interfere with ASAH1 should be considered. Future therapies for cancer should consider not only promoting the elevation of intracellular ceramide to drive cell death but also combination with strategies that prevent ceramide clearance.

ACKNOWLEDGMENTS

Financial support for the authors was provided by grant P01 CA203628. James S. Norris has a financial interest in Sphingogene.

REFERENCES

Ader, I., Gstalder, C., Bouquerel, P., Golzio, M., Andrieu, G., Zalvidea, S., et al. (2015). *Oncotarget, 6*, 13803–13821.

Antoon, J. W., White, M. D., Meacham, W. D., Slaughter, E. M., Muir, S. E., Elliott, S., et al. (2010). *Endocrinology, 151*, 5124–5135.

Aurelio, L., Scullino, C. V., Pitman, M. R., Sexton, A., Oliver, V., Davies, L., et al. (2016). *Journal of Medicinal Chemistry, 59*, 965–984.

Bai, A., Szulc, Z. M., Bielawski, J., Mayroo, N., Liu, X., Norris, J., et al. (2009). *Bioorganic & Medicinal Chemistry, 17*, 1840–1848.

Bai, A., Szulc, Z. M., Bielawski, J., Pierce, J. S., Rembiesa, B., Terzieva, S., et al. (2014). *Bioorganic & Medicinal Chemistry, 22*, 6933–6944.

Beckham, T. H., Cheng, J. C., Lu, P., Marrison, S. T., Norris, J. S., & Liu, X. (2013). *PLoS One, 8*, e76593.

Beckham, T. H., Cheng, J. C., Lu, P., Shao, Y., Troyer, D., Lance, R., et al. (2013). *Oncogenesis, 2*, e49.

Beckham, T. H., Cheng, J. C., Marrison, S. T., Norris, J. S., & Liu, X. (2013). *Advances in Cancer Research, 117*, 1–36.

Beckham, T. H., Lu, P., Cheng, J. C., Zhao, D., Turner, L. S., Zhang, X., et al. (2012). *International Journal of Cancer, 131*, 2034–2043.

Bedia, C., Canals, D., Matabosch, X., Harrak, Y., Casas, J., Llebaria, A., et al. (2008). *Chemistry and Physics of Lipids, 156*, 33–40.

Beljanski, V., Knaak, C., & Smith, C. D. (2010). *The Journal of Pharmacology and Experimental Therapeutics, 333*, 454–464.

Bemis, D. L., Katz, A. E., & Buttyan, R. (2006). *Expert Opinion on Investigational Drugs, 15*, 1191–1200.

Benelli, R., Monteghirfo, S., Vene, R., Tosetti, F., & Ferrari, N. (2010). *Molecular Cancer, 9*, 142.

Bielawska, A., Linardic, C. M., & Hannun, Y. A. (1992). *Journal of Biological Chemistry, 267*, 18493–18497.

Britten, C. D., Garrett-Mayer, E., Chin, S. H., Shirai, K., Ogretmen, B., Bentz, T. A., et al. (2017). *Clinical Cancer Research, 23*, 4642–4650.

Brizuela, L., Martin, C., Jeannot, P., Ader, I., Gstalder, C., Andrieu, G., et al. (2014). *Molecular Oncology, 8*, 1181–1195.

Camacho, L., Meca-Cortes, O., Abad, J. L., Garcia, S., Rubio, N., Diaz, A., et al. (2013). *The Journal of Lipid Research, 54*, 1207–1220.

Camp, E. R., Patterson, L. D., Kester, M., & Voelkel-Johnson, C. (2017). *Cancer Biology & Therapy, 18*, 640–650.

Carter, L. G., D'Orazio, J. A., & Pearson, K. J. (2014). *Endocrine-Related Cancer, 21*, R209–R225.

Casasampere, M., Ordonez, Y. F., Pou, A., & Casas, J. (2016). *Chemistry and Physics of Lipids, 197*, 33–44.

Cheng, J. C., Bai, A., Beckham, T. H., Marrison, S. T., Yount, C. L., Young, K., et al. (2013). *Journal of Clinical Investigation, 123*, 4344–4358.

Cheung, E., Pinski, J., Dorff, T., Groshen, S., Quinn, D. I., Reynolds, C. P., et al. (2009). *Clinical Genitourinary Cancer, 7*, 43–50.

Chua, C. W., Chiu, Y. T., Yuen, H. F., Chan, K. W., Man, K., Wang, X., et al. (2009). *Clinical Cancer Research, 15*, 4322–4335.

Chua, C. W., Lee, D. T., Ling, M. T., Zhou, C., Man, K., Ho, J., et al. (2005). *International Journal of Cancer, 117*, 1039–1048.

Chumanevich, A. A., Poudyal, D., Cui, X., Davis, T., Wood, P. A., Smith, C. D., et al. (2010). *Carcinogenesis, 31*, 1787–1793.

Cingolani, F., Casasampere, M., Sanllehi, P., Casas, J., Bujons, J., & Fabrias, G. (2014). *The Journal of Lipid Research, 55*, 1711–1720.

Coant, N., Sakamoto, W., Mao, C., & Hannun, Y. A. (2017). *Advances in Biological Regulation, 63*, 122–131.

Dai, L., Smith, C. D., Foroozesh, M., Miele, L., & Qin, Z. (2017). *International Journal of Cancer.*

Doan, N. B., Alhajala, H., Al-Gizawiy, M. M., Mueller, W. M., Rand, S. D., Connelly, J. M., et al. (2017). *Oncotarget, 8*, 112662–112674.

Draper, J. M., Xia, Z., Smith, R. A., Zhuang, Y., Wang, W., & Smith, C. D. (2011). *Molecular Cancer Therapeutics, 10*, 2052–2061.

Edmond, V., Dufour, F., Poiroux, G., Shoji, K., Malleter, M., Fouque, A., et al. (2015). *Oncogene, 34*, 996–1005.

Eliyahu, E., Park, J. H., Shtraizent, N., He, X., & Schuchman, E. H. (2007). *The FASEB Journal: Official Publication of the Federation of American Societies for Experimental Biology, 21*, 1403–1409.

Elojeimy, S., Holman, D. H., Liu, X., El-Zawahry, A., Villani, M., Cheng, J. C., et al. (2006). *FEBS Letters, 580*, 4751–4756.

Eto, M., Bennouna, J., Hunter, O. C., Hershberger, P. A., Kanto, T., Johnson, C. S., et al. (2003). *The Prostate, 57*, 66–79.
Fabrias, G., Munoz-Olaya, J., Cingolani, F., Signorelli, P., Casas, J., Gagliostro, V., et al. (2012). *Progress in Lipid Research, 51*, 82–94.
Formelli, F., Clerici, M., Campa, T., Di Mauro, M. G., Magni, A., Mascotti, G., et al. (1993). *Journal of Clinical Oncology, 11*, 2036–2042.
French, K. J., Schrecengost, R. S., Lee, B. D., Zhuang, Y., Smith, S. N., Eberly, J. L., et al. (2003). *Cancer Research, 63*, 5962–5969.
French, K. J., Upson, J. J., Keller, S. N., Zhuang, Y., Yun, J. K., & Smith, C. D. (2006). *The Journal of Pharmacology and Experimental Therapeutics, 318*, 596–603.
French, K. J., Zhuang, Y., Maines, L. W., Gao, P., Wang, W., Beljanski, V., et al. (2010). *The Journal of Pharmacology and Experimental Therapeutics, 333*, 129–139.
Fu, P., Ebenezer, D. L., Ha, A. W., Suryadevara, V., Harijith, A., & Natarajan, V. (2018). *Journal of Cellular Biochemistry*.
Furuya, H., Tamashiro, P. M., Shimizu, Y., Iino, K., Peres, R., Chen, R., et al. (2017). *Carcinogenesis, 38*, 1218–1227.
Gagliostro, V., Casas, J., Caretti, A., Abad, J. L., Tagliavacca, L., Ghidoni, R., et al. (2012). *International Journal of Biochemistry & Cell Biology, 44*, 2135–2143.
Gao, P., Peterson, Y. K., Smith, R. A., & Smith, C. D. (2012). *PLoS One, 7*, e44543.
Gouaze-Andersson, V., Flowers, M., Karimi, R., Fabrias, G., Delgado, A., Casas, J., et al. (2011). *The Prostate, 71*, 1064–1073.
Gouaze-Andersson, V., Yu, J. Y., Kreitenberg, A. J., Bielawska, A., Giuliano, A. E., & Cabot, M. C. (2007). *Biochimica et Biophysica Acta, 1771*, 1407–1417.
Grasso, C. S., Wu, Y. M., Robinson, D. R., Cao, X., Dhanasekaran, S. M., Khan, A. P., et al. (2012). *Nature, 487*, 239–243.
Gupta, S. C., Patchva, S., & Aggarwal, B. B. (2013). *The AAPS Journal, 15*, 195–218.
Haddadi, N., Lin, Y., Simpson, A. M., Nassif, N. T., & McGowan, E. M. (2017). *International Journal of Molecular Sciences, 18*.
Hajj, C., & Haimovitz-Friedman, A. (2013). *Handbook of Experimental Pharmacology*, 115–130.
Hanker, L. C., Karn, T., Holtrich, U., Gatje, R., Rody, A., Heinrich, T., et al. (2013). *International Journal of Gynecological Pathology, 32*, 249–257.
Hilchie, A. L., Furlong, S. J., Sutton, K., Richardson, A., Robichaud, M. R., Giacomantonio, C. A., et al. (2010). *Nutrition and Cancer, 62*, 379–389.
Holman, D. H., Turner, L. S., El-Zawahry, A., Elojeimy, S., Liu, X., Bielawski, J., et al. (2008). *Cancer Chemotherapy and Pharmacology, 61*, 231–242.
Ide, H., Tokiwa, S., Sakamaki, K., Nishio, K., Isotani, S., Muto, S., et al. (2010). *The Prostate, 70*, 1127–1133.
Jasinski, M., Jasinska, L., & Ogrodowczyk, M. (2013). *Central European Journal of Urology, 66*, 144–149.
Kapitonov, D., Allegood, J. C., Mitchell, C., Hait, N. C., Almenara, J. A., Adams, J. K., et al. (2009). *Cancer Research, 69*, 6915–6923.
Koch, J., Gartner, S., Li, C. M., Quintern, L. E., Bernardo, K., Levran, O., et al. (1996). *Journal of Biological Chemistry, 271*, 33110–33115.
Korbelik, M., Banath, J., Zhang, W., Saw, K. M., Szulc, Z. M., Bielawska, A., et al. (2016). *International Journal of Cancer, 139*, 1372–1378.
Kus, G., Kabadere, S., Uyar, R., & Kutlu, H. M. (2015). *In Vitro Cellular & Developmental Biology Animal, 51*, 1056–1063.
Lahiri, S., Park, H., Laviad, E. L., Lu, X., Bittman, R., & Futerman, A. H. (2009). *Journal of Biological Chemistry, 284*, 16090–16098.
Lai, M., Realini, N., La Ferla, M., Passalacqua, I., Matteoli, G., Ganesan, A., et al. (2017). *Scientific Reports, 7*, 7411.

Laviad, E. L., Kelly, S., Merrill, A. H., Jr., & Futerman, A. H. (2012). *Journal of Biological Chemistry, 287*, 21025–21033.
Levy, M., & Futerman, A. H. (2010). *IUBMB Life, 62*, 347–356.
Lewis, C. L., Powell, J. A., & Pitson, S. M. (2017). In A. Pebay, & R. C. B. Wong (Eds.), *Lipidomics of stem cells.*
Lewis, C. S., Voelkel-Johnson, C., & Smith, C. D. (2016). *Oncotarget,* 7, 60181–60192.
Liu, F., Li, X., Lu, C., Bai, A., Bielawski, J., Bielawska, A., et al. (2016). *Oncotarget,* 7, 83907–83925.
Liu, X., Chen, X., Rycaj, K., Chao, H. P., Deng, Q., Jeter, C., et al. (2015). *Oncotarget, 6*, 23959–23986.
Liu, Y. Y., Patwardhan, G. A., Xie, P., Gu, X., Giuliano, A. E., & Cabot, M. C. (2011). *International Journal of Oncology, 39*, 425–431.
Lynch, K. R., Thorpe, S. B., & Santos, W. L. (2016). *Expert Opinion on Therapeutic Patents, 26*, 1409–1416.
Maeng, H. J., Song, J. H., Kim, G. T., Song, Y. J., Lee, K., Kim, J. Y., et al. (2017). *BMB Reports, 50*, 144–149.
Mahdy, A. E., Cheng, J. C., Li, J., Elojeimy, S., Meacham, W. D., Turner, L. S., et al. (2009). *Molecular Therapy: The Journal of the American Society of Gene Therapy, 17*, 430–438.
Malavaud, B., Pchejetski, D., Mazerolles, C., de Paiva, G. R., Calvet, C., Doumerc, N., et al. (2010). *European Journal of Cancer, 46*, 3417–3424.
Marfia, G., Campanella, R., Navone, S. E., Di Vito, C., Riccitelli, E., Hadi, L. A., et al. (2014). *Glia, 62*, 1968–1981.
McNair, C., Urbanucci, A., Comstock, C. E., Augello, M. A., Goodwin, J. F., Launchbury, R., et al. (2017). *Oncogene, 36*, 1655–1668.
Mehta, S., Blackinton, D., Omar, I., Kouttab, N., Myrick, D., Klostergaard, J., et al. (2000). *Cancer Chemotherapy and Pharmacology, 46*, 85–92.
Meshcheryakova, A., Svoboda, M., Tahir, A., Kofeler, H. C., Triebl, A., Mungenast, F., Heinze, G., Gerner, C., Zimmermann, P., Jaritz, M., & Mechtcheriakova, D. (2016). *Oncotarget,* 7, 22295–22323.
Mizutani, N., Inoue, M., Omori, Y., Ito, H., Tamiya-Koizumi, K., Takagi, A., et al. (2015). *Journal of Biochemistry, 158*, 309–319.
Morad, S. A., & Cabot, M. C. (2015). *Biochimica et Biophysica Acta, 1851*, 1134–1145.
Morad, S. A., Levin, J. C., Tan, S. F., Fox, T. E., Feith, D. J., & Cabot, M. C. (2013). *Biochimica et Biophysica Acta, 1831*, 1657–1664.
Morales, A., Paris, R., Villanueva, A., Llacuna, L., Garcia-Ruiz, C., & Fernandez-Checa, J. C. (2007). *Oncogene, 26*, 905–916.
Mullen, T. D., Hannun, Y. A., & Obeid, L. M. (2012). *The Biochemical Journal, 441*, 789–802.
Nagahashi, M., Ramachandran, S., Kim, E. Y., Allegood, J. C., Rashid, O. M., Yamada, A., et al. (2012). *Cancer Research, 72*, 726–735.
Nava, V. E., Cuvillier, O., Edsall, L. C., Kimura, K., Milstien, S., Gelmann, E. P., et al. (2000). *Cancer Research, 60*, 4468–4474.
Neubauer, H. A., Pham, D. H., Zebol, J. R., Moretti, P. A., Peterson, A. L., Leclercq, T. M., et al. (2016). *Oncotarget,* 7, 64886–64899.
Neubauer, H. A., & Pitson, S. M. (2013). *The FEBS Journal, 280*, 5317–5336.
Newton, J., Lima, S., Maceyka, M., & Spiegel, S. (2015). *Experimental Cell Research, 333*, 195–200.
Nieto, M. A., Huang, R. Y., Jackson, R. A., & Thiery, J. P. (2016). *Cell, 166*, 21–45.
Norris, J. S., Bielawska, A., Day, T., El-Zawahri, A., Elojeimy, S., Hannun, Y., et al. (2006). *Cancer Gene Therapy, 13*, 1045–1051.
Ogretmen, B. (2018). *Nature Reviews Cancer, 18*, 33–50.

Park, J. H., & Schuchman, E. H. (2006). *Biochimica et Biophysica Acta, 1758*, 2133–2138.
Park, J. W., Park, W. J., & Futerman, A. H. (2014). *Biochimica et Biophysica Acta, 1841*, 671–681.
Paugh, S. W., Paugh, B. S., Rahmani, M., Kapitonov, D., Almenara, J. A., Kordula, T., et al. (2008). *Blood, 112*, 1382–1391.
Pchejetski, D., Bohler, T., Brizuela, L., Sauer, L., Doumerc, N., Golzio, M., et al. (2010). *Cancer Research, 70*, 8651–8661.
Permpongkosol, S., Wang, J. D., Takahara, S., Matsumiya, K., Nonomura, N., Nishimura, K., et al. (2002). *International Journal of Cancer, 98*, 167–172.
Pienta, K. J., Esper, P. S., Zwas, F., Krzeminski, R., & Flaherty, L. E. (1997). *American Journal of Clinical Oncology, 20*, 36–39.
Pienta, K. J., Nguyen, N. M., & Lehr, J. E. (1993). *Cancer Research, 53*, 224–226.
Pollard, M., Luckert, P. H., & Sporn, M. B. (1991). *Cancer Research, 51*, 3610–3611.
Ponnusamy, S., Selvam, S. P., Mehrotra, S., Kawamori, T., Snider, A. J., Obeid, L. M., et al. (2012). *EMBO Molecular Medicine, 4*, 761–775.
Pruthi, R. S., Derksen, J. E., Moore, D., Carson, C. C., Grigson, G., Watkins, C., et al. (2006). *Clinical Cancer Research, 12*, 2172–2177.
Qin, Z., Dai, L., Trillo-Tinoco, J., Senkal, C., Wang, W., Reske, T., et al. (2014). *Molecular Cancer Therapeutics, 13*, 154–164.
Rahmaniyan, M., Curley, R. W., Jr., Obeid, L. M., Hannun, Y. A., & Kraveka, J. M. (2011). *Journal of Biological Chemistry, 286*, 24754–24764.
Realini, N., Palese, F., Pizzirani, D., Pontis, S., Basit, A., Bach, A., et al. (2016). *Journal of Biological Chemistry, 291*, 2422–2434.
Realini, N., Solorzano, C., Pagliuca, C., Pizzirani, D., Armirotti, A., Luciani, R., et al. (2013). *Scientific Reports, 3*, 1035.
Rex, K., Jeffries, S., Brown, M. L., Carlson, T., Coxon, A., Fajardo, F., et al. (2013). *PLoS One, 8*, e68328.
Rodriguez-Cuenca, S., Barbarroja, N., & Vidal-Puig, A. (2015). *Biochimica et Biophysica Acta, 1851*, 40–50.
Roh, J. L., Park, J. Y., Kim, E. H., & Jang, H. J. (2016). *European Journal of Cancer, 52*, 163–172.
Saad, A. F., Meacham, W. D., Bai, A., Anelli, V., Elojeimy, S., Mahdy, A. E., et al. (2007). *Cancer Biology & Therapy, 6*, 1455–1460.
Sabichi, A. L., Modiano, M. R., Lee, J. J., Peng, Y. M., Xu, M. J., Villar, H., et al. (2003). *Clinical Cancer Research, 9*, 2400–2405.
Saied, E. M., & Arenz, C. (2016). *Chemistry and Physics of Lipids, 197*, 60–68.
Sakamoto, J., Oba, K., Matsui, T., & Kobayashi, M. (2006). *Diseases of the Colon and Rectum, 49*, S82–S91.
Samsel, L., Zaidel, G., Drumgoole, H. M., Jelovac, D., Drachenberg, C., Rhee, J. G., et al. (2004). *The Prostate, 58*, 382–393.
Sanger, N., Ruckhaberle, E., Gyorffy, B., Engels, K., Heinrich, T., Fehm, T., et al. (2014). *Molecular Oncology*.
Schiffmann, S., Sandner, J., Schmidt, R., Birod, K., Wobst, I., Schmidt, H., et al. (2009). *The Journal of Lipid Research, 50*, 32–40.
Schiffmann, S., Ziebell, S., Sandner, J., Birod, K., Deckmann, K., Hartmann, D., et al. (2010). *Biochemical Pharmacology, 80*, 1632–1640.
Schnute, M. E., McReynolds, M. D., Kasten, T., Yates, M., Jerome, G., Rains, J. W., et al. (2012). *The Biochemical Journal, 444*, 79–88.
Schrecengost, R. S., Keller, S. N., Schiewer, M. J., Knudsen, K. E., & Smith, C. D. (2015). *Molecular Cancer Research, 13*, 1591–1601.
Seelan, R. S., Qian, C., Yokomizo, A., Bostwick, D. G., Smith, D. I., & Liu, W. (2000). *Genes Chromosomes & Cancer, 29*, 137–146.

Sharma, A. K. (2011). *Expert Opinion on Therapeutic Patents, 21*, 807–812.
Shin, K. O., Park, N. Y., Seo, C. H., Hong, S. P., Oh, K. W., Hong, J. T., et al. (2012). *Biomolecules & Therapeutics (Seoul), 20*, 470–476.
Siddique, M. M., Li, Y., Chaurasia, B., Kaddai, V. A., & Summers, S. A. (2015). *Journal of Biological Chemistry, 290*, 15371–15379.
Siegel, R. L., Miller, K. D., & Jemal, A. (2018). *CA: A Cancer Journal for Clinicians, 68*, 7–30.
Signorelli, P., Munoz-Olaya, J. M., Gagliostro, V., Casas, J., Ghidoni, R., & Fabrias, G. (2009). *Cancer Letters, 282*, 238–243.
Slawin, K., Kadmon, D., Park, S. H., Scardino, P. T., Anzano, M., Sporn, M. B., et al. (1993). *Cancer Research, 53*, 4461–4465.
Smith, M. R., Manola, J., Kaufman, D. S., Oh, W. K., Bubley, G. J., & Kantoff, P. W. (2006). *Journal of Clinical Oncology, 24*, 2723–2728.
Song, L., Xiong, H., Li, J., Liao, W., Wang, L., Wu, J., et al. (2011). *Clinical Cancer Research, 17*, 1839–1849.
Sooriakumaran, P., Coley, H. M., Fox, S. B., Macanas-Pirard, P., Lovell, D. P., Henderson, A., et al. (2009). *Anticancer Research, 29*, 1483–1488.
Sooriakumaran, P., Macanas-Pirard, P., Bucca, G., Henderson, A., Langley, S. E., Laing, R. W., et al. (2009). *Cancer Genomics Proteomics, 6*, 93–99.
Sugita, M., Willians, M., Dulaney, J. T., & Moser, H. W. (1975). *Biochimica et Biophysica Acta, 398*, 125–131.
Tang, Y., Cao, K., Wang, Q., Chen, J., Liu, R., Wang, S., et al. (2016). *Oncology Reports, 35*, 2907–2915.
Thaller, C., Shalev, M., Frolov, A., Eichele, G., Thompson, T. C., Williams, R. H., et al. (2000). *Journal of Clinical Oncology, 18*, 3804–3808.
Thangavel, S., Yoshitomi, T., Sakharkar, M. K., & Nagasaki, Y. (2015). *Journal of Controlled Release, 209*, 110–119.
Tirodkar, T. S., Lu, P., Bai, A., Scheffel, M. J., Gencer, S., Garrett-Mayer, E., et al. (2015). *Journal of Biological Chemistry, 290*, 13157–13167.
Tirodkar, T. S., & Voelkel-Johnson, C. (2012). *Experimental Oncology, 34*, 231–242.
Tonelli, F., Alossaimi, M., Williamson, L., Tate, R. J., Watson, D. G., Chan, E., et al. (2013). *British Journal of Pharmacology, 168*, 1497–1505.
Tsai, Y. C., Chen, W. Y., Abou-Kheir, W., Zeng, T., Yin, J. J., Bahmad, H., et al. (2018). *Biochimica et Biophysica Acta.*
Turner, L. S., Cheng, J. C., Beckham, T. H., Keane, T. E., Norris, J. S., & Liu, X. (2011). *Prostate Cancer and Prostatic Diseases, 14*, 30–37.
Vejselova, D., Kutlu, H. M., & Kus, G. (2016). *Cytotechnology, 68*, 2721–2728.
Venant, H., Rahmaniyan, M., Jones, E. E., Lu, P., Lilly, M. B., Garrett-Mayer, E., et al. (2015). *Molecular Cancer Therapeutics, 14*, 2744–2752.
Venkata, J. K., An, N., Stuart, R., Costa, L. J., Cai, H., Coker, W., et al. (2014). *Blood, 124*, 1915–1925.
Vethakanraj, H. S., Sesurajan, B. P., Padmanaban, V. P., Jayaprakasam, M., Murali, S., & Sekar, A. K. (2018). *Anti-cancer Drugs, 29*, 50–60.
Wang, J. D., Takahara, S., Nonomura, N., Ichimaru, N., Toki, K., Azuma, H., et al. (1999). *The Prostate, 40*, 50–55.
Watson, D. G., Tonelli, F., Alossaimi, M., Williamson, L., Chan, E., Gorshkova, I., et al. (2013). *Cellular Signalling, 25*, 1011–1017.
Wattenberg, B. W., Pitson, S. M., & Raben, D. M. (2006). *The Journal of Lipid Research, 47*, 1128–1139.
Wegner, M. S., Schiffmann, S., Parnham, M. J., Geisslinger, G., & Grosch, S. (2016). *Progress in Lipid Research, 63*, 93–119.

White-Gilbertson, S., Mullen, T., Senkal, C., Lu, P., Ogretmen, B., Obeid, L., et al. (2009). *Oncogene, 28*, 1132–1141.
Zheng, W., Kollmeyer, J., Symolon, H., Momin, A., Munter, E., Wang, E., et al. (2006). *Biochimica et Biophysica Acta, 1758*, 1864–1884.
Zhou, C., Ling, M. T., Kin-Wah Lee, T., Man, K., Wang, X., & Wong, Y. C. (2006). *Cancer Letters, 233*, 36–47.

CHAPTER ELEVEN

Targeting Sphingosine Kinases for the Treatment of Cancer

Clayton S. Lewis*, Christina Voelkel-Johnson[§] and Charles D. Smith[¶,1]

*Department of Internal Medicine, University of Cincinnati, Cincinnati, OH, United States
[§]Department of Microbiology and Immunology, Medical University of South Carolina, Charleston, SC, United States
[¶]Apogee Biotechnology Corporation, Hummelstown, PA, United States
[1]Corresponding author: E-mail: cdsmith@apogee-biotech.com

Contents

Abstract

Sphingosine kinases (SK1 and SK2) are key, druggable targets within the sphingolipid metabolism pathway that promote tumor growth and pathologic inflammation. A variety of isozyme-selective and dual inhibitors of SK1 and SK2 have been described in the literature, and at least one compound has reached clinical testing in cancer patients. In this chapter, we will review the rationale for targeting SKs and summarize the preclinical and emerging clinical data for ABC294640 as the first-in-class selective inhibitor of SK2.

1. POTENTIAL DRUG TARGETS IN THE SPHINGOLIPID METABOLISM PATHWAY

Sphingolipid metabolism is a key pathway in cancer biology (Hait & Maiti, 2017; Lee & Kolesnick, 2017; Newton, Lima, Maceyka, & Spiegel, 2015; Ogretmen, 2018; Siddique, Li, Chaurasia, Kaddai, & Summers, 2015),

Advances in Cancer Research, Volume 140
ISSN 0065-230X
https://doi.org/10.1016/bs.acr.2018.04.015

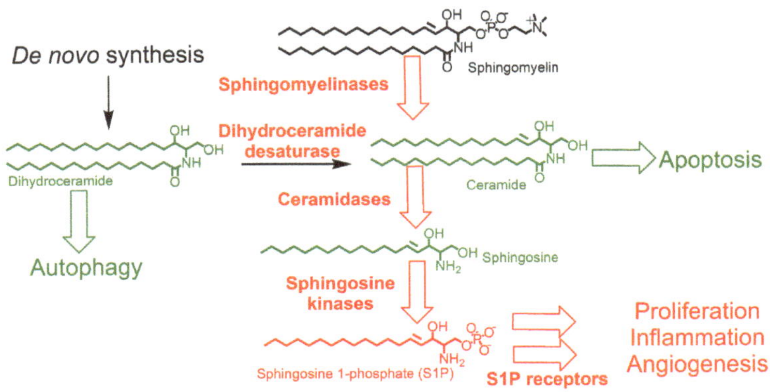

Figure 1 ***A simplified model of sphingolipid metabolism.*** Enzymes and processes that promote tumor growth are shown in red, whereas lipids and processes that inhibit tumor growth are shown in green. Proteins that are currently under consideration as targets for new anticancer drugs include sphingomyelinases, dihydroceramide desaturase (DES1), ceramidases, sphingosine kinases, and sphingosine 1-phosphate (S1P) receptors.

in which ceramides, sphingosine, and sphingosine 1-phosphate (S1P) regulate tumor cell death, proliferation, and drug resistance, as well as host angiogenesis, inflammation, and immunity. As indicated in Fig. 1, ceramide is produced by the hydrolysis of sphingomyelin in response to several growth stimulatory (e.g., growth factors and oncoproteins) and inflammatory (e.g., cytokines and radiation) signals. Alternately, ceramide can be synthesized de novo proceeding through the precursor dihydroceramide, which is converted into ceramide by dihydroceramide desaturase (DES1). Ceramide induces apoptosis in tumor cells without disrupting quiescent normal cells (Kolesnick & Fuks, 2003). Ceramide is hydrolyzed by ceramidases to produce sphingosine, which is phosphorylated by sphingosine kinases (SK1 and SK2) to produce S1P. S1P is dephosphorylated by S1P phosphatase 1 and 2 and degraded by S1P lyase, which cleaves S1P yielding phosphoethanolamine and hexadecenal. In addition to intracellular targets, S1P binds to and activates a family of G protein-coupled receptors, i.e., S1P-receptor 1–5 (S1PR1–5), which mediate at least some of the biological activities of this lipid.

Studies in many cancer cell lines indicate that S1P induces proliferation and protects against ceramide-induced apoptosis. Therefore, a critical balance, i.e., a ceramide/S1P rheostat, has been hypothesized to determine the fate of tumor cells (Spiegel & Milstien, 2002). Sphingolipids also regulate the sensitivities of tumor cells to anticancer drugs (Hendrich & Michalak, 2003;

Sietsma, Veldman, & Kok, 2001). For example, ceramide increases apoptosis induced by paclitaxel (Lucci, Han, Liu, Giuliano, & Cabot, 1999), etoposide (Perry & Kolesnick, 2003), and gemcitabine (Guillermet-Guibert et al., 2009; Modrak, Cardillo, Newsome, Goldenberg, & Gold, 2004). Therefore, inhibition of ceramidase or SK is expected to increase tumor chemosensitivity by elevating ceramide levels in the cells. In addition to their direct effects on tumor cells, SKs regulate deleterious inflammation from cytokines such as tumor necrosis factor-alpha (TNFα) and IL-6 (Aoki, Aoki, Ramanathan, Hait, & Takabe, 2016; Chiurchiu, Leuti, & Maccarrone, 2018; Gomez-Munoz et al., 2016; Pettus, Chalfant, & Hannun, 2004; Snider, Orr Gandy, & Obeid, 2010). In particular, S1P is critical for the activation of granulocytes that escalate inflammatory processes in many cancers, especially during chemo- or radiotherapy. Therefore, manipulation of sphingolipid metabolism to elevate ceramide levels and/or to reduce S1P production is an increasingly important approach to the treatment of hyperproliferative and inflammatory diseases, including cancers.

Among the enzymes and receptors that metabolize or interact with sphingolipids, most drug development efforts have focused on inhibition of ceramidases, SKs, or S1PRs. Recent reviews discuss the roles and pharmacology of ceramidases in detail (Coant, Sakamoto, Mao, & Hannun, 2017; Saied & Arenz, 2016; Tan, Pearson, Feith, & Loughran, 2017). Additionally, S1PR biology and a diverse set of compounds that modulate S1PR signaling have been well discussed in several recent reviews (Hait & Maiti, 2017; Huwiler & Zangemeister-Wittke, 2017; Juif, Kraehenbuehl & Dingemanse, 2016; Mao-Draayer, Sarazin, Fox, & Schiopu, 2017; Patmanathan, Wang, Yap, Herr, & Paterson, 2017; Pyne, El Buri, Adams, & Pyne, 2017). Similarly, a number of excellent recent publications describe the molecular properties and functions of SKs (Haddadi, Lin, Simpson, Nassif, & McGowan, 2017; Pyne, Adams, & Pyne, 2016; Siow & Wattenberg, 2011; Song, Zhou, & Sheng, 2017) and provide comprehensive reviews of SK inhibitors (Aurelio et al., 2016; Cao et al., 2018; Hatoum, Haddadi, Lin, Nassif, & McGowan, 2017; Lynch, Thorpe, & Santos, 2016; Pitman, Costabile, & Pitson, 2016; Plano, Amin, & Sharma, 2014; Pyne, Adams, & Pyne, 2017; Pyne, Bittman, & Pyne, 2011; Sanllehi, Abad, Casas, & Delgado, 2016; Santos & Lynch, 2015). This chapter will not duplicate these contributions but rather will discuss some of the key issues that spotlight the potential utility of inhibiting SK activity in cancer patients and describe the preclinical and early clinical data relating to ABC294640, which is the first SK2-targeted drug to reach clinical testing in cancer patients.

2. SPHINGOSINE KINASES AS TARGETS FOR ANTICANCER DRUGS

SKs are important new targets for anticancer drugs for several reasons. First, conversion of sphingosine into S1P is a key site for manipulation of the ceramide/S1P rheostat that regulates tumor cell proliferation and death. Second, the production of S1P in response to inflammatory cytokines is dependent on SK activity (Billich et al., 2005; Hanna et al., 2001; Maines et al., 2008; Mastrandrea, Sessanna, & Laychock, 2005; Nayak et al., 2010; Radeff-Huang et al., 2007; Snider et al., 2010; Xia et al., 1998), typically through nuclear factor-kappa B (NF-κB) (Xia et al., 1998). Because inflammation is a driving force in many types of cancer, suppression of patient systemic or local inflammation is expected to reduce tumor growth. Finally, sphingolipids are likely to play central roles in regulating tumor immunology, although this potential has been sparsely addressed to date (Molino, Tate, McKillop, & Medin, 2017; Rodriguez et al., 2016; Sensken, Nagarajan, Bode, & Graler, 2011). In particular, we hypothesize that inhibition of SK activity will alter signaling through the PD-1/PD-L1 pathway such that the host immune response to the cancer is improved. This postulate is based on the observations that PD-L1 expression in tumor cells is induced by IFN-γ (Cheng et al., 2007; Dondero et al., 2016; Iwai et al., 2002; Muhlbauer et al., 2006), and IFN-γ signaling is dependent on sphingolipid metabolism (Bajwa et al., 2017; Ottenlinger et al., 2016; Seo, Alexander, & Hahm, 2011; Wakita, Nishimura, Tokura, Furukawa, & Takigawa, 1996). Furthermore, Akt- (Abdelhamed, Ogura, Yokoyama, Saiki, & Hayakawa, 2016; Atefi et al., 2014; Dong et al., 2016; Lastwika et al., 2016; Song et al., 2013; Yang et al., 2017; Zhao et al., 2017), NF-κB- (Gowrishankar et al., 2015), and TNFα- (Wang et al., 2017) signaling induce PD-L1 expression, and all of these pathways are regulated by sphingolipid signaling. Therefore, manipulation of tumor immunology using SK inhibitors is an unexplored potential new approach to improved cancer therapy.

2.1 Sphingosine Kinase 1-Derived S1P

SK1 is activated by numerous growth factors and cytokines including epidermal growth factor (EGF), platelet-derived growth factor, vascular endothelial growth factor, and TNFα (Yester, Tizazu, Harikumar, & Kordula, 2011), and by cross-linking of the high affinity receptor for IgE (FcεRI). Receptor stimulation results in ERK1/2-mediated phosphorylation of Ser225 of SK1, thereby increasing its activity and allowing its translocation to the plasma membrane (Pitson et al., 2003). Calcium and integrin-binding protein 1 bind the activated SK1 and facilitate this translocation (Jarman, Moretti, Zebol, & Pitson, 2010) to the plasma membrane where it interacts with phosphatidylserine (Stahelin et al., 2005). The S1P produced there can be exported from the cell through ATP-binding cassette (ABC) transporters where it can act as an extracellular autocrine or paracrine signaling lipid. S1P binds five different G protein-coupled receptors named S1PR1−5.

S1PR1 couples exclusively to $G\alpha_i$ eliciting a number of responses. Importantly, signaling through S1PR1 is required for lymphocyte egress from lymph nodes and inhibition of SIPR1 function by FTY720 (fingolimod, Gilenya) via internalization sequesters lymphocytes in the lymphoid organs (Matloubian et al., 2004) and provides therapeutic benefit to patients with multiple sclerosis. Cross-linking of FcεRI in response to binding of IgE activates SK1 leading to autocrine signaling of S1P through both S1PR1 and S1PR2. Signaling through S1PR1 in response to FcεRI cross-linking leads to cytoskeleton rearrangement and migration toward the impetus (Jolly et al., 2004).

S1PR2 couples to multiple $G\alpha$ subunits, specifically i, q, and 12/13 (Windh et al., 1999). S1PR2 has frequently been associated with vascularization, and S1PR2 null mice exhibit disturbances in the stria vascularis, which forms the vasculature of the inner ear (Kono et al., 2007). S1PR2 was linked to pathological angiogenesis of the eye following ischemic injury by inhibiting the production of endothelial nitric oxide synthetase (eNOS) and promoting the expression of cyclooxygenase-2, thus driving the inflammatory process (Skoura et al., 2007). S1PR2 has been linked to inflammatory processes in other ways as well. Signaling through S1PR2 leads to mast cell degranulation, which leads to anaphylaxis and resultant pulmonary edema in mice. Pretreatment of mice with an S1PR2 antagonist, JTE-013, drastically attenuated anaphylaxis in response to stimuli due to an inhibition of mast cell degranulation (Oskeritzian et al., 2010). Conversely, S1PR2 also seems to be important for recovery from anaphylaxis. SK1 null mice and S1PR2 null mice have delayed clearance of histamine following anaphylaxis. In addition, they had severe hypotension during that time indicating that SK1-derived S1P signaling through S1PR2 impacts blood pressure control and recovery from anaphylaxis (Olivera et al., 2010).

Like S1PR2, S1PR3 couples to $G\alpha_i$, $G\alpha_q$, and $G\alpha_{12/13}$. Activation of S1PR3 in response to the glucocorticoid dexamethasone in human fibroblasts results in activation of the PI3K/Akt prosurvival pathway, as well as inhibiting apoptosis through increased expression of the antiapoptotic Bcl-2 member Bcl-xL (Nieuwenhuis, Luth, & Kleuser, 2010). Like the other S1P receptors, S1PR3 has been implicated in the inflammatory process. S1P is released from the cell in response to signaling through the protease-activated receptor 1, and this leads to amplification of proinflammatory signaling through S1PR3 (Niessen et al., 2008). S1P signaling through S1PR3 has also been shown to impact vasoconstriction by increasing intracellular Ca^{2+} concentrations and signaling through Rho kinase in human coronary artery smooth muscle cells (Murakami et al., 2010).

S1PR4 couples to $G\alpha_i$ and $G\alpha_{12/13}$ in response to S1P. Less is known about S1PR4, but unlike the other S1P receptors, it has been shown to have immunosuppressive effects. Specifically, signaling through S1PR4 was shown to inhibit the proliferation of T cells and their secretion of effector cytokines while stimulating the release of IL-10, a suppressive cytokine (Wang, Graeler, & Goetzl, 2005).

S1PR5, like S1PR4, couples to both $G\alpha_i$ and $G\alpha_{12/13}$ in response to S1P. S1P signaling through S1PR5 was shown to block the migration of oligodendrocyte precursor cells suggesting that S1P is important in cellular communication events involved in brain development (Novgorodov, El-Alwani, Bielawski, Obeid, & Gudz, 2007). Signaling through S1PR5 also inhibits the motility of esophageal cancer cells (Hu et al., 2010). Conversely, signaling through S1PR5 has been shown to be necessary for the egress of natural killer cells from the bone marrow (Mayol, Biajoux, Marvel, Balabanian, & Walzer, 2011). Furthermore, S1P signaling through S1PR5 has been shown to trigger endoplasmic reticulum (ER) stress in the PC3 prostate cancer cell line leading to autophagic cell death (Huang et al., 2014).

SK1-derived S1P also has distinct intracellular functions. TNFα binding to TNF receptor 1 recruits TNF-associated factor 2 (TRAF2) to the intracellular domain of the receptor. TRAF2 can then bind SK1. The S1P produced by SK1 acts as a cofactor for TRAF2 activating its E3 ubiquitin ligase function. TRAF2 mediates the K63 polyubiquitination of receptor interacting protein 1 (RIP1). The polyubiquitination of RIP1 serves as a scaffold for TGFβ-activated kinase 1 and the IκB kinase complex allowing for the activation of NF-κB (Spiegel & Milstien, 2011). Interestingly, TNFα has also been shown to induce the degradation of SK1 in a cathepsin B-dependent manner (Taha et al., 2005).

2.2 Sphingosine Kinase 2-Derived S1P

Like SK1, SK2 catalyzes the phosphorylation of sphingosine; however, SK2 can also phosphorylate two additional sphingoid bases: dihydrosphingosine and phytosphingosine (Gault, Obeid, & Hannun, 2010). Mice with genetic deletion of SK1 ($SK1^{-/-}$) or SK2 ($SK2^{-/-}$) develop normally indicating that the isozymes are able to compensate for one another to some degree. However, knocking out both genes simultaneously is embryonic lethal (Mizugishi et al., 2005). There are five reported splice variants of SK2, and SK2 has been reported to localize to several subcellular compartments including the cytosol (Okada et al., 2005), the ER (Maceyka et al., 2005), and the mitochondria (Strub et al., 2011). However, the primary localization seems to be the nucleus due to a nuclear localization sequence (NLS) within SK2 (Igarashi et al., 2003). Because of the differences in subcellular localization of SK1 and SK2, pools of S1P produced by SK1 and SK2 are likely to have distinct biological functions.

SK2 has been reported to paradoxically inhibit or promote cell growth. SK2 was first reported to suppress growth and promote apoptosis when it was overexpressed. This activity is attributed to a putative BH3 domain within SK2 that is absent in SK1 and which can bind with Bcl-xL preventing its suppression of proapoptotic Bcl-2 family members such as Bax and Bak (Liu et al., 2003). Mutation of a single leucine residue in the BH3 domain of SK2 drastically decreased apoptosis. Alternately, knockdown of SK2 also results in increased apoptosis in glioblastoma cells (Van Brocklyn et al., 2005), colorectal carcinoma cells, and MCF7 breast cancer cells (Sankala et al., 2007). These seemingly conflicting effects are likely attributed to artifacts from the overexpression of SK2 thereby disrupting the regulation of apoptosis and the subcellular distribution of S1P (Sankala et al., 2007).

Like SK1, SK2 is activated by phosphorylation by Erk1/2 (Hait, Bellamy, Milstien, Kordula, & Spiegel, 2007). In the mitochondria, SK2-derived S1P binds to prohibitin 2, which regulates the assembly of complex IV (cytochrome-*c* oxidase) of the electron transport chain (Strub et al., 2011). S1P derived from SK2 has also been associated with the formation of amyloid plaques in Alzheimer's disease. The major component of these plaques, amyloid-β, is a cleavage product of the amyloid-β precursor protein (APP). Cleavage of APP is carried out by the β-secretase β-site APP cleaving enzyme 1 (BACE1). SK2-derived S1P has been shown to bind the transmembrane domain of BACE1, which allosterically modifies the catalytic domain increasing its activity (Takasugi et al., 2011). Thus, overexpression of SK2 activity could contribute to Alzheimer's disease.

Caspase 1-mediated cleavage of SK2 in cells undergoing apoptosis allows for the translocation of a still catalytically active fragment of SK2 to the plasma membrane where it binds phosphatidylserine and mediates the phosphorylation of sphingosine (Weigert et al., 2010). As with SK1-derived S1P, this S1P can be exported from the cell and serve as a chemoattractant for monocytes and macrophages by interacting with S1P receptors (Gude et al., 2008).

The unique nuclear localization of SK2 provides it with a role in epigenetic regulation of gene expression. For example, SK2 has been shown to bind histone H3 and impact its acetylation in the MCF7 breast cancer cell line (Hait et al., 2009). siRNA knockdown of SK2 was reported to decrease the acetylation of H3—K9, H4—K5, and H2B—K12, and addition of S1P to isolated nuclei of siSK2 cells resulted in restoration of acetylation of these three residues. This same study showed that S1P and dihydro-S1P were able to inhibit histone deacetylases 1 and 2, and that siRNA knockdown of SK2 enhanced histone deacetylase (HDAC) activity. Additionally, the authors showed that activation of SK2 by phorbol 12-myristate 13-acetate (PMA, a potent activator of protein kinase C, upstream of ERK1/2-mediated SK2 phosphorylation) enhanced the colocalization of SK2 with HDAC1. PMA is a known inducer of the transcription of the cyclin-dependent kinase inhibitor p21 and the transcription factor c-fos. When combined with siRNA knockdown of SK2, PMA failed to induce the transcription of p21 and c-fos. This correlated with a reduction in acetylation of H3 associated with the promoters of p21 and c-fos (Hait et al., 2009). This agreed with this lab's earlier findings showing knockdown of SK2 prevented doxycycline-induced p21 expression (Sankala et al., 2007). These results are supported by Igarashi et al. who reported that overexpression of SK2 or SK1 with a fused NLS, but not native SK1, causes an inhibition of DNA synthesis (Igarashi et al., 2003). SK2-derived S1P has, thus, been demonstrated to impact gene transcription through regulation of HDAC activity.

Studies of the role of SK2 in carcinogenesis have also provided discordant results, with some research groups finding SK2 to be a tumor suppressor and others a mediator of tumor growth. For example, treatment of $SK2^{-/-}$ mice with the carcinogen azoxymethane and the inflammation-inducing toxin dextran sodium sulfate was reported to result in more numerous, larger, and more aggressive tumors than wild-type mice receiving the same treatments. This was reported to coincide with an increase in STAT3, NF-κB, and IL-6 expression in hematopoietic cells of the knockout mice, which drove the colitis and associated cancer. Conversely, SK2 has been described as oncogenic in acute lymphoblastic leukemia (ALL) because knockdown of *SK2* or pharmacological inhibition using the inhibitor ABC294640 (discussed below) both attenuated development of leukemia in ALL mouse models (Wallington-Beddoe et al., 2014). This was attributed to a reduction in the expression of c-Myc (and thereby c-Myc target genes) by way of reduced acetylation levels of H3, linking SK2 to the expression this important oncogene. This same study also showed the importance of SK2 in the development of ALL. When BCR/ABL-transduced B-cell progenitor cells were implanted in 29 sublethally irradiated mice, 22 developed ALL with a median survival of 42 days. When the same BCR/ABL translocation was introduced to $SK2^{-/-}$ cells, which were then implanted in 29 sublethally irradiated mice, only 16 developed ALL and these experienced a median survival of 58 days. Cells recovered from the $SK2^{-/-}$ mice had reduced c-Myc expression compared with their wild-type counterparts (Wallington-Beddoe et al., 2014). The expression of SK2 was linked to actin rearrangement in MCF7 cells with expression promoting a structural rearrangement of actin into membrane ruffles/lamellipodia indicative of a more migratory phenotype (Lim, Sun, Bittman, Pyne, & Pyne, 2011). SK2 expression was also shown to negatively correlate with disease-free survival and overall survival in nonsmall cell lung cancer (NSCLC) patients (Wang et al., 2014). Additionally, inhibition of SK2 has delayed tumor growth in mouse xenograft models of pancreatic, kidney, liver, and colon cancers (Beljanski, Knaak, Zhuang, & Smith, 2011; Beljanski, Lewis, & Smith, 2011; Chumanevich et al., 2010).

Overall, the propensity of data indicates that SK2 is a key mediator of enhanced growth of cancer. It is likely that studies using $SK2^{-/-}$ mice do not reflect the normal biology of the critical SK node and clearly do not mimic the pharmacologic inhibition of SK2 because of the absence of the protein, which in and of itself is proapoptotic due to its endogenous BH3 domain. Similarly, studies in which SK2 is overexpressed by transfection are compromised by inappropriate subcellular distribution of SK2 (and its S1P product) and excessive proapoptotic pressure from the elevated presence of BH3 domains.

3. SPHINGOSINE KINASE INHIBITORS

Because of the importance of S1P in growth signaling, a great deal of research has been done with the aim of inhibiting S1P signaling. Numerous therapeutic strategies have been developed that target either one or both SKs, as well as those that inhibit S1PR signaling. As indicated above, these SK inhibitors have been extensively reviewed elsewhere, so only a few compounds are highlighted here. The earliest example of an SK inhibitor is N,N-dimethylsphingosine (DMS), a naturally occurring N-methyl derivative of sphingosine that inhibits both SK1 and SK2 with K_is of ~16 and ~14 μM, respectively (Gao, Peterson, Smith, & Smith, 2012). Dimethylsphingosine inhibits SK activity in human platelets preventing platelet aggregation in response to exogenous S1P (Yatomi et al., 1996) and induces apoptosis in leukemic and colonic carcinoma cell lines (Jendiroba, Klostergaard, Keyhani, Pagliaro, & Freireich, 2002; Sweeney et al., 1996). Two additional sphingosine analogs, L-*threo*-dihydrosphingosine and N,N,N-trimethyl-sphingosine, have also been described as SK inhibitors; however, they also inhibit protein kinase C, which may be responsible for the observed phenotype (Canals & Hannun, 2013). The first nonlipid SK inhibitors were described in 2003, and the compound called SKI-II in particular has been widely used to probe SK involvement in many cell processes (French et al., 2003). SKI-II inhibits both isoforms with K_is of 16 μM (SK1) and 8 μM (SK2) (Gao et al., 2012), but it has also been shown to inhibit DES1 (Cingolani et al., 2014). Treatment of prostate cancer and breast cancer cells with SKI-II induces apoptosis in vitro (Antoon, Meacham, et al., 2011; Leroux et al., 2007) and in vivo (French et al., 2006). The compound PF-543, which is over 100-fold more selective for SK1 than SK2, has a K_i of 3.6 nM for SK1 (Schnute et al., 2012). Though it does not affect the growth of 1483 HNSCC cells, it was able to inhibit the formation of S1P in human whole blood ex vivo by >90% compared to control (Schnute et al., 2012). SK1-I is a specific competitive inhibitor of the ATP-binding site of SK1 with a K_i of 10 μM that has been shown to induce apoptosis in the U937 and Jurkat leukemic cell lines (Paugh et al., 2008). Treatment of these cells with SK1-I decreased ERK1/2 and Akt phosphorylation. Additionally, treatment of AML xenograft bearing mice with SK1-I reduced tumor growth (Paugh et al., 2008). SK1-I also decreased Akt activation and increased apoptosis in glioblastoma cell lines and slowed glioblastoma xenograft growth (Kapitonov et al., 2009). Importantly, in a murine model of breast cancer, treatment with SK1-I reduced metastasis and decreased overall tumor burden (Nagahashi et al., 2012), which was attributed to diminished angiogenesis caused by the suppression of circulating S1P. FTY720 is a structural analog of sphingosine. Following its phosphorylation by SK2, the resultant phospho-FTY720 can inhibit SK1 or exit the cell via ABC transporters to bind S1P receptors (except S1PR2). Binding results in the internalization of the receptor and prolonged receptor downregulation muting extracellular S1P signaling (Nagahashi et al., 2014). Prevention of S1P receptor signaling, mainly that of S1PR1, has a dramatic clinical impact in preventing the release of lymphocytes from lymphoid tissue thereby attenuating the autoimmune response responsible for multiple sclerosis (Allende, Dreier, Mandala, & Proia, 2004; Nagahashi et al., 2014). An analog, (R)-FTY720 methyl ester, was reported to have specific inhibitory activity against SK2, which prevented actin rearrangement in MCF7 cells in response to S1P thereby inhibiting motility (Lim et al., 2011).

3.1 Preclinical Development of ABC294640, the First-in-Class Inhibitor of SK2

Developed by Apogee Biotechnology Corporation, 3-(4-chlorophenyl)-adamantane-1-carboxylic acid (pyridin-4-ylmethyl)amide (ABC294640, Fig. 2) is an inhibitor of SK2 that is competitive with respect to sphingosine and therefore does not have off-target effects on protein kinases

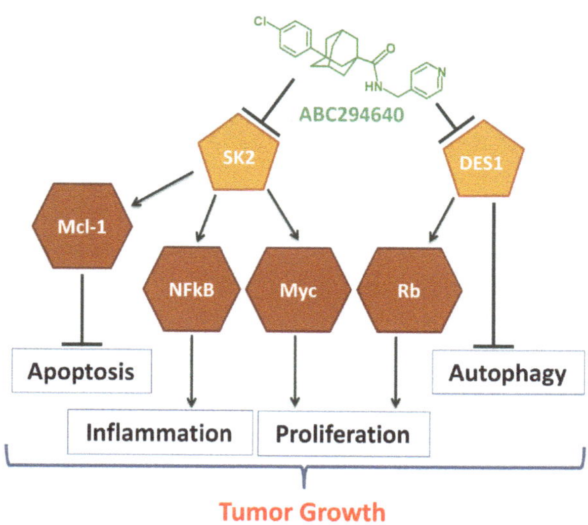

Figure 2 ***Mechanism for Suppression of Tumor Growth by ABC294640.*** ABC294640 has similar potency toward sphingosine kinase (SK2) and dihydroceramide desaturase (DES1). Inhibition of SK2 activity causes downregulation of the expression of myeloid cell leukemia 1 (Mcl-1) and c-Myc and suppresses the activation of nuclear factor-kappa B (NF-κB), thereby promoting apoptosis and inhibiting proliferation and inflammation. In parallel, inhibition of DES1 by ABC294640 promotes autophagy and promotes Rb phosphorylation resulting in suppression of proliferation.

(French et al., 2010; Gao et al., 2012). In tissue culture, ABC294640 reduces S1P levels (French et al., 2010; Gao et al., 2012; Maines, French, Wolpert, Antonetti, & Smith, 2006; Maines et al., 2008), suppresses the proliferation of many tumor cell lines, and inhibits tumor cell migration concomitant with loss of microfilaments. ABC294640 depletes S1P and elevates ceramide in tumor cells, suppresses signaling through pERK, pAkt, and NF-κB, and promotes autophagy and/or apoptosis in a broad range of cancer cells (Antoon, White, et al., 2011; Beljanski, Knaak, & Smith, 2010; Ding et al., 2016; French et al., 2010; Gao et al., 2012; Qin et al., 2014; Schrecengost, Keller, Schiewer, Knudsen, & Smith, 2015). Importantly, ABC294640 downregulates the expression of c-Myc in a variety of tumor cell lines (Lewis, Voelkel-Johnson, & Smith, 2016; Schrecengost et al., 2015; Venant et al., 2015; Venkata et al., 2014; Wallington-Beddoe et al., 2014) and also reduces androgen receptor expression in prostate cancer cells (Schrecengost et al., 2015; Venant et al., 2015). In addition to SK2, ABC294640 inhibits DES1, which accounts for the marked increases in dihydroceramides in cells treated with the drug (French et al., 2010; Venant et al., 2015). ABC294640 is synergistically cytotoxic with gemcitabine toward pancreatic cancer cell lines and results in decreased expression of both RRM2 and MYC, decreased S780 phosphorylation of Rb, and increased acetylation of H3-K9 and p21 (Lewis et al., 2016). The reduction in Rb phosphorylation is consistent with studies by Kraveka et al. who demonstrated that inhibition of DES1 is associated with reduction of Rb phosphorylation and suppression of cell cycle progression (Kraveka et al., 2007). Fig. 2 describes the current model in which ABC294640 dually inhibits SK2 and DES1 resulting in inhibition of multiple signaling pathways that promote tumor growth.

As summarized below, ABC294640 has therapeutic activity in diverse mouse tumor models both alone and in combination with other anticancer drugs (Antoon, White, Driver, Burow, & Beckman, 2012; Antoon et al., 2010; Antoon, White, et al., 2011; Beljanski et al., 2010; Beljanski, Knaak, et al., 2011; Beljanski, Lewis, et al., 2011; Dai et al., 2017; Dai, Smith, Foroozesh, Miele, & Qin, 2018; French et al., 2010; Qin et al., 2014; Schrecengost et al., 2015; Venant et al., 2015; Venkata et al., 2014; Wallington-Beddoe et al., 2014; Xu et al., 2018; Xun et al., 2015; Zhou, Chen, & Yu, 2018), as well as a number of rodent inflammation models (Fitzpatrick, Green, Frauenhoffer, et al., 2011; Fitzpatrick, Green, Maines, & Smith, 2011; Liu et al., 2012; Maines et al., 2008; Maines, Fitzpatrick, Green, Zhuang, & Smith, 2010; Maines et al., 2006; Poti et al., 2012;

Shi et al., 2012) not discussed here. Additionally, ABC294640 diminished tumor incidence and multiplicity in the azoxymethane/dextran sulfate sodium model of colon carcinogenesis (Chumanevich et al., 2010). Antitumor activity is associated with accumulation of ABC294640 in the tumors, reduction of tumor S1P levels, and induction of apoptosis (French et al., 2010). Intratumoral concentrations of ABC294640 can exceed the IC_{50} for tumor cell cytotoxicity by as much as 25-fold (French et al., 2010), and ABC294640 reduces plasma S1P levels ~50% at therapeutically efficacious doses (Beljanski, Lewis, et al., 2011). Preclinical studies demonstrated an excellent oral bioavailability and safety profile for ABC294640, with no hematologic or major organ toxicity (French et al., 2010).

The in vivo anticancer activity of ABC294640 was first shown by inhibition of tumor growth in a syngeneic JC (breast cancer) xenograft model in BALB/c mice (French et al., 2010). In additional studies in breast cancer models, ABC294640 has been shown to bind to the antagonist ligand-binding domain of the estrogen receptor thereby inhibiting the growth of the estrogen receptor positive breast cancer cell line MCF7 both in vitro and in vivo (Antoon et al., 2010). ABC294640 was also effective at inducing apoptotic cell death in triple-negative MCF7 cells (MCF7-TN-R) by diminishing prosurvival signaling through the NF-κB pathway (Antoon, White, et al., 2011). In vitro levels of apoptosis were further increased by addition of either etoposide or doxorubicin after pretreatment with ABC294640. Additionally, Ki67 staining was decreased in MCF7-TN-R orthotopic tumors treated with ABC294640 when compared to control. However, ABC294640 is ineffective against the luminal, endocrine-resistant MDA-MB-361 breast cancer cell line, although the drug did reduce Bcl-2 in MCF7 tumors (Antoon et al., 2012).

A growing body of evidence indicates that ABC294640 may be particularly effective for the treatment of prostate cancer. Initially, ABC294640 was also shown to inhibit proliferation in androgen-resistant PC-3 and LNCaP prostate cancer cell lines (Gestaut, Antoon, Burow, & Beckman, 2014). Schrecengost et al. (2015) then showed that ABC294640 abrogates signaling pathways requisite for prostate cancer growth and proliferation. Of particular importance, ABC294640 treatment of early-stage and advanced prostate cancer models caused downregulation of c-Myc and androgen receptors (both full-length in LNCaP cells and the splice variant in 22Rv1 cells) expression and activity. This corresponded with significant inhibition of growth, proliferation, and cell-cycle progression. Finally, oral administration of ABC294640 was found to dramatically impede xenograft tumor growth. Studies by Venant et al. provided further confirmation of activity toward prostate cancer cells in association with decreased expression of both c-Myc and androgen receptors (Venant et al., 2015). Furthermore, that study confirmed that ABC294640 inhibits DES1 activity resulting in marked increases in dihydroceramides both in tissue culture and in xenografts. Studies by McNaughton et al. examined the effects of SKI-II and ABC294640 in androgen-independent LNCaP-AI prostate cancer cells and demonstrated that these compounds suppress both SK1 and DES1 by inducing their proteolysis (McNaughton, Pitman, Pitson, Pyne, & Pyne, 2016). Overall, the combined effects of ABC294640 on SK1, SK2, DES1, and androgen receptors in androgen-sensitive and castrate-resistant prostate cancer cells and xenografts build a strong rationale for examining the therapeutic effects of ABC294640 in prostate cancer patients.

ABC294640 has been shown to promote autophagic cell death in xenografts of A-498 (kidney cancer) cells in SCID mice as evidenced by increases in Beclin-1 and the cleavage of the microtubule-associated proteins 1A/1B light chain 3A (LC3). This same study demonstrated a decrease of MAPK and Akt pathway activation in response to ABC294640 in vivo (Beljanski et al., 2010). Similarly, ABC294640 caused pronounced apoptosis, cell-cycle arrest, and inhibition of NSCLC growth in vitro and in vivo (Dai et al., 2018). This is consistent with a previous report that ABC294640 enhances the antitumor effects of TRAIL in NSCLC cells, likely by enhancing the translocation of death receptor 4/5 to the cell membrane (Yang et al., 2015). Leili et al. demonstrated that ABC294640 enhances doxorubicin-stimulated apoptosis in NSCLC cells, and that this is associated with suppression of the expression of survivin (Leili, Nasser, Nadereh, Siavoush, & Pouran, 2018). Also using lung cancer cells in vitro, Guan et al. demonstrated that apoptosis induced by ABC294640 can be suppressed by overexpression of glucosylceramide synthase (GCS) or enhanced by the GCS inhibitor PDMP (Guan, Liu, Yan, & Zhou, 2016). Studies by Xun et al. using transformed and primary colorectal cancer cells demonstrated that these cells are also sensitive to growth inhibition by ABC294640 that could be amplified by combination with 5-fluorouracil and cisplatin and attenuated by cotreatment with a JNK inhibitor (Xun et al., 2015). Similarly, ABC294640 promoted apoptosis in skin squamous cell carcinoma cells in vitro and inhibited tumor growth in vivo associated with inhibition of Akt and JNK signaling (Zhou et al., 2018).

The potential use of ABC294640 in treating cancers of hematopoietic origin has also been demonstrated. Treatment of primary effusion lymphoma

(PEL) cells positive for Kaposi's sarcoma-associated herpesvirus (KSHV) (BCBL-1 cell line) with ABC294640 induced dose-dependent apoptosis that corresponded with decreases in phosphorylation of ERK1/2, Akt, and NF-κB (Qin et al., 2014). When these cells were injected into NOD/SCID mice, ABC294640 treatment delayed disease progression as indicated by spleen size, weight gain, and ascites volume. ABC294640 treatment vastly attenuated the pathogenesis of $KSHV^+$ PEL (Qin et al., 2014). Mechanistically, elevation of dihydroceramides was shown to induce apoptosis in KSHV-infected PEL cells through activation of KSHV lytic gene expression (Dai et al., 2015). Finally, impressive activity of ABC294640 toward virally induced cancers both in vitro and in vivo was also demonstrated by Dai, et al. who found that KSHV long-term-infected immortalized endothelial cells are efficiently killed by ABC294640, whereas noninfected cells were not (Dai et al., 2017).

ABC294640 has anticancer activity in models of multiple myeloma (MM). Treatment of both immortalized MM cell lines and freshly isolated $CD138^+$ myeloma cells from myeloma patients resulted in inhibition of proliferation (Venkata et al., 2014). ABC294640 increased caspase 3 activation and caspase 9 cleavage in OPM1 MM cells indicative of apoptotic cell death. Treatment of MM cell lines with ABC294640 also increased mRNA and protein expression of the proapoptotic Bcl-2 family member Noxa, which is regulated by p53. The same study additionally showed that ABC294640 promotes proteosomal degradation of myeloid cell leukemia 1 (Mcl-1), a Bcl-2 family protein overexpressed in the majority of MM patients. Thus, ABC294640 shifts the balance of these apoptotic mediators in the direction of apoptosis. ABC294640 also enhanced the proteosomal degradation of c-Myc in the MM cells. Additionally, the authors demonstrated that ABC294640 induced apoptosis of MM cells even in the presence of bone marrow stromal cells, which support the growth of MM cells. These findings prompted in vivo studies using a luciferase expressing MM.1S cell line xenograft where ABC294640 suppressed tumor growth (Venkata et al., 2014).

ABC294640 has also been evaluated in preclinical testing against ALL (Wallington-Beddoe et al., 2014). ABC294640 inhibited proliferation and induced cell death in ALL cell lines and ALL patient samples but, importantly, not bone marrow mononuclear cells. Cell death in this instance was attributed to a mixed etiology, being caspase-independent but not completely autophagic. Next, a gene expression profile of the impact of ABC294640 on gene expression was compiled, which revealed an inhibition of c-Myc and c-Myc-regulated genes. *MYC* downregulation in response to ABC294640 treatment was confirmed by qRT-PCR. This same study showed that inhibition of SK2 by ABC294640 prevented S1P-mediated inhibition of HDAC activity resulting in a reduction in the association of c-Myc with histone 3 acetylated on lysine 9 and thereby a reduction in c-Myc expression. Xenografts of ALL in NOD/SCID mice treated with ABC294640 had significantly less disease than vehicle control following 21 days of treatment (Wallington-Beddoe et al., 2014).

Combination of ABC294640 with other anticancer drugs has shown promising results. Combination of ABC294640 with sorafenib resulted in synergistic cell killing in A-498 kidney carcinoma and BxPC-3 pancreatic adenocarcinoma cells by means of apoptosis, and the combination attenuated the growth of xenografts of each cell line in SCID mice to a greater degree than either drug alone (Beljanski, Knaak, et al., 2011). This same combination proved effective in two hepatocellular carcinoma xenografts with the mode of cell death being autophagic (Beljanski, Lewis, et al., 2011). As with the A-498 and BxPC-3 cells, ABC294640 decreased the levels of p-Erk1/2 provoking the idea that ABC294640-mediated cell death could come as a result of diminished MAPK signaling. Similarly, combination of ABC294640 with sorafenib caused synergistic apoptosis of cholangiocarcinoma cells, and this was associated with marked suppression of STAT3 phosphorylation (Ding et al., 2016). Combination of ABC294640 with the microtubule stabilizer paclitaxel was shown to greatly increase caspase-9 signaling and apoptosis when compared with either drug alone (White, Chan, Antoon, & Beckman, 2013). ABC294640 has also been combined with doxorubicin, vincristine, imatinib, and bortezomib in ALL cells (Wallington-Beddoe et al., 2014). Only additive effects were observed with the DNA intercalating agent doxorubicin and the mitosis inhibitor vincristine. However, synergistic cell death was produced with the proteasome inhibitor bortezomib and the multityrosine kinase inhibitor (i.e., Bcr-Abl inhibitor) imatinib when either was combined with ABC294640 (Wallington-Beddoe et al., 2014). ABC294640 treatment of cervical carcinoma cells promoted apoptosis that was amplified by combination with the BCL-2 inhibitor GDC-0199 both in vitro and in vivo (Xu et al., 2018).

3.2 Phase I Clinical Trial of ABC294640

A phase I clinical trial of ABC294640 in patients with advanced solid tumors has been completed (Britten et al., 2017). ABC294640 was administered orally on a continuous schedule, with 28 days constituting a Cycle. Tumors were reimaged every two Cycles (8 weeks), and patients were allowed to continue receiving the drug if there was no disease progression by RECIST tumor imaging criteria. Twenty-one patients received the drug in this phase I trial. The first patient on study at 250 mg qd developed grade 4 hyperglycemia in the setting of rapidly progressing pancreatic cancer; however, no other patients experienced possibly drug-related hyperglycemia. Among the four patients enrolled at 750 mg bid, one ovarian cancer patient had dose-limiting grade 3 nausea and vomiting, and two patients were unable to complete Cycle 1 due to diverse possibly drug-related toxicities. One patient at the expanded 500 mg dose level was dose reduced at Cycle 3 due to grade 1 adverse events. The other five patients at 500 mg bid tolerated the drug well, and the 500 mg bid dose level was established as the recommended phase II dose. Across all dose levels, the most common drug-related toxicities were nausea, fatigue, vomiting, and diarrhea. In addition, 2 patients experienced psychiatric disorders including agitation/anxiety, mood changes, and/or hallucinations; and 5 patients experienced grade 1−2 nervous system disorders, including dizziness, dysarthria, dysgeusia, dysesthesia, memory loss, muscle spasms, paresthesia, somnolence, and/or spasticity that resolved on discontinuation of ABC294640.

Among the patients evaluable by RECIST, 1 (6%, a patient with refractory cholangiocarcinoma) had a Partial Response, 6 (38%) had Stable Disease, and 9 (56%) had Progressive Disease as their best response. Subjects included a patient with advanced hepatocellular carcinoma who lived for 10 months after completing 4 cycles; a patient with recurrent metastatic bladder cancer who received 12 cycles; and a patient with advanced cholangiocarcinoma who received 18 cycles.

Pharmacokinetic (PK) profiling was conducted on Days 1 and 28 of Cycle 1, and data from patients treated with 500 mg of ABC294640 are shown in Fig. 3. After oral dosing, plasma levels of ABC294640 typically peaked at 1−2 h, and then declined with a halftime of clearance of ~4 h. The peak plasma concentrations (C_{max}) and AUCs within a cohort were generally similar. Notably, the C_{max} and AUC values for most patients were not significantly different between Day 1 and Day 28, indicating a lack of metabolic adaptation to the drug. Importantly, C_{max} levels of ABC294640 in patients receiving 500 mg were in the range expected to have therapeutic activity. Specifically, antitumor activity in mouse models occurs at ABC294640 doses of 25−50 mg/kg, which results in plasma C_{max} levels of ~3.5 μg/mL (French et al., 2010). Patients receiving a

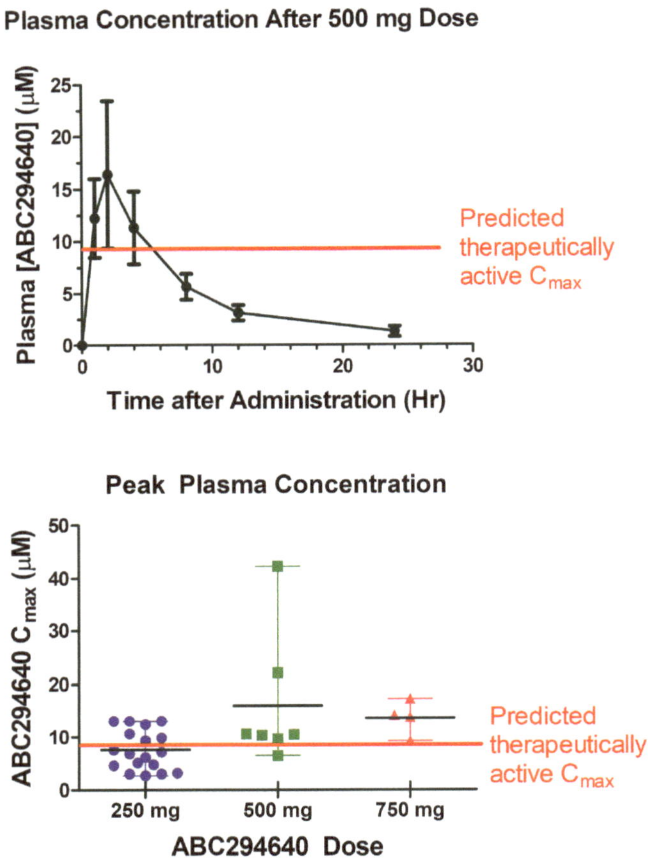

Figure 3 ***Pharmacokinetics of ABC294640 in humans.*** Top Panel. Plasma was isolated from patients given 500 mg of ABC294640 on either Day 1 or Day 28 of Cycle 1 and levels of ABC294640 were quantified using a GLP LC/MS method. Values represent the mean $\pm$ SEM for 7 data sets. Antitumor activity in mouse models occurs at ABC294640 doses $\geq$25 mg/kg, which provides plasma C_{max} levels of 9 μM. The horizontal line indicates the C_{max} predicted to have antitumor activity based on the mouse tumor models. Bottom Panel. C_{max} values are shown for patient samples collected on either Day 1 or Day 28 (there is no change in C_{max} or AUC in this period). The horizontal line indicates the Cmax predicted to have antitumor activity based on the mouse tumor models.

500 mg dose of ABC294640 had C_{max} levels averaging 16.4 μM (Fig. 3), with 9 of 12 profiles exceeding the threshold of 9 μM. Patients given 250 mg of ABC294640 exceeded the C_{max} threshold approximately 50% of the time, whereas all patients receiving 750 mg of ABC294640 reached this level. Because the $t_{1/2}$s for clearance are equal in mice and humans,

the drug exposure in patients at 500 mg and higher dose levels is expected to be sufficient to inhibit SK2 in the tumors.

The most direct pharmacodynamic biomarkers for inhibition of SK2 and DES1 by ABC294640 are S1P and dihydroceramide levels in the tumor. However, tumor biopsies were not obtained in the phase I clinical trial, so sphingolipids in the plasma were measured as a biomarker for ABC294640. In the patients, ABC294640 treatment caused rapid decreases in plasma S1P levels with a parallel increase in plasma dihydroC_{16}-ceramide levels (Fig. 4). In general, plasma S1P reached a minimum approximately 12 h after ABC294640 treatment and recovered to baseline by 24 h. This is consistent with the clearance profile of the drug and supports the BID dosing schedule going forward into phase II clinical trials. The maximum S1P decreases for the 250, 500, and 750 mg cohorts were $51 \pm 6\%$ ($n = 10$), $46 \pm 9\%$ ($n = 5$), and $54 \pm 14\%$ ($n = 2$), respectively, indicating that ABC294640 exposures attained with the 250 mg dose are sufficient to optimally inhibit S1P generation in the patients.

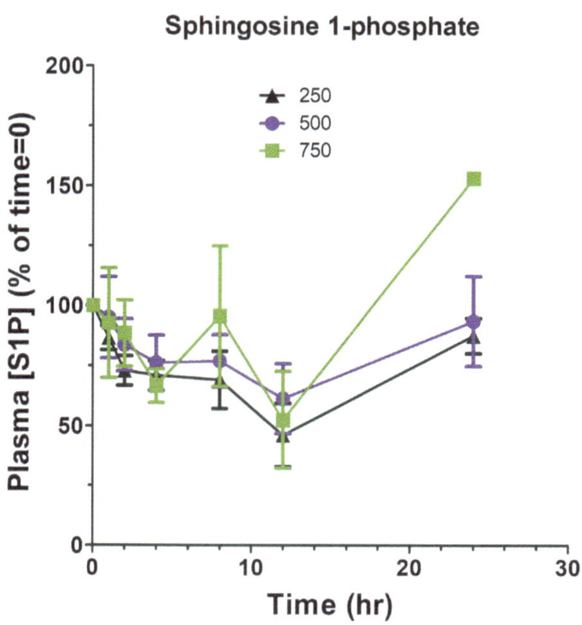

Figure 4 ***Effect of ABC294640 treatment on plasma sphingosine 1-phosphate (S1P) levels.*** S1P concentrations were normalized to the Time 0 levels for individual patients receiving 250 mg (▲), 500 mg (●), or 750 mg (■) of ABC294640.

4. CONCLUSIONS

Continuing research on the sphingolipid metabolism pathway is providing increasing support for targeted disruption of key enzymes in this pathway for the treatment of cancer and inflammatory diseases. Important differences in the biology and pharmacology of SK1 and SK2 have been revealed by insightful basic and translational research and provide rationale for continuing the movement of novel inhibitors into clinical testing. Ultimately, the efficacies and safety of these new drugs will reveal whether or not targeting sphingolipid metabolism provides therapeutic benefit to patients. Clinical testing of ABC294640 is continuing with trials in patients having MM, cholangiocarcinoma, or hepatocellular carcinoma currently underway.

ACKNOWLEDGMENTS

Financial support for the authors was provided by grant P01 CA203628.

REFERENCES

Abdelhamed, S., Ogura, K., Yokoyama, S., Saiki, I., & Hayakawa, Y. (2016). AKT-STAT3 pathway as a downstream target of EGFR signaling to regulate PD-L1 expression on NSCLC cells. *Journal of Cancer, 7*(12), 1579–1586. https://doi.org/10.7150/jca.14713.

Allende, M. L., Dreier, J. L., Mandala, S., & Proia, R. L. (2004). Expression of the sphingosine 1-phosphate receptor, S1P1, on T-cells controls thymic emigration. *Journal of Biological Chemistry, 279*(15), 15396–15401. https://doi.org/10.1074/jbc.M314291200.

Antoon, J. W., Meacham, W. D., Bratton, M. R., Slaughter, E. M., Rhodes, L. V., Ashe, H. B., et al. (2011). Pharmacological inhibition of sphingosine kinase isoforms alters estrogen receptor signaling in human breast cancer. *Journal of Molecular Endocrinology, 46*(3), 205–216. https://doi.org/10.1530/JME-10-0116.

Antoon, J. W., White, M. D., Driver, J. L., Burow, M. E., & Beckman, B. S. (2012). Sphingosine kinase isoforms as a therapeutic target in endocrine therapy resistant luminal and basal-A breast cancer. *Experimental Biology and Medicine (Maywood), 237*(7), 832–844. https://doi.org/10.1258/ebm.2012.012028.

Antoon, J. W., White, M. D., Meacham, W. D., Slaughter, E. M., Muir, S. E., Elliott, S., et al. (2010). Antiestrogenic effects of the novel sphingosine kinase-2 inhibitor ABC294640. *Endocrinology, 151*(11), 5124–5135. https://doi.org/10.1210/en.2010-0420.

Antoon, J. W., White, M. D., Slaughter, E. M., Driver, J. L., Khalili, H. S., Elliott, S., et al. (2011). Targeting NFkB mediated breast cancer chemoresistance through selective inhibition of sphingosine kinase-2. *Cancer Biology & Therapy, 11*(7), 678–689.

Aoki, M., Aoki, H., Ramanathan, R., Hait, N. C., & Takabe, K. (2016). Sphingosine-1-phosphate signaling in immune cells and inflammation: Roles and therapeutic potential. *Mediators of Inflammation, 2016*, 8606878. https://doi.org/10.1155/2016/8606878.

Atefi, M., Avramis, E., Lassen, A., Wong, D. J., Robert, L., Foulad, D., et al. (2014). Effects of MAPK and PI3K pathways on PD-L1 expression in melanoma. *Clinical Cancer Research, 20*(13), 3446–3457. https://doi.org/10.1158/1078-0432.CCR-13-2797.

Aurelio, L., Scullino, C. V., Pitman, M. R., Sexton, A., Oliver, V., Davies, L., et al. (2016). From sphingosine kinase to dihydroceramide desaturase: A structure-activity relationship (SAR) study of the enzyme inhibitory and anticancer activity of 4-((4-(4-Chlorophenyl) thiazol-2-yl)amino)phenol (SKI-II). *Journal of Medicinal Chemistry, 59*(3), 965–984. https://doi.org/10.1021/acs.jmedchem.5b01439.

Bajwa, A., Huang, L., Kurmaeva, E., Ye, H., Dondeti, K. R., Chroscicki, P., et al. (2017). Sphingosine kinase 2 deficiency attenuates kidney fibrosis via IFN-gamma. *Journal of the American Society of Nephrology: JASN, 28*(4), 1145–1161. https://doi.org/10.1681/ASN.2016030306.

Beljanski, V., Knaak, C., & Smith, C. D. (2010). A novel sphingosine kinase inhibitor induces autophagy in tumor cells. *The Journal of Pharmacology and Experimental Therapeutics, 333*(2), 454–464. https://doi.org/10.1124/jpet.109.163337.

Beljanski, V., Knaak, C., Zhuang, Y., & Smith, C. D. (2011). Combined anticancer effects of sphingosine kinase inhibitors and sorafenib. *Investigational New Drugs, 29*(6), 1132–1142. https://doi.org/10.1007/s10637-010-9452-0.

Beljanski, V., Lewis, C. S., & Smith, C. D. (2011). Antitumor activity of sphingosine kinase 2 inhibitor ABC294640 and sorafenib in hepatocellular carcinoma xenografts. *Cancer Biology & Therapy, 11*(5), 524–534.

Billich, A., Bornancin, F., Mechtcheriakova, D., Natt, F., Huesken, D., & Baumruker, T. (2005). Basal and induced sphingosine kinase 1 activity in A549 carcinoma cells: Function in cell survival and IL-1beta and TNF-alpha induced production of inflammatory mediators. *Cellular Signalling, 17*(10), 1203–1217.

Britten, C. D., Garrett-Mayer, E., Chin, S. H., Shirai, K., Ogretmen, B., Bentz, T. A., et al. (2017). A phase I study of ABC294640, a first-in-class sphingosine kinase-2 inhibitor, in patients with advanced solid tumors. *Clinical Cancer Research, 23*(16), 4642–4650. https://doi.org/10.1158/1078-0432.CCR-16-2363.

Canals, D., & Hannun, Y. A. (2013). Novel chemotherapeutic drugs in sphingolipid cancer research. *Handbook of Experimental Pharmacology*, 211–238. https://doi.org/10.1007/978-3-7091-1368-4_12.

Cao, M., Ji, C., Zhou, Y., Huang, W., Ni, W., Tong, X., et al. (2018). Sphingosine kinase inhibitors: A patent review. *International Journal of Molecular Medicine, 41*(5), 2450–2460. https://doi.org/10.3892/ijmm.2018.3505.

Cheng, X., Zhao, Z., Ventura, E., Gran, B., Shindler, K. S., & Rostami, A. (2007). The PD-1/PD-L pathway is up-regulated during IL-12-induced suppression of EAE mediated by IFN-gamma. *Journal of Neuroimmunology, 185*(1–2), 75–86. https://doi.org/10.1016/j.jneuroim.2007.01.012.

Chiurchiu, V., Leuti, A., & Maccarrone, M. (2018). Bioactive lipids and chronic inflammation: Managing the fire within. *Frontiers in Immunology, 9*, 38. https://doi.org/10.3389/fimmu.2018.00038.

Chumanevich, A. A., Poudyal, D., Cui, X., Davis, T., Wood, P. A., Smith, C. D., et al. (2010). Suppression of colitis-driven colon cancer in mice by a novel small molecule inhibitor of sphingosine kinase. *Carcinogenesis, 31*(10), 1787–1793. https://doi.org/10.1093/carcin/bgq158.

Cingolani, F., Casasampere, M., Sanllehi, P., Casas, J., Bujons, J., & Fabrias, G. (2014). Inhibition of dihydroceramide desaturase activity by the sphingosine kinase inhibitor SKI II. *The Journal of Lipid Research, 55*(8), 1711–1720. https://doi.org/10.1194/jlr.M049759.

Coant, N., Sakamoto, W., Mao, C., & Hannun, Y. A. (2017). Ceramidases, roles in sphingolipid metabolism and in health and disease. *Advances in Biological Regulation, 63*, 122–131. https://doi.org/10.1016/j.jbior.2016.10.002.

Dai, L., Bai, A., Smith, C. D., Rodriguez, P. C., Yu, F., & Qin, Z. (2017). ABC294640, a novel sphingosine kinase 2 inhibitor, induces oncogenic virus-infected cell autophagic death and represses tumor growth. *Molecular Cancer Therapeutics, 16*(12), 2724–2734. https://doi.org/10.1158/1535-7163.MCT-17-0485.

Dai, L., Smith, C. D., Foroozesh, M., Miele, L., & Qin, Z. (2018). The sphingosine kinase 2 inhibitor ABC294640 displays anti-non-small cell lung cancer activities in vitro and in vivo. *International Journal of Cancer, 142*(10), 2153–2162. https://doi.org/10.1002/ijc.31234.

Dai, L., Trillo-Tinoco, J., Bai, A., Chen, Y., Bielawski, J., Del Valle, L., et al. (2015). Ceramides promote apoptosis for virus-infected lymphoma cells through induction of ceramide synthases and viral lytic gene expression. *Oncotarget, 6*(27), 24246–24260. https://doi.org/10.18632/oncotarget.4759.

Ding, X., Chaiteerakij, R., Moser, C. D., Shaleh, H., Boakye, J., Chen, G., et al. (2016). Antitumor effect of the novel sphingosine kinase 2 inhibitor ABC294640 is enhanced by inhibition of autophagy and by sorafenib in human cholangiocarcinoma cells. *Oncotarget*. https://doi.org/10.18632/oncotarget.7914.

Dondero, A., Pastorino, F., Della Chiesa, M., Corrias, M. V., Morandi, F., Pistoia, V., et al. (2016). PD-L1 expression in metastatic neuroblastoma as an additional mechanism for limiting immune surveillance. *OncoImmunology, 5*(1), e1064578. https://doi.org/10.1080/2162402X.2015.1064578.

Dong, L., Lv, H., Li, W., Song, Z., Li, L., Zhou, S., et al. (2016). Co-expression of PD-L1 and p-AKT is associated with poor prognosis in diffuse large B-cell lymphoma via PD-1/PD-L1 axis activating intracellular AKT/mTOR pathway in tumor cells. *Oncotarget, 7*(22), 33350–33362. https://doi.org/10.18632/oncotarget.9061.

Fitzpatrick, L. R., Green, C., Frauenhoffer, E. E., French, K. J., Zhuang, Y., Maines, L. W., et al. (2011). Attenuation of arthritis in rodents by a novel orally-available inhibitor of sphingosine kinase. *Inflammopharmacology, 19*(2), 75–87. https://doi.org/10.1007/s10787-010-0060-6.

Fitzpatrick, L. R., Green, C., Maines, L. W., & Smith, C. D. (2011). Experimental osteoarthritis in rats is attenuated by ABC294640, a selective inhibitor of sphingosine kinase-2. *Pharmacology, 87*(3–4), 135–143. https://doi.org/10.1159/000323911.

French, K. J., Schrecengost, R. S., Lee, B. D., Zhuang, Y., Smith, S. N., Eberly, J. L., et al. (2003). Discovery and evaluation of inhibitors of human sphingosine kinase. *Cancer Research, 63*(18), 5962–5969.

French, K. J., Upson, J. J., Keller, S. N., Zhuang, Y., Yun, J. K., & Smith, C. D. (2006). Antitumor activity of sphingosine kinase inhibitors. *The Journal of Pharmacology and Experimental Therapeutics, 318*(2), 596–603. https://doi.org/10.1124/jpet.106.101345.

French, K. J., Zhuang, Y., Maines, L. W., Gao, P., Wang, W., Beljanski, V., et al. (2010). Pharmacology and antitumor activity of ABC294640, a selective inhibitor of sphingosine kinase-2. *The Journal of Pharmacology and Experimental Therapeutics, 333*(1), 129–139. https://doi.org/10.1124/jpet.109.163444.

Gao, P., Peterson, Y. K., Smith, R. A., & Smith, C. D. (2012). Characterization of isoenzyme-selective inhibitors of human sphingosine kinases. *PLoS One, 7*(9), e44543. https://doi.org/10.1371/journal.pone.0044543.

Gault, C. R., Obeid, L. M., & Hannun, Y. A. (2010). An overview of sphingolipid metabolism: From synthesis to breakdown. *Advances in Experimental Medicine and Biology, 688*, 1–23.

Gestaut, M. M., Antoon, J. W., Burow, M. E., & Beckman, B. S. (2014). Inhibition of sphingosine kinase-2 ablates androgen resistant prostate cancer proliferation and survival. *Pharmacological Reports: PR, 66*(1), 174–178. https://doi.org/10.1016/j.pharep.2013.08.014.

Gomez-Munoz, A., Presa, N., Gomez-Larrauri, A., Rivera, I. G., Trueba, M., & Ordonez, M. (2016). Control of inflammatory responses by ceramide, sphingosine 1-phosphate and ceramide 1-phosphate. *Progress in Lipid Research, 61*, 51–62. https://doi.org/10.1016/j.plipres.2015.09.002.

Gowrishankar, K., Gunatilake, D., Gallagher, S. J., Tiffen, J., Rizos, H., & Hersey, P. (2015). Inducible but not constitutive expression of PD-L1 in human melanoma cells is dependent on activation of NF-kappaB. *PLoS One, 10*(4), e0123410. https://doi.org/10.1371/journal.pone.0123410.

Guan, S., Liu, Y. Y., Yan, T., & Zhou, J. (2016). Inhibition of ceramide glucosylation sensitizes lung cancer cells to ABC294640, a first-in-class small molecule SphK2 inhibitor. *Biochemical and Biophysical Research Communications, 476*(4), 230–236. https://doi.org/10.1016/j.bbrc.2016.05.102.

Gude, D. R., Alvarez, S. E., Paugh, S. W., Mitra, P., Yu, J., Griffiths, R., et al. (2008). Apoptosis induces expression of sphingosine kinase 1 to release sphingosine-1-phosphate as a "come-and-get-me" signal. *The FASEB Journal, 22*(8), 2629–2638. https://doi.org/10.1096/fj.08-107169.

Guillermet-Guibert, J., Davenne, L., Pchejetski, D., Saint-Laurent, N., Brizuela, L., Guilbeau-Frugier, C., et al. (2009). Targeting the sphingolipid metabolism to defeat pancreatic cancer cell resistance to the chemotherapeutic gemcitabine drug. *Molecular Cancer Therapeutics, 8*(4), 809–820. https://doi.org/10.1158/1535-7163.MCT-08-1096.

Haddadi, N., Lin, Y., Simpson, A. M., Nassif, N. T., & McGowan, E. M. (2017). "Dicing and splicing" sphingosine kinase and relevance to cancer. *International Journal of Molecular Sciences, 18*(9). https://doi.org/10.3390/ijms18091891.

Hait, N. C., Allegood, J., Maceyka, M., Strub, G. M., Harikumar, K. B., Singh, S. K., et al. (2009). Regulation of histone acetylation in the nucleus by sphingosine-1-phosphate. *Science, 325*(5945), 1254–1257. https://doi.org/10.1126/science.1176709.

Hait, N. C., Bellamy, A., Milstien, S., Kordula, T., & Spiegel, S. (2007). Sphingosine kinase type 2 activation by ERK-mediated phosphorylation. *Journal of Biological Chemistry, 282*(16), 12058–12065. https://doi.org/10.1074/jbc.M609559200.

Hait, N. C., & Maiti, A. (2017). The role of sphingosine-1-phosphate and ceramide-1-phosphate in inflammation and cancer. *Mediators of Inflammation, 2017*, 4806541. https://doi.org/10.1155/2017/4806541.

Hanna, A. N., Berthiaume, L. G., Kikuchi, Y., Begg, D., Bourgoin, S., & Brindley, D. N. (2001). Tumor necrosis factor-alpha induces stress fiber formation through ceramide production: Role of sphingosine kinase. *Molecular Biology of the Cell, 12*(11), 3618–3630.

Hatoum, D., Haddadi, N., Lin, Y., Nassif, N. T., & McGowan, E. M. (2017). Mammalian sphingosine kinase (SphK) isoenzymes and isoform expression: Challenges for SphK as an oncotarget. *Oncotarget, 8*(22), 36898–36929. https://doi.org/10.18632/oncotarget.16370.

Hendrich, A. B., & Michalak, K. (2003). Lipids as a target for drugs modulating multidrug resistance of cancer cells. *Current Drug Targets, 4*(1), 23–30.

Hu, W. M., Li, L., Jing, B. Q., Zhao, Y. S., Wang, C. L., Feng, L., et al. (2010). Effect of S1P5 on proliferation and migration of human esophageal cancer cells. *World Journal of Gastroenterology: WJG, 16*(15), 1859–1866.

Huang, Y. L., Chang, C. L., Tang, C. H., Lin, Y. C., Ju, T. K., Huang, W. P., et al. (2014). Extrinsic sphingosine 1-phosphate activates S1P5 and induces autophagy through generating endoplasmic reticulum stress in human prostate cancer PC-3 cells. *Cellular Signalling, 26*(3), 611–618. https://doi.org/10.1016/j.cellsig.2013.11.024.

Huwiler, A., & Zangemeister-Wittke, U. (2017). The sphingosine 1-phosphate receptor modulator fingolimod as a therapeutic agent: Recent findings and new perspectives. *Pharmacology & Therapeutics*. https://doi.org/10.1016/j.pharmthera.2017.11.001.

Igarashi, N., Okada, T., Hayashi, S., Fujita, T., Jahangeer, S., & Nakamura, S. (2003). Sphingosine kinase 2 is a nuclear protein and inhibits DNA synthesis. *Journal of Biological Chemistry, 278*(47), 46832–46839. https://doi.org/10.1074/jbc.M306577200.

Iwai, Y., Ishida, M., Tanaka, Y., Okazaki, T., Honjo, T., & Minato, N. (2002). Involvement of PD-L1 on tumor cells in the escape from host immune system and tumor immunotherapy by PD-L1 blockade. *Proceedings of the National Academy of Sciences of the United States of America, 99*(19), 12293–12297. https://doi.org/10.1073/pnas.192461099.

Jarman, K. E., Moretti, P. A., Zebol, J. R., & Pitson, S. M. (2010). Translocation of sphingosine kinase 1 to the plasma membrane is mediated by calcium- and integrin-binding protein 1. *Journal of Biological Chemistry, 285*(1), 483–492. https://doi.org/10.1074/jbc.M109.068395.

Jendiroba, D. B., Klostergaard, J., Keyhani, A., Pagliaro, L., & Freireich, E. J. (2002). Effective cytotoxicity against human leukemias and chemotherapy-resistant leukemia cell lines by N-N-dimethylsphingosine. *Leukemia Research, 26*(3), 301–310.

Jolly, P. S., Bektas, M., Olivera, A., Gonzalez-Espinosa, C., Proia, R. L., Rivera, J., et al. (2004). Transactivation of sphingosine-1-phosphate receptors by FcepsilonRI triggering is required for normal mast cell degranulation and chemotaxis. *The Journal of Experimental Medicine, 199*(7), 959–970. https://doi.org/10.1084/jem.20030680.

Juif, P. E., Kraehenbuehl, S., & Dingemanse, J. (2016). Clinical pharmacology, efficacy, and safety aspects of sphingosine-1-phosphate receptor modulators. *Expert Opinion on Drug Metabolism and Toxicology, 12*(8), 879–895. https://doi.org/10.1080/17425255.2016.1196188.

Kapitonov, D., Allegood, J. C., Mitchell, C., Hait, N. C., Almenara, J. A., Adams, J. K., et al. (2009). Targeting sphingosine kinase 1 inhibits Akt signaling, induces apoptosis, and suppresses growth of human glioblastoma cells and xenografts. *Cancer Research, 69*(17), 6915–6923. https://doi.org/10.1158/0008-5472.CAN-09-0664.

Kolesnick, R., & Fuks, Z. (2003). Radiation and ceramide-induced apoptosis. *Oncogene, 22*(37), 5897–5906. https://doi.org/10.1038/sj.onc.1206702.

Kono, M., Belyantseva, I. A., Skoura, A., Frolenkov, G. I., Starost, M. F., Dreier, J. L., et al. (2007). Deafness and stria vascularis defects in S1P2 receptor-null mice. *Journal of Biological Chemistry, 282*(14), 10690–10696. https://doi.org/10.1074/jbc.M700370200.

Kraveka, J. M., Li, L., Szulc, Z. M., Bielawski, J., Ogretmen, B., Hannun, Y. A., et al. (2007). Involvement of dihydroceramide desaturase in cell cycle progression in human neuroblastoma cells. *Journal of Biological Chemistry, 282*(23), 16718–16728. https://doi.org/10.1074/jbc.M700647200.

Lastwika, K. J., Wilson, W., 3rd, Li, Q. K., Norris, J., Xu, H., Ghazarian, S. R., et al. (2016). Control of PD-L1 expression by oncogenic activation of the AKT-mTOR pathway in non-small cell lung cancer. *Cancer Research, 76*(2), 227–238. https://doi.org/10.1158/0008-5472.CAN-14-3362.

Lee, W. K., & Kolesnick, R. N. (2017). Sphingolipid abnormalities in cancer multidrug resistance: Chicken or egg? *Cellular Signalling, 38*, 134–145. https://doi.org/10.1016/j.cellsig.2017.06.017.

Leili, H., Nasser, S., Nadereh, R., Siavoush, D., & Pouran, K. (2018). Sphingosine kinase-2 inhibitor ABC294640 enhances doxorubicin-induced apoptosis of NSCLC cells via altering survivin expression. *Drug Research, 68*(1), 45–53. https://doi.org/10.1055/s-0043-117181.

Leroux, M. E., Auzenne, E., Evans, R., Hail, N., Jr., Spohn, W., Ghosh, S. C., et al. (2007). Sphingolipids and the sphingosine kinase inhibitor, SKI II, induce BCL-2-independent apoptosis in human prostatic adenocarcinoma cells. *The Prostate, 67*(15), 1699–1717. https://doi.org/10.1002/pros.20645.

Lewis, C. S., Voelkel-Johnson, C., & Smith, C. D. (2016). Suppression of c-Myc and RRM2 expression in pancreatic cancer cells by the sphingosine kinase-2 inhibitor ABC294640. *Oncotarget,* 7(37), 60181–60192. https://doi.org/10.18632/oncotarget.11112.

Lim, K. G., Sun, C., Bittman, R., Pyne, N. J., & Pyne, S. (2011). (R)-FTY720 methyl ether is a specific sphingosine kinase 2 inhibitor: Effect on sphingosine kinase 2 expression in HEK 293 cells and actin rearrangement and survival of MCF-7 breast cancer cells. *Cellular Signalling, 23*(10), 1590–1595. https://doi.org/10.1016/j.cellsig.2011.05.010.

Liu, H., Toman, R. E., Goparaju, S. K., Maceyka, M., Nava, V. E., Sankala, H., et al. (2003). Sphingosine kinase type 2 is a putative BH3-only protein that induces apoptosis. *Journal of Biological Chemistry, 278*(41), 40330–40336. https://doi.org/10.1074/jbc.M304455200.

Liu, Q., Rehman, H., Shi, Y., Krishnasamy, Y., Lemasters, J. J., Smith, C. D., et al. (2012). Inhibition of sphingosine kinase-2 suppresses inflammation and attenuates graft injury after liver transplantation in rats. *PLoS One,* 7(7), e41834. https://doi.org/10.1371/journal.pone.0041834.

Lucci, A., Han, T. Y., Liu, Y. Y., Giuliano, A. E., & Cabot, M. C. (1999). Modification of ceramide metabolism increases cancer cell sensitivity to cytotoxics. *International Journal of Oncology, 15*(3), 541–546.

Lynch, K. R., Thorpe, S. B., & Santos, W. L. (2016). Sphingosine kinase inhibitors: A review of patent literature (2006–2015). *Expert Opinion on Therapeutic Patents, 26*(12), 1409–1416. https://doi.org/10.1080/13543776.2016.1226282.

Maceyka, M., Sankala, H., Hait, N. C., Le Stunff, H., Liu, H., Toman, R., et al. (2005). SphK1 and SphK2, sphingosine kinase isoenzymes with opposing functions in sphingolipid metabolism. *Journal of Biological Chemistry, 280*(44), 37118–37129. https://doi.org/10.1074/jbc.M502207200.

Maines, L. W., Fitzpatrick, L. R., French, K. J., Zhuang, Y., Xia, Z., Keller, S. N., et al. (2008). Suppression of ulcerative colitis in mice by orally available inhibitors of sphingosine kinase. *Digestive Diseases and Sciences, 53*(4), 997–1012. https://doi.org/10.1007/s10620-007-0133-6.

Maines, L. W., Fitzpatrick, L. R., Green, C. L., Zhuang, Y., & Smith, C. D. (2010). Efficacy of a novel sphingosine kinase inhibitor in experimental Crohn's disease. *Inflammopharmacology, 18*(2), 73–85. https://doi.org/10.1007/s10787-010-0032-x.

Maines, L. W., French, K. J., Wolpert, E. B., Antonetti, D. A., & Smith, C. D. (2006). Pharmacologic manipulation of sphingosine kinase in retinal endothelial cells: Implications for angiogenic ocular diseases. *Investigative Ophthalmology & Visual Science, 47*(11), 5022–5031.

Mao-Draayer, Y., Sarazin, J., Fox, D., & Schiopu, E. (2017). The sphingosine-1-phosphate receptor: A novel therapeutic target for multiple sclerosis and other autoimmune diseases. *Clinical Immunology, 175,* 10–15. https://doi.org/10.1016/j.clim.2016.11.008.

Mastrandrea, L. D., Sessanna, S. M., & Laychock, S. G. (2005). Sphingosine kinase activity and sphingosine-1 phosphate production in rat pancreatic islets and INS-1 cells: Response to cytokines. *Diabetes, 54*(5), 1429–1436.

Matloubian, M., Lo, C. G., Cinamon, G., Lesneski, M. J., Xu, Y., Brinkmann, V., et al. (2004). Lymphocyte egress from thymus and peripheral lymphoid organs is dependent on S1P receptor 1. *Nature, 427*(6972), 355–360. https://doi.org/10.1038/nature02284.

Mayol, K., Biajoux, V., Marvel, J., Balabanian, K., & Walzer, T. (2011). Sequential desensitization of CXCR4 and S1P5 controls natural killer cell trafficking. *Blood, 118*(18), 4863–4871. https://doi.org/10.1182/blood-2011-06-362574.

McNaughton, M., Pitman, M., Pitson, S. M., Pyne, N. J., & Pyne, S. (2016). Proteasomal degradation of sphingosine kinase 1 and inhibition of dihydroceramide desaturase by the sphingosine kinase inhibitors, SKi or ABC294640, induces growth arrest in androgen-independent LNCaP-AI prostate cancer cells. *Oncotarget, 7*(13), 16663–16675. https://doi.org/10.18632/oncotarget.7693.

Mizugishi, K., Yamashita, T., Olivera, A., Miller, G. F., Spiegel, S., & Proia, R. L. (2005). Essential role for sphingosine kinases in neural and vascular development. *Molecular and Cellular Biology, 25*(24), 11113–11121. https://doi.org/10.1128/MCB.25.24.11113-11121.2005.

Modrak, D. E., Cardillo, T. M., Newsome, G. A., Goldenberg, D. M., & Gold, D. V. (2004). Synergistic interaction between sphingomyelin and gemcitabine potentiates ceramide-mediated apoptosis in pancreatic cancer. *Cancer Research, 64*(22), 8405–8410.

Molino, S., Tate, E., McKillop, W. M., & Medin, J. A. (2017). Sphingolipid pathway enzymes modulate cell fate and immune responses. *Immunotherapy, 9*(14), 1185–1198. https://doi.org/10.2217/imt-2017-0089.

Muhlbauer, M., Fleck, M., Schutz, C., Weiss, T., Froh, M., Blank, C., et al. (2006). PD-L1 is induced in hepatocytes by viral infection and by interferon-alpha and -gamma and mediates T cell apoptosis. *Journal of Hepatology, 45*(4), 520–528. https://doi.org/10.1016/j.jhep.2006.05.007.

Murakami, A., Takasugi, H., Ohnuma, S., Koide, Y., Sakurai, A., Takeda, S., et al. (2010). Sphingosine 1-phosphate (S1P) regulates vascular contraction via S1P3 receptor: Investigation based on a new S1P3 receptor antagonist. *Molecular Pharmacology, 77*(4), 704–713. https://doi.org/10.1124/mol.109.061481.

Nagahashi, M., Hait, N. C., Maceyka, M., Avni, D., Takabe, K., Milstien, S., et al. (2014). Sphingosine-1-phosphate in chronic intestinal inflammation and cancer. *Advances in Biological Regulation, 54*, 112–120. https://doi.org/10.1016/j.jbior.2013.10.001.

Nagahashi, M., Ramachandran, S., Kim, E. Y., Allegood, J. C., Rashid, O. M., Yamada, A., et al. (2012). Sphingosine-1-phosphate produced by sphingosine kinase 1 promotes breast cancer progression by stimulating angiogenesis and lymphangiogenesis. *Cancer Research, 72*(3), 726–735. https://doi.org/10.1158/0008-5472.CAN-11-2167.

Nayak, D., Huo, Y., Kwang, W. X., Pushparaj, P. N., Kumar, S. D., Ling, E. A., et al. (2010). Sphingosine kinase 1 regulates the expression of proinflammatory cytokines and nitric oxide in activated microglia. *Neuroscience, 166*(1), 132–144. https://doi.org/10.1016/j.neuroscience.2009.12.020.

Newton, J., Lima, S., Maceyka, M., & Spiegel, S. (2015). Revisiting the sphingolipid rheostat: Evolving concepts in cancer therapy. *Experimental Cell Research, 333*(2), 195–200. https://doi.org/10.1016/j.yexcr.2015.02.025.

Niessen, F., Schaffner, F., Furlan-Freguia, C., Pawlinski, R., Bhattacharjee, G., Chun, J., et al. (2008). Dendritic cell PAR1-S1P3 signalling couples coagulation and inflammation. *Nature, 452*(7187), 654–658. https://doi.org/10.1038/nature06663.

Nieuwenhuis, B., Luth, A., & Kleuser, B. (2010). Dexamethasone protects human fibroblasts from apoptosis via an S1P3-receptor subtype dependent activation of PKB/Akt and Bcl XL. *Pharmacological Research: The Official Journal of the Italian Pharmacological Society, 61*(5), 449–459. https://doi.org/10.1016/j.phrs.2009.12.005.

Novgorodov, A. S., El-Alwani, M., Bielawski, J., Obeid, L. M., & Gudz, T. I. (2007). Activation of sphingosine-1-phosphate receptor S1P5 inhibits oligodendrocyte progenitor migration. *The FASEB Journal, 21*(7), 1503–1514. https://doi.org/10.1096/fj.06-7420com.

Ogretmen, B. (2018). Sphingolipid metabolism in cancer signalling and therapy. *Nature Reviews Cancer, 18*(1), 33–50. https://doi.org/10.1038/nrc.2017.96.

Okada, T., Ding, G., Sonoda, H., Kajimoto, T., Haga, Y., Khosrowbeygi, A., et al. (2005). Involvement of N-terminal-extended form of sphingosine kinase 2 in serum-dependent regulation of cell proliferation and apoptosis. *Journal of Biological Chemistry, 280*(43), 36318–36325. https://doi.org/10.1074/jbc.M504507200.

Olivera, A., Eisner, C., Kitamura, Y., Dillahunt, S., Allende, L., Tuymetova, G., et al. (2010). Sphingosine kinase 1 and sphingosine-1-phosphate receptor 2 are vital to recovery from anaphylactic shock in mice. *Journal of Clinical Investigation, 120*(5), 1429–1440. https://doi.org/10.1172/JCI40659.

Oskeritzian, C. A., Price, M. M., Hait, N. C., Kapitonov, D., Falanga, Y. T., Morales, J. K., et al. (2010). Essential roles of sphingosine-1-phosphate receptor 2 in human mast cell activation, anaphylaxis, and pulmonary edema. *The Journal of Experimental Medicine, 207*(3), 465–474. https://doi.org/10.1084/jem.20091513.

Ottenlinger, F. M., Mayer, C. A., Ferreiros, N., Schreiber, Y., Schwiebs, A., Schmidt, K. G., et al. (2016). Interferon-Beta increases plasma ceramides of specific chain length in multiple sclerosis patients, unlike fingolimod or natalizumab. *Frontiers in Pharmacology, 7*, 412. https://doi.org/10.3389/fphar.2016.00412.

Patmanathan, S. N., Wang, W., Yap, L. F., Herr, D. R., & Paterson, I. C. (2017). Mechanisms of sphingosine 1-phosphate receptor signalling in cancer. *Cellular Signalling, 34*, 66–75. https://doi.org/10.1016/j.cellsig.2017.03.002.

Paugh, S. W., Paugh, B. S., Rahmani, M., Kapitonov, D., Almenara, J. A., Kordula, T., et al. (2008). A selective sphingosine kinase 1 inhibitor integrates multiple molecular therapeutic targets in human leukemia. *Blood, 112*(4), 1382–1391. https://doi.org/10.1182/blood-2008-02-138958.

Perry, D. K., & Kolesnick, R. N. (2003). Ceramide and sphingosine 1-phosphate in anti-cancer therapies. *Cancer Treatment and Research, 115*, 345–354.

Pettus, B. J., Chalfant, C. E., & Hannun, Y. A. (2004). Sphingolipids in inflammation: Roles and implications. *Current Molecular Medicine, 4*(4), 405–418.

Pitman, M. R., Costabile, M., & Pitson, S. M. (2016). Recent advances in the development of sphingosine kinase inhibitors. *Cellular Signalling, 28*(9), 1349–1363. https://doi.org/10.1016/j.cellsig.2016.06.007.

Pitson, S. M., Moretti, P. A., Zebol, J. R., Lynn, H. E., Xia, P., Vadas, M. A., et al. (2003). Activation of sphingosine kinase 1 by ERK1/2-mediated phosphorylation. *The EMBO Journal, 22*(20), 5491–5500. https://doi.org/10.1093/emboj/cdg540.

Plano, D., Amin, S., & Sharma, A. K. (2014). Importance of sphingosine kinase (SphK) as a target in developing cancer therapeutics and recent developments in the synthesis of novel SphK inhibitors. *Journal of Medicinal Chemistry, 57*(13), 5509–5524. https://doi.org/10.1021/jm4011687.

Poti, F., Bot, M., Costa, S., Bergonzini, V., Maines, L., Varga, G., et al. (2012). Sphingosine kinase inhibition exerts both pro- and anti-atherogenic effects in low-density lipoprotein receptor-deficient (LDL-R(-/-)) mice. *Thrombosis & Haemostasis, 107*(3), 552–561. https://doi.org/10.1160/TH11-08-0583.

Pyne, N. J., Adams, D. R., & Pyne, S. (2017). Sphingosine kinase 2 in autoimmune/inflammatory disease and the development of sphingosine kinase 2 inhibitors. *Trends in Pharmacological Sciences, 38*(7), 581–591. https://doi.org/10.1016/j.tips.2017.04.003.

Pyne, N. J., El Buri, A., Adams, D. R., & Pyne, S. (2017). Sphingosine 1-phosphate and cancer. *Advances in Biological Regulation*. https://doi.org/10.1016/j.jbior.2017.09.006.

Pyne, S., Adams, D. R., & Pyne, N. J. (2016). Sphingosine 1-phosphate and sphingosine kinases in health and disease: Recent advances. *Progress in Lipid Research, 62*, 93–106. https://doi.org/10.1016/j.plipres.2016.03.001.

Pyne, S., Bittman, R., & Pyne, N. J. (2011). Sphingosine kinase inhibitors and cancer: Seeking the golden sword of hercules. *Cancer Research, 71*(21), 6576–6582. https://doi.org/10.1158/0008-5472.can-11-2364.
Qin, Z., Dai, L., Trillo-Tinoco, J., Senkal, C., Wang, W., Reske, T., et al. (2014). Targeting sphingosine kinase induces apoptosis and tumor regression for KSHV-associated primary effusion lymphoma. *Molecular Cancer Therapeutics, 13*(1), 154–164. https://doi.org/10.1158/1535-7163.MCT-13-0466.
Radeff-Huang, J., Seasholtz, T. M., Chang, J. W., Smith, J. M., Walsh, C. T., & Brown, J. H. (2007). Tumor necrosis factor-alpha-stimulated cell proliferation is mediated through sphingosine kinase-dependent Akt activation and cyclin D expression. *Journal of Biological Chemistry, 282*(2), 863–870.
Rodriguez, Y. I., Campos, L. E., Castro, M. G., Aladhami, A., Oskeritzian, C. A., & Alvarez, S. E. (2016). Sphingosine-1 phosphate: A new modulator of immune plasticity in the tumor microenvironment. *Frontiers in Oncologyront Oncol, 6*, 218. https://doi.org/10.3389/fonc.2016.00218.
Saied, E. M., & Arenz, C. (2016). Inhibitors of ceramidases. *Chemistry and Physics of Lipids, 197*, 60–68. https://doi.org/10.1016/j.chemphyslip.2015.07.009.
Sankala, H. M., Hait, N. C., Paugh, S. W., Shida, D., Lepine, S., Elmore, L. W., et al. (2007). Involvement of sphingosine kinase 2 in p53-independent induction of p21 by the chemotherapeutic drug doxorubicin. *Cancer Research, 67*(21), 10466–10474. https://doi.org/10.1158/0008-5472.CAN-07-2090.
Sanllehi, P., Abad, J. L., Casas, J., & Delgado, A. (2016). Inhibitors of sphingosine-1-phosphate metabolism (sphingosine kinases and sphingosine-1-phosphate lyase). *Chemistry and Physics of Lipids, 197*, 69–81. https://doi.org/10.1016/j.chemphyslip.2015.07.007.
Santos, W. L., & Lynch, K. R. (2015). Drugging sphingosine kinases. *ACS Chemical Biology, 10*(1), 225–233. https://doi.org/10.1021/cb5008426.
Schnute, M. E., McReynolds, M. D., Kasten, T., Yates, M., Jerome, G., Rains, J. W., et al. (2012). Modulation of cellular S1P levels with a novel, potent and specific inhibitor of sphingosine kinase-1. *The Biochemical Journal, 444*(1), 79–88. https://doi.org/10.1042/BJ20111929.
Schrecengost, R. S., Keller, S. N., Schiewer, M. J., Knudsen, K. E., & Smith, C. D. (2015). Downregulation of critical oncogenes by the selective SK2 inhibitor ABC294640 hinders prostate cancer progression. *Molecular Cancer Research, 13*(12), 1591–1601. https://doi.org/10.1158/1541-7786.MCR-14-0626.
Sensken, S. C., Nagarajan, M., Bode, C., & Graler, M. H. (2011). Local inactivation of sphingosine 1-phosphate in lymph nodes induces lymphopenia. *The Journal of Immunology: Official Journal of the American Association of Immunologists, 186*(6), 3432–3440. https://doi.org/10.4049/jimmunol.1002169.
Seo, Y. J., Alexander, S., & Hahm, B. (2011). Does cytokine signaling link sphingolipid metabolism to host defense and immunity against virus infections? *Cytokine & Growth Factor Reviews, 22*(1), 55–61. https://doi.org/10.1016/j.cytogfr.2010.12.001.
Shi, Y., Rehman, H., Ramshesh, V. K., Schwartz, J., Liu, Q., Krishnasamy, Y., et al. (2012). Sphingosine kinase-2 inhibition improves mitochondrial function and survival after hepatic ischemia-reperfusion. *Journal of Hepatology, 56*(1), 137–145. https://doi.org/10.1016/j.jhep.2011.05.025.
Siddique, M. M., Li, Y., Chaurasia, B., Kaddai, V. A., & Summers, S. A. (2015). Dihydroceramides: From bit players to lead actors. *Journal of Biological Chemistry, 290*(25), 15371–15379. https://doi.org/10.1074/jbc.R115.653204.
Sietsma, H., Veldman, R. J., & Kok, J. W. (2001). The involvement of sphingolipids in multidrug resistance. *The Journal of Membrane Biology, 181*(3), 153–162.

Siow, D., & Wattenberg, B. (2011). The compartmentalization and translocation of the sphingosine kinases: Mechanisms and functions in cell signaling and sphingolipid metabolism. *Critical Reviews in Biochemistry and Molecular Biology, 46*(5), 365–375. https://doi.org/10.3109/10409238.2011.580097.

Skoura, A., Sanchez, T., Claffey, K., Mandala, S. M., Proia, R. L., & Hla, T. (2007). Essential role of sphingosine 1-phosphate receptor 2 in pathological angiogenesis of the mouse retina. *Journal of Clinical Investigation, 117*(9), 2506–2516. https://doi.org/10.1172/JCI31123.

Snider, A. J., Orr Gandy, K. A., & Obeid, L. M. (2010). Sphingosine kinase: Role in regulation of bioactive sphingolipid mediators in inflammation. *Biochimie, 92*(6), 707–715. https://doi.org/10.1016/j.biochi.2010.02.008.

Song, D. D., Zhou, J. H., & Sheng, R. (2017). Regulation and function of sphingosine kinase 2 in diseases. *Histology & Histopathology*. https://doi.org/10.14670/hh-11-939, 11939.

Song, M., Chen, D., Lu, B., Wang, C., Zhang, J., Huang, L., et al. (2013). PTEN loss increases PD-L1 protein expression and affects the correlation between PD-L1 expression and clinical parameters in colorectal cancer. *PLoS One, 8*(6), e65821. https://doi.org/10.1371/journal.pone.0065821.

Spiegel, S., & Milstien, S. (2002). Sphingosine 1-phosphate, a key cell signaling molecule. *Journal of Biological Chemistry, 277*(29), 25851–25854.

Spiegel, S., & Milstien, S. (2011). The outs and the ins of sphingosine-1-phosphate in immunity. *Nature Reviews Immunology, 11*(6), 403–415. https://doi.org/10.1038/nri2974.

Stahelin, R. V., Hwang, J. H., Kim, J. H., Park, Z. Y., Johnson, K. R., Obeid, L. M., et al. (2005). The mechanism of membrane targeting of human sphingosine kinase 1. *Journal of Biological Chemistry, 280*(52), 43030–43038. https://doi.org/10.1074/jbc.M507574200.

Strub, G. M., Paillard, M., Liang, J., Gomez, L., Allegood, J. C., Hait, N. C., et al. (2011). Sphingosine-1-phosphate produced by sphingosine kinase 2 in mitochondria interacts with prohibitin 2 to regulate complex IV assembly and respiration. *The FASEB Journal, 25*(2), 600–612. https://doi.org/10.1096/fj.10-167502.

Sweeney, E. A., Sakakura, C., Shirahama, T., Masamune, A., Ohta, H., Hakomori, S., et al. (1996). Sphingosine and its methylated derivative N,N-dimethylsphingosine (DMS) induce apoptosis in a variety of human cancer cell lines. *International Journal of Cancer, 66*(3), 358–366. https://doi.org/10.1002/(SICI)1097-0215(19960503)66:3<358::AID-IJC16>3.0.CO;2-7.

Taha, T. A., Kitatani, K., Bielawski, J., Cho, W., Hannun, Y. A., & Obeid, L. M. (2005). Tumor necrosis factor induces the loss of sphingosine kinase-1 by a cathepsin B-dependent mechanism. *Journal of Biological Chemistry, 280*(17), 17196–17202. https://doi.org/10.1074/jbc.M413744200.

Takasugi, N., Sasaki, T., Suzuki, K., Osawa, S., Isshiki, H., Hori, Y., et al. (2011). BACE1 activity is modulated by cell-associated sphingosine-1-phosphate. *The Journal of Neuroscience: The Official Journal of the Society for Neuroscience, 31*(18), 6850–6857. https://doi.org/10.1523/JNEUROSCI.6467-10.2011.

Tan, S. F., Pearson, J. M., Feith, D. J., & Loughran, T. P., Jr. (2017). The emergence of acid ceramidase as a therapeutic target for acute myeloid leukemia. *Expert Opinion on Therapeutic Targets, 21*(6), 583–590. https://doi.org/10.1080/14728222.2017.1322065.

Van Brocklyn, J. R., Jackson, C. A., Pearl, D. K., Kotur, M. S., Snyder, P. J., & Prior, T. W. (2005). Sphingosine kinase-1 expression correlates with poor survival of patients with glioblastoma multiforme: Roles of sphingosine kinase isoforms in growth of glioblastoma cell lines. *Journal of Neuropathology and Experimental Neurology, 64*(8), 695–705.

Venant, H., Rahmaniyan, M., Jones, E. E., Lu, P., Lilly, M. B., Garrett-Mayer, E., et al. (2015). The sphingosine kinase 2 inhibitor ABC294640 reduces the growth of prostate cancer cells and results in accumulation of dihydroceramides in vitro and in vivo. *Molecular Cancer Therapeutics, 14*(12), 2744–2752. https://doi.org/10.1158/1535-7163.MCT-15-0279.

Venkata, J. K., An, N., Stuart, R., Costa, L. J., Cai, H., Coker, W., et al. (2014). Inhibition of sphingosine kinase 2 downregulates the expression of c-Myc and Mcl-1 and induces apoptosis in multiple myeloma. *Blood, 124*(12), 1915–1925. https://doi.org/10.1182/blood-2014-03-559385.

Wakita, H., Nishimura, K., Tokura, Y., Furukawa, F., & Takigawa, M. (1996). Inhibitors of sphingolipid synthesis modulate interferon (IFN)-gamma-induced intercellular adhesion molecule (ICAM)-1 and human leukocyte antigen (HLA)-DR expression on cultured normal human keratinocytes: Possible involvement of ceramide in biologic action of IFN-gamma. *The Journal of Investigative Dermatology, 107*(3), 336–342.

Wallington-Beddoe, C. T., Powell, J. A., Tong, D., Pitson, S. M., Bradstock, K. F., & Bendall, L. J. (2014). Sphingosine kinase 2 promotes acute lymphoblastic leukemia by enhancing MYC expression. *Cancer Research, 74*(10), 2803–2815. https://doi.org/10.1158/0008-5472.CAN-13-2732.

Wang, Q., Li, J., Li, G., Li, Y., Xu, C., Li, M., et al. (2014). Prognostic significance of sphingosine kinase 2 expression in non-small cell lung cancer. *Tumour Biology: The Journal of the International Society for Oncodevelopmental Biology and Medicine, 35*(1), 363–368. https://doi.org/10.1007/s13277-013-1051-1.

Wang, W., Graeler, M. H., & Goetzl, E. J. (2005). Type 4 sphingosine 1-phosphate G protein-coupled receptor (S1P4) transduces S1P effects on T cell proliferation and cytokine secretion without signaling migration. *The FASEB Journal, 19*(12), 1731–1733. https://doi.org/10.1096/fj.05-3730fje.

Wang, X., Yang, L., Huang, F., Zhang, Q., Liu, S., Ma, L., et al. (2017). Inflammatory cytokines IL-17 and TNF-alpha up-regulate PD-L1 expression in human prostate and colon cancer cells. *Immunology Letters, 184*, 7–14. https://doi.org/10.1016/j.imlet.2017.02.006.

Weigert, A., Cremer, S., Schmidt, M. V., von Knethen, A., Angioni, C., Geisslinger, G., et al. (2010). Cleavage of sphingosine kinase 2 by caspase-1 provokes its release from apoptotic cells. *Blood, 115*(17), 3531–3540. https://doi.org/10.1182/blood-2009-10-243444.

White, M. D., Chan, L., Antoon, J. W., & Beckman, B. S. (2013). Targeting ovarian cancer and chemoresistance through selective inhibition of sphingosine kinase-2 with ABC294640. *Anticancer Research, 33*(9), 3573–3579.

Windh, R. T., Lee, M. J., Hla, T., An, S., Barr, A. J., & Manning, D. R. (1999). Differential coupling of the sphingosine 1-phosphate receptors Edg-1, Edg-3, and H218/Edg-5 to the G(i), G(q), and G(12) families of heterotrimeric G proteins. *Journal of Biological Chemistry, 274*(39), 27351–27358.

Xia, P., Gamble, J. R., Rye, K. A., Wang, L., Hii, C. S., Cockerill, P., et al. (1998). Tumor necrosis factor-alpha induces adhesion molecule expression through the sphingosine kinase pathway. *Proceedings of the National Academy of Sciences of the United States of America, 95*(24), 14196–14201.

Xu, L., Jin, L., Yang, B., Wang, L., Xia, Z., Zhang, Q., et al. (2018). The sphingosine kinase 2 inhibitor ABC294640 inhibits cervical carcinoma cell growth. *Oncotarget, 9*(2), 2384–2394. https://doi.org/10.18632/oncotarget.23415.

Xun, C., Chen, M. B., Qi, L., Tie-Ning, Z., Peng, X., Ning, L., et al. (2015). Targeting sphingosine kinase 2 (SphK2) by ABC294640 inhibits colorectal cancer cell growth in vitro and in vivo. *Journal of Experimental & Clinical Cancer Research, 34*, 94. https://doi.org/10.1186/s13046-015-0205-y.

Yang, J., Yang, C., Zhang, S., Mei, Z., Shi, M., Sun, S., et al. (2015). ABC294640, a sphingosine kinase 2 inhibitor, enhances the antitumor effects of TRAIL in non-small cell lung cancer. *Cancer Biology & Therapy, 16*(8), 1194–1204. https://doi.org/10.1080/15384047.2015.1056944.

Yang, L., Huang, F., Mei, J., Wang, X., Zhang, Q., Wang, H., et al. (2017). Posttranscriptional control of PD-L1 expression by 17beta-estradiol via PI3K/Akt signaling pathway in ERalpha-positive cancer cell lines. *International Journal of Gynecological Cancer, 27*(2), 196–205. https://doi.org/10.1097/IGC.0000000000000875.

Yatomi, Y., Ruan, F., Megidish, T., Toyokuni, T., Hakomori, S., & Igarashi, Y. (1996). N,N-dimethylsphingosine inhibition of sphingosine kinase and sphingosine 1-phosphate activity in human platelets. *Biochemistry, 35*(2), 626–633. https://doi.org/10.1021/bi9515533.

Yester, J. W., Tizazu, E., Harikumar, K. B., & Kordula, T. (2011). Extracellular and intracellular sphingosine-1-phosphate in cancer. *Cancer and Metastasis Reviews, 30*(3–4), 577–597. https://doi.org/10.1007/s10555-011-9305-0.

Zhao, L., Li, C., Liu, F., Zhao, Y., Liu, J., Hua, Y., et al. (2017). A blockade of PD-L1 produced antitumor and antimetastatic effects in an orthotopic mouse pancreatic cancer model via the PI3K/Akt/mTOR signaling pathway. *OncoTargets and Therapy, 10*, 2115–2126. https://doi.org/10.2147/OTT.S130481.

Zhou, J., Chen, J., & Yu, H. (2018). Targeting sphingosine kinase 2 by ABC294640 inhibits human skin squamous cell carcinoma cell growth. *Biochemical and Biophysical Research Communications, 497*(2), 535–542. https://doi.org/10.1016/j.bbrc.2018.02.075.

CHAPTER TWELVE

Novel Sphingolipid-Based Cancer Therapeutics in the Personalized Medicine Era

Jeremy Shaw*,[a], Pedro Costa-Pinheiro*,[a], Logan Patterson*, Kelly Drews*, Sarah Spiegel[§] and Mark Kester[¶,||,1]
*Department of Pathology, University of Virginia, Charlottesville, VA, United States
[§]Department of Biochemistry and Molecular Biology, Virginia Commonwealth University School of Medicine, Richmond, VA, United States
[¶]Department of Pharmacology, University of Virginia, Charlottesville, VA, United States
[||]University of Virginia Cancer Center, University of Virginia, Charlottesville, VA, United States
[1]Corresponding author: E-mail: mk5vq@virginia.edu

Contents

[a] Contributed equally.

Advances in Cancer Research, Volume 140
ISSN 0065-230X
https://doi.org/10.1016/bs.acr.2018.04.016

Abstract

Sphingolipids are bioactive lipids that participate in a wide variety of biological mechanisms, including cell death and proliferation. The myriad of pro-death and pro-survival cellular pathways involving sphingolipids provide a plethora of opportunities for dysregulation in cancers. In recent years, modulation of these sphingolipid metabolic pathways has been in the forefront of drug discovery for cancer therapeutics. About two decades ago, researchers first showed that standard of care treatments, e.g., chemotherapeutics and radiation, modulate sphingolipid metabolism to increase endogenous ceramides, which kill cancer cells. Strikingly, resistance to these treatments has also been linked to altered sphingolipid metabolism, favoring lipid species that ultimately lead to cell survival. To this end, many inhibitors of sphingolipid metabolism have been developed to further define not only our understanding of these pathways but also to potentially serve as therapeutic interventions. Therefore, understanding how to better use these new drugs that target sphingolipid metabolism, either alone or in combination with current cancer treatments, holds great potential for cancer control. While sphingolipids in cancer have been reviewed previously (Hannun & Obeid, 2018; Lee & Kolesnick, 2017; Morad & Cabot, 2013; Newton, Lima, Maceyka, & Spiegel, 2015; Ogretmen, 2018; Ryland, Fox, Liu, Loughran, & Kester, 2011) in this chapter, we present a comprehensive review on how standard of care therapeutics affects sphingolipid metabolism, the current landscape of sphingolipid inhibitors, and the clinical utility of sphingolipid-based cancer therapeutics.

1. EFFECTS OF CHEMO- AND RADIATION THERAPIES ON SPHINGOLIPIDS

1.1 Chemotherapy

Chemotherapeutic or cytotoxic agents are drugs commonly used as standard of care for multiple types of cancers due to their ability to kill or damage highly proliferative cells. These agents are usually divided into classes based on their mechanisms of action, and, strikingly, a wide range of these classes have been described as leading to accumulation of sphingolipids (particularly ceramide) in cells. As shown in Fig. 1, we depict the steady state accumulation of endogenous ceramide via chemo- and radiation therapies as a combination of "faucets" and "drains." Enzymes underlying sphingomyelin hydrolysis (sphingomyelinase—SMase), de novo sphingolipid synthesis (serine palmitoyltransferase—SPT, dihydroceramide synthase—CerS, dihydroceramide desaturase—DES), and the sphingolipid salvage pathway (glucosylceramidase—GCase, CerS) are the "faucets" that generate ceramide, whereas enzymes that metabolize ceramide (ceramidase—CDase,

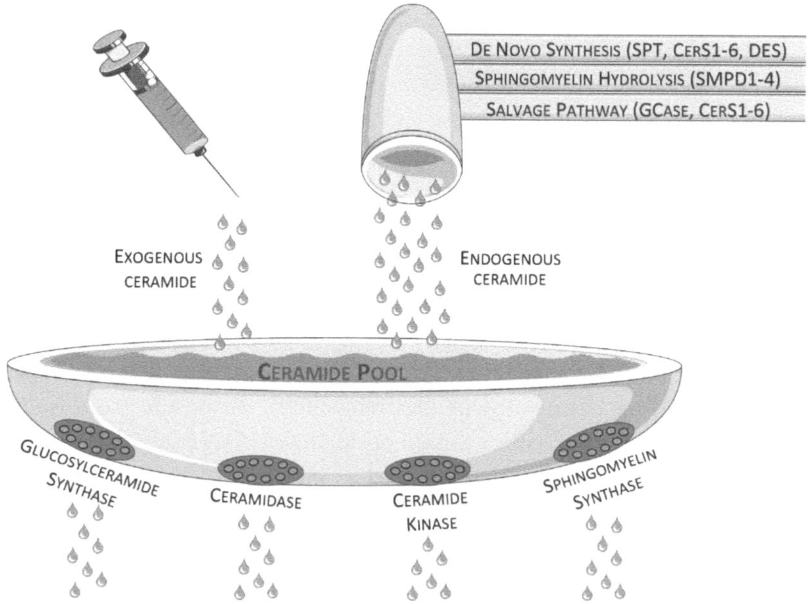

Figure 1 The accumulation of ceramide can be generated from three main pathways (sphingomyelin hydrolysis, de novo synthesis, or salvage pathway) or exogenously added (in the form of short chain ceramides), whereas the degradation of ceramide is facilitated by four main enzymes (SMS, GCS, CDase, and CerK). At a steady state, the level of ceramide is held stable, as flux through the system remains fairly constant. However, similarly to increasing flow through the faucet or clogging the drain, increasing the generation of ceramide or preventing its degradation can lead to an accumulation of ceramide, which can be selectively toxic to cancer cells.

glucosylceramide synthase—GCS, sphingomyelin synthase—SMS, ceramide kinase—CERK) are the "drains." In essence, when too much ceramide accumulates and the metaphorical sink overflows the cell dies. Complicating this system, some of these sphingolipids have separate "faucets" or "drains" due to different isoenzymes of sphingolipid enzymes that generate distinct ceramide species with unique biological and biophysical properties. This diversity in ceramide species is due to distinct ceramide synthases with specific fatty acyl-CoA specificities. As a generalization for the major ceramide species observed in cancer, CerS1 preferentially generates C_{18}-ceramide, CerS2 and CerS4 generate long chain C_{24} species, and CerS5 and CerS6 generate C_{16} species. The "fluid dynamics" of these enzyme activities determines both the efficacy of, and resistance to, chemo- and radiation therapies. In fact, this more interconnected and compensatory metabolic system is just a new version of the "sphingolipid rheostat," by which the pro-death actions of ceramide are counterbalanced by the pro-survival functions of sphingosine-1-phosphate (S1P) (Cuvillier et al., 1996). A detailed metabolic pathway for sphingolipid regulation in cancer is depicted in Fig. 2 (top panel).

Studies performed in yeast showed that under stress conditions, cells responded by generating ceramide, demonstrating that sphingolipid metabolism is an evolutionary response to stress (Cowart & Obeid, 2007). Chemotherapeutics induce stress in cancer cells, and numerous investigations have shown that in response to this stress, ceramide levels are increased, both by sphingomyelin hydrolysis as well as through de novo synthesis of ceramide (Beckham, Cheng, Marrison, Norris, & Liu, 2013). This increase in ceramide levels has been validated in many different classes of chemotherapies including daunorubicin (anthracycline), etoposide (topoisomerase II inhibitor), and gemcitabine (nucleotide analog DNA replication inhibitor). These drugs have all been described as inducers of de novo generation of ceramide, either by activating ceramide synthases or serine palmitoyl transferase (Bose et al., 1995; Chalfant et al., 2002; Jaffrezou et al., 1996; Perry et al., 2000). Inhibition of enzymes important for de novo generation of ceramide decreases the cytotoxicity of these chemotherapeutics, and therefore, decreases their overall efficacy. However, in glioma cells treated with etoposide, the increase in ceramide levels was caused by increased activation of neutral SMase, which is responsible for converting sphingomyelin to ceramide (Sawada et al., 2001). In one clinical trial for head and neck squamous cell carcinoma, administration of a combination of gemcitabine plus doxorubicin increased serum levels of C_{18}-ceramide that correlated with better drug responses (Saddoughi et al., 2011).

Interestingly, altered ceramide levels are not the only biological connection between sphingolipids and chemotherapy; the Cabot laboratory demonstrated that levels of glucosylceramide are increased in breast cancer cell lines (Lavie et al., 1997) and in patients who were resistant to chemotherapy (Lucci et al., 1998). The enzyme that generates glucosylceramide, GCS, is upregulated in several different tumor types such as lung cancer (Zhang et al., 2014), breast cancer (Liu, Patwardhan, Xie, et al., 2011), bladder cancer (Sun et al., 2010), and colorectal cancer (Wang, Liu, Xu, Mu, & Sun, 2014). Given that increased conversion of ceramide to glucosylceramide decreases killing of cancer cells exposed to chemotherapeutic agents and the increased expression/activity of glucosylceramide synthase in several tumor models, focus has shifted to developing inhibitors for this enzyme that could synergize with standard of care drugs.

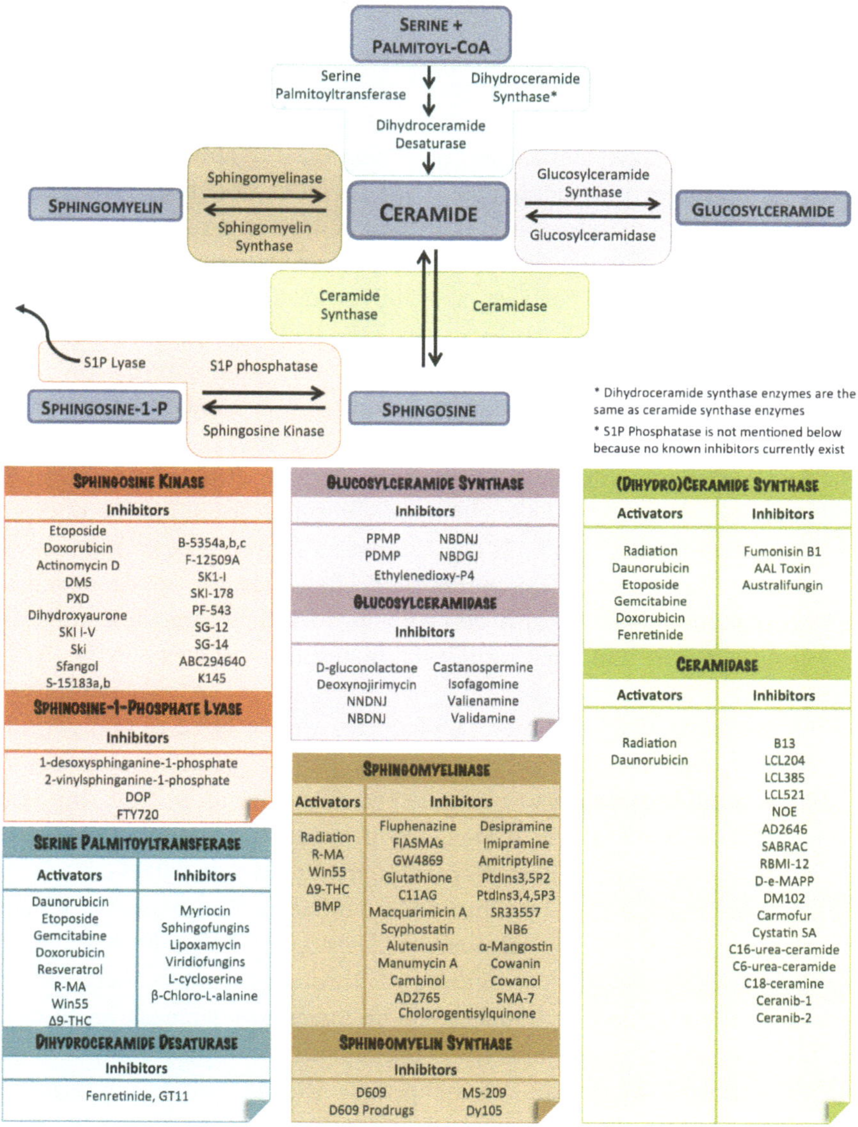

Figure 2 This figure depicts a ceramide-centric view of sphingolipid metabolism. The upper panel outlines the major sphingolipid species (*blue boxes*) and the enzymes that metabolize these lipids (*multicolored boxes*). Inhibitors and activators of specific enzymes in the pathway are listed in the lower panel using enzyme color-coded "sticky notes."

Cancer cells may also decrease ceramide levels by increasing activity of the CDase enzymes that cleave ceramide to sphingosine, which can then be converted to S1P. Due to this conversion from ceramide, a pro-apoptotic lipid, to S1P, a pro-survival lipid, CDase can modulate cell fate under stress conditions. CDase enzymes are divided in three different types according to their optimum pH: neutral, acid, or alkaline. In cancer research, a vast majority of the studies have been devoted to acid ceramidase (ACDase). This enzyme was reported as overexpressed in a myriad of tumor types including prostate (Saad et al., 2007), head and neck squamous cell (Elojeimy et al., 2007), and acute myeloid leukemia (AML) (Tan, Pearson, Feith, & Loughran, 2017). In vitro and in vivo studies have shown that by overexpressing ACDase, tumors are more aggressive and resistant to chemotherapies (Beckham et al., 2012; Saad et al., 2007). In prostate cancer cells, ACDase overexpression promoted resistance to etoposide, cisplatin (platinum-based agent), and gemcitabine by stopping accumulation of ceramide (Beckham, Cheng, et al., 2013). Interestingly, ACDase has been shown in hepatocellular cells to be upregulated after exposure to daunorubicin, a phenomenon that was not observed in normal hepatocytes, suggesting a specific mechanism for cancer cells to resist cytotoxicity (Morales et al., 2007).

Recently, the use of medical marijuana was legalized to mitigate side effects of chemotherapies in 30 states (http://www.governing.com/gov-data/state-marijuana-laws-map-medical-recreational.html). In 1998, the Sanchez group first reported the relationship between sphingolipids and cannabinoids, the active ingredients of medical marijuana (Sanchez, Galve-Roperh, Canova, Brachet, & Guzman, 1998; Sanchez, Galve-Roperh, Rueda, & Guzman, 1998). They demonstrated that THC (tetrahydrocannabinol) caused an increase in ceramide levels through de novo synthesis, which led to apoptosis in C6 glioma cells. This increase of ceramide levels through THC or other cannabinoids was also shown in prostate and colon cancer cells, as well as mantle cell lymphoma (MCL) and multiple myeloma (Barbado et al., 2017; Cianchi et al., 2008; Gustafsson, Christensson, Sander, & Flygare, 2006; Mimeault, Pommery, Wattez, Bailly, & Henichart, 2003). Further work in MCL showed that cannabinoids induced CerS3 and CerS6 expression. Notably, blocking ceramide metabolism into S1P or glucosylceramide also increased the cytotoxic effect of cannabinoids (Gustafsson, Sander, Bielawski, Hannun, & Flygare, 2009).

Increases in S1P levels at the expense of sphingosine and ceramide in cancer are often manifested by an increase in sphingosine kinase (SphK) expression or activity. The conversion of sphingosine to S1P is executed by two SphK isoenzymes that phosphorylate sphingosine. SphK1 is upregulated in paclitaxel-resistant prostate cancer cell lines (Aoyama et al., 2017) as well as in several tumor xenograft models such as stomach, breast, lung, brain, colon, and kidney (Pyne & Pyne, 2010). Moreover, SphK1 expression was shown to be upregulated in AML patient samples (Dick et al., 2015). Although the specific role of SphK2 in cancer is not as well described, specific SphK2 inhibitors are showing efficacy in the clinic (Orr Gandy & Obeid, 2013). In vitro and in vivo studies in prostate cancer and non—small cell lung carcinoma showed that overexpression of SphK1 was responsible for resistance to docetaxel (taxane), camptothecin (topoisomerase inhibitor), and doxorubicin (Akao et al., 2006; Pchejetski et al., 2008; Sauer et al., 2009; Song et al., 2011). However, other studies have also shown that in leukemia cells, treatment with actinomycin D (peptide antibiotic), doxorubicin, or etoposide resulted in reduction of SphK1 due to accumulation of p53 (Taha et al., 2004). SphK1 was also found to be responsible for resistance to cetuximab (a monoclonal antibody chemotherapeutic against Epidermal Growth Factor Receptor (EGFR)) in colorectal cancer both in vitro as well as in patients (Rosa et al., 2013). An interesting mechanism for cell death is to alter the sphingolipid rheostat, not by increasing ceramide levels, but by decreasing S1P levels (French et al., 2003). For example, Oskouian et al. (2006) showed that treating several cancer cell lines with etoposide led to increased levels of S1P lyase and concomitant increases in apoptosis. Taken together, inhibitors of SphK or activators of sphingosine lyase or phosphatases may sensitize patients to conventional therapeutics and prevent drug resistance mechanisms. Although manipulating sphingolipid metabolism as new therapeutic avenue might be enticing, understanding each patient's sphingolipid and enzymatic profile is vital for interfering with this very intricate, complex, and highly regulated process.

1.2 Radiation

Radiotherapy, as currently used in the clinic, was first developed and tested as a treatment for cancer in 1934 by Henri Coutard. As described by Halperin et al., Coutard developed a fractionized protocol and observed that by administrating lower doses of X-rays over a course of days, he could kill cancer cells while allowing normal cells to recover from the effects of radiation (Halperin, Perez, & Brady, 2008). Contrary to chemotherapeutics, radiotherapy is frequently limited to tumor locations and not systemic. By using high-energy liberation of electrons, it is possible to induce cell death specifically in malignant cells (Aureli et al., 2014). This process occurs by induction of DNA damage, as radiation causes double-strand breaks in DNAs that are lethal to cells (Beckham, Cheng, et al., 2013; Carroll, Donaldson, & Obeid, 2015). Cancer cells can attempt to repair these double-strand breaks, but as repair mechanisms are usually dysregulated in cancer, it leads to errors and subsequent mitosis-associated cell death or p53-induced apoptosis (Aureli et al., 2014; Beckham, Cheng, et al., 2013). Strikingly, the work performed by the Kolesnick lab has shown that even p53-independent cell death mechanisms after radiation exposure are regulated by sphingolipid metabolism (Haimovitz-Friedman et al., 1994; Kolesnick & Fuks, 2003). Among sphingolipid species, ceramide was shown to be the major mediator of cellular stress after radiation exposure (Modrak, Gold, & Goldenberg, 2006).

One of the earliest evidences for the role of ceramide in cell death upon radiation led to the discovery of the rapid hydrolysis of sphingomyelin to ceramide by SMase after exposure to ionizing radiation (Haimovitz-Friedman et al., 1994). The identification of different isoforms of SMase that work under different optimal pH conditions, led to the discovery that sphingomyelin metabolism is triggered by different SMase enzymes in a cell-type specific manner (Beckham, Cheng, et al., 2013; Li & Zhang, 2015). In a very elegant set of experiments, the Kolesnick lab showed that by ablating acid SMase in human type I Niemann–Pick disease lymphoblastoid cells, or in p53 knockout mice, ceramide induced apoptosis after radiation was ablated (Lozano et al., 2001; Santana et al., 1996). Ectopic expression of acid SMase as well as by exogenous treatment with ceramide restored sensitivity to irradiation. These results indicated a critical role of ceramide in cell death induced by radiation. Moreover, SMases can be present extracellularly, and in a small group of cancer patients, fractionated high dose radiation was correlated with an increase of SMase extracellular content and ceramide levels (Oskouian & Saba, 2010).

Sphingomyelin hydrolysis is not the only metabolic pathway to generate ceramide in response to radiation. An in vitro study in radiation-resistant DU145 prostate cancer cells showed that pretreatment with resveratrol (an antioxidant agent) resulted in sensitization to radiation. This shift had occurred by induction of de novo synthesis of ceramide, a finding that was validated when sphingolipid synthesis inhibitors blocked sensitization and reverted DU145 cells to radiation-resistant status (Scarlatti et al., 2007).

Different ceramide chain length species are now known to have different roles in cells. Therefore it is not surprising that different responses to radiation treatments have been correlated with changes in various ceramide species or ceramide synthases. CerS5 and CerS6 are activated following radiation and induce cell death. However, C_{24} ceramide generated by CerS2, conferred resistance to radiation in the same human cervical cancer (HeLa) cells (Galadari, Rahman, Pallichankandy, & Thayyullathil, 2015). This paradigm of differential activities and cellular roles between C_{16}/C_{18} versus C_{24} species is actively being investigated in various other cancer therapeutic models. In addition, the role of saturated versus unsaturated ceramide species of similar chain length is also an area deserving further investigation.

In addition to sphingomyelin hydrolysis and de novo synthesis, ceramide synthase activity has long been known to increase in cells after exposure to radiation (Vit & Rosselli, 2003). The current rationale in the field is that ceramide levels are initially increased by activation of sphingomyelin hydrolysis as described above, and then, between 8 and 24 h after irradiation, the increase in ceramide levels shifts to become driven by increased CerS activity (Carroll et al., 2015). Strikingly, the second wave of ceramide increase seems to be dependent on the first one. In the aforementioned type I Niemann–Pick disease cells that lack acid SMase, after radiation treatment, there is an increase in CerS activity but not an increase in ceramide levels (Carroll et al., 2015). This dual wave system of cell death was also validated in Jurkat leukemia cells (Ardail et al., 2009).

Another important sphingolipid pathway is the generation of sphingosine by ceramide cleavage via CDases. Hara et al. (2004) reported that in vitro irradiated glioblastoma cells showed increased expression of ACDase. The authors further showed that by inhibiting this enzyme, glioblastoma cells accumulated ceramide on irradiation and increased levels of apoptosis (Hara et al., 2004). A different laboratory working with prostate cancer cells showed that this upregulation of ACDase after radiation exposure was mediated by c-Jun (Cheng et al., 2013). Similarly to glioblastoma cells, reducing ACDase expression in prostate cancer cells resulted in ceramide accumulation and cell death after radiation treatments (Cheng et al., 2013). Considering the evidence above, conversion of ceramide to sphingosine can be a mechanism to counteract the proapoptotic role of ceramide and lead to cell survival and insensitivity to radiation.

2. CURRENT LANDSCAPE OF INHIBITORS OF SPHINGOLIPID METABOLISM

Sphingolipids are vital to normal cell function and thus the dysregulation of the enzymes responsible for their production and maintenance can lead to a number of diseases, including cancer. As such, the ability to regulate the activity of these enzymes using inhibitors or activators is important for developing novel treatments for diseases in which sphingolipid imbalance occurs. Thus, the search for effective inhibitors for the enzymes that decrease ceramide levels are worth exploring. The following section will discuss in vitro and in vivo anti-cancer effects of some of the best-characterized inhibitors and activators of the main sphingolipid enzymes. The lower panel of Fig. 2 provides a comprehensive list of these drugs, their targets, and their impact in sphingolipid metabolism.

2.1 Targeting Sphingomyelin Synthesis

SMSs generate sphingomyelin from two substrates, ceramide and phosphatidylcholine. Tricyclodecan-9-yl-xanthogenate (D609) is a cytotoxic drug that causes selective tumor cell death by inhibiting SMS leading to increased levels of the pro-apoptotic lipid ceramide (Bai et al., 2004; Meng et al., 2004). However, due to the presence of a xanthate group on D609, cells may metabolize the drug and render it useless.Thus, a series of D609 prodrugs have been developed, which are more stable and effective at inhibiting SMS at lower LD_{50} values (56.6 μM prodrug vs. 117 μM D609) (Bai et al., 2004). Another inhibitor of SMS, MS-209, has been used in numerous studies for its ability to reverse multidrug resistance in tumors (Delgado, Casas, Llebaria, Abad, & Fabrias, 2006; Robert, 2004). Finally, the SMS inhibitor, Dy105, was initially developed as an atherosclerosis treatment but may be useful for future cancer studies (Lou et al., 2014).

2.2 Targeting Ceramide Formation

The first step in the de novo synthesis of ceramide is the condensation of serine and palmitoyl CoA by the enzyme SPT (Hanada, 2003). Myriocin, also known as ISP-1 or thermozymocidin, is the most studied inhibitor of SPT and can inhibit the enzyme at picomolar concentrations (Miyake, Kozutsumi, Nakamura, Fujita, & Kawasaki, 1995; Sauane et al., 2010). More natural products such as sphingofungins, lipoxamycin, and, to a lesser extent, viridiofungins have been shown to inhibit SPT as well (Mandala et al., 1994; Mandala, Thornton, Frommer, Dreikorn, & Kurtz, 1997;

Zweerink, Edison, Wells, Pinto, & Lester, 1992). Other previously used less-specific compounds, L-cycloserine and β-chloro-L-alanine, have received some attention as SPT inhibitors but both are known to have off-target effects. Specifically, L-cycloserine also inhibits 3-ketodihydrosphingosine reductase, which is downstream of SPT in the de novo synthesis pathway, whereas β-chloro-L-alanine has off-target effects inhibiting other pyridoxal-5′phosphate—dependent enzymes (Medlock & Merrill, 1988; Sundaram & Lev, 1984).

The CerS family of six isoenzymes is involved in both the de novo synthesis and the recycling of sphingosine into ceramide in the salvage pathway. One of the most well-known general inhibitors of CerS is Fumonisin B1, a mycotoxin produced by *Fusarium moniliforme* (Bose et al., 1995; Wang, Norred, Bacon, Riley, & Merrill, 1991; Yoo, Norred, Wang, Merrill, & Riley, 1992). A structurally similar toxin, AAL toxin produced by *Alternaria alternata*, has also been shown to inhibit CerS (Delgado et al., 2006; Winter, Gilchrist, Dickman, & Jones, 1996). Australifungin, a mycotoxin from *Sporormiella australis*, potently inhibits CerS but has limited use because of its high chemical reactivity (Delgado et al., 2006; Mandala et al., 1995). Interestingly, despite showing efficacy in preclinical development, N-(4-hydroxyphenyl)retinamide (4-HPR or fenretinide) activates CerS but also inhibits DES, the next enzyme in the de novo pathway (Schulz et al., 2006). Fenretinide is now being tested in clinical trials for its potential as a broad anti-tumor inhibitor in several tumor models.

The sphingolipid-based compound, GT11, is a competitive inhibitor of DES that has been modified to generate more effective versions of the compound (Bedia, Triola, Casas, Llebaria, & Fabrias, 2005; Triola, Fabrias, Casas, & Llebaria, 2003; Triola, Fabrias, & Llebaria, 2001). The utility of these compounds that effect multiple sphingolipid metabolites in cancer are undefined.

Desipramine, imipramine, and amitriptyline, which are all tricyclic antidepressants, are also functional aSMase inhibitors that increase aSMase degradation by lysosomal proteases (Albouz et al., 1981; Arimochi & Morita, 2008; Bhabak & Arenz, 2013; Delgado et al., 2006; Huang et al., 2007; Lu et al., 2009). L-alpha-phosphatidyl-D-myo-inositol-3,5-bisphosphate (PtdIns3,5P2) and the more potent phosphatidyl-myo-inositol 3,4,5-triphosphate (PtdIns3,4,5P3) are both physiological inhibitors of aSMase that are found in humans (Kolzer et al., 2003; Testai, Landek, Goswami, Ahmed, & Dawson, 2004). A calcium channel blocker that sensitized multidrug resistant cells to doxorubicin, SR33557, also functions to inhibit aSMase (Jaffrezou et al., 1991). NB6, a structurally similar molecule to the tricyclic antidepressants, decreases aSMase activity in a different manner, by inhibiting its transcription (Deigner et al., 2001). α-Mangostin, cowanin, and cowanol, compounds isolated from the bark of *Garcinia speciosa*, each inhibit aSMase with varying degrees of potency (14.1, 19.2, and 10.9 μM IC_{50} values, respectively) (Delgado et al., 2006; Okudaira et al., 2000). SMA-7 is another aSMase inhibitor that was shown to inhibit Dextran Sulfate Sodium (DSS)-induced colitis in mice, though other studies have suggested that it may also inhibit nSMase (Canals, Perry, Jenkins, & Hannun, 2011; Sakata et al., 2007; Yokomatsu et al., 2001). A sphingomyelin analogue that was developed to inhibit aSMase, AD2765, increased cellular ceramide levels and cell death in cancer cells in vitro (Darroch et al., 2005). Treatment with fluphenazine, an anti-psychotic drug that induces lysosomal stress and inhibits aSMase activity, led to accumulation of sphingomyelin and subsequent cell death in hypoxic tumors (Klutzny et al., 2017). In contrast to the inhibitors listed above, bismonoacylgycerophopsohate was found to be an activator of aSMase that increased the generation of ceramide (Canals et al., 2011; Linke et al., 2001). Other functional aSMase inhibitors (FIASMAs) that cause aSMase to dissociate from the inner lysosomal membrane and become inactivated include Dextromethorphan, Fluoxetine, Maprotilin, Nortriptyline, Orphenadrine, Sertralin, and Triflupromazine (Kornhuber et al., 2008, 2010).

Numerous studies have explored the effects of inhibition of the SMase family member, neutral SMase (nSMase), utilizing GW4869, a noncompetitive nSMase inhibitor (Luberto et al., 2002; Marchesini, Luberto, & Hannun, 2003; Takahashi, Inanami, Asanuma, & Kuwabara, 2006). The antioxidant glutathione, which is present in mammalian cells at relatively high concentrations (1–20 mM), was reported to almost completely inhibit nSMase at a concentration of 5 mM, suggesting that nSMase is relatively inactive in vivo under normal conditions (Liu & Hannun, 1997). Another nSMase inhibitor, C11AG, which was originally used to inhibit herpes simplex virus-1 replication, was later shown to enhance apoptosis of Jurkat T cell lymphoma cells (Amtmann & Zoller, 2005; Amtmann, Zoller, & Schilling, 2000; Delgado et al., 2006). Macquarimicin A and scyphostatin, two natural products that are primarily nSMase inhibitors, are also weak ACDase inhibitors (Nara, Tanaka, Hosoya, Suzuki-Konagai, & Ogita, 1999; Nara, Tanaka, Masuda-Inoue, et al., 1999; Tanaka et al., 1999). Alutenusin and chlorogentisylquinone are both nSMase inhibitors that were originally purified from culture media of different fungal species

(Uchida, Tomoda, Arai, & Omura, 2001; Uchida, Tomoda, Dong, & Omura, 1999). Manumycin A is an antibiotic produced by the bacterium *Streptomyces parvulus*, which selectively inhibits nSMase, not aSMase, and was also found to be a useful inhibitor of Ras farnesyltransferase (Arenz et al., 2001; Kouchi et al., 1999). Cambinol is another molecule with dual uses, as it was originally discovered as a regulator of SIRT1/2 and was later found to be dramatically more effective for its ability to inhibit nSMase (Figuera-Losada et al., 2015).

2.3 Targeting Glucosylceramide Generation and Degradation

Glucosylceramidase (GCase) hydrolyzes glucosylceramide back to ceramide and glucose. In very early studies, D-gluconolactone was found to inhibit glucosylceramidase, but more potent inhibitors have since been developed (Overkleeft et al., 1998; van Weely, Brandsma, Strijland, Tager, & Aerts, 1993). Deoxynojirimycin is an α-glucosidase inhibitor isolated from mulberry leaves that has been extensively used as a GCase inhibitor, although interestingly, its derivatives, N-nonyl-1-deoxynojirimycin (NNDNJ) and N-Butyldeoxynojirimycin (NBDNJ), have been shown to function as chaperones of GCase that increase its activity (Alfonso et al., 2005; Delgado et al., 2006; Sawkar et al., 2002). Other natural product inhibitors, such as castanospermine, isofagomine, valienamine, and validamine, have also been used to inhibit GCase activity (Delgado et al., 2006; Sasak, Ordovas, Elbein, & Berninger, 1985; Saul, Chambers, Molyneux, & Elbein, 1983).

GCS is the enzyme responsible for adding a glucose molecule to ceramide. A pair of the most widely used GCS inhibitors, D-threo-1-phenyl-2-palmitoylamino-3-morpholino-l-propanol (PPMP) and D-threo-1-phenyl-2-decanoylamino-3-morpholino-propanol (PDMP), are ceramide analogues (Kovacs, Pinter, & Csaba, 2000; Vunnam & Radin, 1980). PPMP has previously been shown to resensitize multidrug-resistant cancer cells to chemotherapies in vitro and in vivo, whereas PDMP has shown cytotoxicity in AML in vitro (Cakir, Saydam, Sahin, & Baran, 2011; Gouaze et al., 2005). A derivative of PDMP, ethylenedioxy-P4, showed specific inhibition of human GCS but did not inhibit other animal, plant, fungal, or bacterial versions of the enzyme (Hillig, Warnecke, & Heinz, 2005). Miglustat, or NBDNJ, has been shown to reduce the effects of Gaucher disease by inhibiting GCS to prevent glucosylceramide buildup (Cox, 2005; Platt, Neises, Dwek, & Butters, 1994). A structurally similar molecule,

N-(n-Butyl)deoxygalactonojirimycin (NBDGJ), also functions to inhibit GCS (Platt, Neises, Karlsson, Dwek, & Butters, 1994).

2.4 Targeting Ceramide-1-Phosphate Formation

CERK, a highly homologous relative of SphK, catalyzes the ATP-dependent specific phosphorylation of ceramide to ceramide-1-phosphate. Two main CERK inhibitors have been developed that block its activity: K1 and NVP-231 (Graf et al., 2008; Kim, Inagaki, et al., 2005; Pastukhov et al., 2014). K1 is an isomer of F-12509A, a SphK inhibitor discussed below, that non-competitively inhibits CERK (Kim, Inagaki, et al., 2005). NVP-231 is a competitive CERK inhibitor (Graf et al., 2008) and inhibits breast and lung-cancer cell growth (Pastukhov et al., 2014).

2.5 Targeting Sphingosine Formation

Similar to the three types of SMase enzymes, CDaseare classified according to their optimal pH and thus are named acid ceramidase (ACDase), neutral ceramidase (nCDase), and alkaline ceramidase (alkCDase). B13 (also referred to as D-NMAPPD) is a ceramide analogue that inhibits ACDase and has been used extensively in vitro and in in vivo models of melanoma, pancreatic cancer, and metastatic colon cancer (Raisova et al., 2002; Samsel et al., 2004; Selzner et al., 2001). This has led to the generation of a number of analogs (LCL204, LCL385, LCL521) that have been designed to target ACDase in the lysosomal compartment that B13 struggles to reach (Bai et al., 2017; Holman et al., 2008; Mahdy et al., 2009). LCL 204 has been shown to inhibit in vitro and in vivo models of AML (Tan et al., 2016). N-oleoylethanolamine (NOE) is a synthetic analogue of ceramide that inhibits ACDase that has been used to sensitize cancer cells to chemotherapies and reduce pro-survival signals (Monick et al., 2004; Morales et al., 2007; Strelow et al., 2000; Sugita, Willians, Dulaney, & Moser, 1975). Like NOE, AD2646, SABRAC, and RBMI-12 are other ceramide analogues that were designed to inhibit ACDase (Camacho et al., 2013; Dagan, Wang, Fibach, & Gatt, 2003). In contrast, D-e-MAPP (or D-MAPP) is a ceramide analogue that inhibits alkCDase (Bielawska et al., 1996). DM102 meanwhile is a specific ACDase inhibitor as it shows no activity toward other CDases and causes cell death in vitro (Gouaze-Andersson et al., 2011). Carmofur, a chemotherapeutic drug used to treat colorectal cancer, was discovered to be an ACDase inhibitor suggesting that its mechanism of action is promoting ceramide buildup in cancer cells (Realini et al., 2013). Cystatin SA, a member of the cystatin family, was identified as one of the first physiological inhibitors of ACDase (Eliyahu et al., 2011). Finally, a very potent ACDase inhibitor, SABRAC (IC_{50} 52 nM) induces ceramide accumulation and reduces cell growth of PC-3 prostate cancer cells in vitro (Camacho et al., 2013).

C_{16}-urea-ceramide, as well as shorter-chain C_6-urea-ceramide, are potent inhibitors of nCDase (Garcia-Barros et al., 2016; Saied & Arenz, 2014; Usta et al., 2001). C_{18}-Ceramine (ceramine), a synthetic ceramide derivative, is also able to inhibit nCDase (Garcia-Barros et al., 2016; Saied & Arenz, 2014). Two additional inhibitors, Ceranib-1 and Ceranib-2, have been identified that broadly inhibit ACDase, nCDase, and alkCDase (Draper et al., 2011).

2.6 Targeting Sphingosine-1-Phosphate Formation and Degradation

Dimethylsphingosine (DMS), a methylated version of sphingosine, is the first SphK inhibitor that was shown to promote apoptosis in cancer cells (Edsall, Van Brocklyn, Cuvillier, Kleuser, & Spiegel, 1998) and to inhibit growth of xenografted breast cancer tumors in mice (Nava et al., 2000).

Based on the lead compound dihydroxyaurone, which was reported to be a SphK inhibitor that has anti-tumor properties in a mouse model of breast cancer, a group of SphK inhibitors (termed SKI I—V) were developed (French et al., 2003). Of these, SKI-II (also called SKi) has found the most use although it inhibits both SphK1 and SphK2 and possibly other targets, complicating understanding of the mechanism by which it promotes cell death in a variety of cancer types (French et al., 2003; Truman, Garcia-Barros, Obeid, & Hannun, 2014). L-threo-dihydrosphingosine, also known as Safingol, was initially reported to be a protein kinase C inhibitor but was later proven to also inhibit SphKs (Buehrer & Bell, 1992; Tsukamoto et al., 2015). Additionally, a number of SphK inhibitors from various bacterial and fungal sources have been discovered. S-15183a and S-15183b were produced by the fungus *Zopfiella inermis*, B-5354a, B-5354b, and B-5454c were all derived from a marine bacteria from the genus *Ruegeria*, and F-12509A was isolated from the fungus *Trichopezizella barbata* (Kono, Sugiura, & Kohama, 2002; Kono, Tanaka, Mizuno, et al., 2000; Kono, Tanaka, Ogita, Hosoya, & Kohama, 2000; Kono et al., 2001).

SK1-I (BML-258), a sphingosine analogue that is an isozyme-specific SphK1 inhibitor, promoted cell death in a number of in vitro cancer models (Paugh et al., 2008). Two other SphK1 inhibitors, SKI-178 and PF-543, also occupy the substrate-binding site of SphK and thus prevent the phosphorylation of sphingosine (Hengst et al., 2017; Schnute et al., 2012; Wang, Knapp, Pyne, Pyne, & Elkins, 2014).

SG-12 and SG-14 were the first two SphK2 inhibitors discovered, but ABC294640 has been used more frequently. ABC294640 displays activity in vitro and in vivo to prevent cancer growth and is currently in phase II clinical trials (Antoon et al., 2010; Beljanski, Knaak, & Smith, 2010; Kim, Kim, et al., 2005). K145 was recently developed as a more specific and potent SphK2 inhibitor and has shown promise in vivo in murine xenografts for suppressing breast tumor growth (K. Liu et al., 2013).

S1P is not only a bioactive mediator; it is also the ultimate sphingolipid metabolite and the only exit point for metabolism of all sphingolipids. S1P lyase is responsible for the irreversible cleavage of S1P into hexadecenal and ethanolamine phosphate, which are substrates for metabolism in various metabolic pathways. 1-deoxysphinganine-1-phosphate, an S1P analogue, is a naturally occurring phosphorylated sphingoid base that inhibits S1P lyase (Stoffel & Grol, 1974). S1P lyase activity can also be inhibited by 2-vinylsphinganine-1-phosphate (Boumendjel & Miller, 1994). S1P lyase is a pyridoxal phosphate (B6)-dependent enzyme but the often used S1P lyase inhibitor 4-deoxypyridoxine, which interferes with many B6-dependent enzymes is generally not used anymore because this lack of specificity results in high levels of toxicity (Serra & Saba, 2010).

In addition, therapeutics can regulate S1P activity by altering S1P receptor expression or activity. As a prime example, FTY720, also known as fingolimod/gilenya, is a myriocin derivative and sphingosine analogue that is phosphorylated in vivo to its active form FTY720-P mainly by SphK2; inhibiting S1P activity through S1P receptor 1, 3, 4, and 5 binding (Brinkmann et al., 2002). In vivo studies have shown that FTY720-P is also able to inhibit S1P lyase activity directly, as S1P lyase mRNA and protein expression were not altered (Bandhuvula, Tam, Oskouian, & Saba, 2005). Thus FTY720, already FDA approved for multiple sclerosis, is now being actively investigated as a cancer therapeutic.

3. DELIVERABLE AND CLINICALLY RELEVANT SPHINGOLIPID-BASED DRUGS FOR CANCER

3.1 Methods for Delivery of Sphingolipids and Mimetics to Cells

Instead of relying on increasing "flow through the faucets or clogging the drains," we can also simply add water to the sink externally to overflow the system (Fig. 1). Over the past 10 to 15 years, a number of approaches have been developed to enhance delivery of sphingolipids and analogues directly to cells in an effort to alter cell physiology, usually to disrupt the growth of tumor cells. Of these delivered sphingolipids, the most common is ceramide, with numerous chemical modifications and synthetic derivatives created to increase cell permeability, efficacy, and chemotoxicity. One method of increasing specificity of ceramide derivatives for negatively charged cellular compartments (such as the mitochondria) is the introduction of a positive charge on the fatty acid residue by the addition of a pyridine structure. Studies with pyridine-ceramides showed that they localized more readily to the mitochondria and induced the release of cytochrome *c*, *a* marker of apoptosis (Novgorodov et al., 2005; Senkal et al., 2006). These positively charged ceramides selectively induced pancreatic cancer cell death (Beckham, Lu, et al., 2013).

Another approach is to shorten the ceramide fatty-acid chain length to C_2, C_6, or C_8 (Liu, Beckman, & Foroozesh, 2013). These synthetic short-chain ceramides have increased cell permeability and have been shown to have potent anti-tumor effects (Obeid, Linardic, Karolak, & Hannun, 1993; Wolff, Dobrowsky, Bielawska, Obeid, & Hannun, 1994). C_6 ceramide has been utilized for the development of a novel nanoliposome formulation termed ceramide nanoliposomes (CNLs). These CNLs can deliver C_6 ceramides effectively as well as synergizing with current chemotherapeuitcs by encapsulating them within the same nanoliposomes (Adiseshaiah et al., 2013; Kester et al., 2015; Ma, Mou, Ding, Zou, & Huang, 2016; Sun, Fox, Adhikary, Kester, & Pearlman, 2008). The efficacy of these CNLs is currently being investigated in a phase 1 clinical trial in patients with solid tumors.

Finally, several sphingosine analogues have been researched as promising sphingolipid drugs, such as safingol (Schwartz et al., 1995, 1997), N,N-DMS (Edsall et al., 1998; Sweeney et al., 1996), fingolimod (Chun & Hartung, 2010; White, Alshaker, Cooper, Winkler, & Pchejetski, 2016), and (R)-FTY720 methyl ether (ROME) (Watson et al., 2013). These sphingosine analogues display anti-tumor effects as inhibitors of SphK enzyme activity and/or S1P receptors signaling.

3.2 Sphingolipid-Based Therapeutics Clinical Pipeline

A search through the clinicaltrials.gov database for various sphingolipid keywords such as S1P, sphingosine, ceramide, and sphingomyelin returned an impressive number of completed, ongoing, or recruiting clinical trials.

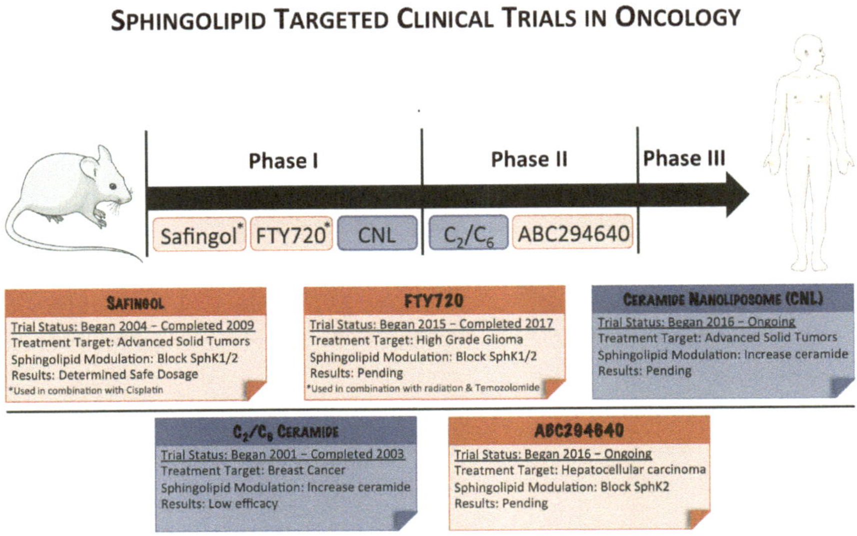

Figure 3 An outline of the sphingolipid-based therapeutics currently undergoing clinical trials for cancer treatment. Although there are few sphingolipid therapies currently in the clinic, many other sphingolipid-based therapeutics are showing promise in preclinical models.

For the sake of brevity and clarity, this review is focused on the three largest classes of sphingolipid-related drugs currently in clinical trials: synthetic ceramides, S1P antagonists, and SphK inhibitors. While clinical trials for other sphingolipid pathway inhibitors exist (notably amitriptyline, an acid SMase inhibitor), they are not as far along in the clinical pipeline and thus will not be discussed here (ClinicalTrials.gov) (Beckmann, Sharma, Gulbins, Becker, & Edelmann, 2014). Fig. 3 depicts various sphingolipid-based therapeutics that have entered clinical oncology trials.

Synthetic ceramides are ceramide analogues that have been modified to increase their tumor-specific toxicity, membrane permeability, or efficacy. As mentioned earlier, numerous synthetic ceramides exist, but the most common involve modifications to the hydrocarbon chain length, rendering it shorter. In 2003, topical administration of C_2 and C_6 ceramides were employed in a phase II study against cutaneous breast cancer, with mixed results (Clinical trial #: NCT00008320) (ClinicalTrials.gov). The trial found that while the C_2 and C_6 ceramides had little to no toxicity, patients yielded only a 4% response rate, shelving topical administration of short chain ceramides as a potential chemotherapeutic (Jatoi et al., 2003). However, by encapsulating a C_6 ceramide into stable nontoxic nanoliposomes, researchers found that they could increase chemotoxicity as compared to free ceramide alone in in vivo models (Kester et al., 2015; Sun et al., 2008). These CNLs entered a phase I study beginning in 2017 for patients with advanced solid tumors at the University of Maryland, the University of Virginia, and the Medical University of South Carolina (Clinical trial #: NCT02834611) (ClinicalTrials.gov).

For the past decade, S1P has been the focus of intense study due to its tumorigenic properties. As a result, numerous inhibitors against the enzyme responsible for its generation, SphK, have been advanced as candidates for clinical trials. Of these, the furthest down the clinical pipeline is ABC294640. ABC294640 acts as a specific SphK2 inhibitor, with numerous studies demonstrating its efficacy and utility as a potential chemotherapeutic across a broad range of cancer types (Schrecengost, Keller, Schiewer, Knudsen, & Smith, 2015; Yang et al., 2015). According to clinicaltrials.gov, six separate trials used ABC294640 against a variety of cancers including pancreatic cancer, cholangiocarcinoma, multiple myeloma, and hepatocellular carcinoma. Trademarked as YELIVA, ABC294640 is currently being employed in a single-arm phase IIA clinical study in the treatment of cholangiocarcinoma at the Mayo Clinic and the MD Anderson Cancer Center (Clinical trial #: NCT03377179).

By far the most common sphingosine analogue in clinical trials today, FTY720 (also known as fingolimod and trademarked as gilenya) has been referenced 94 times on clinicaltrials.gov and is an FDA-approved therapy for multiple sclerosis (Ingwersen et al., 2012). Though reports differ, the most universal mechanism for its potential anti-cancer properties is outcompeting sphingosine for the binding pocket of SphK, thus limiting the conversion of sphingosine to S1P (Billich et al., 2003; Lim et al., 2011; Paugh, Payne, Barbour, Milstien, & Spiegel, 2003). The first trial using FTY720 as a potential treatment for cancer began in 2015, when a phase I safety study of FTY720 with radiation and temozolomide against glioblastoma or anaplastic astrocytoma was conducted at Johns Hopkins University (Clinical trial #: NCT02490930) (ClinicalTrials.gov). For more information regarding FTY720's role in cancer treatment, see the 2016 review by White et al. (2016). FTY720 has shown promise as a cancer therapy partly because of its ability to prevent pro-tumorigenic T cell transport out of the thymus by blocking the S1P receptor (S1PR) on T cells (Matloubian et al., 2004; Segui, Andrieu-Abadie, Jaffrezou, Benoist, & Levade, 2006). In addition, safingol is currently in the clinical setting for a phase 1 clinical trial for metastatic solid tumors in combination with fenretinide. Another clinical trial already completed used safingol combined with cisplatin for metastatic solid tumors. This trial showed that administering safingol with cisplatin was safe for cancer patients and ascertained as a recommended dose for safingol in a future phase 2 trial (Dickson et al., 2011).

Other, glyco- and sphingolipid analogues have also been shown to regulate immune components, which are dysregulated in cancer. α-Galactosylceramide, a sphingolipid produced by marine sponges, activates invariant natural killer T cells, which recruits anti-tumor immune cells (Berkers & Ovaa, 2005) (McEwen-Smith, Salio, & Cerundolo, 2015). A chemical modified α-galactosylceramide, KRN7000, has been used in a phase 1/2 clinical trial for treatment of patients with NCT00698776 myeloma (Berkers & Ovaa, 2005; Ishikawa et al., 2005). Other sphingolipid-targeting drugs are being developed that have immune-regulating properties. The ACDase inhibitor LCL521 decreased viability of the tumor-promoting myeloid-derived suppressor cells (MDSCs) in mice (F. Liu et al., 2016; Umansky & Sevko, 2013). In addition, SphK2-dependent dihydrosphingosine-1-phosphate also ameliorates MDSCs in mice models of solid tumors (Barth et al., 2013). Finally, GM3 ganglioside, overexpressed in human melanomas, was recently targeted with a preventative vaccine in mice models that resulted in rejection of B16 melanoma tumors in a CD8+ dependent manner (Mazorra, Mesa, Fernandez, & Fernandez, 2008).

4. SPHINGOLIPIDS SYNERGIZE WITH CHEMOTHERAPEUTICS

Recent studies have shown that elevating endogenous levels of ceramide by opening metabolic "faucets" or closing some of the enzymatic "drains" can augment the efficacy of conventional treatments in cancer. In addition, exogenous delivery platforms for sphingolipid mimetics can also elevate endogenous ceramide levels (Fig. 1). This section will review underlying combinatorial/synergistic cytotoxic mechanisms between elevated ceramide levels/decreased S1P levels and chemotherapeutic regulation of signal transduction cascades.

4.1 Synergistic Therapeutic Strategies Against Glucosylceramide Synthase

Perhaps the most prevalent area of research regarding synergy of sphingolipid therapies involves inhibiting GCS to limit conversion of pro-apoptotic ceramide into the less toxic glucosylceramide in cancer cells. Additionally, inhibitors of GCS downregulate the multidrug resistance gene 1 (MDR1) and its protein product P-glycoprotein (P-gp) reducing drug resistance (Gouaze et al., 2005; Morad & Cabot, 2013).

In breast cancer cell lines, an antisense-driven GCS knockdown led to higher ceramide levels and increased caspase-3 activity, which resensitized these cells to doxorubicin (Liu, Han, Giuliano, Hansen, & Cabot, 2000). The same lab later showed that using a pharmacological approach to inhibit GCS could also synergize with doxorubicin (an anthracycline chemotherapeutic). However, the novel, mixed backbone oligonucleotide antisense to GCS (MBO-asGCS), showed greater than 10-fold sensitivity to doxorubicin in doxorubicin-resistant lines of breast and ovarian cancer, through increased drug uptake and C_{18}-ceramide—induced apoptosis (Liu et al., 2004; Patwardhan et al., 2009). The MBO-asGCS was also found to decrease levels of MDR-1, cSrc kinase, nuclear β-catenin, and P-gp that sensitized breast, ovarian, cervical, and colon cancer cells to doxorubicin (Liu et al., 2010). Finally, the sensitization of ovarian cancer cells by PPMP or MBO-asGCS to doxorubicin was shown to only be effective in p53 mutants and could be mitigated by CerS inhibition using FB1 (Liu, Patwardhan, Bhinge, et al., 2011).

Using GCS antisense RNA and paclitaxel together enhanced ceramide formation, but, interestingly, also showed 10-fold higher levels of intracellular drug and 80% less P-gp beginning to suggest the relationship between P-gp and GCS. PPMP also restored sensitivity to vinblastine by enhancing its uptake by threefold, and diminished MDR1 expression by 84% in a vinblastine-resistant cell line (Gouaze et al., 2005). Additional studies using shRNAs against GCS confirmed increased apoptosis following treatment with doxorubicin and vinblastine while also displaying synergy with daunorubicin (Sun et al., 2006) and paclitaxel (Sun et al., 2010).

A study in 2001 that showed the GCS inhibitor DL-threo-phenyl-2-hexadecanoylamino-3-pyrrolidino-1-propanol (PPPP) synergizes with vincristine in leukemic cell lines (Olshefski & Ladisch, 2001) paved the way for a more in-depth analysis of 65 human leukemic samples. This study established not only that GCS expression levels correlated with increased multidrug resistance in patients and cell lines but also that PPMP could sensitize leukemic cell lines to doxorubicin (Xie et al., 2008). Counterintuitively, one study in leukemia cell lines showed that by blocking GCS using both PDMP and PPMP, cells were no longer sensitive to daunorubicin due to an anti-apoptotic effect of accumulated galactosylceramides (Grazide et al., 2004).

GCS inhibition using PDMP has also been shown to synergize with new treatment options, including BCL2-inhibitor ABT-737 (Casson et al., 2013) and Imatinib (Baran, Bielawski, Gunduz, & Ogretmen, 2011) in multiple leukemia cell lines, with temozolomide in glioblastoma multiforme (Giussani, Tringali, Riboni, Viani, & Venerando, 2014), with curcumin in melanoma (Yu, Li, Qiu, & Sun, 2012), and with sorafenib in hepatocellular carcinoma (Stefanovic et al., 2016). Efficacy of glucosylceramide synthase inhibition by PPMP and tamoxifen (an anti-estrogen with P-gp antagonistic properties) was also shown to be augmented in combination with C_6-ceramide in neuroepithelioma and ovarian cancer cells, respectively (Chapman, Gouaze-Andersson, Karimi, Messner, & Cabot, 2011; Spinedi, Bartolomeo, & Piacentini, 1998).

4.2 Synergistic Therapeutic Strategies Against Sphingosine Kinase

As mentioned previously, S1P is the most pro-mitogenic sphingolipid. SphK inhibitors decrease S1P levels in cells and increase levels of the more-toxic sphingosine leading frequently to increased ceramides as well.

The first of many synergy studies investigating SphK inhibition in blood cancers was performed in 2006 when a combination of F-12509A, a microbially derived weak inhibitor of SphK and CerK, and doxorubicin or etoposide effectively resensitized AML cell lines, which had developed chemoresistance (Bonhoure et al., 2006). Additionally, decreased SphK1 activity resulting from treatment with etoposide in combination with Perifosine (an Akt inhibitor) was able to increase Fas-receptor—mediated cell death in human T-leukemia—mimicking Jurkat cells (Nyakern, Cappellini, Mantovani, & Martelli, 2006). Another SphK1 inhibitor (Ski), but not a SphK2 inhibitor (ROME), was able to synergistically induce autophagic cell death in T-ALL cells in combination with vincristine (Evangelisti et al., 2014). A different SphK2 inhibitor (Ski-II) was able to synergize with ABT-263 (a BCL2-like protein inhibitor) by causing accumulation of sphingosine and ceramide, leading to in vitro death of multiple human leukemia cell lines (Casson et al., 2013). In myeloma lines, K145 (a specific SphK2 inhibitor) in combination with Bortezomib, a receptor tyrosine kinase (RTK) inhibitor, synergistically induced ER stress and led to apoptosis (Wallington-Beddoe et al., 2017).

This combination of SphK and RTK inhibitors was also successful in solid tumor models. A combination of ABC294640 (a SphK2 inhibitor) and another RTK inhibitor, sorafenib, synergistically induced cytotoxicity in hepatocellular carcinoma (Beljanski, Lewis, & Smith, 2011), pancreatic adenocarcinoma, kidney carcinoma (Beljanski, Knaak, Zhuang, & Smith, 2011), and cholangiocarcinoma (Ding et al., 2016). Finally, there appears to be promise in combining the previously mentioned Ski-II with temozolomide (a monofunctional alkylator) to synergistically increase caspase-dependent cell death in glioblastoma multiforme (Noack, Choi, Richter, Kopp-Schneider, & Regnier-Vigouroux, 2014).

4.3 Synergistic Therapeutic Strategies Against Acid Ceramidase

In multiple prostate cancer cell lines, the ACDase inhibitor NOE synergized with the EGFR inhibitor PD135035 to show drastic increases in both the estimated levels of ceramide as well as the percentage of cells undergoing apoptosis (Mimeault, Pommery, & Henichart, 2003). Interestingly, despite the success of NOE in the previous study, another group studying prostate cancer found that NOE did not synergize when combined with another compound, 4-HPR (fenretinide). However, using a different ACDase inhibitor, DM102, with fenretinide synergistically decreased cell viability, enhanced caspase activity, led to a 30-fold increase in ROS levels, and increased dihydroceramide levels. Although these results suggested activation of de novo ceramide synthesis, myriocin was unable to rescue from this cytotoxic effect, whereas surprisingly, antioxidant vitamin E did convey a partial rescue (Gouaze-Andersson et al., 2011).

Further work in head and neck squamous cell carcinoma showed that while overexpression of ACDase promoted resistance to Fas ligand-induced cell death, Fas-induced death was also potentiated by siRNAs against ACDase. Additionally, inhibition of ACDase using LCL 204 in the presence of Fas ligand led to cellular effects including: decreased survivin protein, increased cleaved caspase-8, and increased caspase-3 levels in vitro. The combination of these same two compounds also decreased tumor volume and promoted survival in in vivo xenograft models (Elojeimy et al., 2007). ACDase inhibition additionally showed synergy with daunorubicin in hepatocarcinoma cells. While NOE or siRNAs against ACDase in combination with daunorubicin increased the ceramide:S1P ratio resulting in increased cell death in vitro, the siRNA and daunorubicin combination was also efficacious in decreasing tumor volume in in vivo xenograft models. This synergistic cell death, which did not occur in healthy mouse hepatocytes, was preceded by markers of mitochondrial-induced apoptosis including ultrastructural changes, generation of ROS, release of cytochrome C, and caspase-3 activation (Morales et al., 2007).

4.4 Synergistic Therapeutic Strategies Using Exogenous Sphingolipids

Short-chain ceramides including C_2 and C_6 species have increased solubility and cell permeability compared with longer-chain, more physiological, ceramides. Short-chain ceramides have proven to be synergistic with docetaxel, doxorubicin, methotrexate and pemetrexed (antifolates) in murine melanoma cell lines, and osteosarcoma (Chapman et al., 2011; Feng, Li, Liu, Yang, & Zhang, 2014; Zhai, Sun, Han, Jin, & Zhang, 2015). Combining C_6-ceramide with ACDase inhibitor (DM102), AZD-8055 (mTORC1/2 dual inhibitor), or AT406 (a small molecule IAP antagonist) was synergistically efficacious in breast cancer cell lines, hepatocellular carcinoma lines, or pancreatic cell lines and primary samples, respectively (Flowers et al., 2012; M. Liu, Gu, Guo, & Fan, 2016; Zhao, Sun, Zhang, Zhang, & Zhang, 2016).

Nanotechnology platforms to improve ceramide delivery including CNL have shown synergy with sorafenib, PPMP, FTY720, gemcitabine, vinblastine, and tamoxifen/verapamil/cyclosporine A in in vitro and in vivo models of melanoma, breast, natural killer cell leukemia, pancreatic, hepatocellular carcinoma, and colorectal cancer, respectively (Adiseshaiah et al., 2013; Chapman et al., 2010; Jiang et al., 2011; Tran, Smith, Kester, & Robertson, 2008; Watters et al., 2013). It has been hypothesized that vinblastine and other microtubule destabilizers such as chloroquine and vincristine synergize with C_6 ceramide by reducing autophagosome formation and switching ceramide-induced autophagy into ceramide-induced cell death (Adiseshaiah et al., 2013; Lima, Milstien, & Spiegel, 2017; Young et al., 2016). A recent study showed that the use of CNL combined with adoptive T cell transfer immunotherapy in mice bearing liver tumors was able to reduce the number of tumor-associated macrophages and decreased the ability of these immune cells to suppress the anti-tumor properties (Li et al., 2018).

5. CONCLUSION

In the ever-growing quest to better understand the basis of cancer, numerous tools and collaborative efforts have emerged as powerful initiatives for research and medicine. The Cancer Genome Atlas represents one of the largest such endeavors, chronicling patient samples and organizing them by cancer type, gene expression, RNA profile, and mutation profile, among others. With the increasing sensitivity and selectivity as well as availability of mass spectrometry devices capable of creating full sphingolipid profiles for patients, we would argue that these vital class of lipids should be included when discussing patient diagnosis and treatment plans, as every year new research is published establishing links between sphingolipids and cancer.

As researchers continue to define the essential cellular functions of sphingolipids, their pivotal role in cancer progression is beginning to come into focus. Discoveries that established links between sphingolipid metabolism and carcinogenesis have cemented the need for a more complete understanding of these bioactive lipids. The medical advances in systems biology, big data, genomics, and epigenetics will synergize with advances in lipidomics to further the precision medicine revolution. All of these tools will further advance the utility of sphingolipid-based chemotherapeutics in oncology. Sphingolipid metabolism inhibitors, sphingolipid-modulated immunotherapies, cannabinoids, and sphingolipid analogues hold the promise for entirely new approach to cancer treatments.

ACKNOWLEDGMENTS

This work was supported by National Institutes of Health Grants 5 P01 CA171983 and R01s CA208396, CA167535 to MK; R01 GM043880 to SS.

Conflict of Interest

Penn State Research Foundation has licensed ceramide nanoliposomes and other sphingolipid-based nanoscale therapeutics to Keystone Nano, Inc (State College, PA). MK is Chief Medical Officer and cofounder of Keystone Nano, Inc.

REFERENCES

Adiseshaiah, P. P., Clogston, J. D., McLeland, C. B., Rodriguez, J., Potter, T. M., Neun, B. W., et al. (2013). Synergistic combination therapy with nanoliposomal C6-ceramide and vinblastine is associated with autophagy dysfunction in hepatocarcinoma and colorectal cancer models. *Cancer Letters, 337*(2), 254–265.

Akao, Y., Banno, Y., Nakagawa, Y., Hasegawa, N., Kim, T. J., & Murate, T. (2006). High expression of sphingosine kinase 1 and S1P receptors in chemotherapy-resistant prostate cancer PC3 cells and their camptothecin-induced up-regulation. *Biochemical and Biophysical Research Communications, 342*(4), 1284–1290.

Albouz, S., Hauw, J. J., Berwald-Netter, Y., Boutry, J. M., Bourdon, R., & Baumann, N. (1981). Tricyclic antidepressants induce sphingomyelinase deficiency in fibroblast and neuroblastoma cell cultures. *Biomedicine, 35*(7–8), 218–220.

Alfonso, P., Pampin, S., Estrada, J., Rodriguez-Rey, J. C., Giraldo, P., Sancho, J., et al. (2005). Miglustat (NB-DNJ) works as a chaperone for mutated acid beta-glucosidase in cells transfected with several Gaucher disease mutations. *Blood Cells, Molecules & Diseases, 35*(2), 268–276.

Amtmann, E., & Zoller, M. (2005). Stimulation of CD95-induced apoptosis in T-cells by a subtype specific neutral sphingomyelinase inhibitor. *Biochemical Pharmacology, 69*(8), 1141–1148.

Amtmann, E., Zoller, M., & Schilling, G. (2000). Neutral sphingomyelinase-inhibiting guanidines prevent herpes simplex virus-1 replication. *Drugs Under Experimental and Clinical Research, 26*(2), 57–65.

Antoon, J. W., White, M. D., Meacham, W. D., Slaughter, E. M., Muir, S. E., Elliott, S., et al. (2010). Antiestrogenic effects of the novel sphingosine kinase-2 inhibitor ABC294640. *Endocrinology, 151*(11), 5124–5135.

Aoyama, Y., Sobue, S., Mizutani, N., Inoue, C., Kawamoto, Y., Nishizawa, Y., et al. (2017). Modulation of the sphingolipid rheostat is involved in paclitaxel resistance of the human prostate cancer cell line PC3-PR. *Biochemical and Biophysical Research Communications, 486*(2), 551–557.

Ardail, D., Maalouf, M., Boivin, A., Chapet, O., Bodennec, J., Rousson, R., et al. (2009). Diversity and complexity of ceramide generation after exposure of jurkat leukemia cells to irradiation. *International Journal of Radiation Oncology, Biology, Physics, 73*(4), 1211–1218.

Arenz, C., Thutewohl, M., Block, O., Waldmann, H., Altenbach, H. J., & Giannis, A. (2001). Manumycin A and its analogues are irreversible inhibitors of neutral sphingomyelinase. *Chembiochem, 2*(2), 141–143.

Arimochi, H., & Morita, K. (2008). Desipramine induces apoptotic cell death through nonmitochondrial and mitochondrial pathways in different types of human colon carcinoma cells. *Pharmacology, 81*(2), 164–172.

Aureli, M., Murdica, V., Loberto, N., Samarani, M., Prinetti, A., Bassi, R., et al. (2014). Exploring the link between ceramide and ionizing radiation. *Glycoconjugate Journal, 31*(6–7), 449–459.

Bai, A., Mao, C., Jenkins, R. W., Szulc, Z. M., Bielawska, A., & Hannun, Y. A. (2017). Anticancer actions of lysosomally targeted inhibitor, LCL521, of acid ceramidase. *PLoS One, 12*(6), e0177805.

Bai, A., Meier, G. P., Wang, Y., Luberto, C., Hannun, Y. A., & Zhou, D. (2004). Prodrug modification increases potassium tricyclo[5.2.1.0(2,6)]-decan-8-yl dithiocarbonate (D609) chemical stability and cytotoxicity against U937 leukemia cells. *The Journal of Pharmacology and Experimental Therapeutics, 309*(3), 1051–1059.

Bandhuvula, P., Tam, Y. Y., Oskouian, B., & Saba, J. D. (2005). The immune modulator FTY720 inhibits sphingosine-1-phosphate lyase activity. *Journal of Biological Chemistry, 280*(40), 33697–33700.

Baran, Y., Bielawski, J., Gunduz, U., & Ogretmen, B. (2011). Targeting glucosylceramide synthase sensitizes imatinib-resistant chronic myeloid leukemia cells via endogenous ceramide accumulation. *Journal of Cancer Research and Clinical Oncology, 137*(10), 1535–1544.

Barbado, M. V., Medrano, M., Caballero-Velazquez, T., Alvarez-Laderas, I., Sanchez-Abarca, L. I., Garcia-Guerrero, E., et al. (2017). Cannabinoid derivatives exert a potent anti-myeloma activity both in vitro and in vivo. *International Journal of Cancer, 140*(3), 674–685.

Barth, B. M., Shanmugavelandy, S. S., Kaiser, J. M., McGovern, C., Altinoglu, E. I., Haakenson, J. K., et al. (2013). PhotoImmunoNanoTherapy reveals an anticancer role for sphingosine kinase 2 and dihydrosphingosine-1-phosphate. *ACS Nano, 7*(3), 2132–2144.

Beckham, T. H., Cheng, J. C., Marrison, S. T., Norris, J. S., & Liu, X. (2013). Interdiction of sphingolipid metabolism to improve standard cancer therapies. *Advances in Cancer Research, 117*, 1–36.

Beckham, T. H., Lu, P., Cheng, J. C., Zhao, D., Turner, L. S., Zhang, X., et al. (2012). Acid ceramidase-mediated production of sphingosine 1-phosphate promotes prostate cancer invasion through upregulation of cathepsin B. *International Journal of Cancer, 131*(9), 2034–2043.

Beckham, T. H., Lu, P., Jones, E. E., Marrison, T., Lewis, C. S., Cheng, J. C., et al. (2013). LCL124, a cationic analog of ceramide, selectively induces pancreatic cancer cell death by accumulating in mitochondria. *The Journal of Pharmacology and Experimental Therapeutics, 344*(1), 167–178.

Beckmann, N., Sharma, D., Gulbins, E., Becker, K. A., & Edelmann, B. (2014). Inhibition of acid sphingomyelinase by tricyclic antidepressants and analogons. *Frontiers in Physiology, 5*, 331.

Bedia, C., Triola, G., Casas, J., Llebaria, A., & Fabrias, G. (2005). Analogs of the dihydroceramide desaturase inhibitor GT11 modified at the amide function: Synthesis and biological activities. *Organic & Biomolecular Chemistry, 3*(20), 3707–3712.

Beljanski, V., Knaak, C., & Smith, C. D. (2010). A novel sphingosine kinase inhibitor induces autophagy in tumor cells. *The Journal of Pharmacology and Experimental Therapeutics, 333*(2), 454–464.

Beljanski, V., Knaak, C., Zhuang, Y., & Smith, C. D. (2011). Combined anticancer effects of sphingosine kinase inhibitors and sorafenib. *Investigational New Drugs, 29*(6), 1132–1142.

Beljanski, V., Lewis, C. S., & Smith, C. D. (2011). Antitumor activity of sphingosine kinase 2 inhibitor ABC294640 and sorafenib in hepatocellular carcinoma xenografts. *Cancer Biology & Therapy, 11*(5), 524–534.

Berkers, C. R., & Ovaa, H. (2005). Immunotherapeutic potential for ceramide-based activators of iNKT cells. *Trends in Pharmacological Sciences, 26*(5), 252–257.

Bhabak, K. P., & Arenz, C. (2013). Novel drugs targeting sphingolipid metabolism. *Handbook of Experimental Pharmacology, 215*, 187–196.

Bielawska, A., Greenberg, M. S., Perry, D., Jayadev, S., Shayman, J. A., McKay, C., et al. (1996). (1S,2R)-D-erythro-2-(N-myristoylamino)-1-phenyl-1-propanol as an inhibitor of ceramidase. *Journal of Biological Chemistry, 271*(21), 12646–12654.

Billich, A., Bornancin, F., Devay, P., Mechtcheriakova, D., Urtz, N., & Baumruker, T. (2003). Phosphorylation of the immunomodulatory drug FTY720 by sphingosine kinases. *Journal of Biological Chemistry, 278*(48), 47408–47415.

Bonhoure, E., Pchejetski, D., Aouali, N., Morjani, H., Levade, T., Kohama, T., et al. (2006). Overcoming MDR-associated chemoresistance in HL-60 acute myeloid leukemia cells by targeting sphingosine kinase-1. *Leukemia, 20*(1), 95–102.

Bose, R., Verheij, M., Haimovitz-Friedman, A., Scotto, K., Fuks, Z., & Kolesnick, R. (1995). Ceramide synthase mediates daunorubicin-induced apoptosis: An alternative mechanism for generating death signals. *Cell, 82*(3), 405–414.

Boumendjel, A., & Miller, S. P. (1994). Synthesis of sphingosine-1-phosphate and dihydrosphingosine-1-phosphate. *The Journal of Lipid Research, 35*(12), 2305–2311.

Brinkmann, V., Davis, M. D., Heise, C. E., Albert, R., Cottens, S., Hof, R., et al. (2002). The immune modulator FTY720 targets sphingosine 1-phosphate receptors. *Journal of Biological Chemistry, 277*(24), 21453–21457.

Buehrer, B. M., & Bell, R. M. (1992). Inhibition of sphingosine kinase in vitro and in platelets. Implications for signal transduction pathways. *Journal of Biological Chemistry, 267*(5), 3154–3159.

Cakir, Z., Saydam, G., Sahin, F., & Baran, Y. (2011). The roles of bioactive sphingolipids in resveratrol-induced apoptosis in HL60: Acute myeloid leukemia cells. *J Cancer Res Clin Oncol, 137*(2), 279–286.

Camacho, L., Meca-Cortes, O., Abad, J. L., Garcia, S., Rubio, N., Diaz, A., et al. (2013). Acid ceramidase as a therapeutic target in metastatic prostate cancer. *The Journal of Lipid Research, 54*(5), 1207–1220.

Canals, D., Perry, D. M., Jenkins, R. W., & Hannun, Y. A. (2011). Drug targeting of sphingolipid metabolism: Sphingomyelinases and ceramidases. *British Journal of Pharmacology, 163*(4), 694–712.

Carroll, B., Donaldson, J. C., & Obeid, L. (2015). Sphingolipids in the DNA damage response. *Advances in Biological Regulation, 58*, 38–52.

Casson, L., Howell, L., Mathews, L. A., Ferrer, M., Southall, N., Guha, R., et al. (2013). Inhibition of ceramide metabolism sensitizes human leukemia cells to inhibition of BCL2-like proteins. *PLoS One, 8*(1), e54525.

Chalfant, C. E., Rathman, K., Pinkerman, R. L., Wood, R. E., Obeid, L. M., Ogretmen, B., et al. (2002). De novo ceramide regulates the alternative splicing of caspase 9 and Bcl-x in A549 lung adenocarcinoma cells. Dependence on protein phosphatase-1. *Journal of Biological Chemistry, 277*(15), 12587–12595.

Chapman, J. V., Gouaze-Andersson, V., Karimi, R., Messner, M. C., & Cabot, M. C. (2011). P-glycoprotein antagonists confer synergistic sensitivity to short-chain ceramide in human multidrug-resistant cancer cells. *Experimental Cell Research, 317*(12), 1736–1745.

Chapman, J. V., Gouaze-Andersson, V., Messner, M. C., Flowers, M., Karimi, R., Kester, M., et al. (2010). Metabolism of short-chain ceramide by human cancer cells—implications for therapeutic approaches. *Biochemical Pharmacology, 80*(3), 308–315.

Cheng, J. C., Bai, A., Beckham, T. H., Marrison, S. T., Yount, C. L., Young, K., et al. (2013). Radiation-induced acid ceramidase confers prostate cancer resistance and tumor relapse. *Journal of Clinical Investigation, 123*(10), 4344–4358.

Chun, J., & Hartung, H. P. (2010). Mechanism of action of oral fingolimod (FTY720) in multiple sclerosis. *Clinical Neuropharmacology, 33*(2), 91–101.

Cianchi, F., Papucci, L., Schiavone, N., Lulli, M., Magnelli, L., Vinci, M. C., et al. (2008). Cannabinoid receptor activation induces apoptosis through tumor necrosis factor alpha-mediated ceramide de novo synthesis in colon cancer cells. *Clinical Cancer Research, 14*(23), 7691–7700.

Cowart, L. A., & Obeid, L. M. (2007). Yeast sphingolipids: Recent developments in understanding biosynthesis, regulation, and function. *Biochimica et Biophysica Acta, 1771*(3), 421–431.

Cox, T. M. (2005). Substrate reduction therapy for lysosomal storage diseases. *Acta Paediatrica Supplement, 94*(447), 69–75. discussion 57.

Cuvillier, O., Pirianov, G., Kleuser, B., Vanek, P. G., Coso, O. A., Gutkind, S., et al. (1996). Suppression of ceramide-mediated programmed cell death by sphingosine-1-phosphate. *Nature, 381*(6585), 800–803.

Dagan, A., Wang, C., Fibach, E., & Gatt, S. (2003). Synthetic, non-natural sphingolipid analogs inhibit the biosynthesis of cellular sphingolipids, elevate ceramide and induce apoptotic cell death. *Biochimica et Biophysica Acta, 1633*(3), 161–169.

Darroch, P. I., Dagan, A., Granot, T., He, X., Gatt, S., & Schuchman, E. H. (2005). A lipid analogue that inhibits sphingomyelin hydrolysis and synthesis, increases ceramide, and leads to cell death. *The Journal of Lipid Research, 46*(11), 2315–2324.

Deigner, H. P., Claus, R., Bonaterra, G. A., Gehrke, C., Bibak, N., Blaess, M., et al. (2001). Ceramide induces aSMase expression: Implications for oxLDL-induced apoptosis. *The FASEB Journal, 15*(3), 807–814.

Delgado, A., Casas, J., Llebaria, A., Abad, J. L., & Fabrias, G. (2006). Inhibitors of sphingolipid metabolism enzymes. *Biochimica et Biophysica Acta, 1758*(12), 1957–1977.

Dick, T. E., Hengst, J. A., Fox, T. E., Colledge, A. L., Kale, V. P., Sung, S. S., et al. (2015). The apoptotic mechanism of action of the sphingosine kinase 1 selective inhibitor SKI-178 in human acute myeloid leukemia cell lines. *The Journal of Pharmacology and Experimental Therapeutics, 352*(3), 494–508.

Dickson, M. A., Carvajal, R. D., Merrill, A. H., Jr., Gonen, M., Cane, L. M., & Schwartz, G. K. (2011). A phase I clinical trial of safingol in combination with cisplatin in advanced solid tumors. *Clinical Cancer Research, 17*(8), 2484–2492.

Ding, X., Chaiteerakij, R., Moser, C. D., Shaleh, H., Boakye, J., Chen, G., et al. (2016). Antitumor effect of the novel sphingosine kinase 2 inhibitor ABC294640 is enhanced by inhibition of autophagy and by sorafenib in human cholangiocarcinoma cells. *Oncotarget, 7*(15), 20080–20092.

Draper, J. M., Xia, Z., Smith, R. A., Zhuang, Y., Wang, W., & Smith, C. D. (2011). Discovery and evaluation of inhibitors of human ceramidase. *Molecular Cancer Therapeutics, 10*(11), 2052–2061.

Edsall, L. C., Van Brocklyn, J. R., Cuvillier, O., Kleuser, B., & Spiegel, S. (1998). N,N-dimethylsphingosine is a potent competitive inhibitor of sphingosine kinase but not of protein kinase C: Modulation of cellular levels of sphingosine 1-phosphate and ceramide. *Biochemistry, 37*(37), 12892–12898.

Eliyahu, E., Shtraizent, N., He, X., Chen, D., Shalgi, R., & Schuchman, E. H. (2011). Identification of cystatin SA as a novel inhibitor of acid ceramidase. *Journal of Biological Chemistry, 286*(41), 35624–35633.

Elojeimy, S., Liu, X., McKillop, J. C., El-Zawahry, A. M., Holman, D. H., Cheng, J. Y., et al. (2007). Role of acid ceramidase in resistance to FasL: Therapeutic approaches based on acid ceramidase inhibitors and FasL gene therapy. *Molecular Therapy, 15*(7), 1259–1263.

Evangelisti, C., Evangelisti, C., Teti, G., Chiarini, F., Falconi, M., Melchionda, F., et al. (2014). Assessment of the effect of sphingosine kinase inhibitors on apoptosis,unfolded protein response and autophagy of T-cell acute lymphoblastic leukemia cells; indications for novel therapeutics. *Oncotarget, 5*(17), 7886–7901.

Feng, L. X., Li, M., Liu, Y. J., Yang, S. M., & Zhang, N. (2014). Synergistic enhancement of cancer therapy using a combination of ceramide and docetaxel. *International Journal of Molecular Sciences, 15*(3), 4201–4220.

Figuera-Losada, M., Stathis, M., Dorskind, J. M., Thomas, A. G., Bandaru, V. V., Yoo, S. W., et al. (2015). Cambinol, a novel inhibitor of neutral sphingomyelinase 2 shows neuroprotective properties. *PLoS One, 10*(5), e0124481.

Flowers, M., Fabrias, G., Delgado, A., Casas, J., Abad, J. L., & Cabot, M. C. (2012). C6-ceramide and targeted inhibition of acid ceramidase induce synergistic decreases in breast cancer cell growth. *Breast Cancer Research and Treatment, 133*(2), 447–458.

French, K. J., Schrecengost, R. S., Lee, B. D., Zhuang, Y., Smith, S. N., Eberly, J. L., et al. (2003). Discovery and evaluation of inhibitors of human sphingosine kinase. *Cancer Research, 63*(18), 5962–5969.

Galadari, S., Rahman, A., Pallichankandy, S., & Thayyullathil, F. (2015). Tumor suppressive functions of ceramide: Evidence and mechanisms. *Apoptosis, 20*(5), 689–711.

Garcia-Barros, M., Coant, N., Kawamori, T., Wada, M., Snider, A. J., Truman, J. P., et al. (2016). Role of neutral ceramidase in colon cancer. *The FASEB Journal, 30*(12), 4159–4171.

Giussani, P., Tringali, C., Riboni, L., Viani, P., & Venerando, B. (2014). Sphingolipids: Key regulators of apoptosis and pivotal players in cancer drug resistance. *International Journal of Molecular Sciences, 15*(3), 4356–4392.

Gouaze-Andersson, V., Flowers, M., Karimi, R., Fabrias, G., Delgado, A., Casas, J., et al. (2011). Inhibition of acid ceramidase by a 2-substituted aminoethanol amide synergistically sensitizes prostate cancer cells to N-(4-hydroxyphenyl) retinamide. *The Prostate, 71*(10), 1064–1073.

Gouaze, V., Liu, Y. Y., Prickett, C. S., Yu, J. Y., Giuliano, A. E., & Cabot, M. C. (2005). Glucosylceramide synthase blockade down-regulates P-glycoprotein and resensitizes multidrug-resistant breast cancer cells to anticancer drugs. *Cancer Research, 65*(9), 3861–3867.

Graf, C., Klumpp, M., Habig, M., Rovina, P., Billich, A., Baumruker, T., et al. (2008). Targeting ceramide metabolism with a potent and specific ceramide kinase inhibitor. *Molecular Pharmacology, 74*(4), 925–932.

Grazide, S., Terrisse, A. D., Lerouge, S., Laurent, G., & Jaffrezou, J. P. (2004). Cytoprotective effect of glucosylceramide synthase inhibition against daunorubicin-induced apoptosis in human leukemic cell lines. *Journal of Biological Chemistry, 279*(18), 18256–18261.

Gustafsson, K., Christensson, B., Sander, B., & Flygare, J. (2006). Cannabinoid receptor-mediated apoptosis induced by R(+)-methanandamide and Win55,212-2 is associated with ceramide accumulation and p38 activation in mantle cell lymphoma. *Molecular Pharmacology, 70*(5), 1612–1620.

Gustafsson, K., Sander, B., Bielawski, J., Hannun, Y. A., & Flygare, J. (2009). Potentiation of cannabinoid-induced cytotoxicity in mantle cell lymphoma through modulation of ceramide metabolism. *Molecular Cancer Research, 7*(7), 1086–1098.

Haimovitz-Friedman, A., Kan, C. C., Ehleiter, D., Persaud, R. S., McLoughlin, M., Fuks, Z., et al. (1994). Ionizing radiation acts on cellular membranes to generate ceramide and initiate apoptosis. *The Journal of Experimental Medicine, 180*(2), 525–535.

Halperin, E. C., Perez, C. A., & Brady, L. W. (2008). *Perez and Brady's principles and practice of radiation oncology* (5th ed.). Philadelphia: Wolters Kluwer Health/Lippincott Williams & Wilkins.

Hanada, K. (2003). Serine palmitoyltransferase, a key enzyme of sphingolipid metabolism. *Biochimica et Biophysica Acta, 1632*(1–3), 16–30.

Hannun, Y. A., & Obeid, L. M. (2018). Sphingolipids and their metabolism in physiology and disease. *Nature Reviews. Molecular Cell Biology, 19*(3), 175–191.

Hara, S., Nakashima, S., Kiyono, T., Sawada, M., Yoshimura, S., Iwama, T., et al. (2004). p53-Independent ceramide formation in human glioma cells during gamma-radiation-induced apoptosis. *Cell Death and Differentiation, 11*(8), 853–861.

Hengst, J. A., Dick, T. E., Sharma, A., Doi, K., Hegde, S., Tan, S. F., et al. (2017). SKI-178: A multitargeted inhibitor of sphingosine kinase and microtubule dynamics demonstrating therapeutic efficacy in acute myeloid leukemia models. *Cancer Translational Medicine, 3*(4), 109–121.

Hillig, I., Warnecke, D., & Heinz, E. (2005). An inhibitor of glucosylceramide synthase inhibits the human enzyme, but not enzymes from other organisms. *Bioscience, Biotechnology, and Biochemistry, 69*(9), 1782–1785.

Holman, D. H., Turner, L. S., El-Zawahry, A., Elojeimy, S., Liu, X., Bielawski, J., et al. (2008). Lysosomotropic acid ceramidase inhibitor induces apoptosis in prostate cancer cells. *Cancer Chemotherapy and Pharmacology, 61*(2), 231–242.

Huang, C. J., Cheng, H. H., Chou, C. T., Kuo, C. C., Lu, Y. C., Tseng, L. L., et al. (2007). Desipramine-induced Ca2+ movement and cytotoxicity in PC3 human prostate cancer cells. *Toxicology in Vitro, 21*(3), 449–456.

Ingwersen, J., Aktas, O., Kuery, P., Kieseier, B., Boyko, A., & Hartung, H. P. (2012). Fingolimod in multiple sclerosis: Mechanisms of action and clinical efficacy. *Clinical Immunology, 142*(1), 15–24.

Ishikawa, A., Motohashi, S., Ishikawa, E., Fuchida, H., Higashino, K., Otsuji, M., et al. (2005). A phase I study of alpha-galactosylceramide (KRN7000)-pulsed dendritic cells in patients with advanced and recurrent non-small cell lung cancer. *Clinical Cancer Research, 11*(5), 1910–1917.

Jaffrezou, J. P., Herbert, J. M., Levade, T., Gau, M. N., Chatelain, P., & Laurent, G. (1991). Reversal of multidrug resistance by calcium channel blocker SR33557 without photoaffinity labeling of P-glycoprotein. *Journal of Biological Chemistry, 266*(29), 19858–19864.

Jaffrezou, J. P., Levade, T., Bettaieb, A., Andrieu, N., Bezombes, C., Maestre, N., et al. (1996). Daunorubicin-induced apoptosis: Triggering of ceramide generation through sphingomyelin hydrolysis. *The EMBO Journal, 15*(10), 2417–2424.

Jatoi, A., Suman, V. J., Schaefer, P., Block, M., Loprinzi, C., Roche, P., et al. (2003). A phase II study of topical ceramides for cutaneous breast cancer. *Breast Cancer Research and Treatment, 80*(1), 99–104.

Jiang, Y., DiVittore, N. A., Kaiser, J. M., Shanmugavelandy, S. S., Fritz, J. L., Heakal, Y., et al. (2011). Combinatorial therapies improve the therapeutic efficacy of nanoliposomal ceramide for pancreatic cancer. *Cancer Biology & Therapy, 12*(7), 574–585.

Kester, M., Bassler, J., Fox, T. E., Carter, C. J., Davidson, J. A., & Parette, M. R. (2015). Preclinical development of a C6-ceramide NanoLiposome, a novel sphingolipid therapeutic. *Biological Chemistry, 396*(6–7), 737–747.

Kim, J. W., Inagaki, Y., Mitsutake, S., Maezawa, N., Katsumura, S., Ryu, Y. W., et al. (2005). Suppression of mast cell degranulation by a novel ceramide kinase inhibitor, the F-12509A olefin isomer K1. *Biochimica et Biophysica Acta, 1738*(1–3), 82–90.

Kim, J. W., Kim, Y. W., Inagaki, Y., Hwang, Y. A., Mitsutake, S., Ryu, Y. W., et al. (2005). Synthesis and evaluation of sphingoid analogs as inhibitors of sphingosine kinases. *Bioorganic & Medicinal Chemistry, 13*(10), 3475–3485.

Klutzny, S., Lesche, R., Keck, M., Kaulfuss, S., Schlicker, A., Christian, S., et al. (2017). Functional inhibition of acid sphingomyelinase by Fluphenazine triggers hypoxia-specific tumor cell death. *Cell Death & Disease, 8*(3), e2709.

Kolesnick, R., & Fuks, Z. (2003). Radiation and ceramide-induced apoptosis. *Oncogene, 22*(37), 5897–5906.

Kolzer, M., Arenz, C., Ferlinz, K., Werth, N., Schulze, H., Klingenstein, R., et al. (2003). Phosphatidylinositol-3,5-bisphosphate is a potent and selective inhibitor of acid sphingomyelinase. *Biological Chemistry, 384*(9), 1293–1298.

Kono, K., Sugiura, M., & Kohama, T. (2002). Inhibition of recombinant sphingosine kinases by novel inhibitors of microbial origin, F-12509A and B-5354c. *Journal of Antibiotics, 55*(1), 99–103.

Kono, K., Tanaka, M., Mizuno, T., Kodama, K., Ogita, T., & Kohama, T. (2000). B-535a, b and c, new sphingosine kinase inhibitors, produced by a marine bacterium; taxonomy, fermentation, isolation, physico-chemical properties and structure determination. *Journal of Antibiotics, 53*(8), 753–758.

Kono, K., Tanaka, M., Ogita, T., Hosoya, T., & Kohama, T. (2000). F-12509A, a new sphingosine kinase inhibitor, produced by a discomycete. *Journal of Antibiotics, 53*(5), 459–466.

Kono, K., Tanaka, M., Ono, Y., Hosoya, T., Ogita, T., & Kohama, T. (2001). S-15183a and b, new sphingosine kinase inhibitors, produced by a fungus. *Journal of Antibiotics, 54*(5), 415–420.

Kornhuber, J., Tripal, P., Reichel, M., Muhle, C., Rhein, C., Muehlbacher, M., et al. (2010). Functional inhibitors of acid sphingomyelinase (FIASMAs): A novel pharmacological group of drugs with broad clinical applications. *Cellular Physiology and Biochemistry: International Journal of Experimental Cellular Physiology, Biochemistry, and Pharmacology, 26*(1), 9–20.

Kornhuber, J., Tripal, P., Reichel, M., Terfloth, L., Bleich, S., Wiltfang, J., et al. (2008). Identification of new functional inhibitors of acid sphingomyelinase using a structure-property-activity relation model. *Journal of Medicinal Chemistry, 51*(2), 219–237.

Kouchi, H., Nakamura, K., Fushimi, K., Sakaguchi, M., Miyazaki, M., Ohe, T., et al. (1999). Manumycin A, inhibitor of ras farnesyltransferase, inhibits proliferation and migration of rat vascular smooth muscle cells. *Biochemical and Biophysical Research Communications, 264*(3), 915–920.

Kovacs, P., Pinter, M., & Csaba, G. (2000). Effect of glucosphingolipid synthesis inhibitor (PPMP and PDMP) treatment on *Tetrahymena pyriformis*: Data on the evolution of the signaling system. *Cell Biochemistry and Function, 18*(4), 269–280.

Lavie, Y., Cao, H., Volner, A., Lucci, A., Han, T. Y., Geffen, V., et al. (1997). Agents that reverse multidrug resistance, tamoxifen, verapamil, and cyclosporin A, block glycosphingolipid metabolism by inhibiting ceramide glycosylation in human cancer cells. *Journal of Biological Chemistry, 272*(3), 1682–1687.

Lee, W. K., & Kolesnick, R. N. (2017). Sphingolipid abnormalities in cancer multidrug resistance: Chicken or egg? *Cellular Signalling, 38*, 134–145.
Li, F., & Zhang, N. (2015). Ceramide: Therapeutic potential in combination therapy for cancer treatment. *Current Drug Metabolism, 17*(1), 37–51.
Li, G., Liu, D., Kimchi, E. T., Kaifi, J. T., Qi, X., Manjunath, Y., et al. (2018). Nanoliposome C6-ceramide increases the anti-tumor immune response and slows growth of liver tumors in mice. *Gastroenterology, 154*(4), 1024–1036. e1029.
Lim, K. G., Tonelli, F., Li, Z., Lu, X., Bittman, R., Pyne, S., et al. (2011). FTY720 analogues as sphingosine kinase 1 inhibitors: Enzyme inhibition kinetics, allosterism, proteasomal degradation, and actin rearrangement in MCF-7 breast cancer cells. *Journal of Biological Chemistry, 286*(21), 18633–18640.
Lima, S., Milstien, S., & Spiegel, S. (2017). Sphingosine and sphingosine kinase 1 involvement in endocytic membrane trafficking. *Journal of Biological Chemistry, 292*(8), 3074–3088.
Linke, T., Wilkening, G., Lansmann, S., Moczall, H., Bartelsen, O., Weisgerber, J., et al. (2001). Stimulation of acid sphingomyelinase activity by lysosomal lipids and sphingolipid activator proteins. *Biological Chemistry, 382*(2), 283–290.
Liu, B., & Hannun, Y. A. (1997). Inhibition of the neutral magnesium-dependent sphingomyelinase by glutathione. *Journal of Biological Chemistry, 272*(26), 16281–16287.
Liu, F., Li, X., Lu, C., Bai, A., Bielawski, J., Bielawska, A., et al. (2016). Ceramide activates lysosomal cathepsin B and cathepsin D to attenuate autophagy and induces ER stress to suppress myeloid-derived suppressor cells. *Oncotarget, 7*(51), 83907–83925.
Liu, J., Beckman, B. S., & Foroozesh, M. (2013). A review of ceramide analogs as potential anticancer agents. *Future Medicinal Chemistry, 5*(12), 1405–1421.
Liu, K., Guo, T. L., Hait, N. C., Allegood, J., Parikh, H. I., Xu, W., et al. (2013). Biological characterization of 3-(2-amino-ethyl)-5-[3-(4-butoxyl-phenyl)-propylidene]-thiazolidine-2,4-dione (K145) as a selective sphingosine kinase-2 inhibitor and anticancer agent. *PLoS One, 8*(2), e56471.
Liu, M., Gu, P., Guo, W., & Fan, X. (2016). C6 ceramide sensitizes the anti-hepatocellular carcinoma (HCC) activity by AZD-8055, a novel mTORC1/2 dual inhibitor. *Tumour Biology, 37*(8), 11039–11048.
Liu, Y. Y., Gupta, V., Patwardhan, G. A., Bhinge, K., Zhao, Y., Bao, J., et al. (2010). Glucosylceramide synthase upregulates MDR1 expression in the regulation of cancer drug resistance through cSrc and beta-catenin signaling. *Molecular Cancer, 9*, 145.
Liu, Y. Y., Han, T. Y., Giuliano, A. E., Hansen, N., & Cabot, M. C. (2000). Uncoupling ceramide glycosylation by transfection of glucosylceramide synthase antisense reverses adriamycin resistance. *Journal of Biological Chemistry, 275*(10), 7138–7143.
Liu, Y. Y., Han, T. Y., Yu, J. Y., Bitterman, A., Le, A., Giuliano, A. E., et al. (2004). Oligonucleotides blocking glucosylceramide synthase expression selectively reverse drug resistance in cancer cells. *The Journal of Lipid Research, 45*(5), 933–940.
Liu, Y. Y., Patwardhan, G. A., Bhinge, K., Gupta, V., Gu, X., & Jazwinski, S. M. (2011). Suppression of glucosylceramide synthase restores p53-dependent apoptosis in mutant p53 cancer cells. *Cancer Research, 71*(6), 2276–2285.
Liu, Y. Y., Patwardhan, G. A., Xie, P., Gu, X., Giuliano, A. E., & Cabot, M. C. (2011). Glucosylceramide synthase, a factor in modulating drug resistance, is overexpressed in metastatic breast carcinoma. *International Journal of Oncology, 39*(2), 425–431.
Lou, B., Dong, J., Li, Y., Ding, T., Bi, T., Li, Y., et al. (2014). Pharmacologic inhibition of sphingomyelin synthase (SMS) activity reduces apolipoprotein-B secretion from hepatocytes and attenuates endotoxin-mediated macrophage inflammation. *PLoS One, 9*(7), e102641.
Lozano, J., Menendez, S., Morales, A., Ehleiter, D., Liao, W. C., Wagman, R., et al. (2001). Cell autonomous apoptosis defects in acid sphingomyelinase knockout fibroblasts. *Journal of Biological Chemistry, 276*(1), 442–448.

Luberto, C., Hassler, D. F., Signorelli, P., Okamoto, Y., Sawai, H., Boros, E., et al. (2002). Inhibition of tumor necrosis factor-induced cell death in MCF7 by a novel inhibitor of neutral sphingomyelinase. *Journal of Biological Chemistry, 277*(43), 41128–41139.

Lucci, A., Cho, W. I., Han, T. Y., Giuliano, A. E., Morton, D. L., & Cabot, M. C. (1998). Glucosylceramide: A marker for multiple-drug resistant cancers. *Anticancer Research, 18*(1B), 475–480.

Lu, T., Huang, C. C., Lu, Y. C., Lin, K. L., Liu, S. I., Wang, B. W., et al. (2009). Desipramine-induced Ca-independent apoptosis in Mg63 human osteosarcoma cells: Dependence on P38 mitogen-activated protein kinase-regulated activation of caspase 3. *Clinical and Experimental Pharmacology & Physiology, 36*(3), 297–303.

Ma, Y. Y., Mou, X. Z., Ding, Y. H., Zou, H., & Huang, D. S. (2016). Delivery systems of ceramide in targeted cancer therapy: Ceramide alone or in combination with other antitumor agents. *Expert Opinion on Drug Delivery, 13*(10), 1397–1406.

Mahdy, A. E., Cheng, J. C., Li, J., Elojeimy, S., Meacham, W. D., Turner, L. S., et al. (2009). Acid ceramidase upregulation in prostate cancer cells confers resistance to radiation: AC inhibition, a potential radiosensitizer. *Molecular Therapy, 17*(3), 430–438.

Mandala, S. M., Frommer, B. R., Thornton, R. A., Kurtz, M. B., Young, N. M., Cabello, M. A., et al. (1994). Inhibition of serine palmitoyl-transferase activity by lipoxamycin. *Journal of Antibiotics, 47*(3), 376–379.

Mandala, S. M., Thornton, R. A., Frommer, B. R., Curotto, J. E., Rozdilsky, W., Kurtz, M. B., et al. (1995). The discovery of australifungin, a novel inhibitor of sphinganine N-acyltransferase from *Sporormiella australis*. Producing organism, fermentation, isolation, and biological activity. *Journal of Antibiotics, 48*(5), 349–356.

Mandala, S. M., Thornton, R. A., Frommer, B. R., Dreikorn, S., & Kurtz, M. B. (1997). Viridiofungins, novel inhibitors of sphingolipid synthesis. *Journal of Antibiotics, 50*(4), 339–343.

Marchesini, N., Luberto, C., & Hannun, Y. A. (2003). Biochemical properties of mammalian neutral sphingomyelinase 2 and its role in sphingolipid metabolism. *Journal of Biological Chemistry, 278*(16), 13775–13783.

Matloubian, M., Lo, C. G., Cinamon, G., Lesneski, M. J., Xu, Y., Brinkmann, V., et al. (2004). Lymphocyte egress from thymus and peripheral lymphoid organs is dependent on S1P receptor 1. *Nature, 427*(6972), 355–360.

Mazorra, Z., Mesa, C., Fernandez, A., & Fernandez, L. E. (2008). Immunization with a GM3 ganglioside nanoparticulated vaccine confers an effector CD8(+) T cells-mediated protection against melanoma B16 challenge. *Cancer Immunology, Immunotherapy, 57*(12), 1771–1780.

McEwen-Smith, R. M., Salio, M., & Cerundolo, V. (2015). The regulatory role of invariant NKT cells in tumor immunity. *Cancer Immunology Research, 3*(5), 425–435.

Medlock, K. A., & Merrill, A. H., Jr. (1988). Inhibition of serine palmitoyltransferase in vitro and long-chain base biosynthesis in intact Chinese hamster ovary cells by beta-chloroalanine. *Biochemistry, 27*(18), 7079–7084.

Meng, A., Luberto, C., Meier, P., Bai, A., Yang, X., Hannun, Y. A., et al. (2004). Sphingomyelin synthase as a potential target for D609-induced apoptosis in U937 human monocytic leukemia cells. *Experimental Cell Research, 292*(2), 385–392.

Mimeault, M., Pommery, N., & Henichart, J. P. (2003). Synergistic antiproliferative and apoptotic effects induced by epidermal growth factor receptor and protein kinase a inhibitors in human prostatic cancer cell lines. *International Journal of Cancer, 106*(1), 116–124.

Mimeault, M., Pommery, N., Wattez, N., Bailly, C., & Henichart, J. P. (2003). Antiproliferative and apoptotic effects of anandamide in human prostatic cancer cell lines: Implication of epidermal growth factor receptor down-regulation and ceramide production. *The Prostate, 56*(1), 1–12.

Miyake, Y., Kozutsumi, Y., Nakamura, S., Fujita, T., & Kawasaki, T. (1995). Serine palmitoyltransferase is the primary target of a sphingosine-like immunosuppressant, ISP-1/myriocin. *Biochemical and Biophysical Research Communications, 211*(2), 396–403.

Modrak, D. E., Gold, D. V., & Goldenberg, D. M. (2006). Sphingolipid targets in cancer therapy. *Molecular Cancer Therapeutics, 5*(2), 200–208.

Monick, M. M., Mallampalli, R. K., Bradford, M., McCoy, D., Gross, T. J., Flaherty, D. M., et al. (2004). Cooperative prosurvival activity by ERK and Akt in human alveolar macrophages is dependent on high levels of acid ceramidase activity. *The Journal of Immunology, 173*(1), 123–135.

Morad, S. A., & Cabot, M. C. (2013). Ceramide-orchestrated signalling in cancer cells. *Nature Reviews Cancer, 13*(1), 51–65.

Morales, A., Paris, R., Villanueva, A., Llacuna, L., Garcia-Ruiz, C., & Fernandez-Checa, J. C. (2007). Pharmacological inhibition or small interfering RNA targeting acid ceramidase sensitizes hepatoma cells to chemotherapy and reduces tumor growth in vivo. *Oncogene, 26*(6), 905–916.

Nara, F., Tanaka, M., Hosoya, T., Suzuki-Konagai, K., & Ogita, T. (1999). Scyphostatin, a neutral sphingomyelinase inhibitor from a discomycete, Trichopeziza mollissima: Taxonomy of the producing organism, fermentation, isolation, and physico-chemical properties. *Journal of Antibiotics, 52*(6), 525–530.

Nara, F., Tanaka, M., Masuda-Inoue, S., Yamasato, Y., Doi-Yoshioka, H., Suzuki-Konagai, K., et al. (1999). Biological activities of scyphostatin, a neutral sphingomyelinase inhibitor from a discomycete, Trichopeziza mollissima. *Journal of Antibiotics, 52*(6), 531–535.

Nava, V. E., Cuvillier, O., Edsall, L. C., Kimura, K., Milstien, S., Gelmann, E. P., et al. (2000). Sphingosine enhances apoptosis of radiation-resistant prostate cancer cells. *Cancer Research, 60*(16), 4468–4474.

Newton, J., Lima, S., Maceyka, M., & Spiegel, S. (2015). Revisiting the sphingolipid rheostat: Evolving concepts in cancer therapy. *Experimental Cell Research, 333*(2), 195–200.

Noack, J., Choi, J., Richter, K., Kopp-Schneider, A., & Regnier-Vigouroux, A. (2014). A sphingosine kinase inhibitor combined with temozolomide induces glioblastoma cell death through accumulation of dihydrosphingosine and dihydroceramide, endoplasmic reticulum stress and autophagy. *Cell Death & Disease, 5*, e1425.

Novgorodov, S. A., Szulc, Z. M., Luberto, C., Jones, J. A., Bielawski, J., Bielawska, A., et al. (2005). Positively charged ceramide is a potent inducer of mitochondrial permeabilization. *Journal of Biological Chemistry, 280*(16), 16096–16105.

Nyakern, M., Cappellini, A., Mantovani, I., & Martelli, A. M. (2006). Synergistic induction of apoptosis in human leukemia T cells by the Akt inhibitor perifosine and etoposide through activation of intrinsic and Fas-mediated extrinsic cell death pathways. *Molecular Cancer Therapeutics, 5*(6), 1559–1570.

Obeid, L. M., Linardic, C. M., Karolak, L. A., & Hannun, Y. A. (1993). Programmed cell death induced by ceramide. *Science, 259*(5102), 1769–1771.

Ogretmen, B. (2018). Sphingolipid metabolism in cancer signalling and therapy. *Nature Reviews Cancer, 18*(1), 33–50.

Okudaira, C., Ikeda, Y., Kondo, S., Furuya, S., Hirabayashi, Y., Koyano, T., et al. (2000). Inhibition of acidic sphingomyelinase by xanthone compounds isolated from Garcinia speciosa. *Journal of Enzyme Inhibition, 15*(2), 129–138.

Olshefski, R. S., & Ladisch, S. (2001). Glucosylceramide synthase inhibition enhances vincristine-induced cytotoxicity. *International Journal of Cancer, 93*(1), 131–138.

Orr Gandy, K. A., & Obeid, L. M. (2013). Targeting the sphingosine kinase/sphingosine 1-phosphate pathway in disease: Review of sphingosine kinase inhibitors. *Biochimica et Biophysica Acta, 1831*(1), 157–166.

Oskouian, B., & Saba, J. D. (2010). Cancer treatment strategies targeting sphingolipid metabolism. *Advances in Experimental Medicine and Biology, 688*, 185–205.

Oskouian, B., Sooriyakumaran, P., Borowsky, A. D., Crans, A., Dillard-Telm, L., Tam, Y. Y., et al. (2006). Sphingosine-1-phosphate lyase potentiates apoptosis via p53- and p38-dependent pathways and is down-regulated in colon cancer. *Proceedings of the National Academy of Sciences of the United States of America, 103*(46), 17384–17389.

Overkleeft, H. S., Renkema, G. H., Neele, J., Vianello, P., Hung, I. O., Strijland, A., et al. (1998). Generation of specific deoxynojirimycin-type inhibitors of the non-lysosomal glucosylceramidase. *Journal of Biological Chemistry, 273*(41), 26522–26527.

Pastukhov, O., Schwalm, S., Zangemeister-Wittke, U., Fabbro, D., Bornancin, F., Japtok, L., et al. (2014). The ceramide kinase inhibitor NVP-231 inhibits breast and lung cancer cell proliferation by inducing M phase arrest and subsequent cell death. *British Journal of Pharmacology, 171*(24), 5829–5844.

Patwardhan, G. A., Zhang, Q. J., Yin, D., Gupta, V., Bao, J., Senkal, C. E., et al. (2009). A new mixed-backbone oligonucleotide against glucosylceramide synthase sensitizes multidrug-resistant tumors to apoptosis. *PLoS One, 4*(9), e6938.

Paugh, S. W., Paugh, B. S., Rahmani, M., Kapitonov, D., Almenara, J. A., Kordula, T., et al. (2008). A selective sphingosine kinase 1 inhibitor integrates multiple molecular therapeutic targets in human leukemia. *Blood, 112*(4), 1382–1391.

Paugh, S. W., Payne, S. G., Barbour, S. E., Milstien, S., & Spiegel, S. (2003). The immunosuppressant FTY720 is phosphorylated by sphingosine kinase type 2. *FEBS Letters, 554*(1–2), 189–193.

Pchejetski, D., Doumerc, N., Golzio, M., Naymark, M., Teissie, J., Kohama, T., et al. (2008). Chemosensitizing effects of sphingosine kinase-1 inhibition in prostate cancer cell and animal models. *Molecular Cancer Therapeutics, 7*(7), 1836–1845.

Perry, D. K., Carton, J., Shah, A. K., Meredith, F., Uhlinger, D. J., & Hannun, Y. A. (2000). Serine palmitoyltransferase regulates de novo ceramide generation during etoposide-induced apoptosis. *Journal of Biological Chemistry, 275*(12), 9078–9084.

Platt, F. M., Neises, G. R., Dwek, R. A., & Butters, T. D. (1994). N-butyldeoxynojirimycin is a novel inhibitor of glycolipid biosynthesis. *Journal of Biological Chemistry, 269*(11), 8362–8365.

Platt, F. M., Neises, G. R., Karlsson, G. B., Dwek, R. A., & Butters, T. D. (1994). N-butyldeoxygalactonojirimycin inhibits glycolipid biosynthesis but does not affect N-linked oligosaccharide processing. *Journal of Biological Chemistry, 269*(43), 27108–27114.

Pyne, N. J., & Pyne, S. (2010). Sphingosine 1-phosphate and cancer. *Nature Reviews Cancer, 10*(7), 489–503.

Raisova, M., Goltz, G., Bektas, M., Bielawska, A., Riebeling, C., Hossini, A. M., et al. (2002). Bcl-2 overexpression prevents apoptosis induced by ceramidase inhibitors in malignant melanoma and HaCaT keratinocytes. *FEBS Letters, 516*(1–3), 47–52.

Realini, N., Solorzano, C., Pagliuca, C., Pizzirani, D., Armirotti, A., Luciani, R., et al. (2013). Discovery of highly potent acid ceramidase inhibitors with in vitro tumor chemosensitizing activity. *Scientific Reports, 3*, 1035.

Robert, J. (2004). MS-209 schering. *Current Opinion in Investigational Drugs, 5*(12), 1340–1347.

Rosa, R., Marciano, R., Malapelle, U., Formisano, L., Nappi, L., D'Amato, C., et al. (2013). Sphingosine kinase 1 overexpression contributes to cetuximab resistance in human colorectal cancer models. *Clinical Cancer Research, 19*(1), 138–147.

Ryland, L. K., Fox, T. E., Liu, X., Loughran, T. P., & Kester, M. (2011). Dysregulation of sphingolipid metabolism in cancer. *Cancer Biology & Therapy, 11*(2), 138–149.

Saad, A. F., Meacham, W. D., Bai, A., Anelli, V., Elojeimy, S., Mahdy, A. E., et al. (2007). The functional effects of acid ceramidase overexpression in prostate cancer progression and resistance to chemotherapy. *Cancer Biology & Therapy, 6*(9), 1455–1460.

Saddoughi, S. A., Garrett-Mayer, E., Chaudhary, U., O'Brien, P. E., Afrin, L. B., Day, T. A., et al. (2011). Results of a phase II trial of gemcitabine plus doxorubicin in patients with recurrent head and neck cancers: Serum C(1)(8)-ceramide as a novel biomarker for monitoring response. *Clinical Cancer Research, 17*(18), 6097–6105.

Saied, E. M., & Arenz, C. (2014). Small molecule inhibitors of ceramidases. *Cellular Physiology and Biochemistry, 34*(1), 197–212.

Sakata, A., Ochiai, T., Shimeno, H., Hikishima, S., Yokomatsu, T., Shibuya, S., et al. (2007). Acid sphingomyelinase inhibition suppresses lipopolysaccharide-mediated release of inflammatory cytokines from macrophages and protects against disease pathology in dextran sulphate sodium-induced colitis in mice. *Immunology, 122*(1), 54–64.

Samsel, L., Zaidel, G., Drumgoole, H. M., Jelovac, D., Drachenberg, C., Rhee, J. G., et al. (2004). The ceramide analog, B13, induces apoptosis in prostate cancer cell lines and inhibits tumor growth in prostate cancer xenografts. *The Prostate, 58*(4), 382–393.

Sanchez, C., Galve-Roperh, I., Canova, C., Brachet, P., & Guzman, M. (1998). Delta9-tetrahydrocannabinol induces apoptosis in C6 glioma cells. *FEBS Letters, 436*(1), 6–10.

Sanchez, C., Galve-Roperh, I., Rueda, D., & Guzman, M. (1998). Involvement of sphingomyelin hydrolysis and the mitogen-activated protein kinase cascade in the Delta9-tetrahydrocannabinol-induced stimulation of glucose metabolism in primary astrocytes. *Molecular Pharmacology, 54*(5), 834–843.

Santana, P., Pena, L. A., Haimovitz-Friedman, A., Martin, S., Green, D., McLoughlin, M., et al. (1996). Acid sphingomyelinase-deficient human lymphoblasts and mice are defective in radiation-induced apoptosis. *Cell, 86*(2), 189–199.

Sasak, V. W., Ordovas, J. M., Elbein, A. D., & Berninger, R. W. (1985). Castanospermine inhibits glucosidase I and glycoprotein secretion in human hepatoma cells. *The Biochemical Journal, 232*(3), 759–766.

Sauane, M., Su, Z. Z., Dash, R., Liu, X., Norris, J. S., Sarkar, D., et al. (2010). Ceramide plays a prominent role in MDA-7/IL-24-induced cancer-specific apoptosis. *Journal of Cellular Physiology, 222*(3), 546–555.

Sauer, L., Nunes, J., Salunkhe, V., Skalska, L., Kohama, T., Cuvillier, O., et al. (2009). Sphingosine kinase 1 inhibition sensitizes hormone-resistant prostate cancer to docetaxel. *International Journal of Cancer, 125*(11), 2728–2736.

Saul, R., Chambers, J. P., Molyneux, R. J., & Elbein, A. D. (1983). Castanospermine, a tetrahydroxylated alkaloid that inhibits beta-glucosidase and beta-glucocerebrosidase. *Archives of Biochemistry and Biophysics, 221*(2), 593–597.

Sawada, M., Nakashima, S., Kiyono, T., Nakagawa, M., Yamada, J., Yamakawa, H., et al. (2001). p53 regulates ceramide formation by neutral sphingomyelinase through reactive oxygen species in human glioma cells. *Oncogene, 20*(11), 1368–1378.

Sawkar, A. R., Cheng, W. C., Beutler, E., Wong, C. H., Balch, W. E., & Kelly, J. W. (2002). Chemical chaperones increase the cellular activity of N370S beta -glucosidase: A therapeutic strategy for Gaucher disease. *Proceedings of the National Academy of Sciences of the United States of America, 99*(24), 15428–15433.

Scarlatti, F., Sala, G., Ricci, C., Maioli, C., Milani, F., Minella, M., et al. (2007). Resveratrol sensitization of DU145 prostate cancer cells to ionizing radiation is associated to ceramide increase. *Cancer Letters, 253*(1), 124–130.

Schnute, M. E., McReynolds, M. D., Kasten, T., Yates, M., Jerome, G., Rains, J. W., et al. (2012). Modulation of cellular S1P levels with a novel, potent and specific inhibitor of sphingosine kinase-1. *The Biochemical Journal, 444*(1), 79–88.

Schrecengost, R. S., Keller, S. N., Schiewer, M. J., Knudsen, K. E., & Smith, C. D. (2015). Downregulation of critical oncogenes by the selective SK2 inhibitor ABC294640 hinders prostate cancer progression. *Molecular Cancer Research, 13*(12), 1591–1601.

Schulz, A., Mousallem, T., Venkataramani, M., Persaud-Sawin, D. A., Zucker, A., Luberto, C., et al. (2006). The CLN9 protein, a regulator of dihydroceramide synthase. *Journal of Biological Chemistry, 281*(5), 2784–2794.

Schwartz, G. K., Haimovitz-Friedman, A., Dhupar, S. K., Ehleiter, D., Maslak, P., Lai, L., et al. (1995). Potentiation of apoptosis by treatment with the protein kinase C-specific inhibitor safingol in mitomycin C-treated gastric cancer cells. *Journal of the National Cancer Institute, 87*(18), 1394–1399.

Schwartz, G. K., Ward, D., Saltz, L., Casper, E. S., Spiess, T., Mullen, E., et al. (1997). A pilot clinical/pharmacological study of the protein kinase C-specific inhibitor safingol alone and in combination with doxorubicin. *Clinical Cancer Research, 3*(4), 537–543.

Segui, B., Andrieu-Abadie, N., Jaffrezou, J. P., Benoist, H., & Levade, T. (2006). Sphingolipids as modulators of cancer cell death: Potential therapeutic targets. *Biochimica et Biophysica Acta, 1758*(12), 2104–2120.

Selzner, M., Bielawska, A., Morse, M. A., Rudiger, H. A., Sindram, D., Hannun, Y. A., et al. (2001). Induction of apoptotic cell death and prevention of tumor growth by ceramide analogues in metastatic human colon cancer. *Cancer Research, 61*(3), 1233–1240.

Senkal, C. E., Ponnusamy, S., Rossi, M. J., Sundararaj, K., Szulc, Z., Bielawski, J., et al. (2006). Potent antitumor activity of a novel cationic pyridinium-ceramide alone or in combination with gemcitabine against human head and neck squamous cell carcinomas in vitro and in vivo. *The Journal of Pharmacology and Experimental Therapeutics, 317*(3), 1188–1199.

Serra, M., & Saba, J. D. (2010). Sphingosine 1-phosphate lyase, a key regulator of sphingosine 1-phosphate signaling and function. *Advances in Enzyme Regulation, 50*(1), 349–362.

Song, L., Xiong, H., Li, J., Liao, W., Wang, L., Wu, J., et al. (2011). Sphingosine kinase-1 enhances resistance to apoptosis through activation of PI3K/Akt/NF-kappaB pathway in human non-small cell lung cancer. *Clinical Cancer Research, 17*(7), 1839–1849.

Spinedi, A., Bartolomeo, S. D., & Piacentini, M. (1998). Apoptosis induced by N-hexanoylsphingosine in CHP-100 cells associates with accumulation of endogenous ceramide and is potentiated by inhibition of glucocerebroside synthesis. *Cell Death and Differentiation, 5*(9), 785–791.

Stefanovic, M., Tutusaus, A., Martinez-Nieto, G. A., Barcena, C., de Gregorio, E., Moutinho, C., et al. (2016). Targeting glucosylceramide synthase upregulation reverts sorafenib resistance in experimental hepatocellular carcinoma. *Oncotarget, 7*(7), 8253–8267.

Stoffel, W., & Grol, M. (1974). Chemistry and biochemistry of 1-desoxysphinganine 1-phosphonate (dihydrosphingosine-1-phosphonate). *Chemistry and Physics of Lipids, 13*(4), 372–388.

Strelow, A., Bernardo, K., Adam-Klages, S., Linke, T., Sandhoff, K., Kronke, M., et al. (2000). Overexpression of acid ceramidase protects from tumor necrosis factor-induced cell death. *The Journal of Experimental Medicine, 192*(5), 601–612.

Sugita, M., Willians, M., Dulaney, J. T., & Moser, H. W. (1975). Ceramidase and ceramide synthesis in human kidney and cerebellum. Description of a new alkaline ceramidase. *Biochimica et Biophysica Acta, 398*(1), 125–131.

Sun, Y., Fox, T., Adhikary, G., Kester, M., & Pearlman, E. (2008). Inhibition of corneal inflammation by liposomal delivery of short-chain, C-6 ceramide. *Journal of Leukocyte Biology, 83*(6), 1512–1521.

Sun, Y., Zhang, T., Gao, P., Meng, B., Gao, Y., Wang, X., et al. (2010). Targeting glucosylceramide synthase downregulates expression of the multidrug resistance gene MDR1 and sensitizes breast carcinoma cells to anticancer drugs. *Breast Cancer Research and Treatment, 121*(3), 591–599.

Sun, Y. L., Zhou, G. Y., Li, K. N., Gao, P., Zhang, Q. H., Zhen, J. H., et al. (2006). Suppression of glucosylceramide synthase by RNA interference reverses multidrug resistance in human breast cancer cells. *Neoplasma, 53*(1), 1–8.

Sundaram, K. S., & Lev, M. (1984). Inhibition of sphingolipid synthesis by cycloserine in vitro and in vivo. *Journal of Neurochemistry, 42*(2), 577–581.

Sweeney, E. A., Sakakura, C., Shirahama, T., Masamune, A., Ohta, H., Hakomori, S., et al. (1996). Sphingosine and its methylated derivative N,N-dimethylsphingosine (DMS) induce apoptosis in a variety of human cancer cell lines. *International Journal of Cancer, 66*(3), 358–366.

Taha, T. A., Osta, W., Kozhaya, L., Bielawski, J., Johnson, K. R., Gillanders, W. E., et al. (2004). Down-regulation of sphingosine kinase-1 by DNA damage: Dependence on proteases and p53. *Journal of Biological Chemistry, 279*(19), 20546–20554.

Takahashi, E., Inanami, O., Asanuma, T., & Kuwabara, M. (2006). Effects of ceramide inhibition on radiation-induced apoptosis in human leukemia MOLT-4 cells. *Journal of Radiation Research, 47*(1), 19–25.

Tan, S. F., Liu, X., Fox, T. E., Barth, B. M., Sharma, A., Turner, S. D., et al. (2016). Acid ceramidase is upregulated in AML and represents a novel therapeutic target. *Oncotarget,* 7(50), 83208–83222.

Tan, S. F., Pearson, J. M., Feith, D. J., & Loughran, T. P., Jr. (2017). The emergence of acid ceramidase as a therapeutic target for acute myeloid leukemia. *Expert Opinion on Therapeutic Targets, 21*(6), 583–590.

Tanaka, M., Nara, F., Yamasato, Y., Masuda-Inoue, S., Doi-Yoshioka, H., Kumakura, S., et al. (1999). Macquarimicin A inhibits membrane-bound neutral sphingomyelinase from rat brain. *Journal of Antibiotics, 52*(7), 670–673.

Testai, F. D., Landek, M. A., Goswami, R., Ahmed, M., & Dawson, G. (2004). Acid sphingomyelinase and inhibition by phosphate ion: Role of inhibition by phosphatidyl-myo-inositol 3,4,5-triphosphate in oligodendrocyte cell signaling. *Journal of Neurochemistry, 89*(3), 636–644.

Tran, M. A., Smith, C. D., Kester, M., & Robertson, G. P. (2008). Combining nanoliposomal ceramide with sorafenib synergistically inhibits melanoma and breast cancer cell survival to decrease tumor development. *Clinical Cancer Research, 14*(11), 3571–3581.

Triola, G., Fabrias, G., Casas, J., & Llebaria, A. (2003). Synthesis of cyclopropene analogues of ceramide and their effect on dihydroceramide desaturase. *The Journal of Organic Chemistry, 68*(26), 9924–9932.

Triola, G., Fabrias, G., & Llebaria, A. (2001). Synthesis of a cyclopropene analogue of ceramide, a potent inhibitor of dihydroceramide desaturase. This work was supported by the Direccion General de Ensenanza Superior e Investigacion Cientifica (grant PB97-1171) and the Departament d'Universitats, Recerca i Societat de la Informacio, Generalitat de Catalunya (grant 1999-SGR 00187 and a Predoctoral fellowship to G.T.). We thank Dr. J. Casas, Dr. A. Delgado, and Dr. J. Joglar for their help in different aspects of this work *Angewandte Chemie, 40*(10), 1960–1962.

Truman, J. P., Garcia-Barros, M., Obeid, L. M., & Hannun, Y. A. (2014). Evolving concepts in cancer therapy through targeting sphingolipid metabolism. *Biochimica et Biophysica Acta, 1841*(8), 1174–1188.

Tsukamoto, S., Huang, Y., Kumazoe, M., Lesnick, C., Yamada, S., Ueda, N., et al. (2015). Sphingosine kinase-1 protects multiple myeloma from apoptosis driven by cancer-specific inhibition of RTKs. *Molecular Cancer Therapeutics, 14*(10), 2303–2312.

Uchida, R., Tomoda, H., Arai, M., & Omura, S. (2001). Chlorogentisylquinone, a new neutral sphingomyelinase inhibitor, produced by a marine fungus. *Journal of Antibiotics, 54*(11), 882–889.

Uchida, R., Tomoda, H., Dong, Y., & Omura, S. (1999). Alutenusin, a specific neutral sphingomyelinase inhibitor, produced by Penicillium sp. FO-7436. *Journal of Antibiotics, 52*(6), 572–574.

Umansky, V., & Sevko, A. (2013). Tumor microenvironment and myeloid-derived suppressor cells. *Cancer Microenvironment, 6*(2), 169–177.

Usta, J., El Bawab, S., Roddy, P., Szulc, Z. M., Hannun, Y. A., & Bielawska, A. (2001). Structural requirements of ceramide and sphingosine based inhibitors of mitochondrial ceramidase. *Biochemistry, 40*(32), 9657–9668.

Vit, J. P., & Rosselli, F. (2003). Role of the ceramide-signaling pathways in ionizing radiation-induced apoptosis. *Oncogene, 22*(54), 8645–8652.

Vunnam, R. R., & Radin, N. S. (1980). Analogs of ceramide that inhibit glucocerebroside synthetase in mouse brain. *Chemistry and Physics of Lipids, 26*(3), 265–278.

Wallington-Beddoe, C. T., Bennett, M. K., Vandyke, K., Davies, L., Zebol, J. R., Moretti, P. A. B., et al. (2017). Sphingosine kinase 2 inhibition synergises with bortezomib to target myeloma by enhancing endoplasmic reticulum stress. *Oncotarget, 8*(27), 43602–43616.

Wang, C., Liu, J. N., Xu, L., Mu, Y. L., & Sun, P. (2014). Expression and significance of glucosylceramide synthase in colorectal carcinoma tissues. *European Review for Medical and Pharmacological Sciences, 18*(23), 3632–3637.

Wang, E., Norred, W. P., Bacon, C. W., Riley, R. T., & Merrill, A. H., Jr. (1991). Inhibition of sphingolipid biosynthesis by fumonisins. Implications for diseases associated with Fusarium moniliforme. *Journal of Biological Chemistry, 266*(22), 14486–14490.

Wang, J., Knapp, S., Pyne, N. J., Pyne, S., & Elkins, J. M. (2014). Crystal structure of sphingosine kinase 1 with PF-543. *ACS Medicinal Chemistry Letters, 5*(12), 1329–1333.

Watson, D. G., Tonelli, F., Alossaimi, M., Williamson, L., Chan, E., Gorshkova, I., et al. (2013). The roles of sphingosine kinases 1 and 2 in regulating the Warburg effect in prostate cancer cells. *Cellular Signalling, 25*(4), 1011–1017.

Watters, R. J., Fox, T. E., Tan, S. F., Shanmugavelandy, S., Choby, J. E., Broeg, K., et al. (2013). Targeting glucosylceramide synthase synergizes with C6-ceramide nanoliposomes to induce apoptosis in natural killer cell leukemia. *Leukemia and Lymphoma, 54*(6), 1288–1296.

van Weely, S., Brandsma, M., Strijland, A., Tager, J. M., & Aerts, J. M. (1993). Demonstration of the existence of a second, non-lysosomal glucocerebrosidase that is not deficient in Gaucher disease. *Biochimica et Biophysica Acta, 1181*(1), 55–62.

White, C., Alshaker, H., Cooper, C., Winkler, M., & Pchejetski, D. (2016). The emerging role of FTY720 (Fingolimod) in cancer treatment. *Oncotarget, 7*(17), 23106–23127.

Winter, C. K., Gilchrist, D. G., Dickman, M. B., & Jones, C. (1996). Chemistry and biological activity of AAL toxins. *Advances in Experimental Medicine and Biology, 392*, 307–316.

Wolff, R. A., Dobrowsky, R. T., Bielawska, A., Obeid, L. M., & Hannun, Y. A. (1994). Role of ceramide-activated protein phosphatase in ceramide-mediated signal transduction. *Journal of Biological Chemistry, 269*(30), 19605–19609.

Xie, P., Shen, Y. F., Shi, Y. P., Ge, S. M., Gu, Z. H., Wang, J., et al. (2008). Overexpression of glucosylceramide synthase in associated with multidrug resistance of leukemia cells. *Leukemia Research, 32*(3), 475–480.

Yang, J., Yang, C., Zhang, S., Mei, Z., Shi, M., Sun, S., et al. (2015). ABC294640, a sphingosine kinase 2 inhibitor, enhances the antitumor effects of TRAIL in non-small cell lung cancer. *Cancer Biology & Therapy, 16*(8), 1194–1204.

Yokomatsu, T., Takechi, H., Akiyama, T., Shibuya, S., Kominato, T., Soeda, S., et al. (2001). Synthesis and evaluation of a difluoromethylene analogue of sphingomyelin as an inhibitor of sphingomyelinase. *Bioorganic & Medicinal Chemistry Letters, 11*(10), 1277–1280.

Yoo, H. S., Norred, W. P., Wang, E., Merrill, A. H., Jr., & Riley, R. T. (1992). Fumonisin inhibition of de novo sphingolipid biosynthesis and cytotoxicity are correlated in LLC-PK1 cells. *Toxicology and Applied Pharmacology, 114*(1), 9–15.

Young, M. M., Takahashi, Y., Fox, T. E., Yun, J. K., Kester, M., & Wang, H. G. (2016). Sphingosine kinase 1 cooperates with autophagy to maintain endocytic membrane trafficking. *Cell Reports, 17*(6), 1532–1545.

Yu, T., Li, J., Qiu, Y., & Sun, H. (2012). 1-phenyl-2-decanoylamino-3-morpholino-1-propanol (PDMP) facilitates curcumin-induced melanoma cell apoptosis by enhancing ceramide accumulation, JNK activation, and inhibiting PI3K/AKT activation. *Molecular and Cellular Biochemistry, 361*(1–2), 47–54.

Zhai, L., Sun, N., Han, Z., Jin, H. C., & Zhang, B. (2015). Liposomal short-chain C6 ceramide induces potent anti-osteosarcoma activity in vitro and in vivo. *Biochemical and Biophysical Research Communications, 468*(1–2), 274–280.

Zhang, C., Lin, X., Song, Y., Zhang, X., Li, H., & Wang, Q. (2014). Overexpression of glucosylceramide synthase and its significance in the clinical outcome of non-small cell lung cancer. *Chinese Medical Journal, 127*(17), 3071–3076.

Zhao, X., Sun, B., Zhang, J., Zhang, R., & Zhang, Q. (2016). Short-chain C6 ceramide sensitizes AT406-induced anti-pancreatic cancer cell activity. *Biochemical and Biophysical Research Communications, 479*(2), 166–172.

Zweerink, M. M., Edison, A. M., Wells, G. B., Pinto, W., & Lester, R. L. (1992). Characterization of a novel, potent, and specific inhibitor of serine palmitoyltransferase. *Journal of Biological Chemistry, 267*(35), 25032–25038.

CHAPTER THIRTEEN

Side Effects in Cancer Therapy: Are Sphingolipids to Blame?

Falak Patel and Stefka D. Spassieva[1]
Department of Physiology, University of Kentucky, Lexington, KY, United States
[1]Corresponding author: E-mail: Stefanka.Spassieva@uky.edu

Contents

Abstract

Cancer patients' quality of life is greatly dependent on the efficacy of treatments and their associated side effects, which can significantly reduce the overall quality of life. Although the effectiveness of cancer treatments has improved over time, adverse effects persist with each treatment. Some side effects, such as paclitaxel-induced peripheral neuropathy, can be dose limiting, thus further reducing the potential of paclitaxel chemotherapy treatment. Premature ovarian failure in young female patients due to radiation and chemotherapy therapy can have devastating infertility consequences. In recent years, a class of lipids known as sphingolipids has been identified as playing a role in the side effects of cancer therapies. Advanced analytical technologies, such as mass spectrometry, have provided great aid in detecting and distinguishing individual sphingolipids at low concentrations. Sphingolipids play an important role in cell proliferation and apoptosis and, importantly, sphingolipid metabolism has been shown to be dysregulated in cancer. The goal of this review is to summarize the latest findings of the role of sphingolipids in the injurious side effects in various cancer treatments. A better understanding of the molecular mechanisms driving these sphingolipid-induced side effects can help develop new drugs and treatments for cancer that have fewer side effects, thus improving treatment efficacy and quality of life.

Advances in Cancer Research, Volume 140
ISSN 0065-230X
https://doi.org/10.1016/bs.acr.2018.04.017

1. INTRODUCTION

Cancer is a disease with significant societal burden. According to the World Health Organization, there were 14 million new cases diagnosed and 8.2 million cancer-related deaths worldwide in 2012. In the United States, cancer is the second leading cause of death after heart disease. Indeed, it is estimated that over 600,000 deaths were attributable to cancer in 2017 alone (Siegel, Miller, & Jemal, 2017). In the last few decades, advances in cancer research have led to new innovative treatments that lowered the cancer death rate by about 25% (Siegel et al., 2017). With improved treatments, more patients overcome cancer and live longer; however, cancer treatments come at a cost. They present a wide range of side effects, which can reduce the treatments' efficacies, and, in some cases, can have a lasting negative impact on patients' quality of life (Maguire & Selby, 1989; Mansano-Schlosser & Ceolim, 2012). For example, some of the side effects associated with chemotherapy, such as nausea and vomiting, are only apparent during the course of the treatment, whereas others, such as peripheral neuropathy, can persist for months or even years after the treatment is complete (Seretny et al., 2014). Another example is female infertility, a life-long negative consequence of radiation or chemotherapy (Perez, Knudson, Leykin, Korsmeyer, & Tilly, 1997; Tilly & Kolesnick, 2002; Yucebilgin et al., 2004). In light of these substantial health risks, patients' diminished quality of life resulting from cancer treatment becomes an important health concern. It is critical, therefore, to elucidate the mechanisms causing these side effects to improve treatment efficacy and patients' quality of life.

It is well documented that sphingolipid metabolism is dysregulated in cancer cells (reviewed in Hannun & Obeid, 2018; Ogretmen, 2018; Ryland, Fox, Liu, Loughran, & Kester, 2011 and discussed in the previous chapters of this compendium). Sphingolipids are structural components of biological membranes and, recently, have been widely studied due to their importance as intra- and extracellular signaling molecules (reviewed in Hannun & Obeid, 2018; Merrill, 2011). The main structural component of sphingolipids is a sphingoid base, an aliphatic amino alcohol that can be amide-linked to a fatty acid moiety of a different chain length to form a ceramide (Fig. 1). Ceramides can be O-linked at the C1 position to different types of head groups to form complex sphingolipids. For example, association with a phosphocholine head group forms sphingomyelin. Meanwhile, a sugar

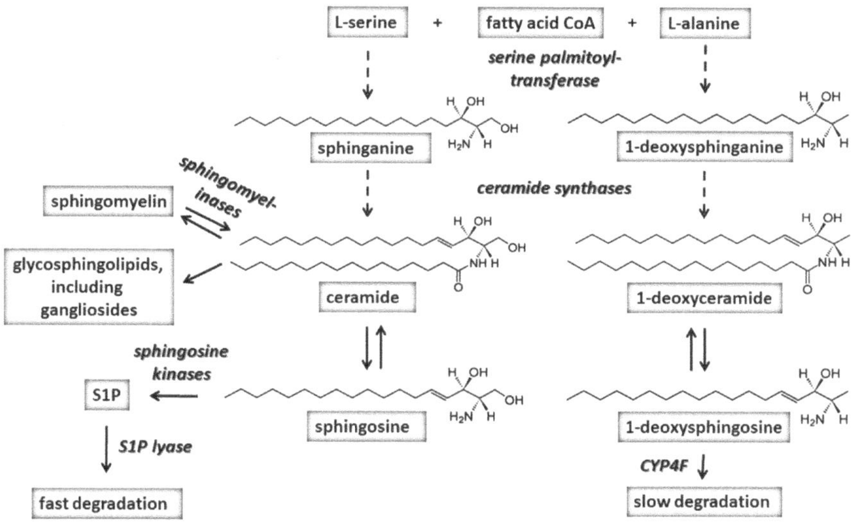

Figure 1 ***Schematic representation of the sphingolipid metabolic pathway.*** The enzymes discussed in this review are highlighted in italic. *CYP4F*, cytochrome P450 family 4 subfamily F; *S1P*, sphingosine-1-phosphate.

(glucose or galactose) head group forms a hexosylceramide or, as in the case of gangliosides, the head group is an oligosaccharide containing one or more sialic acids (*N*-acetylneuraminic acid and/or *N*-glycolylneuraminic acid) (Groux-Degroote, Guerardel, Julien, & Delannoy, 2015; Merrill, 2011; Merrill, Sullards, Wang, Voss, & Riley, 2001; Yu, Tsai, Ariga, & Yanagisawa, 2011). Many sphingolipid metabolites, e.g., ceramide, dihydroceramide, sphinganine, sphingosine-1-phosphate (S1P), gangliosides, and sphingomyelin, have been shown to play a role in cancer (reviewed in Ogretmen, 2018). Of these, ceramide and S1P are the most extensively studied. They exert opposite effects on cell proliferation and death. Their ratio, the so-called "ceramide/S1P rheostat" is a well-established concept wherein ceramide is associated with apoptosis and cell death, and S1P is associated with cell proliferation and survival pathways (Maceyka, Payne, Milstien, & Spiegel, 2002; Selvam & Ogretmen, 2013; Van Brocklyn & Williams, 2012). An example for a sphingolipid enzyme involved in cancer is sphingosine kinase-1 (SK1). Elevated expression of SK1 has been observed in many types of cancer such as colon, breast, uterine, kidney, and lung cancers (reviewed in Ponnusamy et al., 2010). SK1 phosphorylates sphingosine into S1P. Therefore, overexpression of SK1 increases the amount of S1P, leading to proliferation of tumor cells. By contrast, cancer therapy focuses on increasing the de novo synthesis of ceramide, which has apoptotic properties (Ogretmen & Hannun, 2004). Cancer cells are known to avoid apoptosis by preventing the accumulation of ceramide (Ponnusamy et al., 2010).

Although the role of sphingolipids in cancer biology and their potential as targets for cancer treatment is undisputable and intensively studied for the last few decades, their role in the side effects of treatment became evident only recently. This review will focus on the emerging role of the sphingolipid pathway, enzymes, and metabolites in cancer treatment side effects.

2. SPHINGOLIPIDS AND CHEMOTHERAPY-INDUCED PERIPHERAL NEUROPATHY

Chemotherapy uses drugs that generally target the regulation of the cell cycle (Chan, Koh, & Li, 2012). Because tumor cells have dysregulated cell cycle resulting in aberrantly high cell division rates, chemotherapy predominantly affects tumor cells. However, it can also affect other nontransformed cells types with a rapid rate of cell division such as hair follicles, gastrointestinal cells, and bone marrow cells. Consequently, alopecia, nausea and vomiting, and reduced red blood cell production, respectively, can result (Gerber, 2008; Luanpitpong & Rojanasakul, 2012). Notably, chemotherapeutics such as taxanes, vinca alkaloids, thalidomide, bortezomib, or platinum-based drugs, can additionally, affect differentiated, nondividing cells such as neurons, resulting in chemotherapy-induced peripheral neuropathy (CIPN) (Addington & Freimer, 2016; Janes et al., 2014). CIPN is a progressive and dose-dependent side effect of these drugs (Addington & Freimer, 2016). The symptoms of CIPN include shooting pain, tingling, numbness, and in severe cases, burning sensations in the extremities (Miltenburg & Boogerd, 2014). The recovery from peripheral neuropathy caused by chemotherapy varies among patients and can take anywhere from weeks to a year after completion of a treatment to abate. According to a comprehensive review of multiple studies containing total of 2085 patients, the range of prevalence for CIPN varied from 57.7% to 78.4% (average of 68.1%) 1 month after the cessation of chemotherapy (Seretny et al., 2014). The prevalence range was reduced to 6.4%–53.5% 6 months after completion of chemotherapy. Because CIPN is a dose-limiting side effect of chemotherapy, it requires the tolerable dosage of the chemotherapy drug to be decreased, thus reducing its overall effectiveness. A study with breast cancer patients receiving taxane therapy showed that dose reduction or termination due to peripheral neuropathy lowered the received cumulative dose by 9.4% and 28.4%, respectively (Speck et al., 2013). While the clinical manifestation of CIPN is well documented, the underlying mechanisms are poorly understood and currently there are no treatments or prevention strategies clinically available to address this side effect. Interestingly, a couple of recent studies provided evidence that sphingolipid metabolites and signaling could be involved in CIPN due to paclitaxel (Janes et al., 2014; Kramer et al., 2015).

Paclitaxel is commonly used to treat a variety of cancers such as breast, ovarian, head and neck, prostate, and non–small cell lung carcinomas (Rowinsky, Eisenhauer, Chaudhry, Arbuck, & Donehower, 1993). Paclitaxel is a semisynthetic hydrophobic compound, which binds to tubulin, inhibiting microtubule depolymerization. Active microtubule depolymerization is essential for the proper function of the mitotic spindle. When perturbed, an activation of the mitotic checkpoint can result, causing cell cycle arrest, and ultimately, the death of tumor cells (Weaver, 2014). While it is well established that the anticancer activity of paclitaxel is due to its interference with mitotic spindle dynamics, the molecular mechanisms leading to neuropathy are less clear (Gornstein & Schwarz, 2014). Therefore, currently, the only options available to address paclitaxel-induced neuropathy in the clinic are to reduce the dose or terminate the treatment.

2.1 1-Deoxysphingolipids as Intermediates of Paclitaxel-Induced Peripheral Neuropathy

In a recent clinical study from our laboratory with breast cancer patients receiving standard paclitaxel chemotherapy, we found significant associations between the plasma levels of 1-deoxysphingolipids, a minor class of sphingolipids, and the incidence and severity of patients' peripheral neuropathy symptoms (Kramer et al., 2015). 1-Deoxysphignolipids are produced when the first enzyme of the de novo sphingolipid pathway, serine palmitoyltransferase (SPT), utilizes L-alanine rather than its preferred amino acid substrate L-serine (Fig. 1) (Merrill, 2011). As a result, L-alanine–derived 1-deoxysphingolipids lack the important functional hydroxyl group at the C_1 position. The hydroxyl group at C_1 is present in L-serine–derived sphingolipids, which is required for their metabolic conversion and degradation. Until recently, 1-deoxysphingolipids were considered metabolic "dead ends" because they cannot be degraded by S1P lyase, the last catabolic enzyme of the sphingolipid pathway. However, in a very recent study, it has been found that the 1-deoxysphingolipid (1-deoxysphingosine) can be metabolized by the cytochrome P450 family 4 subfamily F (CYP4F) enzyme (Alecu, Othman, et al., 2017). The CYP4F enzyme is expressed in the liver and is responsible for metabolizing toxic compounds and fatty acids through hydroxylation and desaturation. However, using isotope-labeling experiments, the authors showed that the degradation of 1-deoxysphingolipids is very slow compared to the degradation of L-serine—derived sphingolipids. Therefore, it is likely that the downstream pathway of 1-deoxysphingosine is a natural liver detoxification pathway through cytochrome P450 enzymes (Alecu, Othman, et al., 2017). At normal physiological conditions, the affinity of SPT toward L-alanine is low, resulting in low levels of 1-deoxysphingolipids (Gable et al., 2010). On the other hand, when produced in excess, 1-deoxysphingolipids accumulate, likely due to their slow degradation through an atypical liver-detoxification pathway (Alecu, Othman, et al., 2017).

In hereditary sensory and autonomic neuropathy type I (HSAN1), the *SPTLC1* and *SPTLC2* genes, which are responsible for encoding SPT's two major subunits, are mutated. The SPT encoded from these mutated genes has increased specificity for L-alanine (Gable et al., 2010; Penno et al., 2010). These mutations eventually lead to increased levels of neurotoxic 1-deoxysphingolipds in the plasma of HSAN1 patients'. The study from our laboratory demonstrated that when cells are treated in vitro with paclitaxel, the levels of 1-deoxysphingolipids are significantly increased in a dose-dependent manner (Kramer et al., 2015). It is not likely that paclitaxel causes a mutation to the SPT-encoding genes. Interestingly, in a rat model of paclitaxel-induced neuropathy, it has been shown that in the dorsal root ganglion, paclitaxel can transiently decrease L-serine levels through decreased expression of 3-phosphoglycerate dehydrogenase, the rate-limiting enzyme of de novo L-serine synthesis (Kiya, Kawamata, Namiki, & Yamakage, 2011). However, results from our study revealed that the increased levels of 1-deoxysphingolipids are not due to reduced levels of L-serine or a higher availability of L-alanine; L-alanine/L-serine ratio was not significantly altered by paclitaxel (Kramer et al., 2015). Instead, the results showed that paclitaxel treatment resulted in elevated expression of the SPTLC1 subunit and, in some cases (U87 astroglioma cell line), the SPTLC2 subunit of SPT. Our study also revealed a significant dose-dependent increase in de novo production of 1-deoxysphinganine (L-alanine as substrate) and sphinganine (L-serine as substrate), which indicates that paclitaxel treatment increases SPT activity (Kramer et al., 2015). This finding suggests that treatments such as paclitaxel, which significantly elevate the expression of the major SPT subunits and/or SPT activity, likely increase the production of 1-deoxysphingolipids. As discussed above, 1-deoxysphingolipids have a slow degradation rate compared to the L-serine–derived sphingolipids. Therefore, when 1-deoxysphingolipids' production is increased due to SPT upregulation, they can accumulate over time to pathophysiological levels.

1-Deoxysphingolipids were recently characterized as the neurotoxic molecular intermediates in HSAN1 and have been implicated in diabetic neuropathy (Penno et al., 2010; Wei et al., 2014). Of note, diabetes mellitus, which by itself can cause peripheral neuropathy, is a risk factor for developing CIPN (Bao et al., 2016). The neurotoxicity of 1-deoxysphingolipids was directly shown in phase I clinical trials where the 1-deoxysphingolipid, 1-deoxysphinganine (spisulosine/E-285), was tested for its anticancer potential. The clinical trials were terminated because patients developed severe, and in some cases fatal, neurotoxicity (Baird et al., 2009; Massard et al., 2012; Schoffski et al., 2011). In addition, the neurotoxicity of 1-deoxysphingolipids has been shown in neuronal cultures in vitro (Güntert et al., 2016; Penno et al., 2010; Spassieva et al., 2016).

Although the exact cause for neurotoxicity due to accumulation of 1-deoxysphingolipids is unknown, recent studies have proposed some possible mechanisms involving mitochondria or the cytoskeleton (Alecu, Tedeschi, et al., 2017; Güntert et al., 2016). One of the studies showed that an exogenously applied alkyne analogue of 1-deoxysphinganine localizes to mitochondria and causes mitochondrial dysfunction (Alecu, Tedeschi, et al., 2017). The alkyne analogue of 1-deoxysphiganine is a traceable probe with very similar properties to the naturally occurring 1-deoxysphiganine. In primary sensory neurons, the exogenously applied 1-deoxysphiganine analogue caused swollen and fragmentated mitochondria, as well as reduced mitochondria respiratory capacity and ATP levels. The levels of the 1-deoxysphinganine probe were very prominent in the mitochondria during the early few minutes of the treatment at a nontoxic concentration (0.1 μM). After a longer incubation (1h), detectable amounts of the probe were seen in the endoplasmic reticulum (ER) and the Golgi. When the cells were treated with a toxic concentration (0.5–1.0 μM) for a longer period of time (10–24 h), the ER became enlarged and the cells showed signs of an ER stress, as evidenced by increased levels of X-box binding protein 1's product, which is a marker of ER stress(Alecu, Tedeschi, et al., 2017). ER stress is known to activate caspase 3 and caspase 12, which subsequently activate a general cytotoxic mechanism leading to programmed cell death (Hitomi et al., 2004). An earlier study that tested spisulosine/ES 285 (1-deoxysphinganine) in vitro (NIH-3T3 and RH7777 cell lines) corroborates these findings, showing that 1-deoxysphinganine can activate caspase 3 and caspase 12 (Salcedo et al., 2007). The results from these studies agree with other published research that suggests that paclitaxel neurotoxicity results from adversely affecting mitochondria function and/or activating ER stress pathways (Andoh et al., 2017; Canta, Pozzi, & Carozzi, 2015; Xiao & Bennett, 2012).

Another proposed mechanism for 1-deoxysphingolipids neurotoxicity suggests that they disrupt the cytoskeleton (Güntert et al., 2016). In neurons, the cytoskeleton consists of microtubules, actin fibers, and neurofilaments, which are responsible for maintaining cell's structure, shape, and compartmentalization, all essential for neuronal survival (Fletcher & Mullins, 2010). Analyses of primary cortical neurons treated with high concentration of 1-deoxysphinganine (2 μM) for 24 h or 48 h revealed decreased viability and reduced actin filaments and microtubules (Güntert et al., 2016). In addition, the study showed that 1-deoxysphinganine treatment resulted in decreased RhoA and Rac1 activity. RhoA and Rac1 are small GTPases involved in cytoskeleton regulation and neuronal survival (Stankiewicz & Linseman, 2014). It is worth noting that this result corroborates earlier findings that 1-deoxysphinganine (spisulosine) treatment of Vero cells disrupts actin stress fibers through RhoA signaling (Cuadros et al., 2000). Moreover, the clinical trial study (above) that demonstrated 1-deoxysphinganine (ES-285/spisulosine) neurotoxicity in humans, revealed an altered expression of genes involved in the actin cytoskeleton (Massard et al., 2012). The study of Güntert et al. (2016) suggests that 1-deoxysphinganine—induced cytoskeleton abnormalities and neuronal death are due to overactivation of the N-methyl-D-aspartate (NMDA) receptor. The NMDA receptor plays an important role in synapse formation; however, its overactivation can cause excitotoxicity, leading to neurodegeneration (Dore, Aow, & Malinow, 2016).

The in vitro studies discussed above tested neurotoxicity of only 1-deoxysphingoid bases, i.e., 1-deoxysphingosine and 1-deoxysphinganine. Both 1-deoxysphingoid bases are easily available from commercial sources, facilitating their study. However, our clinical study with breast cancer patients receiving paclitaxel chemotherapy revealed an association between the plasma levels of certain 1-deoxydihydroceramides ($C_{22:1}$ and $C_{18:1}$) and 1-deoxyceramides ($C_{22:1}$ and C_{24}) with the incidence and severity of patients' neuropathy symptoms (Kramer et al., 2015). These results raise the question whether particular 1-deoxydihydroceramide or 1-deoxyceramide species are neurotoxic as well. The recent finding that ceramide synthase inhibitor fumonisin B1 can reduce the neurotoxic effect of 1-deoxysphinganine in vitro supports this idea (Güntert et al., 2016). Ceramide synthase uses 1-deoxysphingoid and L-serine–derived sphingoid bases as substrates to generate 1-deoxyceramides and ceramides, respectively (Levy & Futerman, 2010; Merrill, 2011). Although our study did not show a direct association between 1-deoxysphingoid bases and patients' neuropathy symptoms during paclitaxel treatment, the baseline plasma level of 1-deoxysphinganine was indicative of which patient developed neuropathy (Kramer et al., 2015). This is an exciting result because the baseline plasma level of 1-deoxysphinganine can be used as a diagnostic marker for susceptibility to paclitaxel-induced peripheral neuropathy.

As stated above, there is no prevention strategy available to address paclitaxel-induced peripheral neuropathy. The emerging evidence that 1-deoxysphingolipids likely play a role in the neurotoxicity of paclitaxel treatment provides the basis for a prevention strategy. Recently, a 52-week clinical case study tested the effects of an oral L-serine supplementation on the plasma levels of 1-deoxysphingolipids in an HSAN1 patient (Auranen et al., 2017). This dose-escalation study showed that L-serine supplementation significantly lowered the plasma levels of 1-deoxysphingolipids of the HSAN1 patient with no report of major side effects. The Auranen et al. study extended the findings of an earlier study that demonstrated the ability of L-serine supplementation to reduce 1-deoxysphingolipid plasma levels in a preclinical animal model of HSAN1, in HSAN1 patients, and in vitro (Garofalo et al., 2011). Interestingly, a study in a rat model of painful paclitaxel-induced peripheral neuropathy tested L-serine intraperitoneal injections as a treatment option and found that L-serine injections decreased neuropathic pain behavior (Kiya et al., 2011). L-serine is the preferred amino acid substrate for SPT (Gable et al., 2010). When supplemented, L-serine can be used to outcompete the alternative L-alanine substrate, thereby reducing the overproduction of 1-deoxysphingolipids that results from the upregulation of SPT. This is the premise for using L-serine supplementation as a prevention strategy for paclitaxel-induced peripheral neuropathy. L-serine can be applied orally and, as the study of Auranen et al. shows, has no major side effects at dose escalation. This warrants further investigation in preclinical models and in cancer patients suffering from neuropathy due to paclitaxel chemotherapy.

2.2 Neuroinflammation, S1P Receptor Subtype 1 Signaling, and Paclitaxel-Induced Peripheral Neuropathy

Another proposed mechanism by which the sphingolipid pathway can contribute to paclitaxel-induced peripheral neuropathy is by activating multiple neuroinflammatory responses in the dorsal horn of the spinal cord (Janes et al., 2014). The authors focused on S1P and its proinflammatory role through S1P receptors, specifically S1P receptor 1. Of note, in rats, it has been shown previously that paclitaxel-induced neuroinflammation can sensitize the nociceptors, which was suggested as a possible cause for painful neuropathic behavior (Wang & Lehky, 2012). S1P is the final metabolite of the canonical sphingolipid pathway, which is produced by the phosphorylation of sphingosine by SK1 and SK2 (Fig. 1) (Merrill, 2011). Extracellular S1P acts by binding to and activating the S1P receptors (Brinkmann, 2007; Lee et al., 1998). There are five S1P receptors, four of which are expressed in the nervous system, both in the glia and neurons (Brinkmann, 2007; Toman & Spiegel, 2002). Notably, S1P-receptor signaling has been implicated in pain (summarized in Welch, Sim-Selley, & Selley, 2012). The results from Janes et al. (2014) showed that in the spinal cord of rats, paclitaxel can increase the activity of several enzymes from the sphingolipid pathway: SPT, sphingomyelinase (total), and SK1, leading to increased levels of S1P and subsequently to upregulation of the proinflammatory cytokines TNFα and IL-1β. Importantly, the authors showed that systemic administration of fingolimod (FTY720), together with paclitaxel, decreased the production of TNFα and IL-1β, concomitantly increasing the production of the antiinflammatory cytokines IL-10 and IL-4 and decreasing the neuropathic pain behavior due to paclitaxel. Fingolimod (FTY720) is an FDA-approved drug for multiple sclerosis, which is a functional antagonist of S1P receptor 1 (Mandala et al., 2002), suggesting an exciting possibility for treating of paclitaxel-induced peripheral neuropathy.

3. CERAMIDE IN CHEMOTHERAPY OR RADIATION THERAPY–INDUCED FEMALE INFERTILITY

Radiation therapy affects tumor cells' proliferation by introducing double strand breaks, leading to a DNA damage and genomic instability

(Borrego-Soto, Ortiz-López, & Rojas-Martínez, 2015; Jackson, 2002). Radiation activates the DNA damage response pathways and the irradiated tumor cells ultimately undergo apoptosis (Deckbar, Jeggo, & Lobrich, 2011). Radiation can also cause the formation of reactive oxygen species, which can indirectly lead to DNA damage by generating apurinic and apyrimidinic sites (Borrego-Soto et al., 2015). The side effects of radiation therapy are localized to the vicinity of the tumor (Bentzen, 2006). Because radiation therapy is provided directly to the skin, the most prevalent side effects are skin erythema, dryness, easy bruising, blistering, and itching and are most commonly experienced within a few weeks of treatment. High doses of radiation can cause DNA damage of the adjacent healthy tissues, resulting in late side effects that include fibrosis, atrophy, and vascular damage. Premature ovarian failure is one of the most common long-term side effects in premenopausal female patients who received radiation or chemotherapy therapy (Blumenfeld & Eckman, 2005). Radiation therapy—associated ovarian failure depends on the amount of the radiation received by the ovaries. For example, the ovaries are significantly exposed to radiation when the pelvic and the abdominal regions are treated for cervical and rectal cancers (Meirow & Nugent, 2001; Wallace, Thomson, Saran, & Kelsey, 2005). In addition to radiation therapy, some chemotherapeutic drugs used to treat lymphoma, such as cyclophosphamide, oncovin, dacarbazine, and procarbazine, can cause premature ovarian failure in 50%—75% of female patients (Blumenfeld & Eckman, 2005). Because each female is born with a finite number of primary oocytes and ovarian primordial follicular cells lack the ability to regenerate, ovarian failure is almost always irreversible (Perez et al., 1997; Tilly & Kolesnick, 2002; Yucebilgin et al., 2004). Moreover, younger patients experiencing infertility are also shown to have altered emotions and relationships that greatly affect their quality of life (Carter et al., 2005).

Various stress stimuli including chemotherapeutic agents and ionizing radiation can activate one of the enzymes from the sphingolipid pathway that generates ceramide, acid sphingomyelinase (ASMase) (Hua & Kolesnick, 2013; Verheij, van Blitterswijk, & Bartelink, 1998; Watters, 1999). The ASMase knockout mouse has a reduced ability to convert sphingomyelin into ceramide. Interestingly, due to reduced ceramide production, the female ASMase knockout mice showed less apoptosis in the oocytes (Hua & Kolesnick, 2013; Tilly & Kolesnick, 2002). ASMase knockout mice have 1.6—2.0 times more oocytes compared to wild-type mice at birth. An earlier study that focused on how disruption of the gene responsible for encoding ASMase, SMPD1, affects the rate of oocyte apoptosis had similar findings (Morita et al., 2000). The data from that study suggested that SMPD1-deficient female mice had ovarian hyperplasia and 11,000 more primordial follicles. It is worth investigating further to understand the role of ASMase-derived ceramides in radiation therapy—induced ovarian apoptosis. Specifically, evaluating whether inhibiting ASMase during radiation therapy can reduce the rate of ovarian failure without compromising the radiation therapy itself may provide particularly valuable information.

In addition to radiation therapy, chemotherapy can also increase the ceramide level, which can then lead to apoptosis in oocytes. In mice, treatment with doxorubicin, a chemotherapeutic drug, can lead to apoptosis of unfertilized oocytes and cell cycle arrest in fertilized oocytes (Perez et al., 1997). Ceramide, along with several proapoptotic proteins such as p53, bcl-2, and CASP, shown to be responsible for the ovarian apoptosis. Doxorubicin was shown in a later study to both upregulate ASMase and increase ceramide in cells (Dumitru, Carpinteiro, Trarbach, Hengge, & Gulbins, 2007). In an earlier study in cells treated with daunorubicin, a chemotherapeutic similar to doxorubicin, it was shown that ceramide generation through a different sphingomyelinase activation, neutral SMase (nSMase) can initiate apoptosis (Jaffrézou et al., 1996). Research that is more recent showed that doxorubicin could activate nSMase2 through the p53 pathway, increase ceramide, and induce cell growth arrest (Shamseddine et al., 2015). Taken together, these studies suggest that ceramide generation can lead to apoptosis through doxorubicin-activation of SMases (acid and neutral), which likely plays a role in ovarian toxicity in doxorubicin chemotherapy.

4. ANTIGANGLIOSIDE IMMUNOTHERAPY SIDE EFFECTS

Once a tumor releases micrometastatic cells into the circulation, it becomes very difficult to detect them due to their microscopic size. Therefore, these micrometastases cannot be destroyed using radiation therapy (Galili, 2018). Although chemotherapy is a systemic treatment, some of the metastatic cells travel further from the blood circulation and the concentration of the chemotherapy drug may not always be sufficient to kill all the metastatic cancer cells (Galili, 2018). In the last few decades, a new therapeutic approach has emerged, which involves immunotherapy using monoclonal antibodies against cancer-associated antigens.

Gangliosides, a subclass of complex glycosphingolipids, are highly expressed in a variety of tumors: neuroblastoma, retinoblastoma, melanoma, small cell lung cancer, osteosarcoma, and rhabdomyosarcoma and are therefore considered excellent targets in cancer immunotherapy (Agostino, Yuriev, & Ramsland, 2012; Ahmed & Cheung, 2014; Navid, Santana, & Barfield, 2010). Gangliosides are glycosphingolipids that contain one or more sialic acid residues. Sphingoid bases and ceramide precursors of gangliosides are synthesized in ER and then additional carbohydrate moieties are added in the Golgi apparatus by glycosyltransferases to generate gangliosides (Yu et al., 2011). Although in cells gangliosides are primarily anchored in the outer leaflet of the plasma membrane, outside of cells, they are found floating in body fluids in monomers or loosely associated micelles (Agostino et al., 2012; Antoine, Camdessanché, Ferraud, & Caudie, 2004; Krengel & Bousquet, 2014; Yu et al., 2011). Gangliosides play a role in cell adhesion, differentiation, and proliferation, in both normal and tumor cells (Groux-Degroote et al., 2015; Krengel & Bousquet, 2014).

As stated above, gangliosides are highly expressed in tumor cells (Daniotti, Vilcaes, Torres Demichelis, Ruggiero, & Rodriguez-Walker, 2013; Hakomori, 1994). The expression of the ganglioside GD3 is higher in melanoma, glioblastoma, and breast cancer (Daniotti et al., 2013). The disialoganglioside, GD2, is expressed in higher amounts on the cell surface of neuroectoderm-derived tumors such as neuroblastoma, retinoblastoma, melanoma, osteosarcoma, brain tumor, small cell lung cancer, glioma, and breast cancer (Cazet et al., 2012; Modak & Cheung, 2007). Another ganglioside, GM2, is highly expressed in T-cell acute lymphocytic leukemia, whereas GM3 is highly expressed in non–small cell lung cancer (Daniotti et al., 2013). These gangliosides have been identified as tumor-associated carbohydrate antigens and are excellent targets for cancer treatments (Ahmed & Cheung, 2014). Moreover, abnormal glycosylation patterns have been known to be a hallmark of cancer, e.g., the acetylation of sialic acid on GD3 prevents apoptosis and promotes tumor proliferation in the case of glioblastoma (Daniotti et al., 2013; Munkley & Elliott, 2016; Tuccillo et al., 2014). In addition, experiments in MDA-MB-231 breast cancer cells expressing GD3 synthase showed that increased expression of GD2 ganglioside enhances tumor proliferation by activating c-Met proto-oncogene (Cazet et al., 2012). Importantly, monoclonal antibodies against

GD2 inhibited cell proliferation through decreased phosphorylation of c-Met (Cazet et al., 2012).

Because gangliosides are embedded in the plasma membrane with the glycan group exposed on the outer side, antiganglioside antibodies can bind to gangliosides on the tumor cell surfaces. The binding of the antiganglioside antibody flags the tumor cells so the immune system can recognize and destroy them. The heterogeneous tumor blood supply limits the uniform delivery of most antibodies to solid bulky tumors compromising the antibody treatment of large solid tumors (Ahmed & Cheung, 2014; Weiner & Adams, 2000). The phase 2 trial of an anti-GD2 antibody treatment in melanoma patients revealed that antitumor activity of the antibody was greater in less bulky tumors (Albertini et al., 2012). Therefore, antibody treatment was suggested for patients with relatively small tumors but with high risk for relapse.

Although promising, antiganglioside immunotherapy has its limitations. The GD2 ganglioside is expressed in healthy brain tissue, peripheral nerves, and melanocytes (Ahmed & Cheung, 2014). The adverse effects observed with antiganglioside administration include hypertension, severe pain, fever, and urticaria (Ahmed & Cheung, 2014; Navid et al., 2010). The pain associated with peripheral nerve damage is the most significant limitation of antiganglioside immunotherapy (Ahmed & Cheung, 2014).

Although the exact mechanism behind neuropathic pain in the anti-GD2 immunotherapy is not known, a mechanism has been proposed in the field of autoimmune disease, where the immune system generates autoantibodies against some gangliosides such as GM1, GD1a, and GD1b, which causes painful neuropathy. Autoantibodies against GM1 and GD1a are associated with motor axonal neuropathy and motor—sensory axonal neuropathy, whereas the neuropathy due to antibodies against GD1b is described as sensory ataxic (Susuki et al., 2012). The proposed mechanism for these immune-mediated neuropathies involves disruption of the nodes of Ranvier of motor nerves (Susuki et al., 2012). The nodes of Ranvier are the only nonmyelinated parts of the axolemma (the plasma membrane of axons); therefore, the antibodies against gangliosides have access to and can recognize the gangliosides there. The nodes of Ranvier are enriched with voltage-gated sodium iron channels and play a very important role in saltatory propagation of action potential signaling along the axon (Susuki & Rasband, 2008). This disruption in the function of the nodes of Ranvier causes nerve dysfunction due to reduced conduction of nerve impulses. Several studies determined that antiganglioside antibodies cause disruption in the nodes of Ranvier via the complement pathway (McGonigal et al., 2010; Susuki et al., 2007, 2012). The results from the study of Susuki et al. (2012) showed that higher amounts of the complement component C3 was deposited at the nodes in nerve sites injected with the anti-GD1a/GT1b antibody. The complement pathway is a part of the immune system that helps remove pathogens by depositing complement fragments on the pathogens' surface, making them easier targets for clearance (Ramaglia, Daha, & Baas, 2008). A recent study in neuroblastoma patients tested this concept by comparing the pain outcome of two anti-GD2 antibodies (Anghelescu et al., 2015). One of the anti-GD2 antibodies was designed to diminish the activation of the complement pathway and the result of the study showed that comparted to the established anti-GD2 antibody elicited less pain.

5. CONCLUDING REMARKS

The recent studies in the fields of sphingolipid and cancer, discussed above, have shown the involvement of sphingolipids in the side effects of cancer treatments, such as peripheral neuropathy, pain, and infertility (summarized in Fig. 2). Some of the mechanisms behind the adverse effects due to sphingolipids, such as paclitaxel-induced peripheral neuropathy, are currently under investigation. Others, such as radiation and chemotherapy-induced infertility and antiganglioside painful neuropathy, have less attention and are so far underinvestigated. Understanding these mechanisms

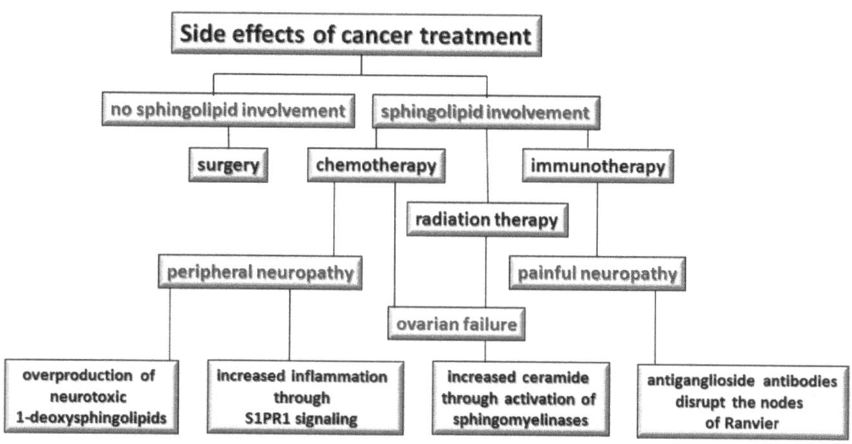

Figure 2 Schematic view of the side effects in cancer therapy involving sphingolipids.

can provide possibilities for improving the existing therapies by reducing their side effects. Such strategies can target the enzymes of the sphingolipid biosynthetic pathway or increase the degradation of toxic sphingolipids by enhancing the activity of their exit metabolic pathway. Further studies, however, are needed to fully understand how adverse effects due to sphingolipids during cancer treatments can be eliminated. For example, the model for the mechanisms behind toxic pain associated with antiganglioside antibodies is not specifically based on research in antiganglioside immunotherapy, but on ganglioside autoimmune diseases. Therefore, the involvement of activation of the complement pathway in neuropathy needs to be further investigated in cancer patients receiving antiganglioside immunotherapy. In summary, current studies have shown significant evidence that in some cases, sphingolipids can be blamed for patients' diminished quality of life due to various side effects of cancer treatments.

REFERENCES

Addington, J., & Freimer, M. (2016). Chemotherapy-induced peripheral neuropathy: An update on the current understanding. *F1000Research, 5*. https://doi.org/10.12688/f1000research.8053.1. F1000 Faculty Rev-1466.

Agostino, M., Yuriev, E., & Ramsland, P. A. (2012). Antibody recognition of cancer-related gangliosides and their mimics investigated using in silico site mapping. *PLoS One,* 7(4), e35457. https://doi.org/10.1371/journal.pone.0035457.

Ahmed, M., & Cheung, N.-K. V. (2014). Engineering anti-GD2 monoclonal antibodies for cancer immunotherapy. *FEBS Letters, 588*(2), 288–297. https://doi.org/10.1016/j.febslet.2013.11.030.

Albertini, M. R., Hank, J. A., Gadbaw, B., Kostlevy, J., Haldeman, J., Schalch, H., et al. (2012). Phase II trial of hu14.18-IL2 for patients with metastatic melanoma. *Cancer Immunology, Immunotherapy, 61*(12), 2261–2271. https://doi.org/10.1007/s00262-012-1286-5.

Alecu, I., Othman, A., Penno, A., Saied, E. M., Arenz, C., von Eckardstein, A., et al. (2017). Cytotoxic 1-deoxysphingolipids are metabolized by a cytochrome P450-dependent pathway. *Journal of Lipid Research, 58*(1), 60–71. https://doi.org/10.1194/jlr.M072421.

Alecu, I., Tedeschi, A., Behler, N., Wunderling, K., Lamberz, C., Lauterbach, M. A., et al. (2017). Localization of 1-deoxysphingolipids to mitochondria induces mitochondrial dysfunction. *The Journal of Lipid Research, 58*(1), 42–59. https://doi.org/10.1194/jlr.M068676.

Andoh, T., Uta, D., Kato, M., Toume, K., Komatsu, K., & Kuraishi, Y. (2017). Prophylactic administration of aucubin inhibits paclitaxel-induced mechanical allodynia via the inhibition of endoplasmic reticulum stress in peripheral schwann cells. *Biological and Pharmaceutical Bulletin, 40*(4), 473–478. https://doi.org/10.1248/bpb.b16-00899.

Anghelescu, D. L., Goldberg, J. L., Faughnan, L. G., Wu, J., Mao, S., Furman, W. L., et al. (2015). Comparison of pain outcomes between two anti-GD2 antibodies in patients with neuroblastoma. *Pediatric Blood and Cancer, 62*(2), 224–228. https://doi.org/10.1002/pbc.25280.

Antoine, J. C., Camdessanché, J. P., Ferraud, K., & Caudie, C. (2004). Antiganglioside antibodies in paraneoplastic peripheral neuropathies. *Journal of Neurology, Neurosurgery & Psychiatry, 75*(12), 1765.

Auranen, M., Toppila, J., Suriyanarayanan, S., Lone, M. A., Paetau, A., Tyynismaa, H., et al. (2017). Clinical and metabolic consequences of L-serine supplementation in hereditary sensory and autonomic neuropathy type 1C. *Cold Spring Harbor Molecular Case Studies, 3*(6), a002212. https://doi.org/10.1101/mcs.a002212.

Baird, R. D., Kitzen, J., Clarke, P. A., Planting, A., Reade, S., Reid, A., et al. (2009). Phase I safety, pharmacokinetic, and pharmacogenomic trial of ES-285, a novel marine cytotoxic agent, administered to adult patients with advanced solid tumors. *Molecular Cancer Therapeutics, 8*(6), 1430−1437. https://doi.org/10.1158/1535-7163.MCT-08-1167.

Bao, T., Basal, C., Seluzicki, C., Li, S. Q., Seidman, A. D., & Mao, J. J. (2016). Long-term chemotherapy-induced peripheral neuropathy among breast cancer survivors: Prevalence, risk factors, and fall risk. *Breast Cancer Research and Treatment, 159*(2), 327−333. https://doi.org/10.1007/s10549-016-3939-0.

Bentzen, S. M. (2006). Preventing or reducing late side effects of radiation therapy: Radiobiology meets molecular pathology. *Nature Reviews Cancer, 6*, 702. https://doi.org/10.1038/nrc1950.

Blumenfeld, Z., & Eckman, A. (2005). Preservation of fertility and ovarian function and minimization of chemotherapy-induced gonadotoxicity in young women by GnRH-a. *Journal of the National Cancer Institute Monographs, 34*, 40−43. https://doi.org/10.1093/jncimonographs/lgi015.

Borrego-Soto, G., Ortiz-López, R., & Rojas-Martínez, A. (2015). Ionizing radiation-induced DNA injury and damage detection in patients with breast cancer. *Genetics and Molecular Biology, 38*(4), 420−432. https://doi.org/10.1590/S1415-475738420150019.

Brinkmann, V. (2007). Sphingosine 1-phosphate receptors in health and disease: Mechanistic insights from gene deletion studies and reverse pharmacology. *Pharmacology & Therapeutics, 115*(1), 84−105. https://doi.org/10.1016/j.pharmthera.2007.04.006.

Canta, A., Pozzi, E., & Carozzi, V. A. (2015). Mitochondrial dysfunction in chemotherapy-induced peripheral neuropathy (CIPN). *Toxics, 3*(2), 198−223. https://doi.org/10.3390/toxics3020198.

Carter, J., Rowland, K., Chi, D., Brown, C., Abu-Rustum, N., Castiel, M., et al. (2005). Gynecologic cancer treatment and the impact of cancer-related infertility. *Gynecologic Oncology, 97*(1), 90−95. https://doi.org/10.1016/j.ygyno.2004.12.019.

Cazet, A., Bobowski, M., Rombouts, Y., Lefebvre, J., Steenackers, A., Popa, I., et al. (2012). The ganglioside GD2 induces the constitutive activation of c-Met in MDA-MB-231 breast cancer cells expressing the GD3 synthase. *Glycobiology, 22*(6), 806−816. https://doi.org/10.1093/glycob/cws049.

Chan, K. S., Koh, C. G., & Li, H. Y. (2012). Mitosis-targeted anti-cancer therapies: Where they stand. *Cell Death & Disease, 3*(10), e411. https://doi.org/10.1038/cddis.2012.148.

Cuadros, R., Montejo de Garcini, E., Wandosell, F., Faircloth, G., Fernandez-Sousa, J. M., & Avila, J. (2000). The marine compound spisulosine, an inhibitor of cell proliferation, promotes the disassembly of actin stress fibers. *Cancer Letters, 152*(1), 23−29.

Daniotti, J., Vilcaes, A., Torres Demichelis, V., Ruggiero, F., & Rodriguez-Walker, M. (2013). Glycosylation of glycolipids in cancer: Basis for development of novel therapeutic approaches. *Frontiers in Oncology, 3*(306). https://doi.org/10.3389/fonc.2013.00306.

Deckbar, D., Jeggo, P. A., & Lobrich, M. (2011). Understanding the limitations of radiation-induced cell cycle checkpoints. *Critical Reviews in Biochemistry and Molecular Biology, 46*(4), 271−283. https://doi.org/10.3109/10409238.2011.575764.

Dore, K., Aow, J., & Malinow, R. (2016). The emergence of NMDA receptor metabotropic function: Insights from imaging. *Frontiers in Synaptic Neuroscience, 8*, 20. https://doi.org/10.3389/fnsyn.2016.00020.

Dumitru, C. A., Carpinteiro, A., Trarbach, T., Hengge, U. R., & Gulbins, E. (2007). Doxorubicin enhances TRAIL-induced cell death via ceramide-enriched membrane platforms. *Apoptosis, 12*(8), 1533–1541. https://doi.org/10.1007/s10495-007-0081-9.

Fletcher, D. A., & Mullins, R. D. (2010). Cell mechanics and the cytoskeleton. *Nature, 463*(7280), 485–492. https://doi.org/10.1038/nature08908.

Gable, K., Gupta, S. D., Han, G., Niranjanakumari, S., Harmon, J. M., & Dunn, T. M. (2010). A disease-causing mutation in the active site of serine palmitoyltransferase causes catalytic promiscuity. *Journal of Biological Chemistry, 285*(30), 22846–22852. https://doi.org/10.1074/jbc.M110.122259.

Galili, U. (2018). Chapter 10-cancer immunotherapy by anti-Gal-mediated in situ conversion of tumors into autologous vaccines. In *The natural anti-Gal antibody as foe turned friend in medicine* (pp. 171–198). Academic Press.

Garofalo, K., Penno, A., Schmidt, B. P., Lee, H. J., Frosch, M. P., von Eckardstein, A., et al. (2011). Oral L-serine supplementation reduces production of neurotoxic deoxysphingolipids in mice and humans with hereditary sensory autonomic neuropathy type 1. *Journal of Clinical Investigation, 121*(12), 4735–4745. https://doi.org/10.1172/JCI57549.

Gerber, D. E. (2008). Targeted therapies: A new generation of cancer treatments. *American Family Physician, 77*(3), 311–319.

Gornstein, E., & Schwarz, T. L. (2014). The paradox of paclitaxel neurotoxicity: Mechanisms and unanswered questions. *Neuropharmacology, 76*(Pt A), 175–183. https://doi.org/10.1016/j.neuropharm.2013.08.016.

Groux-Degroote, S., Guerardel, Y., Julien, S., & Delannoy, P. (2015). Gangliosides in breast cancer: New perspectives. *Biochemistry (Mosc), 80*(7), 808–819. https://doi.org/10.1134/s0006297915070020.

Güntert, T., Hänggi, P., Othman, A., Suriyanarayanan, S., Sonda, S., Zuellig, R. A., et al. (2016). 1-Deoxysphingolipid-induced neurotoxicity involves N-methyl-D-aspartate receptor signaling. *Neuropharmacology, 110*(Part A), 211–222. https://doi.org/10.1016/j.neuropharm.2016.03.033.

Hakomori, S-i. (1994). Chapter 18 Role of gangliosides in tumor progression. In L. Svennerholm, A. K. Asbury, R. A. Reisfeld, K. Sandhoff, K. Suzuki, G. Tettamanti, et al. (Eds.), *Vol. 101. Progress in brain research* (pp. 241–250). Elsevier.

Hannun, Y. A., & Obeid, L. M. (2018). Sphingolipids and their metabolism in physiology and disease. *Nature Reviews. Molecular Cell Biology, 19*(3), 175–191. https://doi.org/10.1038/nrm.2017.107.

Hitomi, J., Katayama, T., Taniguchi, M., Honda, A., Imaizumi, K., & Tohyama, M. (2004). Apoptosis induced by endoplasmic reticulum stress depends on activation of caspase-3 via caspase-12. *Neuroscience Letters, 357*(2), 127–130. https://doi.org/10.1016/j.neulet.2003.12.080.

Hua, G., & Kolesnick, R. (2013). Using ASMase knockout mice to model human diseases. *Handbook of Experimental Pharmacology*, (216), 29–54. https://doi.org/10.1007/978-3-7091-1511-4_2.

Jackson, S. P. (2002). Sensing and repairing DNA double-strand breaks. *Carcinogenesis, 23*(5), 687–696. https://doi.org/10.1093/carcin/23.5.687.

Jaffrézou, J. P., Levade, T., Bettaïeb, A., Andrieu, N., Bezombes, C., Maestre, N., et al. (1996). Daunorubicin-induced apoptosis: Triggering of ceramide generation through sphingomyelin hydrolysis. *The EMBO Journal, 15*(10), 2417–2424.

Janes, K., Little, J. W., Li, C., Bryant, L., Chen, C., Chen, Z., et al. (2014). The development and maintenance of paclitaxel-induced neuropathic pain require activation of the sphingosine 1-phosphate receptor subtype 1. *Journal of Biological Chemistry, 289*(30), 21082–21097. https://doi.org/10.1074/jbc.M114.569574.

Kiya, T., Kawamata, T., Namiki, A., & Yamakage, M. (2011). Role of satellite cell-derived L-serine in the dorsal root ganglion in paclitaxel-induced painful peripheral neuropathy. *Neuroscience, 174*, 190–199. https://doi.org/10.1016/j.neuroscience.2010.11.046.

Kramer, R., Bielawski, J., Kistner-Griffin, E., Othman, A., Alecu, I., Ernst, D., et al. (2015). Neurotoxic 1-deoxysphingolipids and paclitaxel-induced peripheral neuropathy. *The FASEB Journal, 29*(11), 4461–4472. https://doi.org/10.1096/fj.15-272567.

Krengel, U., & Bousquet, P. A. (2014). Molecular recognition of gangliosides and their potential for cancer immunotherapies. *Frontiers in Immunology, 5*(325). https://doi.org/10.3389/fimmu.2014.00325.

Lee, M. J., Van Brocklyn, J. R., Thangada, S., Liu, C. H., Hand, A. R., Menzeleev, R., et al. (1998). Sphingosine-1-phosphate as a ligand for the G protein-coupled receptor EDG-1. *Science, 279*(5356), 1552–1555.

Levy, M., & Futerman, A. H. (2010). Mammalian ceramide synthases. *IUBMB Life, 62*(5), 347–356. https://doi.org/10.1002/iub.319.

Luanpitpong, S., & Rojanasakul, Y. (2012). *Chemotherapy-induced alopecia*.

Maceyka, M., Payne, S. G., Milstien, S., & Spiegel, S. (2002). Sphingosine kinase, sphingosine-1-phosphate, and apoptosis. *Biochimica et Biophysica Acta, 1585*(2–3), 193–201.

Maguire, P., & Selby, P. (1989). Assessing quality of life in cancer patients. *British Journal of Cancer, 60*(3), 437–440.

Mandala, S., Hajdu, R., Bergstrom, J., Quackenbush, E., Xie, J., Milligan, J., et al. (2002). Alteration of lymphocyte trafficking by sphingosine-1-phosphate receptor agonists. *Science, 296*(5566), 346–349. https://doi.org/10.1126/science.1070238.

Mansano-Schlosser, T. C., & Ceolim, M. F. (2012). Quality of life of cancer patients during the chemotherapy period. *Texto & Contexto - Enfermagem, 21*, 600–607.

Massard, C., Salazar, R., Armand, J. P., Majem, M., Deutsch, E., Garcia, M., et al. (2012). Phase I dose-escalating study of ES-285 given as a three-hour intravenous infusion every three weeks in patients with advanced malignant solid tumors. *Investigational New Drugs, 30*(6), 2318–2326. https://doi.org/10.1007/s10637-011-9772-8.

McGonigal, R., Rowan, E. G., Greenshields, K. N., Halstead, S. K., Humphreys, P. D., Rother, R. P., et al. (2010). Anti-GD1a antibodies activate complement and calpain to injure distal motor nodes of Ranvier in mice. *Brain, 133*(Pt 7), 1944–1960. https://doi.org/10.1093/brain/awq119.

Meirow, D., & Nugent, D. (2001). The effects of radiotherapy and chemotherapy on female reproduction. *Human Reproduction Update, 7*(6), 535–543.

Merrill, A. H., Jr. (2011). Sphingolipid and glycosphingolipid metabolic pathways in the era of sphingolipidomics. *Chemical Reviews, 111*(10), 6387–6422. https://doi.org/10.1021/cr2002917.

Merrill, A. H., Jr., Sullards, M. C., Wang, E., Voss, K. A., & Riley, R. T. (2001). Sphingolipid metabolism: Roles in signal transduction and disruption by fumonisins. *Environmental Health Perspectives, 109*(Suppl. 2), 283–289.

Miltenburg, N. C., & Boogerd, W. (2014). Chemotherapy-induced neuropathy: A comprehensive survey. *Cancer Treatment Reviews, 40*(7), 872–882. https://doi.org/10.1016/j.ctrv.2014.04.004.

Modak, S., & Cheung, N. K. (2007). Disialoganglioside directed immunotherapy of neuroblastoma. *Cancer Investigation, 25*(1), 67–77. https://doi.org/10.1080/07357900601130763.

Morita, Y., Perez, G. I., Paris, F., Miranda, S. R., Ehleiter, D., Haimovitz-Friedman, A., et al. (2000). Oocyte apoptosis is suppressed by disruption of the acid sphingomyelinase gene or by sphingosine -1-phosphate therapy. *Nature Medicine, 6*, 1109. https://doi.org/10.1038/80442.

Munkley, J., & Elliott, D. J. (2016). Hallmarks of glycosylation in cancer. *Oncotarget, 7*(23), 35478–35489. https://doi.org/10.18632/oncotarget.8155.

Navid, F., Santana, V. M., & Barfield, R. C. (2010). Anti-GD2 antibody therapy for GD2-expressing tumors. *Current Cancer Drug Targets, 10*(2), 200–209.

Ogretmen, B. (2018). Sphingolipid metabolism in cancer signalling and therapy. *Nature Reviews Cancer, 18*(1), 33–50. https://doi.org/10.1038/nrc.2017.96.

Ogretmen, B., & Hannun, Y. A. (2004). Biologically active sphingolipids in cancer pathogenesis and treatment. *Nature Reviews Cancer, 4*, 604. https://doi.org/10.1038/nrc1411.

Penno, A., Reilly, M. M., Houlden, H., Laura, M., Rentsch, K., Niederkofler, V., et al. (2010). Hereditary sensory neuropathy type 1 is caused by the accumulation of two neurotoxic sphingolipids. *Journal of Biological Chemistry, 285*(15), 11178–11187. https://doi.org/10.1074/jbc.M109.092973.

Perez, G. I., Knudson, C. M., Leykin, L., Korsmeyer, S. J., & Tilly, J. L. (1997). Apoptosis-associated signaling pathways are required for chemotherapy-mediated female germ cell destruction. *Nature Medicine, 3*, 1228. https://doi.org/10.1038/nm1197-1228.

Ponnusamy, S., Meyers-Needham, M., Senkal, C. E., Saddoughi, S. A., Sentelle, D., Selvam, S. P., et al. (2010). Sphingolipids and cancer: Ceramide and sphingosine-1-phosphate in the regulation of cell death and drug resistance. *Future Oncology, 6*(10), 1603–1624. https://doi.org/10.2217/fon.10.116.

Ramaglia, V., Daha, M. R., & Baas, F. (2008). The complement system in the peripheral nerve: Friend or foe? *Molecular Immunology, 45*(15), 3865–3877. https://doi.org/10.1016/j.molimm.2008.06.018.

Rowinsky, E. K., Eisenhauer, E. A., Chaudhry, V., Arbuck, S. G., & Donehower, R. C. (1993). Clinical toxicities encountered with paclitaxel (Taxol). *Seminars in Oncology, 20*(4 Suppl. 3), 1–15.

Ryland, L. K., Fox, T. E., Liu, X., Loughran, T. P., & Kester, M. (2011). Dysregulation of sphingolipid metabolism in cancer. *Cancer Biology & Therapy, 11*(2), 138–149.

Salcedo, M., Cuevas, C., Alonso, J. L., Otero, G., Faircloth, G., Fernandez-Sousa, J. M., et al. (2007). The marine sphingolipid-derived compound ES 285 triggers an atypical cell death pathway. *Apoptosis, 12*(2), 395–409. https://doi.org/10.1007/s10495-006-0573-z.

Schoffski, P., Dumez, H., Ruijter, R., Miguel-Lillo, B., Soto-Matos, A., Alfaro, V., et al. (2011). Spisulosine (ES-285) given as a weekly three-hour intravenous infusion: Results of a phase I dose-escalating study in patients with advanced solid malignancies. *Cancer Chemotherapy and Pharmacology, 68*(6), 1397–1403. https://doi.org/10.1007/s00280-011-1612-1.

Selvam, S. P., & Ogretmen, B. (2013). Sphingosine kinase/sphingosine 1-phosphate signaling in cancer therapeutics and drug resistance. *Handbook of Experimental Pharmacology, 216*, 3–27. https://doi.org/10.1007/978-3-7091-1511-4_1.

Seretny, M., Currie, G. L., Sena, E. S., Ramnarine, S., Grant, R., MacLeod, M. R., et al. (2014). Incidence, prevalence, and predictors of chemotherapy-induced peripheral neuropathy: A systematic review and meta-analysis. *Pain, 155*(12), 2461–2470. https://doi.org/10.1016/j.pain.2014.09.020.

Shamseddine, A. A., Clarke, C. J., Carroll, B., Airola, M. V., Mohammed, S., Rella, A., et al. (2015). P53-dependent upregulation of neutral sphingomyelinase-2: Role in doxorubicin-induced growth arrest. *Cell Death & Disease, 6*, e1947. https://doi.org/10.1038/cddis.2015.268.

Siegel, R. L., Miller, K. D., & Jemal, A. (2017). Cancer statistics, 2017. *CA: A Cancer Journal for Clinicians, 67*(1), 7–30. https://doi.org/10.3322/caac.21387.

Spassieva, S. D., Ji, X., Liu, Y., Gable, K., Bielawski, J., Dunn, T. M., et al. (2016). Ectopic expression of ceramide synthase 2 in neurons suppresses neurodegeneration induced by ceramide synthase 1 deficiency. *Proceedings of the National Academy of Sciences of the United States of America, 113*(21), 5928–5933. https://doi.org/10.1073/pnas.1522071113.

Speck, R. M., Sammel, M. D., Farrar, J. T., Hennessy, S., Mao, J. J., Stineman, M. G., & DeMichele, A. (2013). Impact of chemotherapy-induced peripheral neuropathy on treatment delivery in nonmetastatic breast cancer. *Journal of Oncology Practice, 9*(5), e234–e240. https://doi.org/10.1200/JOP.2012.000863.

Stankiewicz, T. R., & Linseman, D. A. (2014). Rho family GTPases: Key players in neuronal development, neuronal survival, and neurodegeneration. *Frontiers in Cellular Neuroscience, 8*, 314. https://doi.org/10.3389/fncel.2014.00314.

Susuki, K., & Rasband, M. N. (2008). Molecular mechanisms of node of Ranvier formation. *Current Opinion in Cell Biology, 20*(6), 616–623. https://doi.org/10.1016/j.ceb.2008.09.007.

Susuki, K., Rasband, M. N., Tohyama, K., Koibuchi, K., Okamoto, S., Funakoshi, K., et al. (2007). Anti-GM1 antibodies cause complement-mediated disruption of sodium channel clusters in peripheral motor nerve fibers. *The Journal of Neuroscience, 27*(15), 3956–3967. https://doi.org/10.1523/JNEUROSCI.4401-06.2007.

Susuki, K., Yuki, N., Schafer, D. P., Hirata, K., Zhang, G., Funakoshi, K., et al. (2012). Dysfunction of nodes of Ranvier: A mechanism for anti-ganglioside antibody-mediated neuropathies. *Experimental Neurology, 233*(1), 534–542. https://doi.org/10.1016/j.expneurol.2011.11.039.

Tilly, J. L., & Kolesnick, R. N. (2002). Sphingolipids, apoptosis, cancer treatments and the ovary: Investigating a crime against female fertility. *Biochimica et Biophysica Acta, 1585*(2–3), 135–138. https://doi.org/10.1016/S1388-1981(02)00333-5.

Toman, R. E., & Spiegel, S. (2002). Lysophospholipid receptors in the nervous system. *Neurochemical Research, 27*(7–8), 619–627.

Tuccillo, F. M., de Laurentiis, A., Palmieri, C., Fiume, G., Bonelli, P., Borrelli, A., et al. (2014). Aberrant glycosylation as biomarker for cancer: Focus on CD43. *BioMed Research International, 2014*, 13. https://doi.org/10.1155/2014/742831.

Van Brocklyn, J. R., & Williams, J. B. (2012). The control of the balance between ceramide and sphingosine-1-phosphate by sphingosine kinase: Oxidative stress and the seesaw of cell survival and death. *Comparative Biochemistry and Physiology. Part B, Biochemistry & Molecular Biology, 163*(1), 26–36. https://doi.org/10.1016/j.cbpb.2012.05.006.

Verheij, M., van Blitterswijk, W. J., & Bartelink, H. (1998). Radiation-induced apoptosis—the ceramide-SAPK signaling pathway and clinical aspects. *Acta Oncologica, 37*(6), 575–581.

Wallace, W. H. B., Thomson, A. B., Saran, F., & Kelsey, T. W. (2005). Predicting age of ovarian failure after radiation to a field that includes the ovaries. *International Journal of Radiation Oncology, Biology, Physics, 62*(3), 738–744. https://doi.org/10.1016/j.ijrobp.2004.11.038.

Wang, X.-M., & Lehky, T. J. (2012). Discovering cytokines as targets for chemotherapy-induced painful peripheral neuropathy. *Cytokine, 59*(1), 3–9. https://doi.org/10.1016/j.cyto.2012.03.027.

Watters, D. (1999). Molecular mechanisms of ionizing radiation-induced apoptosis. *Immunology and Cell Biology, 77*, 263. https://doi.org/10.1046/j.1440-1711.1999.00824.x.

Weaver, B. A. (2014). How Taxol/paclitaxel kills cancer cells. *Molecular Biology of the Cell, 25*(18), 2677–2681. https://doi.org/10.1091/mbc.E14-04-0916.

Weiner, L. M., & Adams, G. P. (2000). New approaches to antibody therapy. *Oncogene, 19*, 6144. https://doi.org/10.1038/sj.onc.1204000.

Wei, N., Pan, J., Pop-Busui, R., Othman, A., Alecu, I., Hornemann, T., et al. (2014). Altered sphingoid base profiles in type 1 compared to type 2 diabetes. *Lipids in Health and Disease, 13*, 161. https://doi.org/10.1186/1476-511X-13-161.

Welch, S. P., Sim-Selley, L. J., & Selley, D. E. (2012). Sphingosine-1-phosphate receptors as emerging targets for treatment of pain. *Biochemical Pharmacology, 84*(12), 1551–1562. https://doi.org/10.1016/j.bcp.2012.08.010.

Xiao, W. H., & Bennett, G. J. (2012). Effects of mitochondrial poisons on the neuropathic pain produced by the chemotherapeutic agents, paclitaxel and oxaliplatin. *Pain, 153*(3), 704–709. https://doi.org/10.1016/j.pain.2011.12.011.

Yucebilgin, M. S., Terek, M. C., Ozsaran, A., Akercan, F., Zekioglu, O., Isik, E., et al. (2004). Effect of chemotherapy on primordial follicular reserve of rat: An animal model of premature ovarian failure and infertility. *The Australian & New Zealand Journal of Obstetrics & Gynaecology, 44*(1), 6–9. https://doi.org/10.1111/j.1479-828X.2004.00143.x.

Yu, R. K., Tsai, Y.-T., Ariga, T., & Yanagisawa, M. (2011). Structures, biosynthesis, and functions of gangliosides—An overview. *Journal of Oleo Science, 60*(10), 537–544.

FSC
www.fsc.org
MIX
Papier | Fördert gute Waldnutzung
FSC® C083411

Kolibri 360 GmbH
Ferdinand-Jühlke-Straße 7
99095 Erfurt, Deutschland
produktsicherheit@kolibri360.de